T0324968

An Introduction to the
# THEORY OF WAVE MAPS AND RELATED GEOMETRIC PROBLEMS

An Introduction to the

# THEORY OF WAVE MAPS AND RELATED GEOMETRIC PROBLEMS

**Dan-Andrei Geba**

*University of Rochester, USA*

**Manoussos G Grillakis**

*University of Maryland, College Park, USA*

NEW JERSEY · LONDON · SINGAPORE · BEIJING · SHANGHAI · HONG KONG · TAIPEI · CHENNAI · TOKYO

*Published by*

World Scientific Publishing Co. Pte. Ltd.

5 Toh Tuck Link, Singapore 596224

*USA office:* 27 Warren Street, Suite 401-402, Hackensack, NJ 07601

*UK office:* 57 Shelton Street, Covent Garden, London WC2H 9HE

**Library of Congress Cataloging-in-Publication Data**
Names: Geba, Dan-Andrei, 1973–  | Grillakis, Manoussos G.
Title: An introduction to the theory of wave maps and related geometric problems /
    by Dan-Andrei Geba (University of Rochester, USA),
    Manoussos G. Grillakis (University of Maryland, College Park, USA).
Description: New Jersey : World Scientific, 2016. | Includes bibliographical references and index.
Identifiers: LCCN 2016026890 | ISBN 9789814713900 (hardcover : alk. paper)
Subjects: LCSH: Nonlinear wave equations. | Differential equations, Hyperbolic. |
    Solitons. | Geometry, Differential.
Classification: LCC QA927 .G37 2016 | DDC 516.9--dc23
LC record available at https://lccn.loc.gov/2016026890

**British Library Cataloguing-in-Publication Data**
A catalogue record for this book is available from the British Library.

Printed in Singapore

Dan-Andrei Geba dedicates this work to his family, mentors, and friends, to whom he owes everything.

Manoussos Grillakis dedicates this book to his parents, Marika and Georgios.

# Preface

The wave map problem is one of the most beautiful and challenging nonlinear hyperbolic problems, which has kept the attention of mathematicians for more than thirty years now. The study of the problem involves diverse issues, e.g., well-posedness, regularity, formation of singularities, and stability of solitons, and combines intricate tools from analysis, geometry, and topology. Moreover, the wave map system has a natural formulation as the Euler-Lagrange system for a map between manifolds, a special case being the nonlinear sigma model, which is one of the fundamental objects in classical field theory. One of the goals of this book is to offer an up-to-date and self-contained overview of the main regularity theory for wave maps. Another goal is to introduce, to a wide mathematical audience, physically motivated generalizations of the wave map system (e.g., the Skyrme model), which are extremely interesting and pose challenging new questions in their own right.

The topic of wave maps has experienced an incredible advancement in the past ten years. This is precisely the time passed from the moment when the last monograph (i.e., Tao [171]) which tried to give a state-of-the-art for this topic appeared. Our book tries to fill this gap by presenting the most recent developments in the field, e.g., the resolution of the large data global regularity theory for wave maps. These results are very technical, being accessible only to experts in their current format. Our goal is to try to explain them to a wider group, which includes advanced graduate students, in the hope of stimulating new research ideas. Moreover, this book is the first one which discusses, from a mathematical point of view, the time evolution for the models which are extensions of the nonlinear sigma model: the Skyrme, Faddeev, and Adkins-Nappi theories.

Our book starts by introducing the reader to the physical motivation and the mathematical formulation of the wave map problem and its generalizations. This is followed by developing the analytic background needed in the investigation of these problems, which include Strichartz estimates and hyperbolic Sobolev spaces. The third chapter is devoted to the study of the local and small data global well-posedness theories for the wave map equation, where one can see the motivation for the emergence of more and more powerful analytic techniques needed in handling the challenging nature of these topics. This chapter also includes detailed expositions for two results, due to Tao [169] and Shatah-Struwe [146], respectively, which make the case for the important role played by the intrinsic geometric aspect of the wave map problem. Next, we discuss the resolution of the large data regularity theory for wave maps in the energy-critical case by Sterbenz-Tataru's program [160, 161]. We focus on the second part (i.e., [161]) of this work, which provides a complete description of the regimes when a large data, finite energy wave map blows up and when it is global-in-time. Our presentation of this topic reworks Sterbenz-Tataru's argument, including a significant number of refinements and extra details. The fifth chapter addresses well-posedness questions for the classical Skyrme model and its extensions. There, we present Wong's result [191] on the Skyrme model using Christodoulou's regular hyperbolicity framework and Lei-Lin-Zhou's global regularity theorem [103] for the $2 + 1$-dimensional Faddeev problem, which relies on an adaptation of Klainerman's vector field method. Following this, we discuss equivariant results for all the equations considered in this book, which include non-concentration of energy, small data global well-posedness in critical Besov spaces, and global regularity for sufficiently smooth large data. Finally, we turn our attention to the phenomenon of collapse for wave maps and examine Raphaël-Rodnianski's result [136] on the topic. We put forth a novel approach relying on the associated Hodge system, which makes the structure of the problem more transparent. The book also includes an appendix detailing the basic differential geometry concepts needed for following the presentation of the material.

*Dan-Andrei Geba and Manoussos Georgios Grillakis*

# Acknowledgments

Dan-Andrei Geba expresses his gratitude to Viorel Barbu, Sergiu Klainerman, Igor Rodnianski, and especially Daniel Tataru, his mentors in the area of partial differential equations, for introducing him to the subject, for supervising his doctoral dissertation, and for inspiration, support, and friendship throughout his academic career. He also thanks his colleagues at University of Rochester, particularly Sarada Rajeev, for introducing him to the Skyrme model and related problems and for countless hours spent on explaining ideas from theoretical physics, and Allan Greenleaf and Alex Iosevich, for many valuable discussions, comments, and professional advice. He owes a great deal of appreciation to the students and participants to his lectures, especially to his doctoral students, Matthew Creek, Daniel da Silva, and Xiang Zhang. Extra thanks go to Matthew for typing a version of some course notes that were partly included in this book.

Dan-Andrei Geba acknowledges support from the College of Arts and Sciences at University of Rochester for allowing him to take a sabbatical leave in the spring of 2015, which significantly helped in the completion of this project. He is also grateful for support to the National Science Foundation, through Career grant DMS-0747656, and to the Simons Foundation, through grant # 359727. Finally, he expresses infinite appreciation toward his family for everything, especially his wife, Dana, for her immense love, patience, and strong belief in him, his older daughter, Maria, for providing first-class in-house editorship, and his younger one, Ioana, for motivating him throughout the writing process by frequently asking "How much do you have left?"

Manoussos Grillakis would like to thank Constantine Gryllakis, Walter A. Strauss, Constantine Dafermos and Louis Nirenberg for inspiring him and his wife Kristi Dobrovolski for her love and support.

Together, they would like to thank the Chairman of World Scientific Publishing, Dr. K. K. Phua, for the invitation to write this book, and their editors, Mrs. Jessica Barrows and Ms. E. H. Chionh, for their patience, understanding, and support in working with them.

# Contents

# List of Figures

# Chapter 1

# Introduction

The goal of this chapter is to introduce and motivate from various perspectives the problems that are discussed in this book. First, we describe the physical background of these problems and the precise motivation for the physical study of each one of them. Following this, we focus on their mathematical formulation and we develop some of their basic properties, such as energy bounds, various symmetries, and static solutions. The underlying theme for the mathematical exposition is the prominent role played by the rich geometric structure of these problems.

## 1.1 Physical description and motivation

In this section, we include a brief description from the physical point of view for a number of field theories that will make the subject of our mathematical investigations. At the center of this description lie the concepts of *nonlinear sigma model* and *topological soliton*, which are essential in formulating the framework introducing the aforementioned theories. We strived for a concise and direct presentation of these concepts and we refer the avid reader to the monographs of Lee [100] and Manton and Sutcliffe [113] for a more complete exposition of them.

### 1.1.1 *Nonlinear sigma models*

In physics terminology, a *nonlinear sigma model* is a scalar field theory describing maps $\phi : M \to N$, also called *field configurations*, from a spacetime $(M, g)$, the *domain manifold*, to a complete Riemannian manifold $(N, h)$,

1

the *target manifold*, which are formal critical points of the action

$$S = \int_M \frac{1}{2} \langle \partial^\mu \phi, \partial_\mu \phi \rangle_h \, dg = \int_M \frac{1}{2} g^{\mu\nu} \partial_\mu \phi^i \partial_\nu \phi^j h_{ij}(\phi) \, dg. \qquad (1.1)$$

A nonlinear sigma model was first introduced by Gell-Mann and Lévy [55], though similar results appeared at the same time in works by Gürsey [64, 65]. The Gell-Mann-Lévy model, which corresponds to

$$(M, g) = (\mathbb{R}^{3+1}, \operatorname{diag}(-1, 1, 1, 1)) \qquad \text{and} \qquad N = \mathbb{S}^3 \subset \mathbb{R}^4$$

endowed with the induced Riemannian metric, is of particular interest in high energy physics. It describes three subatomic particles, called $\pi$ *mesons* or *pions* and denoted by $\pi = (\pi_x, \pi_y, \pi_z)$, and the interactions between them. If these interactions were ignored, the field $\pi : \mathbb{R}^{3+1} \to \mathbb{R}^3$ would verify the wave equation $\square \pi = 0$. The importance of the $\pi$ mesons comes from the fact that forces they mediate are responsible for holding the nuclei of atoms together.

The sigma terminology originates from a linear model, also featured in [55], where, apart from the $\pi$ mesons, a fourth independent particle, called a $\sigma$ *meson* and denoted by $\sigma$, was postulated to exist. This was done even if, at the time, there was no supporting experimental evidence for its existence. The notation used by Gell-Mann and Lévy for this theory,

$$\phi = (\pi, \sigma) = (\pi_x, \pi_y, \pi_z, \sigma) \in \mathbb{R}^4,$$

was also employed for the nonlinear model, in which case, due to $\phi \in \mathbb{S}^3$, $\sigma$ is no longer an independent degree of freedom. This is why the term *nonlinear sigma model* is a bit confusing, as the original $\sigma$ meson has disappeared. The book by Lee [100] explains these matters thoroughly.

A very interesting question to be studied in connection to nonlinear field theories, which include the nonlinear sigma models, is the existence of *topological solitons*. According to the book by Manton and Sutcliffe [113], which has a complete and beautiful exposition of this topic, *a topological soliton is a localized, finite-energy, exact solution of the associated field equations, which is topologically distinct from the vacuum*[1]. As such, a topological soliton is a stable, particle-like object.

In what follows, we discuss this concept in the context of the nonlinear sigma model (1.1) with $(M, g) = (\mathbb{R}^{n+1}, \operatorname{diag}(-1, 1, \ldots, 1))$, the Minkowski spacetime, and $N = \mathbb{S}^n$, the unit sphere of $\mathbb{R}^{n+1}$; this model is usually called the $O(n + 1)$ sigma model, due to the symmetries present in the

---

[1]The *vacuum* or *vacuum configuration* is any field configuration that is constant both in time and in space, having zero topological charge.

target. As we shall see further, field configurations for this model satisfy the following system of semilinear wave equations,

$$\partial_\alpha \partial^\alpha \phi + \langle \partial^\alpha \phi, \partial_\alpha \phi \rangle_h \, \phi = 0, \tag{1.2}$$

and have an a priori conserved energy, which is given by

$$\mathcal{E}(\phi)(t) = \frac{1}{2} \int_{\mathbb{R}^n} |\nabla \phi(t)|_h^2 \, dx = \mathcal{E}(\phi)(0).$$

This implies that the gradient of any instantaneous soliton (i.e., $x \to \phi(t,x)$) vanishes at spatial infinity and, modulo a compactification of the space, we can identify $\phi(t)$ with a map $\tilde{\phi}(t) : \mathbb{S}^n \to \mathbb{S}^n$. This has a winding number, or *topological charge* in physical terminology, associated to it, which is described by

$$Q(\phi)(t) = \frac{1}{4\pi} \int_{\mathbb{R}^2} \phi \cdot (\partial_1 \phi \wedge \partial_2 \phi)(t,x) \, dx \tag{1.3}$$

when $n = 2$, and by

$$Q(\phi)(t) = \frac{1}{2\pi^2} \int_{\mathbb{R}^3} \det(\phi, \partial_1 \phi, \partial_2 \phi, \partial_3 \phi)(t,x) \, dx \tag{1.4}$$

when $n = 3$. The specific constants in the above integrals are chosen such that $Q$ has integer values. If the time evolution is continuous, the topological charge is a conserved quantity of the soliton.

In what concerns the Gell-Mann-Lévy model (i.e., the $O(4)$ sigma model), it was Tony Skyrme's revolutionary idea in the 1960s to suggest that $Q$ is nothing but the *baryon number* of the nucleus, which is the total number of neutrons and protons. This was later confirmed by studies in quantum chromodynamics (e.g., Balachandran-Nair-Rajeev-Stern [8,9], Witten [189,190]), the fundamental theory of nuclear interactions, and also by numerical calculations (Battye-Sutcliffe [10]). Furthermore, based on the attractive nature of the forces between $\pi$ mesons, the physical intuition postulated that, as time evolves, a soliton with topological charge equal to one would shrink to a point (i.e., a vacuum configuration), emitting its energy as radiation carried to infinity. Thus, one expected this model to develop singularities in finite time, a fact which was corroborated by the blow-up solution constructed by Shatah [143]. A few years after Shatah's result, Turok and Spergel [181] were able to find this solution in closed form.

Next, we focus on static topological solitons for the $O(n+1)$ sigma model and observe that (1.2) reduces to

$$\Delta \phi + |\nabla_x \phi|_h^2 \, \phi = 0,$$

which is invariant under the scaling transformation

$$\phi \to \phi_\lambda = \phi_\lambda(x) \overset{\text{def}}{=} \phi\left(\frac{x}{\lambda}\right).$$

The topological charge is also scale-invariant, i.e.,

$$Q(\phi_\lambda) = Q(\phi), \tag{1.5}$$

whereas the energy changes according to

$$\mathcal{E}(\phi_\lambda) = \lambda^{n-2}\mathcal{E}(\phi). \tag{1.6}$$

Relevant to our discussion is a classical non-existence result, which can be formulated as follows (see also [113]):

**Proposition 1.1 (Derrick [37]).** *If the static energy functional associated to a field theory does not have a stationary point with respect to a spatial rescaling, then that theory admits no static topological solitons.*

What is meant above by *the static energy functional* is precisely the map

$$\lambda \mapsto \mathcal{E}(\phi_\lambda),$$

where $\phi$ is a static field of the respective model.

Due to (1.6), for the $O(n+1)$ sigma model with $n \geq 3$, this functional is monotone in $\lambda$ and thus does not have stationary points. As a consequence of Derrick's result, these nonlinear sigma models (which include the Gell-Mann-Lévy model) do not admit static topological solitons. On the contrary, Derrick's result does not apply to the $O(3)$ sigma model, as its static energy functional is constant and, accordingly, has infinitely many stationary points. In fact, the $O(3)$ sigma model admits static topological solitons called *lumps*. This terminology has to do with an apparent instability manifested by the fact that they do not have a prescribed scale.

### 1.1.2 *Skyrme and Skyrme-like models*

The previous discussion of nonlinear sigma models raises the issue of whether there are physically-inspired generalizations for the $O(n+1)$ sigma model, with $n \geq 3$, that have static topological solitons. This served as the main motivation that led Skyrme to the discovery of his model [152–154], which can be seen as a generalization of the $O(4)$ sigma model that has static topological solitons called *skyrmions*. Historically speaking, the skyrmions are the first topological solitons to model a particle. The *Skyrme model* has also been found to be an effective model for quantum chromodynamics (QCD), Witten [189, 190] being the first one to show the profound connections between the Skyrme model and the topological anomalies of QCD.

The action of the Skyrme model is given by[2]

$$S = \int_{\mathbb{R}^{3+1}} \left[ \frac{1}{2} \langle \partial^\mu \phi, \partial_\mu \phi \rangle_h \right.$$
$$\left. - \frac{\Lambda^2}{4} \left( \langle \partial^\mu \phi, \partial^\nu \phi \rangle_h \langle \partial_\mu \phi, \partial_\nu \phi \rangle_h - \langle \partial^\mu \phi, \partial_\mu \phi \rangle_h^2 \right) \right] dg, \tag{1.7}$$

with $\phi : \mathbb{R}^{3+1} \to \mathbb{S}^3$ and $\Lambda$ being a constant having the dimension of length. In this case, the a priori conserved energy has the schematic form

$$\mathcal{E}(\phi)(t) = \frac{1}{2} \int_{\mathbb{R}^3} |\nabla \phi(t)|_h^2 + \text{“} |\nabla \phi(t)|_h^4 \text{”} \, dx.$$

This implies that the static energy functional satisfies

$$\mathcal{E}(\phi_\lambda) = \lambda \, E_2(\phi) + \frac{1}{\lambda} E_4(\phi),$$

where $E_2(\phi)$ and $E_4(\phi)$ denote the quadratic and quartic terms, respectively, in the corresponding static energy. Hence, it has the stationary point

$$\lambda = \sqrt{\frac{E_4(\phi)}{E_2(\phi)}}.$$

From this computation, it is clear that any other term which is of order four or higher in spatial derivatives would have worked in place of the one introduced by Skyrme in (1.7) to evade Derrick's result. However, Skyrme's term is the only quartic term which is conformally-invariant and for which the resulting Euler-Lagrange equations remain of second order in the time derivative. For $\Lambda = 1$, these equations can be written as

$$(1 + \langle \partial^\mu \phi, \partial_\mu \phi \rangle_h) \, \partial_\alpha \partial^\alpha \phi - \langle \partial^\mu \phi, \partial^\alpha \phi \rangle_h \partial_\mu \partial_\alpha \phi$$
$$+ (\langle \partial^\alpha \phi, \partial_\alpha \partial_\mu \phi \rangle_h - \langle \partial_\mu \phi, \partial_\alpha \partial^\alpha \phi \rangle_h) \partial^\mu \phi$$
$$- \{ \langle \partial_\mu \phi, \partial_\alpha \phi \rangle_h \langle \partial^\mu \phi, \partial^\alpha \phi \rangle_h$$
$$- (1 + \langle \partial^\mu \phi, \partial_\mu \phi \rangle_h) \langle \partial^\alpha \phi, \partial_\alpha \phi \rangle_h \} \phi = 0, \tag{1.8}$$

which is a system of quasilinear partial differential equations.

The presence of stationary points for the static energy functional associated to the Skyrme model does not guarantee by itself the existence of

---

[2]The original model also contained a term that gives the $\pi$ mesons a mass. However, the qualitative features of the field theory are unchanged by its inclusion (see also the comments in [113]). Hence, in order to simplify the exposition, we decided to work with the massless Lagrangian.

skyrmions. However, their existence was first proven rigorously by Kapitanskiĭ and Ladyzenskaya [77] through variational methods. For the skyrmion
of unit topological charge, a later argument using only ODE techniques was
provided by McLeod and Troy [114].

Next, we discuss three important field theories, which are intimately
related to the Skyrme model and have static topological solitons. The
first one is the *Faddeev model* [43, 44], a nonlinear field theory modeling
elementary heavy particles, whose action is described by

$$S = \int_{\mathbb{R}^{3+1}} \frac{1}{2} \partial_\mu \mathbf{n} \cdot \partial^\mu \mathbf{n} + \frac{1}{4} (\partial_\mu \mathbf{n} \wedge \partial_\nu \mathbf{n}) \cdot (\partial^\mu \mathbf{n} \wedge \partial^\nu \mathbf{n}) \, dg. \qquad (1.9)$$

In the above, $v_1 \wedge v_2$ denotes the cross product of the vectors $v_1$ and $v_2$ in
$\mathbb{R}^3$ and $\mathbf{n} : \mathbb{R}^{3+1} \to \mathbb{S}^2$. The associated Euler-Lagrange equations take the
form

$$\mathbf{n} \wedge \partial_\mu \partial^\mu \mathbf{n} + (\partial_\mu [\mathbf{n} \cdot (\partial^\mu \mathbf{n} \wedge \partial^\nu \mathbf{n})]) \partial_\nu \mathbf{n} = 0, \qquad (1.10)$$

yet another system of quasilinear equations. We notice that the action (1.9)
is also well-defined for maps $\mathbf{n} : \mathbb{R}^{n+1} \to \mathbb{S}^2$, with an identical variational
formulation.

The connection between the Faddeev and Skyrme models is revealed by
restricting the image of a field configuration $\phi : \mathbb{R}^{3+1} \to \mathbb{S}^3$ in (1.7) to an
equatorial 2-sphere of $\mathbb{S}^3$ (identified in this case with $\mathbb{S}^2$). Indeed, if we
write the metric on $\mathbb{S}^3$ as

$$h = du^2 + \sin^2 u \, d\mathbf{n}^2,$$

where $0 \le u \le \pi$ and $\mathbf{n} \in \mathbb{S}^2$, the Faddeev action (1.9) is obtained from
(1.7) by prescribing

$$\phi = (u, \mathbf{n}) = \left( \frac{\pi}{2}, \mathbf{n} \right).$$

By comparison to the Skyrme model, the Faddeev model has a remark-
able new feature: it admits *knotted solitons*, as verified numerically by Battye and Sutcliffe [11, 12]. The topological structure of the Faddeev theory
matches this new feature. It can be described by introducing the 2-form

$$F_{\mu\nu}(\mathbf{n}) \stackrel{\mathrm{def}}{=} \mathbf{n} \cdot (\partial_\mu \mathbf{n} \wedge \partial_\nu \mathbf{n}),$$

which is exact. This implies the existence of a gauge potential $A = A_\mu \, dx^\mu$
satisfying $F = dA$ and a simple computation shows that the topological
current

$$J^\mu \stackrel{\mathrm{def}}{=} \epsilon^{\mu\alpha\beta\gamma} F_{\alpha\beta}(\mathbf{n}) A_\gamma(\mathbf{n})$$

is divergence-free, i.e.,

$$\partial_\mu J^\mu = 0.$$

In the above, $\epsilon^{\mu\alpha\beta\gamma}$ denotes the 4-dimensional Levi-Civita symbol. As a consequence, the Faddeev map $\mathbf{n}$ has a conserved topological charge,

$$Q(\mathbf{n})(t) = \frac{1}{32\pi^2} \int_{\mathbb{R}^3} \epsilon^{\alpha\beta\gamma} F_{\alpha\beta}(\mathbf{n}) A_\gamma(\mathbf{n})(t, x) \, dx, \qquad (1.11)$$

which is also called the *Hopf index* of $\mathbf{n}$.

The second Skyrme-like theory we present here is a $2 + 1$-dimensional model appearing first in work by Leese, Peyrard, and Zakrzewski [102] (see also Piette-Schroers-Zakrzewski [132] and Weidig [184]), which has since been found to have applications in condensed matter physics [101] and cosmology [15]. It is usually referred to as the *baby Skyrme model*, being an extension of the $O(3)$ sigma model. As mentioned before, the $O(3)$ sigma model has static topological solitons, but Derrick's argument fails to prescribe a certain scale for them. The main goal of the baby Skyrme model is to rectify this issue in the context of a Lagrangian that inherits the Skyrme Lagrangian.

Thus, we are naturally led to consider actions of the type

$$\begin{aligned} S = \int_{\mathbb{R}^{2+1}} \Big[ & \frac{1}{2} \langle \partial^\mu \phi, \partial_\mu \phi \rangle_h \\ & - \frac{\Lambda^2}{4} \left( \langle \partial^\mu \phi, \partial^\nu \phi \rangle_h \langle \partial_\mu \phi, \partial_\nu \phi \rangle_h - \langle \partial^\mu \phi, \partial_\mu \phi \rangle_h^2 \right) \qquad (1.12) \\ & + \tilde\Lambda^2 V(\phi) \Big] \, dg. \end{aligned}$$

In the above, $\phi : \mathbb{R}^{2+1} \to \mathbb{S}^2$ and $\tilde\Lambda$ is a constant having the dimension of inverse length. The static energy functional is given by

$$\mathcal{E}(\phi_\lambda) = \lambda^2 E_0(\phi) + E_2(\phi) + \frac{1}{\lambda^2} E_4(\phi),$$

where $E_0(\phi)$ is the energy component due to the potential term $V(\phi)$, and $E_2(\phi)$ and $E_4(\phi)$ have the same significance as for the Skyrme model. The potential term is needed to ensure that

$$\lambda = \sqrt[4]{\frac{E_4(\phi)}{E_0(\phi)}}$$

is the only stationary point for $\mathcal{E}(\phi_\lambda)$.

The choices for $V(\phi)$ vary in literature. Initially, Leese-Peyrard-Zakrzewski [102] chose

$$V(\phi) = (1 + \phi_3)^4, \qquad \phi = (\phi_1, \phi_2, \phi_3) \in \mathbb{S}^2.$$

Using the stereographic projection

$$W = \frac{\phi_1 - i\phi_2}{1 + \phi_3}$$

from $\mathbb{S}^2$ to $\mathbb{C}$, they were able to find a static solution for the Euler-Lagrange equations described by

$$W = W(x_1, x_2) = \mu(x_1 + ix_2)$$

with $\mu$ being a parameter depending solely on $\Lambda$ and $\tilde{\Lambda}$. For this reason, this theory is sometimes called the *holomorphic* model. In [132], the authors worked with $V(\phi) = 1 - \phi_3$ and obtained non-radial minimal energy solitons. On the contrary, Weidig [184] chose $V(\phi) = 1 - \phi_3^2$ and derived minimal energy solitons that are radially-symmetric.

Finally, we introduce the *Adkins-Nappi model* [1], which can be regarded as a semilinear Skyrme model. It is a generalization of the $O(4)$ sigma model having static topological solitons, but it does so without introducing higher-order derivative terms. Instead, this theory describes the nonlinear coupling of the scalar field from the sigma model with a gauge field. From the physical point of view, the idea is to add a short range repulsion among the $\pi$ mesons, which is created by their interactions with an $\omega$ vector meson. Later, a $2+1$-dimensional Adkins-Nappi-like model was proposed by Foster and Sutcliffe [47].

The action of the Adkins-Nappi model is given by[3]

$$S = \int_{\mathbb{R}^{3+1}} \left[ \frac{1}{2} \langle \partial^\mu \phi, \partial_\mu \phi \rangle_h + \frac{1}{4} F^{\mu\nu} F_{\mu\nu} - A_\mu J^\mu \right] dg, \qquad (1.13)$$

where

$$A = (A_\mu)_\mu : \mathbb{R}^{3+1} \to \mathbb{R}^4$$

is the gauge field representing the $\omega$ meson, the 2-form

$$F_{\mu\nu} \overset{\text{def}}{=} \partial_\mu A_\nu - \partial_\nu A_\mu$$

is its associated field tensor, and

$$J^\mu \overset{\text{def}}{=} c\,\epsilon^{\mu\nu\rho\sigma}\,\partial_\nu \phi^i\,\partial_\rho \phi^j\,\partial_\sigma \phi^k\,\epsilon_{ijk}$$

is the topological current for the scalar field $\phi : \mathbb{R}^{3+1} \to \mathbb{S}^3$. In the last formula, $\epsilon$ stands for the Levi-Civita symbol and $c$ is a normalizing constant.

---

[3]The original theory also involved two terms introducing masses for both the $\pi$ and $\omega$ mesons. As was the case for the Skyrme model, these terms do not change the qualitative properties of the Adkins-Nappi theory and, as a consequence, we give them no further consideration.

In the static case, we easily obtain that $J^1 = J^2 = J^3 = 0$, and, as the topological current provides the "source term" for the gauge field, it follows that we can assume $A_1 = A_2 = A_3 = 0$. A calculation for the energy in this regime shows that

$$\mathcal{E}(\phi, A) \simeq \int_{\mathbb{R}^3} |\nabla_x \phi|_h^2 + |\nabla_x A_0|^2 \, dx.$$

As $A$ is a gauge field, its associated scaling transformation is

$$A \to A_\lambda = A_\lambda(x) \overset{\text{def}}{=} \frac{1}{\lambda} A\left(\frac{x}{\lambda}\right),$$

which implies that the static energy functional has the form

$$\mathcal{E}(\phi_\lambda, A_\lambda) = \lambda \mathcal{E}(\phi) + \frac{1}{\lambda} \mathcal{E}(A),$$

where $\mathcal{E}(\phi)$ and $\mathcal{E}(A)$ are the parts of energy due to the scalar and gauge fields, respectively. Thus, we derive that this functional has the unique stationary point

$$\lambda = \sqrt{\frac{\mathcal{E}(A)}{\mathcal{E}(\phi)}},$$

a fact which allows one to bypass Derrick's result.

The Euler-Lagrange equations for the Adkins-Nappi model are given by

$$\partial_\alpha \partial^\alpha \phi^i + \langle \partial^\alpha \phi, \partial_\alpha \phi \rangle_h \, \phi^i - \frac{3c}{2} F_{\alpha\beta} \, \epsilon^{\alpha\beta\gamma\delta} \, h^{il}(\phi) \, \partial_\gamma \phi^j \, \partial_\delta \phi^k \, \epsilon_{ljk} = 0 \quad (1.14)$$

and

$$\partial_\alpha F^{\alpha\beta} = J^\beta, \tag{1.15}$$

which form a coupled system of semilinear equations. This justifies calling this theory a semilinear Skyrme model.

## 1.2 Mathematical formulation and basic properties

We focus here on presenting the mathematical background for the nonlinear sigma and Skyrme models, which includes among others a thorough discussion of the associated Euler-Lagrange equations and of the energy-momentum tensor. At the end, to further motivate the importance of these problems, we describe the reduction, in the presence of symmetries, of the Einstein vacuum equations to a system which includes a wave map type of problem.

### 1.2.1   *The wave map problem: initial formulation and the Euler-Lagrange equations*

We will consider first the simplest field theory introduced so far, the nonlinear sigma model, for which we make the extra assumption that the domain manifold is Lorentzian. This is called the *wave map problem* by the mathematics community. The terminology has to do with this problem being the hyperbolic analogue of what is known as the *harmonic map problem*, which is the nonlinear sigma model for the case when the domain is Riemannian.

Precisely, the basic object of study for this problem is a map

$$\phi : (M, g_{\mu\nu}) \to (N, h_{ab}),$$

where the domain or base $M^{n+1}$ is an $n+1$-dimensional Lorentzian manifold, that is the signature of the metric $g_{\mu\nu}$ is $(n,1)$, and the target $N^k$ is a $k$-dimensional Riemannian manifold. The map $\phi$ is called a *wave map* if it is a stationary point for the Lagrangian functional

$$\mathbf{L}[\phi] \stackrel{\text{def}}{=} \int_M \left\{ \frac{1}{2} g^{\mu\nu}(x) h_{ab}(\phi) \nabla_\mu \phi^a \nabla_\nu \phi^b \right\} d\mu_g. \tag{1.16}$$

The Lagrangian is written in local coordinates on the target, for which we use the notation $\phi^a = \phi^a(x^\mu)$ to indicate their dependence on the coordinates of the base, that are labeled by $x^\mu$. As such, the indices $a$ and $\mu$ take the values $\{1, 2, \dots, k\}$ and $\{0, 1, \dots, n\}$, respectively. We denoted by $d\mu_g$ the measure with respect to the metric $g_{\mu\nu}$ on the base manifold, which is given by

$$d\mu_g \stackrel{\text{def}}{=} \sqrt{-\det(g_{\mu\nu})} \prod_{\mu=0}^{n} dx^\mu .$$

We applied the standard convention to write $g^{\mu\nu} := (g_{\mu\nu})^{-1}$ and $h^{ab} := (h_{ab})^{-1}$ for the inverses of the two metric tensors, which will be used in raising indices. Needless to say, in the formula for the Lagrangian, we relied on the Einstein summation convention, which is used throughout the book. Moreover, in order to simplify the notation, we will also employ the notational conventions

$$\langle \nabla \phi^a, \nabla \phi^b \rangle \stackrel{\text{def}}{=} g^{\mu\nu} \nabla_\mu \phi^a \nabla_\nu \phi^b, \tag{1.17}$$

$$\langle \nabla_\mu \phi, \nabla_\nu \phi \rangle \stackrel{\text{def}}{=} h_{ab} \nabla_\mu \phi^a \nabla_\mu \phi^b, \tag{1.18}$$

$$\langle \nabla \phi, \nabla \phi \rangle \stackrel{\text{def}}{=} h_{ab} \, g^{\mu\nu} \nabla_\mu \phi^a \nabla_\nu \phi^b, \tag{1.19}$$

to designate the partial and total contractions of the tensor $\nabla_\mu \phi^a \nabla_\nu \phi^b$, respectively. Thus, the Lagrangian becomes then $\langle \nabla \phi, \nabla \phi \rangle$.

The Euler-Lagrange equations associated to (1.16) can be derived via a simple variation argument. Namely, we consider a small arbitrary variation of the target coordinates, which is denoted by $\delta\phi$, and we compute the change in value of the Lagrangian functional up to first order in $\delta\phi$. This is done using integration by parts and taking advantage of $\delta\phi$ vanishing on the boundary of $M$. We arrive thus at the formula

$$\mathbf{L}[\phi + \delta\phi] - \mathbf{L}[\phi] = \int_M \frac{\delta\mathbf{L}}{\delta\phi^a}\,\delta\phi^a\,d\mu_g,$$

where

$$\frac{\delta\mathbf{L}}{\delta\phi^a} = h_{ab}\Box\phi^b - \frac{\partial h_{ab}}{\partial\phi^c}\langle\nabla\phi^b, \nabla\phi^c\rangle + \frac{1}{2}\frac{\partial h_{bc}}{\partial\phi^a}\langle\nabla\phi^b, \nabla\phi^c\rangle$$

$$= h_{ab}\left\{\Box\phi^b - \Gamma^b_{cd}(\phi)\langle\nabla\phi^c, \nabla\phi^d\rangle\right\}.$$

In the above, we introduced the d'Alembertian or the wave operator

$$\Box \stackrel{\text{def}}{=} -\nabla_\mu\nabla^\mu \qquad (1.20)$$

and the Christoffel symbols on the target manifold, which are given by

$$\Gamma_{cd;a} \stackrel{\text{def}}{=} \frac{1}{2}\left\{\partial_c h_{ad} + \partial_d h_{ac} - \partial_a h_{cd}\right\},$$

whereas $\Gamma^b_{cd} := h^{ba}\Gamma_{cd;a}$. A stationary point for the Lagrangian functional is defined to satisfy

$$\frac{\delta\mathbf{L}}{\delta\phi^a} = 0,$$

which leads to the following system of equations:

$$\Box\phi^a - \Gamma^a_{bc}(\phi)\langle\nabla\phi^b, \nabla\phi^c\rangle = 0. \qquad (1.21)$$

This is called in literature the *wave map system*.

It is easy to see that, if the target manifold is flat[4], then the wave map system reduces to $\Box\phi^a = 0$, which is a system of uncoupled linear wave equations. Hence, one can think of (1.21) as a nonlinear generalization of a system of linear wave equations and, if we adopt this point of view, then the wave map system describes the time evolution of the map $\phi$. In order to have a complete mathematical problem, we need to prescribe appropriate initial data on a spacelike hypersurface[5] in $M$ and subsequently try to solve the problem either forward or backward in "time".

---

[4]This means that, in some appropriate local coordinate system, the Christoffel symbols vanish. Such is the case when $N = \mathbb{R}^k$ and one chooses a Cartesian coordinate system on $N$.

[5]We refer the reader to the appendix for complete details on the timelike/spacelike/null terminology.

This is more transparent when we take the base manifold to be the Minkowski spacetime, i.e., $M = \mathbb{R}^{n+1}$ and $g_{\mu\nu} = m_{\mu\nu} = \text{diag}(-1, 1, \ldots, 1)$, for which the time variable is labeled by $t := x^0$ and $\mathbf{x} := (x^1, \ldots, x^n) \in \mathbb{R}^n$ collectively denotes the space variables. In this case, the inverse metric is given by $m^{\mu\nu} = m_{\mu\nu} = \text{diag}(-1, 1, \ldots, 1)$ and the wave operator takes the familiar form $\Box = \partial_t^2 - \Delta_{\mathbf{x}}$. The simplest way to prescribe initial data in this setting is to let $\phi^a(0, \mathbf{x}) := \phi_0^a(\mathbf{x})$ and $\partial_t \phi^a(0, \mathbf{x}) := \phi_1^a(\mathbf{x})$, where $\phi_0$ and $\phi_1$ are given functions on $\mathbb{R}^n$. Thus, the complete mathematical problem to be solved is the following Cauchy problem:

$$\begin{cases} \Box \phi^a - \Gamma_{bc}^a(\phi) \langle \nabla \phi^b, \nabla \phi^c \rangle = 0, & \phi^a = \phi^a(t, \mathbf{x}), \quad 1 \le a \le k, \\ \phi^a(0, \mathbf{x}) = \phi_0^a(\mathbf{x}), & \partial_t \phi^a(0, \mathbf{x}) = \phi_1^a(\mathbf{x}). \end{cases} \tag{1.22}$$

From either (1.21) or the above system, it would seem that the nonlinearity in the wave map problem is due to the presence of the Christoffel symbols. This is somewhat misleading, for if we take $N = \mathbb{R}^2$ and use polar coordinates instead of Cartesian ones, then we see that the Christoffel symbols do not vanish, yet the equations of the wave map system are linear. We will see later that a better way to think about the wave map problem is that the nonlinearity is due in fact to the curvature of the target manifold. It is both desirable and beneficial to rewrite the equations in such a manner that the true nature of the nonlinearity is revealed. This goal is achieved by deriving a Hodge system which, under reasonable assumptions, is equivalent to the original system of equations.

### 1.2.2 The wave map problem: symmetries of the base and the energy-momentum tensor

The symmetries of the wave map problem play a prominent role in its analysis, in the sense that they generate conserved quantities that can be used as a priori estimates. The central idea here is that these bounds, when combined with other analytic estimates, allow us to obtain local and sometimes global existence results for the Cauchy problem (1.22). In this section, we start covering the essentials of this aspect for the wave map problem, namely the relation between its symmetries and conserved quantities. There are two distinct types of symmetries to be considered: symmetries generated by the base and symmetries coming from the target. In what follows, we focus on the former.

The essential point that we wish to make here is that the Lagrangian is coordinate independent. As a consequence, it is invariant under an infinitesimal change of coordinates on the base manifold. We can then investigate the corresponding symmetries by constructing the energy-momentum tensor

$$T_{\mu\nu} \overset{\text{def}}{=} \langle \nabla_\mu \phi, \nabla_\nu \phi \rangle - \frac{1}{2} g_{\mu\nu} \langle \nabla\phi, \nabla\phi \rangle, \tag{1.23}$$

which is symmetric in the indices $\mu$ and $\nu$ and satisfies the structure equations

$$\nabla_\mu \{ T^\mu_\nu \} = 0. \tag{1.24}$$

The derivation of these equations in general is entirely similar to the case when $\phi$ is a scalar field, the reason being that the coordinates $\phi^a = \phi^a(x)$ on the target behave like scalar fields with respect to the variables on the base manifold.

A sketch for the argument proving (1.24) goes as follows, with a more thorough discussion in the appendix. We consider an arbitrary vector field $\xi$ on the base manifold with compact support and argue that the derivative of the Lagrangian along the flow generated by $\xi$ is zero, due to the invariances mentioned above. To compute this derivative, we rely on the Lie derivative in the direction of $\xi$, which is denoted by $\mathcal{L}_\xi$, and the basic properties

$$\mathcal{L}_\xi g_{\mu\nu} = \nabla_\mu \xi_\nu + \nabla_\nu \xi_\mu, \qquad \mathcal{L}_\xi d\mu_g = (\text{div } \xi) d\mu_g, \qquad \mathcal{L}_\xi \nabla_\nu \phi = \nabla_\nu \mathcal{L}_\xi \phi.$$

Using also (1.16), integration by parts, and (1.21), we derive

$$\begin{aligned}
0 &= \frac{d}{ds}\Big|_{s=0} \{ \mathbf{L}[\phi(s)] \} \\
&= \int_M \left\{ -\frac{1}{2}(\nabla^\mu \xi^\nu + \nabla^\nu \xi^\mu)\langle \nabla_\mu \phi, \nabla_\nu \phi \rangle + \frac{1}{2}(\nabla_\mu \xi^\mu)\langle \nabla\phi, \nabla\phi \rangle \right\} d\mu_g \\
&\quad + \int_M \left\{ g^{\mu\nu} h_{ab} \nabla_\mu \phi^a \mathcal{L}_\xi \nabla_\nu \phi^b + \frac{1}{2}\langle \nabla\phi^a, \nabla\phi^b \rangle \frac{\partial h_{ab}}{\partial \phi^c} \mathcal{L}_\xi \phi^c \right\} d\mu_g \\
&= \int_M \left\{ \xi^\nu \nabla_\mu \Big( \langle \nabla^\mu \phi, \nabla_\nu \phi \rangle - \frac{1}{2}\delta^\mu_\nu \langle \nabla\phi, \nabla\phi \rangle \Big) \right\} d\mu_g \\
&\quad + \int_M \left\{ \Big( -\nabla_\mu(h_{ab} \nabla^\mu \phi^a) + \frac{1}{2}\frac{\partial h_{ac}}{\partial \phi^b} \langle \nabla\phi^a, \nabla\phi^c \rangle \Big) \mathcal{L}_\xi \phi^b \right\} d\mu_g \\
&= \int_M \{ \xi^\nu \nabla_\mu \{ T^\mu_\nu \} \} d\mu_g,
\end{aligned}$$

which implies (1.24) since $\xi$ is arbitrary.

The standard way in which we use (1.24) is by contracting it first with an arbitrary vector field $X$ to deduce

$$-\nabla_\mu \{ T^\mu_\nu X^\nu \} + (\nabla_\mu X_\nu) T^{\mu\nu} = 0. \tag{1.25}$$

Following this, if we define the vector field $P^\mu := T^\mu_\nu X^\nu$ and the scalar $R := (\nabla_\mu X_\nu) T^{\mu\nu}$, then the previous identity takes the form $-\nabla_\mu P^\mu + R = 0$, which we claim provides useful a priori estimates when integrated over appropriate domains. As an example, in the particular case when $X$ is a Killing vector field, i.e.,

$$\mathcal{L}_X g_{\mu\nu} = \nabla_\mu X_\nu + \nabla_\nu X_\mu = 0,$$

by relying on the symmetry of $T$, we infer that $R = 0$ and, subsequently, $\nabla_\mu P^\mu = 0$. This leads to conserved quantities by integration on specific spacelike hypersurfaces of $M$.

Let us discuss now how we can derive useful bounds from (1.25) in the general case. A first remark is that, when integrating (1.25) over certain domains, it would be beneficial to obtain quantities with a definite sign, which we ask to be positive by convention. Assuming the boundary of the domain of integration contains level sets for some specific function $u$, an application of Gauss's theorem results in surface integrals of the form

$$\int_{u=\text{const}} \{T_{\mu\nu} n^\mu X^\nu\} \, d\sigma, \tag{1.26}$$

where $\mathbf{n} := c\nabla u$ is the normal unit vector to the hypersurface $\{u = \text{const}\}$, with $c$ being an appropriate normalizing constant. We argue that if the vector field $X$ is timelike, $\mathbf{n}$ is either timelike or null, and they have similar orientation[6], then the integrand of the previous surface integral is nonnegative.

First, we pursue this when $\mathbf{n}$ is timelike and, as such, $\langle \mathbf{n}, \mathbf{n} \rangle = -1$. Due to the signature of the metric $g$, we can find a collection of vectors $\{e_j\}$, with $1 \le j \le n$, such that

$$\langle e_j, e_k \rangle = \delta_{jk}, \qquad \langle \mathbf{n}, e_j \rangle = 0.$$

We decompose both $X$ and $\nabla\phi$ with respect to the orthonormal frame formed by $\mathbf{n}$ and $\{e_1, \dots, e_n\}$ to obtain

$$X = c_0 \mathbf{n} + \sum_j c_j e_j$$

and

$$\nabla\phi = -(\nabla_{\mathbf{n}}\phi)\mathbf{n} + \sum_j (\nabla_{e_j}\phi)e_j,$$

---

[6] In the particular case of the Minkowski spacetime, this would amount to both $X$ and $\mathbf{n}$ being future or past-oriented.

where $\nabla_\xi \phi := \xi^\mu \nabla_\mu \phi$. Since $X$ is timelike, we derive

$$\langle X, X \rangle = -c_0^2 + \sum_j c_j^2 < 0. \tag{1.27}$$

Moreover, what we meant above by $X$ and $\mathbf{n}$ having similar orientation is that $c_0 = -\langle X, \mathbf{n} \rangle > 0$. We have now all the prerequisites to show that

$$T_{\mu\nu} \, \mathbf{n}^\mu X^\nu \geq 0. \tag{1.28}$$

Indeed, using the formula (1.23), the Cauchy-Schwarz inequality, and (1.27), we derive[7]

$$
\begin{aligned}
T_{\mu\nu} \, \mathbf{n}^\mu X^\nu &= \frac{c_0}{2} \left\{ \langle \nabla_\mathbf{n} \phi, \nabla_\mathbf{n} \phi \rangle + \sum_j \langle \nabla_{e_j} \phi, \nabla_{e_j} \phi \rangle \right\} + \sum_j c_j \langle \nabla_{e_j} \phi, \nabla_\mathbf{n} \phi \rangle \\
&\geq \frac{c_0}{2} \left\{ \langle \nabla_\mathbf{n} \phi, \nabla_\mathbf{n} \phi \rangle + \sum_j \langle \nabla_{e_j} \phi, \nabla_{e_j} \phi \rangle \right\} \\
&\quad - \left( \sum_j c_j^2 \right)^{1/2} \langle \nabla_\mathbf{n} \phi, \nabla_\mathbf{n} \phi \rangle^{1/2} \left( \sum_j \langle \nabla_{e_j} \phi, \nabla_{e_j} \phi \rangle \right)^{1/2} \\
&\geq \frac{c_0 - \left( \sum_j c_j^2 \right)^{1/2}}{2} \left\{ \langle \nabla_\mathbf{n} \phi, \nabla_\mathbf{n} \phi \rangle + \sum_j \langle \nabla_{e_j} \phi, \nabla_{e_j} \phi \rangle \right\} \geq 0.
\end{aligned}
$$

It is worth noticing that this yields control over all derivatives of $\phi$.

Next, we consider the case when $\mathbf{n}$ is a null vector, meaning that $\langle \mathbf{n}, \mathbf{n} \rangle = 0$. We follow an approach similar to the previous one, yet with an important, distinct, step. Here, we cannot construct as above an orthonormal frame containing $\mathbf{n}$. Instead, we rely on building a *null frame*[8], which, apart from $\mathbf{n}$, contains a dual null vector $\mathbf{n}^\dagger$, i.e.,

$$\langle \mathbf{n}^\dagger, \mathbf{n}^\dagger \rangle = 0, \qquad \langle \mathbf{n}, \mathbf{n}^\dagger \rangle = -1,$$

and $n - 1$ spacelike vectors $\{e_j\}$ $(1 \leq j \leq n - 1)$ such that

$$\langle e_j, e_k \rangle = \delta_{jk}, \qquad \langle \mathbf{n}, e_j \rangle = \langle \mathbf{n}^\dagger, e_j \rangle = 0.$$

---

[7]We keep in mind that $\{\phi^a\}$ are coordinates on the target manifold, on which we have a Riemannian metric, and, as a result, $\langle \nabla_\xi \phi, \nabla_\xi \phi \rangle = h_{ab} \nabla_\xi \phi^a \nabla_\xi \phi^b \geq 0$ holds for an arbitrary vector $\xi$.

[8]In the particular case when the base is the Minkowski spacetime, an example of a null frame is

$$\left\{ (\nabla_t - \nabla_r)/\sqrt{2}, (\nabla_t + \nabla_r)/\sqrt{2}, \nabla_\omega \right\},$$

where $\nabla_r$ stands for the radial derivative and $\nabla_\omega$ collectively denotes the angular derivatives.

As before, we decompose $X$ and $\nabla\phi$ with respect to the null frame to deduce

$$X = c\mathbf{n} + c^\dagger \mathbf{n}^\dagger + \sum_j c_j e_j,$$

where $c = -\langle \mathbf{n}^\dagger, X \rangle$ and $c^\dagger = -\langle \mathbf{n}, X \rangle$, and

$$\nabla\phi = -(\nabla_{\mathbf{n}^\dagger}\phi)\mathbf{n} - (\nabla_{\mathbf{n}}\phi)\mathbf{n}^\dagger + \sum_j (\nabla_{e_j}\phi)e_j,$$

with the shorthand notation $\nabla_\xi \phi$ having the same meaning as when $\mathbf{n}$ was timelike. The fact that $X$ is timelike implies

$$\langle X, X \rangle = -2cc^\dagger + \sum_j c_j^2 < 0, \tag{1.29}$$

and the orientation convention is such that $c^\dagger > 0$ (hence, $c > 0$ holds too). Using the above decompositions, the formula for the energy-momentum tensor, the Cauchy-Schwarz inequality, and (1.29), we infer

$$\begin{aligned}
T_{\mu\nu} \mathbf{n}^\mu X^\nu &= c\langle \nabla_{\mathbf{n}}\phi, \nabla_{\mathbf{n}}\phi \rangle + \frac{c^\dagger}{2} \sum_j \langle \nabla_{e_j}\phi, \nabla_{e_j}\phi \rangle + \sum_j c_j \langle \nabla_{e_j}\phi, \nabla_{\mathbf{n}}\phi \rangle \\
&\geq c\langle \nabla_{\mathbf{n}}\phi, \nabla_{\mathbf{n}}\phi \rangle + \frac{c^\dagger}{2} \sum_j \langle \nabla_{e_j}\phi, \nabla_{e_j}\phi \rangle \\
&\quad - \left( \sum_j c_j^2 \right)^{1/2} \langle \nabla_{\mathbf{n}}\phi, \nabla_{\mathbf{n}}\phi \rangle^{1/2} \left( \sum_j \langle \nabla_{e_j}\phi, \nabla_{e_j}\phi \rangle \right)^{1/2} \\
&\geq C \left\{ \langle \nabla_{\mathbf{n}}\phi, \nabla_{\mathbf{n}}\phi \rangle + \sum_j \langle \nabla_{e_j}\phi, \nabla_{e_j}\phi \rangle \right\} \geq 0,
\end{aligned}$$

where $C > 0$ is a sufficiently small constant. This proves that (1.28) is also true in this setting. However, an important distinction when compared to the case when $\mathbf{n}$ is timelike is that the previous bound yields control over all derivatives of $\phi$, except for the one in the direction of $\mathbf{n}^\dagger$. As we shall see later on, this information will lead to control over tangential derivatives, but not transversal derivatives to characteristic hypersurfaces.

In order to use the previous argument which showed, under reasonable assumptions, that $T_{\mu\nu}\mathbf{n}^\mu X^\nu$ has a definite sign, and to finally see the usefulness of (1.25) in deriving a priori bounds, we specialize our discussion to the case when the base manifold $M$ can be foliated by the level sets of a certain "time function" $u = u(x)$. This means that

$$M = \bigcup_t \{u = t\},$$

where $t$ is a parameter that we can think of as the "time variable" and, accordingly, the level sets of $u$ can be interpreted as "time slices" for $M$. Moreover, we ask for conditions to exist on $M$ such that integrating (1.25) on the slice $\{u = t\}$ yields

$$\frac{d}{dt}E(t) = \int\limits_{u=t} \{\nabla_\mu X_\nu\, T^{\mu\nu}\}\, d\sigma. \tag{1.30}$$

In the above, we denoted

$$E(t) \overset{\text{def}}{:=} \int\limits_{u=t} \{\mathbf{n}_\mu T^\mu_{\ \nu} X^\nu\}\, d\sigma$$

and we call this *the stored energy on* $\{u = t\}$, with $\mathbf{n}$ being related to $u$ as in (1.26). This terminology can be motivated by choosing $X$ to be timelike, the level sets to be spacelike, and $X$ and $n$ to share the same orientation, which implies that $E(t)$ is nonnegative. Given this, it is not implausible to claim that, for suitable $X$'s, one derives from (1.30) that

$$\frac{d}{dt}E(t) \leq C\|\nabla_\mu X_\nu\|_{L^\infty(\{u=t\})}\, E(t).$$

Following an application of Grönwall's inequality, we obtain control on the growth of the energy if $t \mapsto \|\nabla_\mu X_\nu\|_{L^\infty(\{u=t\})}$ is integrable with respect to the $t$-"coordinate". Most of the time, this control is referred to as an *energy-type estimate*.

What we have just finished is, of course, a very informal argument. However, it describes all the key steps behind the strategy of using the energy-momentum tensor to obtain relevant a priori bounds. Moreover, one should really appreciate the generality of the last estimate, which serves both as a starting point and a guide in studying analytic properties of wave maps.

**Remark 1.1.** Besides energy-type bounds, there is another set of estimates which has the potential of being useful. These are generated by conformal Killing vector fields[9], which are identified by the condition

$$\mathcal{L}_X g_{\mu\nu} = c_M g_{\mu\nu},$$

where $c_M = c_M(x)$ is the conformal factor on the base manifold. For such vector fields, the identity (1.25) becomes

$$\nabla_\mu \{T^\mu_{\ \nu} X^\nu\} = \frac{1}{2}\mathcal{L}_X g_{\mu\nu}\, T^{\mu\nu} = -\frac{(n-1)c_M}{4}\langle \nabla\phi, \nabla\phi\rangle. \tag{1.31}$$

---

[9] An example of a conformal Killing vector field is the scaling one, $S = x^\mu \nabla_\mu$, for which, when $M = \mathbb{R}^{n+1}$ is the Minkowski spacetime, we have $\mathcal{L}_S g_{\mu\nu} = 2g_{\mu\nu}$.

It is an unfortunate fact that the right-hand side does not have a definite sign unless $n = 1$, in which case we have a conformally invariant problem that is easier to handle. Despite this, for $n \neq 1$, conformal estimates have been proven extremely useful starting with the works of Friedrichs [49], Morawetz [120, 121], Morawetz-Strauss [122]. The reason for this will be explained later and is related to symmetries on the target manifold.

### 1.2.3   The wave map problem: energy conservation and the finite speed of propagation property

In this section, we discuss one of the most important properties for wave maps defined on Minkowski spacetimes, which is the conservation of energy, both in its global and local versions. As a consequence of this, we also establish a finite speed of propagation property, which, among other things, implies that compactly supported initial data for the Cauchy problem (1.22) evolve into solutions that are compactly supported when restricted on time slices in the domain of existence.

We start by recalling the notational conventions for the Minkowski spacetime $\left(\mathbb{R}^{n+1}, g = \mathrm{diag}(-1, 1, \ldots, 1)\right)$, where $t = x^0$ and $\mathbf{x} = (x^1, \ldots, x^n)$. We also write $\vec{\nabla} := (\nabla_1, \ldots, \nabla_n)$ to denote the spatial gradient. We work with the timelike Killing vector field $X = \nabla_t$, which, due to the discussion in the previous section, generates[10] the identity

$$\nabla_\mu P^\mu = 0, \tag{1.32}$$

with

$$P^0 \overset{\text{def}}{=} -\frac{1}{2}\left\{\langle \nabla_t \phi, \nabla_t \phi \rangle + \langle \vec{\nabla}\phi, \vec{\nabla}\phi \rangle\right\}, \tag{1.33}$$

$$P^j \overset{\text{def}}{=} \langle \nabla_j \phi, \nabla_t \phi \rangle, \qquad 1 \leq j \leq n. \tag{1.34}$$

If we assume that the wave map $\phi$ is defined on a domain including the spacetime slab $[0, t] \times \mathbb{R}^n$ and that it decays sufficiently fast to some constant at spatial infinity, then integrating the previous differential identity on $[0, t] \times \mathbb{R}^n$ yields the conservation law

$$\mathcal{E}(\phi(t)) = \mathcal{E}(\phi(0)), \tag{1.35}$$

where

$$\mathcal{E}(\phi(t)) \overset{\text{def}}{=} \frac{1}{2} \int_{\mathbb{R}^n} \left\{\langle \nabla_t \phi(t), \nabla_t \phi(t) \rangle + \langle \vec{\nabla}\phi(t), \vec{\nabla}\phi(t) \rangle\right\} d\mathbf{x} \tag{1.36}$$

---

[10]This can also be motivated by the time-translation invariance of the Lagrangian.

denotes the total energy of the wave map. The equality (1.35) is known as the *general energy identity* and it will later play a crucial role as a basic a priori estimate for the evolution of wave maps we will be studying. This identity implies that if we start with initial data of finite energy, then the solution of (1.22) must be in $\dot{H}^1(\mathbb{R}^n)$ for all fixed times. Unfortunately, this is not strong enough to guarantee that the solution is continuous for $n \geq 1$.

Next, we would like to derive a local version of the energy identity and then explain how this implies that the wave map system (1.21) has the finite speed of propagation property (with the speed being normalized to be 1). For this purpose, we fix a point $(t_0, \mathbf{x}_0) \in \mathbb{R}^{n+1}$ and a radius $r > 0$, and we define the conical regions

$$\mathcal{C}^+(t_0 - r, \mathbf{x}_0) \overset{\text{def}}{=} \big\{ (t, \mathbf{x}) \mid |\mathbf{x} - \mathbf{x}_0| \leq t - t_0 + r, \ t \geq t_0 - r \big\},$$

$$\mathcal{C}^-(t_0 + r, \mathbf{x}_0) \overset{\text{def}}{=} \big\{ (t, \mathbf{x}) \mid |\mathbf{x} - \mathbf{x}_0| \leq t_0 + r - t, \ t \leq t_0 + r \big\},$$

which are the forward characteristic cone with tip at $(t_0 - r, \mathbf{x}_0)$ and the backward characteristic cone with tip at $(t_0 + r, \mathbf{x}_0)$, respectively. Moreover, we denote by

$$\mathcal{C}_r(t_0, \mathbf{x}_0) \overset{\text{def}}{=} \mathcal{C}^+(t_0 - r, \mathbf{x}_0) \cap \mathcal{C}^-(t_0 + r, \mathbf{x}_0)$$

the diamond-shaped region obtained by intersecting the two cones. For the linear wave equation, this represents the domain of influence for the ball centered at $\mathbf{x}_0$, of radius $r$, which lies on the time slice $t = t_0$.

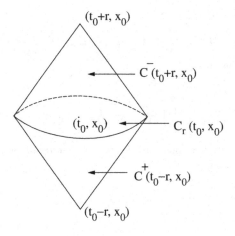

Fig. 1.1   The diamond-shaped region $\mathcal{C}_r(t_0, \mathbf{x}_0)$.

We also introduce subsets of $\mathcal{C}_r(t_0, \mathbf{x}_0)$, which are defined for $|s - t_0| \leq r$, as follows:

$$C_r^{[t_0, s]}(t_0, \mathbf{x}_0) \overset{\text{def}}{=} C(t_0, \mathbf{x}_0) \cap \left\{ (t, \mathbf{x}) \,\middle|\, |t - t_0| \leq |s - t_0|, \ \mathbf{x} \in \mathbb{R}^n \right\},$$

$$B_{r-|s-t_0|}^s(\mathbf{x}_0) \overset{\text{def}}{=} \mathcal{C}_r(t_0, \mathbf{x}_0) \cap (\{s\} \times \mathbb{R}^n),$$

$$\partial C_r^{[t_0, s]}(t_0, \mathbf{x}_0) \overset{\text{def}}{=} C_r^{[t_0, s]}(t_0, \mathbf{x}_0) \cap \left\{ (t, \mathbf{x}) \,\middle|\, |\mathbf{x} - \mathbf{x}_0| = |t - t_0| \right\}.$$

Following this geometric setup, we integrate the differential identity (1.32) on the frustum $C_r^{[t_0, s]}(t_0, \mathbf{x}_0)$ to obtain

$$0 = \int_{\partial\left(C_r^{[t_0, s]}(t_0, \mathbf{x}_0)\right)} \{P \cdot \mathbf{n}\} \, d\sigma = \int_{B_r^{t_0}(\mathbf{x}_0)} P \cdot (\text{sgn}(t_0 - s), 0, \ldots, 0) \, d\sigma$$

$$+ \int_{B_{r-|s-t_0|}^s(\mathbf{x}_0)} \{P \cdot (\text{sgn}(s - t_0), 0, \ldots, 0)\} \, d\sigma$$

$$+ \int_{\partial C_r^{[t_0, s]}(t_0, \mathbf{x}_0)} \left\{ P \cdot \left( \text{sgn}(s - t_0), \frac{\mathbf{x} - \mathbf{x}_0}{|\mathbf{x} - \mathbf{x}_0|} \right) \right\} \, d\sigma,$$

where $P := (P^0, P^1, \ldots, P^n)$ and $\mathbf{n}$ is the outward unit normal to the frustum. Substituting the values for $P^0$ and $P^j$ from (1.33) and (1.34), respectively, we arrive at

$$\int_{B_r^{t_0}(\mathbf{x}_0)} \{e(\phi)\} \, dx = \int_{B_{r-|s-t_0|}^s(\mathbf{x}_0)} \{e(\phi)\} \, dx$$

$$+ \frac{1}{\sqrt{2}} \int_{\partial C_r^{[t_0, s]}(t_0, \mathbf{x}_0)} \left\{ e(\phi) + \text{sgn}(t_0 - s)\langle \frac{\mathbf{x} - \mathbf{x}_0}{|\mathbf{x} - \mathbf{x}_0|} \cdot \vec{\nabla}\phi, \nabla_t\phi \rangle \right\} \, d\sigma,$$

where

$$e(\phi) \overset{\text{def}}{=} \frac{1}{2} \left\{ \langle \nabla_t\phi, \nabla_t\phi \rangle + \langle \vec{\nabla}\phi, \vec{\nabla}\phi \rangle \right\} \tag{1.37}$$

is usually called, on the basis of (1.36), the *energy density*. A careful computation shows that

$$e(\phi) + \text{sgn}(t_0 - s)\langle \frac{\mathbf{x} - \mathbf{x}_0}{|\mathbf{x} - \mathbf{x}_0|} \cdot \vec{\nabla}\phi, \nabla_t\phi \rangle$$

$$= \frac{1}{2} \left\{ \left| \left( \nabla_t + \text{sgn}(t_0 - s)\frac{\mathbf{x} - \mathbf{x}_0}{|\mathbf{x} - \mathbf{x}_0|} \cdot \vec{\nabla} \right) \phi \right|^2 + \frac{1}{|\mathbf{x} - \mathbf{x}_0|^2} |\vec{\nabla}_\theta\phi|^2 \right\} \geq 0,$$

where we denoted above by $\vec{\nabla}_\theta$ the collection of angular derivatives. We collect all these facts in the following simple lemma.

**Lemma 1.1.** *With the notational conventions introduced in this section, let $\phi$ be a classical wave map (i.e., a $C^2$ solution of (1.21)) defined on a domain including $C_r(t_0, \mathbf{x}_0)$. For all $|s - t_0| \leq r$, the following integral identity holds:*

$$\int_{B_r^{t_0}(\mathbf{x}_0)} \{e(\phi)\} dx = \int_{B_{r-|s-t_0|}^s(\mathbf{x}_0)} \{e(\phi)\} dx$$

$$+ \frac{1}{2\sqrt{2}} \int_{\partial C_r^{[t_0,s]}(t_0,\mathbf{x}_0)} \left\{ \left| \left( \nabla_t + \operatorname{sgn}(t_0 - s) \frac{\mathbf{x} - \mathbf{x}_0}{|\mathbf{x} - \mathbf{x}_0|} \cdot \vec{\nabla} \right) \phi \right|^2 \right.$$

$$\left. + \frac{1}{|\mathbf{x} - \mathbf{x}_0|^2} |\vec{\nabla}_\theta \phi|^2 \right\} d\sigma.$$

*This is usually called in the literature the* local version of the energy identity. *As a consequence, we have*

$$\int_{B_{r-|s-t_0|}^s(\mathbf{x}_0)} \{e(\phi)\} dx \leq \int_{B_r^{t_0}(\mathbf{x}_0)} \{e(\phi)\} dx, \qquad \forall |s - t_0| \leq r. \qquad (1.38)$$

An immediate consequence of the energy inequality in the lemma is that the evolution of wave maps has finite speed of propagation. Concretely, let us assume that $\phi$ is trivial in the ball $B_r^{t_0}(\mathbf{x}_0)$, meaning

$$\phi(t_0, \mathbf{x}) = \phi_0 = \text{constant} \qquad \text{and} \qquad \nabla_t \phi(t_0, \mathbf{x}) = 0, \qquad \forall |\mathbf{x} - \mathbf{x}_0| \leq r.$$

Hence, $e(\phi) = 0$ on $B_r^{t_0}(\mathbf{x}_0)$ and, due to (1.38), we also obtain that $e(\phi) = 0$ on $B_{r-|s-t_0|}^s(\mathbf{x}_0)$, for all $t_0 - r \leq s \leq t_0 + r]$. This implies $\phi$ is constant on $C_r(t_0, \mathbf{x}_0)$ and is equal to $\phi_0$.

On the other hand, consider now a wave map $\psi$ which is trivial outside the ball $B_r^{t_0}(\mathbf{x}_0)$, i.e.,

$$\psi(t_0, \mathbf{x}) = \psi_0 = \text{constant} \qquad \text{and} \qquad \nabla_t \psi(t_0, \mathbf{x}) = 0, \qquad \forall |\mathbf{x} - \mathbf{x}_0| \geq r.$$

We claim that the interested reader can verify, through a similar argument based on (1.38), that the equality $\psi = \psi_0$ also holds in the region

$$\{(t, \mathbf{x}) | |\mathbf{x} - \mathbf{x}_0| \geq r + |t - t_0|\}.$$

One way to interpret this is that a "disturbance" localized at time $t_0$ inside the ball $B_r^{t_0}(\mathbf{x}_0)$ cannot travel forward or backward in time with speed larger than 1. Both $\phi$'s and $\psi$'s behaviors are usually referred to as manifestations of the *finite speed of propagation property* for wave maps.

As a final remark, not even the local version of the energy identity is robust enough to handle questions regarding the regularity or the blow-up of wave maps. This motivates the need for finer analytic methods and a substantial part of the present book is devoted to this task. Nevertheless, energy and energy-type estimates remain an essential ingredient for any investigation concerning the Cauchy problem (1.22).

### 1.2.4 *The wave map problem: scaling invariance and energy criticality*

Here, we follow up on considerations made in Section 1.1 concerning scaling transformations. It is easy to see that if $\phi$ solves (1.21), then so does the scaled field

$$\phi_\lambda(t, \mathbf{x}) \overset{\text{def}}{=} \phi\left(\frac{t}{\lambda}, \frac{\mathbf{x}}{\lambda}\right), \tag{1.39}$$

where $\lambda$ is an arbitrary real parameter. As a function of $\lambda$, the total energy defined by (1.36) changes according to

$$\mathcal{E}(\phi_\lambda) = \lambda^{n-2}\mathcal{E}(\phi), \tag{1.40}$$

which implies that the energy is scale-invariant precisely when $n = 2$. In relation to this formula, we adopt the following terminological convention.

*The wave map problem with $\mathbb{R}^{n+1}$ as its base manifold is called:*

- *energy-critical if $n = 2$;*
- *energy-subcritical if $n < 2$;*
- *energy-supercritical if $n > 2$.*

This distinction can be motivated by the analytic techniques developed later in the book that offer complete answers to critical and subcritical problems and only partial answers to supercritical problems, for which many questions remain unanswered. In fact, one of our major goals is to present the most recent advances for the energy-critical problem. The above terminology will be further addressed in Chapter 3, where we introduce the concept of well-posedness for the Cauchy problem (1.22).

### 1.2.5 *The wave map problem: general discussion of symmetries on the target*

We start here by discussing the effect on the wave map problem of the symmetries on the target manifold $N$. It is a useful idea to see the derivatives of $\phi$ as vector fields on $N$. If we define the "deformation tensor"

$$S_\mu^a(\phi) \overset{\text{def}}{=} \nabla_\mu \phi^a = \partial_\mu \phi^a, \tag{1.41}$$

then we can think of $S_\mu^a$, for $0 \le \mu \le n$, as contravariant vector fields[11] on the target.

First, let us recall the formulas for the covariant and contravariant derivatives on the target. We use the notation $\widehat{\nabla}$ to distinguish them from the ones on the base. One has

$$\widehat{\nabla}_b X^a = \frac{\partial X^a}{\partial \phi^b} + \Gamma_{bc}^a X^c, \tag{1.42}$$

$$\widehat{\nabla}_b X_a = \frac{\partial X_a}{\partial \phi^b} - \Gamma_{ba}^c X_c, \tag{1.43}$$

where $\Gamma_{ab}^c$ denotes the Christoffel symbols on the target. If we contract the second formula with the deformation tensor, then we obtain

$$S_\mu^b \widehat{\nabla}_b X_a = S_\mu^b \frac{\partial X_a}{\partial \phi^b} - S_\mu^b \Gamma_{ba}^c X_c = \partial_\mu X_a - S_\mu^b \Gamma_{ba}^c X_c.$$

Introducing the definitions

$$\widehat{\nabla}_{S_\mu} X^{a,\nu} \overset{\text{def}}{=} S_\mu^b \widehat{\nabla}_b X^{a,\nu} + \Gamma_{\mu\alpha}^\nu X^{a,\alpha}, \tag{1.44}$$

$$\widehat{\nabla}_{S_\mu} X_\nu^a \overset{\text{def}}{=} S_\mu^b \widehat{\nabla}_b X_\nu^a - \Gamma_{\mu\nu}^\alpha X_\alpha^a, \tag{1.45}$$

where $X^{a,\mu} := g^{\mu\nu} X_\nu^a$, it follows that the Euler-Lagrange equations (1.21) can be written in the concise form

$$\widehat{\nabla}_{S_\mu} S^{a,\mu} = 0. \tag{1.46}$$

This can be seen as a generalization of the geodesic equation.

One way to see the connection between the wave map system and the symmetries on the target is by considering a vector field $X^a(\phi)$ on the target and the associated Lie derivative coming from the group of transformations

$$\frac{d\phi^a}{ds} = X^a(\phi).$$

The resulting structure equations are nothing but (1.21). Furthermore, contracting the wave map system with an arbitrary covector $X_a(\phi)$ yields

$$-\nabla_\mu \{X_a(\phi)\nabla^\mu \phi^a\} + \frac{1}{2} g^{\mu\nu} S_\mu^a S_\nu^b (\mathcal{L}_X h_{ab}) = 0, \tag{1.47}$$

---

[11]They are also covariant vector fields on the base.

where the Lie derivative of the metric $h$ is given by $\mathcal{L}_X h_{ab} = \widehat{\nabla}_a X_b + \widehat{\nabla}_b X_a$. If we consider now $X$ to be Killing (i.e., $\mathcal{L}_X h_{ab} = 0$), then the current $J^\mu := X_a(\phi) \nabla^\mu \phi^a$ obeys the conservation law

$$\nabla_\mu J^\mu = 0.$$

In the more general case when $X$ is conformal Killing, meaning $\mathcal{L}_X h_{ab} = c_N(\phi) h_{ab}$ with $c_N = c_N(\phi)$ being the conformal factor on the target, we derive the differential identity

$$\nabla_\mu J^\mu = \frac{c_N(\phi)}{2} \langle \nabla\phi, \nabla\phi \rangle. \tag{1.48}$$

**Remark 1.2.** It makes sense now to ask whether the previous conformal identity or the identity (1.31) in Remark 1.1 provides better results. In both cases, we are dealing with an error term on the right-hand side which has an indefinite sign. To make these identities useful, the idea is to combine them such that these error terms cancel out. Unfortunately, this can only happen when the conformal factor $c_N(\phi)$ is constant, which restricts the area of applicability for conformal a priori estimates. Nevertheless, conformal fields are useful when the wave maps have extra symmetries, as in the case of the equivariant ansatz.

### 1.2.6　The wave map problem: discussion of symmetries on the target when $N = \mathbb{S}^2$ or $N = \mathbb{H}^2$

In this subsection, we specialize the discussion about symmetries on the target to the simplest nontrivial examples, which serve as prototypes for our investigation. These are the extreme cases when $N = \mathbb{S}^2$, the 2-dimensional unit sphere, and $N = \mathbb{H}^2$, the 2-dimensional unit hyperboloid. When written in spherical and pseudo-spherical coordinates, respectively, the corresponding metrics take the form

$$h = d\phi^2 + \sin^2\phi \, d\theta^2,$$
$$h = d\phi^2 + \sinh^2\phi \, d\theta^2.$$

It is perhaps more instructive to see both these manifolds as Riemann surfaces with conformal metrics of the type $h(w, \overline{w}) \, dw \, d\overline{w}$, where $w \in D \subset \mathbb{C}$ and the Riemannian metric is given by $h_{ab} = h\delta_{ab}$. Under a stereographic projection, $\mathbb{S}^2$ is associated to

$$h(w, \overline{w}) = \frac{4}{(1 + |w|^2)^2}, \qquad w \in \mathbb{C}, \tag{1.49}$$

while for $\mathbb{H}^2$ we have

$$h(w,\overline{w}) = \frac{4}{(1-|w|^2)^2}, \qquad w \in D \stackrel{\text{def}}{=} \{z; |z| < 1\}. \qquad (1.50)$$

We recall here that the stereographic projection for our two model cases is the change of variables

$$\cos\phi = \frac{1-|w|^2}{1+|w|^2}, \qquad \sin\phi\, e^{i\theta} = \frac{2w}{1+|w|^2}, \qquad N = \mathbb{S}^2,$$

and

$$\cosh\phi = \frac{1+|w|^2}{1-|w|^2}, \qquad \sinh\phi\, e^{i\theta} = \frac{2w}{1-|w|^2}, \qquad N = \mathbb{H}^2.$$

This can be seen for $\mathbb{S}^2$ by denoting $|w| := p$ and writing $\phi = \arctan\left(2p/(1-p^2)\right)$. A simple computation shows that

$$d\phi^2 + \sin^2\phi\, d\theta^2 = 4\left(\frac{1}{(1+p^2)^2}dp^2 + \frac{p^2}{(1+p^2)^2}d\theta^2\right).$$

For $\mathbb{H}^2$, one works with $\phi = \tanh^{-1}(2p/(1+p^2))$ and the calculation runs along similar lines.

In this context, the Lagrangian takes the form

$$\mathbf{L}(w,\overline{w}) \stackrel{\text{def}}{=} \int_M \{g^{\mu\nu}h(w,\overline{w})\nabla_\mu w \nabla_\nu \overline{w}\}\, d\mu_g$$

and one computes its variation with respect to $\overline{w}$ by viewing $(w,\overline{w})$ as independent variables. The resulting Euler-Lagrange equation is

$$\Box w - \frac{\partial \log(h)}{\partial w}\langle \nabla w, \nabla w\rangle = 0, \qquad (1.51)$$

which, for our two cases, reads as

$$\Box w + \frac{2\overline{w}}{1+|w|^2}\langle \nabla w, \nabla w\rangle = 0 \qquad \text{for } \mathbb{S}^2 \qquad (1.52)$$

and

$$\Box w - \frac{2\overline{w}}{1-|w|^2}\langle \nabla w, \nabla w\rangle = 0 \qquad \text{for } \mathbb{H}^2. \qquad (1.53)$$

These are simple looking equations, whose study will occupy a substantial portion of the present work and from which we can learn a lot in preparation to analyze more general problems.

With these prerequisites out of the way, we now turn our attention to the symmetries on the target in these two cases. As a matter of fact, we already know what they are: rotations when $N = \mathbb{S}^2$ and hyperbolic rotations when

$N = \mathbb{H}^2$. What we want to understand is how they manifest themselves under the stereographic projection. The key observation for (1.52) is that it is invariant under fractional linear transformations of the type

$$f(w) = \frac{aw + b}{cw + d},$$

where

$$U \stackrel{\text{def}}{=} \begin{pmatrix} a & b \\ c & d \end{pmatrix} = \begin{pmatrix} e^{i\alpha}\cos\beta & e^{i\gamma}\sin\beta \\ -e^{-i\gamma}\sin\beta & e^{-i\alpha}\cos\beta \end{pmatrix}.$$

The invariance of the Lagrangian is equivalent to $UU^* = I$, which means that we can write $U$ in the form indicated above. Indeed, it is easy to show by a direct computation that

$$\frac{df\,d\overline{f}}{(1 + |f|^2)^2} = \frac{dw\,d\overline{w}}{(1 + |w|^2)^2}.$$

This invariance represents a three-parameter group and its generators give rise to three conserved currents, namely

$$Q^\mu \stackrel{\text{def}}{=} \frac{1}{(1 + |w|^2)^2}\left(\overline{w}\nabla^\mu w - w\nabla^\mu\overline{w}\right),$$

$$J^\mu \stackrel{\text{def}}{=} \frac{1}{(1 + |w|^2)^2}\left(\nabla^\mu w + w^2\nabla^\mu\overline{w}\right).$$

The second current is complex-valued and so, its real and imaginary parts yield two conserved currents.

The analog for the case of (1.53) is represented by

$$\begin{pmatrix} a & b \\ c & d \end{pmatrix} = \begin{pmatrix} e^{i\alpha}\cosh\beta & e^{i\gamma}\sinh\beta \\ e^{-i\gamma}\sinh\beta & e^{-i\alpha}\cosh\beta \end{pmatrix}$$

and the derivation of the resulting conserved currents are left to the reader as an exercise. We will see shortly that there is one more conserved current, which is of a topological nature.

### 1.2.7 *The wave map problem: initial considerations on the associated Hodge system*

Here, we want to show that the Euler-Lagrange equation (1.51) can be written as a first-order system coupled with a set of compatibility conditions. Together, they form a Hodge-type system which has the benefit that the nonlinear nature of the problem is more transparent, in the sense that the curvature of the target appears explicitly in the equations. Moreover, in the case when $N$ is 2-dimensional, we will see that a $U(1)$ gauge field appears.

The complex analog of the "deformation tensor" (1.41) is $S_\mu := \nabla_\mu w$ and we would like to employ complex differentiation on the target manifold. For this purpose, we recall that, for $h_{ab} = h\delta_{ab}$, the Christoffel symbols are given by

$$\Gamma^1_{11} = -\Gamma^1_{22} = \Gamma^2_{12} = \frac{\partial_1 h}{2h}, \qquad \Gamma^1_{12} = -\Gamma^2_{11} = \Gamma^2_{22} = \frac{\partial_2 h}{2h},$$

while for a vector field $X$, routine computations lead to

$$\widehat\nabla_1 X^1 = \partial_1 X^1 + \Gamma^1_{11} X^1 + \Gamma^1_{12} X^2 = \partial_1 X^1 + \frac{1}{2h}\left(\partial_1 h\, X^1 + \partial_2 h\, X^2\right),$$

$$\widehat\nabla_2 X^1 = \partial_2 X^1 + \Gamma^1_{21} X^1 + \Gamma^1_{22} X^2 = \partial_2 X^1 + \frac{1}{2h}\left(\partial_2 h\, X^1 - \partial_1 h\, X^2\right),$$

$$\widehat\nabla_1 X^2 = \partial_1 X^2 + \Gamma^2_{11} X^1 + \Gamma^2_{12} X^2 = \partial_1 X^2 + \frac{1}{2h}\left(-\partial_2 h\, X^1 + \partial_1 h\, X^2\right),$$

$$\widehat\nabla_2 X^2 = \partial_2 X^2 + \Gamma^2_{21} X^1 + \Gamma^2_{22} X^2 = \partial_2 X^2 + \frac{1}{2h}\left(\partial_1 h\, X^1 + \partial_2 h\, X^2\right).$$

If we consider now the complex vector field $X^1 + iX^2$ and use the complex differentiation notation $\partial_w h := (1/2)(\partial_1 - i\partial_2)h$, then the previous formulas yield the identity

$$Y^b \widehat\nabla_b (X^1 + iX^2) = Y^b \partial_b (X^1 + iX^2) + \frac{\partial_w h}{h}(X^1 + iX^2)(Y^1 + iY^2),$$

for an arbitrary vector field $Y$. It follows that we can write

$$\widehat\nabla_{S_\mu} X = \nabla_\mu X + \frac{\partial_w h}{h} S_\mu X,$$

which implies

$$\widehat\nabla_{S_\mu} S_\nu = \nabla_\mu S_\nu + \frac{\partial_w h}{h} S_\mu S_\nu. \tag{1.54}$$

Next, we multiply this equation by $\sqrt{h}$ to deduce

$$\sqrt{h}\,\widehat\nabla_{S_\mu} S_\nu = (\nabla_\mu + iA_\mu)\,\psi_\nu,$$

where we introduced

$$\psi_\nu \stackrel{\text{def}}{=} \sqrt{h}\, S_\nu \tag{1.55}$$

and the real gauge field

$$A_\mu \stackrel{\text{def}}{=} \frac{1}{2i}\left(\partial_w \log(h)\nabla_\mu w - \partial_{\overline w}\log(h)\nabla_\mu \overline w\right), \tag{1.56}$$

which can be thought of as a real gauge field. Given that

$$\widehat\nabla_{S_\mu} S_\nu = \widehat\nabla_{S_\nu} S_\mu \qquad \text{and} \qquad \widehat\nabla_{S_\mu} S^\mu = 0,$$

we derive

$$\left(\nabla_\mu + iA_\mu\right)\psi_\nu = \left(\nabla_\nu + iA_\nu\right)\psi_\mu \qquad \text{and} \qquad \left(\nabla_\mu + iA_\mu\right)\psi^\mu = 0.$$

Furthermore, if we define

$$F_{\mu\nu} \overset{\text{def}}{=} \nabla_\mu A_\nu - \nabla_\nu A_\mu \qquad \text{and} \qquad G_{\mu\nu} \overset{\text{def}}{=} \frac{1}{2i}\left(\psi_\mu\overline{\psi}_\nu - \overline{\psi}_\mu\psi_\nu\right), \qquad (1.57)$$

then we can infer that

$$
\begin{aligned}
F_{\mu\nu} &= 2\frac{\partial^2 \log(h)}{\partial w \partial \overline{w}}\frac{1}{2i}\left(\nabla_\nu w \nabla_\mu \overline{w} - \nabla_\nu \overline{w}_\mu \nabla_\mu w\right) \\
&= (-K)\frac{1}{2i}\left(\psi_\nu \overline{\psi}_\mu - \overline{\psi}_\nu \psi_\mu\right) \\
&= K G_{\mu\nu},
\end{aligned}
\qquad (1.58)
$$

where $K$ is the Gaussian curvature. This is due to the Gauss equation

$$\frac{\partial^2 \log(h)}{\partial w \partial \overline{w}} = \frac{-K}{2}h.$$

Summarizing what we have so far, we obtain the Hodge system

$$(p_\mu + A_\mu)\psi^\mu = 0, \qquad (1.59)$$

$$(p_\mu + A_\mu)\psi_\nu = (p_\nu + A_\nu)\psi_\mu, \qquad (1.60)$$

$$F_{\mu\nu} - K G_{\mu\nu} = 0, \qquad (1.61)$$

with $p_\mu := \frac{1}{i}\nabla_\mu$ designating the momentum operator.

**Remark 1.3.** First, we notice that (1.60) and (1.61) are compatibility equations, which hold for arbitrary maps. The evolution part of the map is described by (1.59). Second, the Hodge system is manifestly gauge-invariant under the gauge transformation

$$\left(\psi_\mu, A_\mu\right) \mapsto \left(e^{i\theta}\psi_\mu, A_\mu - \partial_\mu\theta\right),$$

where $\theta$ is an arbitrary phase function. Third, if the curvature $K$ is equal to zero, then $F_{\mu\nu} = 0$ and, as a result, we can "gauge-away" $A_\mu$. In this case, the Hodge system reduces to a set of linear equations for each $\psi_\mu$ (i.e., $\Box\psi_\mu = 0$). Finally, the gauge invariance gives us a degree of freedom, since we can impose a "gauge fixing" on the field $A_\mu$. The most common choice is the Lorentz gauge, in which we impose $\nabla_\mu A^\mu = 0$. Later, we will see that for the case when the base manifold is $\mathbb{R}^{n+1}$, with $n \geq 4$, the Coulomb gauge is better suited for certain estimates.

In the context of this subsection, let us discuss next the topological conserved current, which was previously alluded to. We consider the base manifold to be $M = \mathbb{R}^{2+1}$ and we introduce

$$E^\mu \stackrel{\text{def}}{=} \frac{1}{2}\epsilon^{\mu\alpha\beta}F_{\alpha\beta} = \frac{K}{2}\epsilon^{\mu\alpha\beta}G_{\alpha\beta}, \qquad (1.62)$$

which is easily seen to satisfy $\nabla_\mu E^\mu = 0$. We work under the assumption that the map is such that $E^1$ and $E^2$ decay to zero at spatial infinity and we integrate the previous differential identity over the spacetime slab $[0, t] \times \mathbb{R}^2$. The result is that

$$\deg(\phi(t)) \stackrel{\text{def}}{=} \int_{\mathbb{R}^2} \left\{ E^0(t, \mathbf{x}) \right\} d\mathbf{x}$$

is a conserved quantity, which is related to the degree of the map $\mathbf{x} \mapsto \phi(t, \mathbf{x})$. This can be explained in a simple way by taking $N = \mathbb{S}^2$. We dissolve the complex notation by writing $\psi_\mu = \psi_\mu^1 + i\psi_\mu^2$ and we recall that $\psi_\mu^a = \sqrt{h}\,\nabla_\mu\phi^a$, where $h = 4/(1 + |w|^2)^2$, and the measure of integration on the target manifold is given by $h\,d\overline{w} \wedge dw/2i$. Using the notation convention that Roman letters stand for spatial indices, we can write

$$E_0 = \frac{K}{2!}\epsilon^{jk}\nabla_j\phi^a\nabla_k\phi^b\epsilon_{ab} = K\det(\nabla_j\phi^a). \qquad (1.63)$$

Since the degree takes discrete values, we can split the evolution of smooth maps into distinct sectors, each one identified by a fixed degree dictated by the initial data. Smooth solutions preserve the degree of the map, while blow-up scenarios force the degree of map to drop.

### 1.2.7.1 *An equivalent form for the Hodge system*

In studying the Hodge system (1.59)-(1.61), it is common to derive from it equations involving wave operators applied to the complex field $\psi_\mu$. For this purpose, let us define first the $A$-covariant d'Alembertian

$$\Box^A \stackrel{\text{def}}{=} (p_\mu + A_\mu)(p^\mu + A^\mu) = \Box + \langle p, A \rangle + \langle A, p \rangle + \langle A, A \rangle. \qquad (1.64)$$

Following this, one applies $(p_\mu + A_\mu)$ to the right side of (1.60) and uses the commutation relation

$$\left[p_\mu + A_\mu, p_\nu + A_\nu\right]\psi^\mu = -R_{\mu\nu\alpha}{}^\mu\psi^\alpha + \frac{1}{i}F_{\mu\nu}\psi^\mu = R_{\nu\alpha}\psi^\alpha + \frac{1}{i}F_{\mu\nu}\psi^\mu,$$

where

$$R_{\nu\alpha} \stackrel{\text{def}}{=} R_{\nu\mu\alpha}{}^\mu$$

is the Ricci tensor on the base. If we also factor in (1.59), then the end result is

$$\Box^A \psi_\nu - R_{\nu\mu}\psi^\mu + i\psi_\mu F^\mu_{\ \nu} = 0. \qquad (1.65)$$

Thus, one now studies the system formed by the previous equation and (1.61).

At this point, a natural question is whether the new system is equivalent to the original Hodge system. The answer is, in principle, positive and, in order to simplify the explanation of why this is the case, we restrict ourselves to the setting when $M = \mathbb{R}^{2+1}$. This implies that $R_{\mu\nu} = 0$ and, as a consequence, (1.65) reduces itself to

$$\Box^A \psi_\nu + i\psi^\alpha F_{\alpha\nu} = 0. \qquad (1.66)$$

Next, based on $\nabla^A_\mu := \nabla_\mu + iA_\mu$, we introduce:

$$\Lambda_\mu \overset{\text{def}}{=} \epsilon_\mu^{\ \alpha\beta} \nabla^A_\alpha \psi_\beta, \qquad \tau \overset{\text{def}}{=} \nabla^A_\mu \psi^\mu = \text{tr}(\nabla_\mu \psi_\nu),$$

$$E_\mu \overset{\text{def}}{=} \frac{1}{2}\epsilon_\mu^{\ \alpha\beta} F_{\alpha\beta}, \qquad H_\mu \overset{\text{def}}{=} \frac{1}{2}\epsilon_\mu^{\ \alpha\beta} G_{\alpha\beta}, \qquad (1.67)$$

$$\left(d^A \wedge \Lambda\right)_\mu \overset{\text{def}}{=} \epsilon_{\mu\alpha}^{\ \ \beta} \nabla^{A,\alpha} \Lambda_\beta.$$

On the account of both (1.66) and (1.61), we derive that $E = KH$ and

$$\nabla^A_\mu \Lambda^\mu = \frac{i}{2}\epsilon^{\mu\alpha\beta} F_{\mu\alpha}\psi_\beta = i\langle E, \psi\rangle = 0.$$

A straightforward calculation using the identity

$$\epsilon_{\mu\alpha}^{\ \ \beta} \epsilon_\beta^{\ \gamma\delta} = \delta_\mu^{\ \gamma}\delta_\alpha^{\ \delta} - \delta_\mu^{\ \delta}\delta_\alpha^{\ \gamma}$$

yields

$$\left(d^A \wedge \Lambda\right)_\mu = \Box^A \psi_\mu + i\psi^\alpha F_{\alpha\mu} + \nabla^A_\mu \tau,$$

which, due to (1.66), implies

$$\left(d^A \wedge \Lambda\right)_\mu = \nabla^A_\mu \tau.$$

By computing the divergence and the curl of this equation, we discover that

$$\Box^A \Lambda_\mu - i\left(E \wedge \Lambda\right)_\mu = 2iE_\mu \tau, \qquad (1.68)$$

$$\Box^A \tau = 2i\langle E, \Lambda\rangle. \qquad (1.69)$$

If we make the basic assumption that $H$ is uniformly bounded, then a standard energy-type argument for the system formed by the previous two equations combined with an application of Grönwall's inequality proves that $\Lambda(0) = \tau(0) = 0$ implies $\Lambda(t) = \tau(t) = 0$ for later times. Thus, smooth solutions of (1.66) are equivalent to solutions of the Hodge system (1.59)-(1.61). We are going to see an illustration of this simple argument later, for the particular case of the wave map system on $\mathbb{R}^{n+1}$, with $n \geq 4$.

**Remark 1.4.** In solving the system formed by (1.66) and (1.61), one can see the former as a nonlinear wave equation with a cubic nonlinearity, for which a caricature problem is

$$\Box\psi + \psi^3 = 0.$$

Here, the novelty is that the d'Alembertian is $A$-covariant and we have the choice of fixing the gauge field at our disposal.

### 1.2.7.2 *An interesting observation*

It is an interesting and curious fact that one can view the Hodge system as being connected to a more general Lagrangian, which is defined for the pair of fields $(\psi_\mu, A_\mu)$. In order to formulate this Lagrangian, we construct the tensor

$$\omega_{\mu\nu} \stackrel{\text{def}}{=} i(p_\mu + A_\mu)\psi_\nu \tag{1.70}$$

and, in order to avoid unnecessary complications, we specialize to the case of constant curvature. The Lagrangian takes the form

$$\mathbf{L}[\psi, A] \stackrel{\text{def}}{=} \int_{\mathbb{R}^{n+1}} \left\{ \left(\overline{\omega}_{\mu\nu}\,\omega^{\mu\nu} - \overline{\omega}_\mu{}^\mu \omega_\nu{}^\nu\right) \right.$$
$$\left. - \frac{1}{2K} F_{\mu\nu}\,F^{\mu\nu} - \frac{K}{2} G_{\mu\nu}\,G^{\mu\nu} \right\} dx. \tag{1.71}$$

We notice that the first two terms in the Lagrangian form an expression of the type $\mathrm{tr}(\omega^2) - (\mathrm{tr}(\omega))^2$, which is similar to what will appear later in the context of the Skyrme model.

For computing the associated Euler-Lagrange equations, we first define the antisymmetric tensor $\Lambda_{\mu\nu} := \omega_{\mu\nu} - \omega_{\nu\mu}$, which is consistent with (1.67). A straightforward calculation of the variational derivatives $\delta\mathbf{L}/\delta\overline{\psi}^\nu$ and $\delta\mathbf{L}/\delta A^\nu$ yields the system

$$\Box^A\psi_\nu + \nabla_\nu^A\{\nabla_\mu^A\psi^\mu\} + iK\psi^\mu G_{\mu\nu} = 0,$$
$$\nabla_\mu\{F^\mu{}_\nu - KG^\mu{}_\nu\} - i\frac{K}{2}\left(\overline{\psi}_\mu\Lambda_\nu{}^\mu - \psi_\mu\overline{\Lambda}_\nu{}^\mu\right) = 0. \tag{1.72}$$

It is easy to see that if $\Lambda = 0$ and $\omega_\mu{}^\mu = 0$, then the above system reduces to the Hodge system formed by (1.66) and (1.61).

One may wonder if the above Lagrangian formulation leads to more conserved quantities that may be useful. Unfortunately, this does not seem to be the case. It suffices to notice the following identity,

$$\overline{\omega}_{\mu\nu}\,\omega^{\mu\nu} - \overline{\omega}_\mu{}^\mu \omega_\nu{}^\nu = \nabla_\mu J^\mu + \frac{1}{2}\overline{\Lambda}_{\mu\nu}\,\Lambda^{\mu\nu} + F_{\mu\nu}\,G^{\mu\nu},$$

where the current is given by

$$J^\mu \stackrel{\text{def}}{=} \nabla_\nu \left\{ \mathcal{R}e(\overline{\psi}^\nu\,\psi^\mu) \right\} - 2\mathcal{R}e(\overline{\psi}^\mu\,\omega_\nu{}^\nu).$$

This means that the Lagrangian (1.71) takes the form

$$\mathbf{L}(\psi, A) = \int_{\mathbb{R}^{n+1}} \left\{ \nabla_\mu J^\mu + \frac{1}{2}\overline{\Lambda}_{\mu\nu}\Lambda^{\mu\nu} \right.$$
$$\left. - \frac{1}{2K}\left(F_{\mu\nu} - KG_{\mu\nu}\right)\left(F^{\mu\nu} - KG^{\mu\nu}\right) \right\} dx.$$

In fact, by using $\Lambda_\mu = \epsilon_\mu{}^{\alpha\beta}\omega_{\alpha\beta}$ and slightly abusing the notation, we can rewrite the Euler-Lagrange equations (1.72) in the suggestive form

$$d^A \wedge \Lambda = i\psi \wedge (E - KG), \qquad \nabla \wedge (E - KH) + \mathcal{I}m(\overline{\psi} \wedge \Lambda) = 0,$$

where $E$, $H$, and $d^A \wedge \Lambda$ have the meaning from (1.67). These computations point to the fact that the energy-momentum tensor associated to the Lagrangian (1.71) does not lead to useful conserved quantities.

## 1.2.8 *The wave map problem: the Hodge system in the general case of a parallelizable target manifold*

The Hodge system associated to the wave map problem can be derived in the more general setting of a parallelizable target, which admits a smooth orthonormal frame system at all of its points. Thus, if we denote this frame system by $\{e_j\}_{j=1}^d$, where $\dim N = d$, then we have

$$\langle e_j, e_k \rangle_h \stackrel{\text{def}}{=} h_{ab}\,e_j^a e_k^b = \delta_{jk}.$$

We start by decomposing the deformation tensor $S_\mu^a$ with respect to this frame, i.e.,

$$S_\mu = \sum_j \langle e_j, S_\mu \rangle_h e_j.$$

Introducing $\psi_\mu^j := \delta^{jk}\langle e_k, S_\mu \rangle_h$, we deduce

$$S_\mu = \psi_\mu^j e_j,$$

where summation over the index $j$ is assumed in the last formula.

Next, we compute the mixed derivative

$$\widehat{\nabla}_{S_\mu} S_\nu = \widehat{\nabla}_{S_\mu}(\psi_\nu^j e_j) = (\nabla_\mu \psi_\nu^j) e_j + \psi_\nu^j (\widehat{\nabla}_{S_\mu} e_j)$$

and, connected to this expression, we write the decomposition

$$\widehat{\nabla}_{S_\mu} e_j = \sum_k \langle e_k, \widehat{\nabla}_{S_\mu} e_j \rangle_h e_k.$$

If we define the matrix

$$A_{\mu, kj} \overset{\text{def}}{=} \langle e_k, \widehat{\nabla}_{S_\mu} e_j \rangle_h,$$

then we notice that it is antisymmetric due to $\nabla_\mu \langle e_k, e_j \rangle_h = \widehat{\nabla}_{S_\mu} \langle e_k, e_j \rangle_h = 0$. We raise the index $k$ according to

$$A_{\mu, j}{}^k \overset{\text{def}}{=} \delta^{kl} \langle e_l, \widehat{\nabla}_{S_\mu} e_j \rangle,$$

which implies

$$\widehat{\nabla}_{S_\mu} e_j = A_{\mu, j}{}^k e_k.$$

With these notational conventions, the formula for the mixed derivative becomes

$$\widehat{\nabla}_{S_\mu} S_\nu = \left\{ \nabla_\mu \psi_\nu^j + A_{\mu, k}{}^j \psi_\nu^k \right\} e_j.$$

It follows now that the identity $\widehat{\nabla}_{S_\mu} S_\nu = \widehat{\nabla}_{S_\nu} S_\mu$ and the evolution equation $\widehat{\nabla}_{S_\mu} S^\mu = 0$ become

$$\nabla_\mu \psi_\nu^j + A_{\mu, k}{}^j \psi_\nu^k = \nabla_\nu \psi_\mu^j + A_{\nu, k}{}^j \psi_\mu^k \tag{1.73}$$

and

$$\nabla_\mu \psi^{j, \mu} + A_{\mu, k}{}^j \psi^{k, \mu} = 0, \tag{1.74}$$

respectively.

Next, we turn our attention to the derivation of the curvature equation, for which we introduce the antisymmetric tensor

$$F_{\mu\nu, kj} \overset{\text{def}}{=} \nabla_\mu A_{\nu, kj} - \nabla_\nu A_{\mu, kj} + [A_\mu, A_\nu]_{kj}.$$

To simplify the calculations, we adopt the antisymmetrizing notational convention with respect to the indices $\mu$ and $\nu$, and thus derive

$$\widehat{\nabla}_{S_{[\mu}} \langle e_k, \widehat{\nabla}_{S_{\nu]}} e_j \rangle_h = \langle \widehat{\nabla}_{S_{[\mu}} e_k, \widehat{\nabla}_{S_{\nu]}} e_j \rangle_h + \langle e_k, \widehat{\nabla}_{S_{[\mu}} \widehat{\nabla}_{S_{\nu]}} e_j \rangle,$$

or, equivalently,

$$\nabla_{[\mu} A_{\nu], kj} + \frac{1}{2} [A_\mu, A_\nu]_{kj} = \frac{1}{2} \psi_\mu^m \psi_\nu^n R_{mnkj}, \tag{1.75}$$

with $R_{mnkj}$ denoting the curvature tensor. Finally, if we define

$$G_{\mu\nu}^{mn} \overset{\text{def}}{=} \psi_\mu^{[m} \psi_\nu^{n]}$$

and use the $A$-covariant derivative notation $\nabla_\mu^A := \nabla_\mu + A_\mu$, then (1.73)-(1.75) can be written as the Hodge system

$$\nabla_\mu^A \psi^\mu = 0, \tag{1.76}$$

$$\nabla_\mu^A \psi_\nu = \nabla_\nu^A \psi_\mu, \tag{1.77}$$

$$F_{\mu\nu,kj} = G_{\mu\nu}^{mn} R_{mnkj}. \tag{1.78}$$

The unknown field is $\psi_\mu^j$, with $0 \le \mu \le n$ and $1 \le j \le d$, and the equations are invariant under the transformations of the orthogonal group $O(d)$. The above computations can be easily adapted to the more general case of a target manifold which admits a uniform isometric embedding into a Euclidean space.

### 1.2.8.1    *The special case of a 3-dimensional target manifold with conformal structure*

It is perhaps more instructive to experience the previous calculations for a 3-dimensional target endowed with a conformal structure. Thus, one can identify the antisymmetric matrices $A_{\mu,kj}$ with vectors and can regard the expression $A_{\mu,k}^{\;\;j} \psi_\nu^k$ as an exterior/cross product. These considerations will help us better understand the Skyrme model, where the target manifold is $\mathbb{S}^3$, and, in particular, the Skyrme modification of the Hodge system. In this case, things are even simpler, as the gauge group is either $SU(2)$ or $SO(3)$.

In the present setting, the metric on the target has the form[12] $h_{ab} = h\,\delta_{ab}$, the coordinates are labeled as $(\phi^a)_{1 \le a \le 3}$, and we introduce the notations $u := \log(h)$ and $u_a := \widehat{\nabla}_a u$. An easy computation shows that the Christoffel symbols are given by

$$\Gamma_{ab}^c = \frac{1}{2} \left( u_a \delta_b^{\;c} + u_b \delta_a^{\;c} - u^c \delta_{ab} \right),$$

with $u^c := \delta^{cd} u_d$. For the deformation tensor $S_\mu^a := \nabla_\mu \phi^a$, using notational conventions such as $\widehat{\nabla} u \cdot S_\mu := \delta_b^{\;a} u_a S_\mu^b$, it follows that

$$\widehat{\nabla}_{S_\mu} S_\nu^a = \nabla_\mu S_\nu^a + \Gamma_{bc}^a S_\mu^b S_\nu^c$$

$$= \nabla_\mu S_\nu + \frac{1}{2} \left[ \left( \widehat{\nabla} u \cdot S_\mu \right) S_\nu^a + \left( \widehat{\nabla} u \cdot S_\nu \right) S_\mu^a - u^a \left( S_\mu \cdot S_\nu \right) \right].$$

---

[12]Here, we have in mind the cases $h(\phi) = 4/(1 \pm |\phi|^2)^2$, modeling the unit sphere $\mathbb{S}^3$ and the 3-dimensional hyperboloid $\mathbb{H}^3$, respectively.

If we define as before $\psi_\mu^a := \sqrt{h}\, S_\mu^a$, then, by multiplying the previous equation by $\sqrt{h}$, we obtain

$$
\sqrt{h}\, \widehat{\nabla}_{S_\mu} S_\nu^a = \nabla_\mu \psi_\nu^a + \frac{1}{2}\left\{ (\widehat{\nabla} u \cdot \psi_\nu) S_\mu^a - u^a (S_\mu \cdot \psi_\nu) \right\}
$$
$$
= \nabla_\mu \psi_\nu^a + \frac{1}{2}\left( (\widehat{\nabla} u \wedge S_\mu) \wedge \psi_\nu \right)^a. \tag{1.79}
$$

In the above, we also relied on $S_\nu \cdot \psi_\nu := \delta_{ab} S_\mu^a \psi_\nu^b$, whereas the last term can be interpreted as a triple exterior product.

With the help of the totally antisymmetric tensor $\widehat{\epsilon}_{abc}$, we construct the gauge field

$$
A_\mu^a \overset{\text{def}}{=} \frac{1}{2}\widehat{\epsilon}^a_{\ bc} u^b S_\mu^c = \frac{1}{2}\left( \widehat{\nabla} u \wedge S_\mu \right)^a
$$

and, subsequently, the tensors

$$
F_{\mu\nu}^a \overset{\text{def}}{=} \nabla_\mu A_\nu^a - \nabla_\nu A_\mu^a + \left( A_\mu \wedge A_\nu \right)^a,
$$
$$
G_{\mu\nu}^a \overset{\text{def}}{=} \left( \psi_\mu \wedge \psi_\nu \right)^a,
$$

where $1 \le a \le 3$. Our goal is to arrive at the gauge system

$$
\nabla_\mu \psi^{a,\mu} + \left( A_\mu \wedge \psi^\mu \right)^a = 0, \tag{1.80}
$$
$$
\nabla_\mu \psi_\nu^a + \left( A_\mu \wedge \psi_\nu \right)^a = \nabla_\nu \psi_\mu^a + \left( A_\nu \wedge \psi_\mu \right)^a, \tag{1.81}
$$
$$
F_{\mu\nu}^a = K G_{\mu\nu}^a. \tag{1.82}
$$

Based on (1.79), it is easy to see that the first two equations are restatements of the equation $\widehat{\nabla}_{S_\mu} S^\mu = 0$ and the identity $\widehat{\nabla}_{S_\mu} S_\nu = \widehat{\nabla}_{S_\nu} S_\mu$, respectively. Hence, we are left to derive the curvature equation, for which we need to further compute $F_{\mu\nu}$. Direct calculations yield

$$
\nabla_\mu A_\nu^a - \nabla_\nu A_\mu^a = \frac{1}{2}\widehat{\epsilon}^{ab}_{\ \ c}\, \partial_d \partial_b u \left( S_\mu^d S_\nu^c - S_\nu^d S_\mu^c \right)
$$
$$
= \frac{1}{2}\widehat{\epsilon}^{ab}_{\ \ c}\, \widehat{\epsilon}^c_{\ de} \left( S_\mu \wedge S_\nu \right)^d \partial^e \partial_b u
$$
$$
= \frac{1}{2}\left( \delta^a_{\ d}\delta^b_{\ e} - \delta^a_{\ e}\delta^b_{\ d} \right) \left( S_\mu \wedge S_\nu \right)^d \partial^e \partial_b u
$$
$$
= \frac{1}{2}\left[ \left( \psi_\mu \wedge \psi_\nu \right)^a \left( h^{bc}\, \partial_b \partial_d u \right) - \left( \psi_\mu \wedge \psi_\nu \right)^b (\partial_b \partial_c u) h^{ca} \right]
$$

and

$$
\left( A_\mu \wedge A_\nu \right)^a = \frac{1}{4}\left( (\widehat{\nabla} u \wedge S_\mu) \wedge (\widehat{\nabla} u \wedge S_\nu) \right)^a
$$
$$
= -\frac{1}{4}\left( (S_\mu \cdot (\widehat{\nabla} u \wedge S_\nu)) \widehat{\nabla} u \right)^a
$$
$$
= \frac{1}{4}\left( (\widehat{\nabla} u \cdot (S_\mu \wedge S_\nu)) \widehat{\nabla} u \right)^a
$$
$$
= \frac{1}{4}\left( \psi_\mu \wedge \psi_\nu \right)^b \partial_b u\, \partial_c u\, h^{ca}.
$$

Combining these two expressions with the definition of $F_{\mu\nu}$, we deduce (1.82), i.e.,

$$F^a_{\mu\nu} = G^b_{\mu\nu} K_b{}^a,$$

where the curvature tensor has the formula

$$K_b{}^a = \frac{1}{2}\left[(h^{dc}\partial_d\partial_c u)\delta_b{}^a - (\partial_b\partial_c u)h^{ca} + \frac{1}{2}\partial_b u\,\partial_c u\,h^{ca}\right]$$

$$= \frac{1}{2}\left(\delta_b{}^a \Delta_h u - \widehat{\nabla}_b\widehat{\nabla}^a u + \widehat{\nabla}_b u\widehat{\nabla}^a u\right).$$

For the cases we are interested in, namely $N = \mathbb{S}^3$ and $N = \mathbb{H}^3$, we have that $K_b{}^a = K\delta_b{}^a$, with $K = 1$ and $K = -1$, respectively.

As a final remark, if we adopt the short-hand notation $\nabla^A_\mu := \nabla_\mu + A\wedge$, then we can write the system (1.80)-(1.82) in the more compact form

$$\nabla^A_\mu \psi^\mu = 0, \qquad \nabla^A_\mu \psi_\nu = \nabla^A_\nu \psi_\mu, \qquad F_{\mu\nu} = KG_{\mu\nu}. \tag{1.83}$$

This coincides with (1.76)-(1.78), the only difference being that $\psi_\mu$ and $A_\mu$ are now vectors in $\mathbb{R}^3$.

## 1.2.9    *The wave map problem: static solutions for the 2+1-dimensional case*

In this subsection, we would like to investigate the question of static solutions to the 2+1-dimensional wave map problem, for which the base is $\mathbb{R}^{2+1}$ and the target is either $\mathbb{S}^2$ or $\mathbb{H}^2$. Using the elegant language of complex analysis, if we write $z := x^1 + ix^2$ for $\mathbf{x} = (x^1, x^2) \in \mathbb{R}^2$ and introduce the complex derivatives

$$\partial_{\bar{z}} \overset{\text{def}}{=} \frac{1}{2}(\partial_1 + i\partial_2) \qquad \text{and} \qquad \partial_z \overset{\text{def}}{=} \frac{1}{2}(\partial_1 - i\partial_2),$$

then we have that $|\partial_1|^2 + |\partial_2|^2 = 4\partial_{\bar{z}}\cdot\partial_z$ and $\Delta = 4\partial_{\bar{z}}\partial_z$. Therefore, according to (1.52)-(1.53), static solutions correspond to $w = w(z, \bar{z})$ satisfying

$$-\partial_{\bar{z}}\partial_z w \pm \frac{2\bar{w}}{1 \pm |w|^2}\partial_{\bar{z}}w\,\partial_z w = 0. \tag{1.84}$$

It is clear that if $w$ is either analytic or conjugate analytic, then it is a static solution.

Let us focus now on analytic solutions and, in particular, polynomials

$$w(z) = \prod_{j=1}^{p}(z - z_j).$$

In the case when the target manifold is $\mathbb{H}^2$ (i.e., $h = 4/(1 - |w|^2)^2$), none of these are admissible solutions for obvious reasons. However, when the target is $\mathbb{S}^2$ (i.e., $h = 4/(1 + |w|^2)^2$), these solutions are energy minimizers within a certain topological class determined by the degree of the polynomial. To explain this, we rely on (1.56) and (1.57) to define the one-form $a := A_\mu dx^\mu$ and the two-form $g := G_{\mu\nu} dx^\mu dx^\nu$, such that, according to (1.58), the relation $da = Kg$ holds. For the static problem, we have $a = A_1 dx^1 + A_2 dx^2$ and $g = G_{12} dx^1 dx^2$. Moreover, one can check easily the identity

$$4h\left|\partial_{\bar{z}} w\right|^2 = h\left(|\partial_1 w|^2 + |\partial_2 w|^2\right) + 2G_{12},$$

which, integrated over the complex plane, yields

$$\int_{\mathbb{C}} h\left(|\partial_1 w|^2 + |\partial_2 w|^2\right) dx^1 dx^2 = \int_{\mathbb{C}} 4h\left|\partial_{\bar{z}} w\right|^2 dx^1 dx^2 - \frac{2}{K} \int_{\mathbb{C}} da.$$

The left-hand side represents the energy of the map and thus is minimized if $\partial_{\bar{z}} w = 0$. If we choose $w = z^k$, then the last integral on the right-hand side picks only the singularities of $w$ at $\infty$, i.e.,

$$-\frac{2}{K} \int_{\mathbb{C}} da = 4\pi k.$$

Even if we found meaningful solutions to the static wave map problem, we would like to take a closer look at its associated Hodge system,

$$\nabla_j^A \psi^j = 0, \qquad \nabla_j^A \psi_k = \nabla_k^A \psi_j, \qquad \partial_1 A_2 - \partial_2 A_1 = KG_{12}, \qquad (1.85)$$

with $1 \le j \le 2$. It is well-known that this system is completely integrable (e.g., Hélein [68]) and we want to discuss its main features. For this purpose, we notice that we can write

$$A_1 = \partial_2 \omega + \partial_1 v, \qquad A_2 = -\partial_1 \omega + \partial_2 v,$$

where $\omega$ and $v$ are some real potentials. We work again with complex variables and we introduce the operators

$$\mathcal{L}_{\bar{z}} \stackrel{\text{def}}{=} \partial_{\bar{z}} + \omega_{,\bar{z}} + iv_{,\bar{z}}, \qquad \mathcal{L}_z \stackrel{\text{def}}{=} -\partial_z + \omega_{,z} - iv_{,z},$$

for which we have $\left(\mathcal{L}_{\bar{z}}\right)^* = \mathcal{L}_z$. Using $\psi_1$ and $\psi_2$, we also construct

$$\psi_{\bar{z}} \stackrel{\text{def}}{=} \frac{1}{2}\left(\psi_1 + i\psi_2\right), \qquad \psi_z \stackrel{\text{def}}{=} \frac{1}{2}\left(\psi_1 - i\psi_2\right).$$

**Remark 1.5.** In the above definitions, $\bar{z}$ and $z$ are indices for $\psi_{\bar{z}}$ and $\psi_z$, whereas for $\omega_{,\bar{z}} := \partial_{\bar{z}} \omega$ and $\omega_{,z} := \partial_z \omega$ they designate complex differentiation. Moreover, the operators $\mathcal{L}_{\bar{z}}$ and $\mathcal{L}_z$ can be cast as weighted differentiations:

$$\mathcal{L}_{\bar{z}} = e^{-\omega - iv} \partial_{\bar{z}} e^{\omega + iv}, \qquad \mathcal{L}_z = -e^{\omega - iv} \partial_z e^{-\omega + iv}.$$

With these notational conventions, the Hodge system (1.85) can be written in complex form

$$\mathcal{L}_{\bar{z}}\psi_z = 0, \qquad \mathcal{L}_z\psi_{\bar{z}} = 0, \qquad -4\partial_{\bar{z}}\partial_z\omega = K\big(|\psi_{\bar{z}}|^2 - |\psi_z|^2\big). \qquad (1.86)$$

The operators $\mathcal{L}_{\bar{z}}$ and $\mathcal{L}_z$ do not commute; instead, they satisfy

$$[\mathcal{L}_{\bar{z}}, \mathcal{L}_z] = \frac{K}{2}\big(|\psi_z|^2 - |\psi_{\bar{z}}|^2\big).$$

The energy of the map is given by the integral

$$E \overset{\text{def}}{=} \int_{\mathbb{C}} \big\{|\psi_{\bar{z}}|^2 + |\psi_z|^2\big\}\, dm_z, \qquad dm_z \overset{\text{def}}{=} \frac{1}{2i}d\bar{z}\wedge dz = dx^1 dx^2.$$

In this setting, the energy-momentum tensor is traceless and can be complexified by writing

$$T \overset{\text{def}}{=} T_{11} + iT_{12} = 4\psi_{\bar{z}}\overline{\psi_z}.$$

This is due to the fact that $T_{11} = |\psi_1|^2 - |\psi_2|^2$ and $T_{12} = \langle\psi_2, \psi_1\rangle = 2\mathcal{R}e\{\psi_1\overline{\psi_2}\}$. In addition, the structure equations $\nabla_j T^j{}_k = 0$ reduce to $\partial_z T = 0$, as

$$\partial_z\big\{\psi_{\bar{z}}\overline{\psi_z}\big\} = \big(\mathcal{L}_z\psi_{\bar{z}}\big)\overline{\psi_z} + \psi_{\bar{z}}\big(\overline{\mathcal{L}_{\bar{z}}\psi_z}\big) = 0.$$

Finally, the complex two-form $\overline{T}\, dx^1 dx^2$ is the Hopf differential of the harmonic map.

Now, we can address the question of how to find solutions for (1.86). A simple idea is to pick $\psi_{\bar{z}} = 0$ and fix the gauge by assuming $v = 0$. Using the previous remark, the first equation of (1.86) simplifies to $\partial_{\bar{z}}\big(e^\omega\psi_z\big) = 0$, which implies $\psi_z = 2f'(z)e^{-\omega}$ for some analytic function $f = f(z)$. Plugging this profile in the last equation of (1.86), we deduce the Liouville-type equation

$$\partial_{\bar{z}}\partial_z\omega = K|f'|^2 e^{-2\omega}.$$

One recognizes that $\omega = \log\big(1 + |f|^2\big)$ is a solution if $K = 1$. For $K = -1$, we can find solutions to this equation, but they are not acceptable since they have infinite energy. The simple choice $f(z) = z^n$, or any polynomial of degree $n$, provides us with a family of solutions having topological degree $n$. The case $f(z) = z$ is of particular interest to us, as the invariance of the equation under rotations and internal dilations means that the family

$$w = \frac{Acz + B}{Ccz + D}, \qquad AB - CD = 1,$$

also represents solutions.

### 1.2.10 *The wave map problem: the completely integrable* $1 + 1$-*dimensional case*

Here, we will briefly comment on the $1 + 1$-dimensional problem, for which the wave map system (1.21) is completely integrable (e.g., Pohlmeyer [133]). In this context, the energy-momentum tensor has the form

$$T_{\mu\nu} = \frac{1}{2} \left\{ \psi_\mu \overline{\psi}_\nu + \overline{\psi}_\mu \psi_\nu \right\} - \frac{1}{2} g_{\mu\nu} \langle \overline{\psi}, \psi \rangle, \tag{1.87}$$

where, as before, $\psi_\mu = \sqrt{h} \nabla_\mu \phi$, $h$ is the conformal factor of the target metric, and $\phi$ is complex-valued.

If we consider the base manifold to be the $1 + 1$-dimensional Minkowski spacetime, then it is easily checked that $T_{11} = T_{00}$. The structure equations of the energy-momentum tensor become

$$\nabla_0 T_{00} - \nabla_1 T_{10} = 0, \qquad \nabla_0 T_{10} - \nabla_1 T_{00} = 0,$$

from which we infer

$$\Box T_{00} = 0.$$

Hence, the energy density is a solution of the wave equation. We can use this equation to obtain either a priori uniform-in-time bounds for $\phi$ or energy estimates for $T_{00}$, with the end conclusion being that the solution is global in time (e.g., Keel-Tao [79] and Machihara-Nakanishi-Tsugawa [111]). An alternative way to analyze the above system is by adding and subtracting its equations to derive

$$\left( \nabla_0 - \nabla_1 \right) \left\{ T_{00} + T_{10} \right\} = 0, \qquad \left( \nabla_0 + \nabla_1 \right) \left\{ T_{00} - T_{10} \right\} = 0.$$

Now, we can use the null coordinates $x^0 \pm x^1$ to integrate these equations (e.g., see the construction of solitary waves by Shatah-Strauss [144]).

### 1.2.11 *The Skyrme model: its energy-momentum tensor and associated Hodge system*

Our goal in this subsection is to describe the mathematical formulation of the Skyrme model, which is a generalization of the $3 + 1$-dimensional wave map problem admitting static topological solitons. We recall that the Skyrme model is concerned with formal critical points $\phi : \mathbb{R}^{3+1} \to \mathbb{S}^3$ for the Lagrangian

$$\mathbf{L}_S[\phi] = \int_{\mathbb{R}^{3+1}} \left\{ \frac{1}{2} \langle \nabla^\mu \phi, \nabla_\mu \phi \rangle + \frac{\alpha^2}{4} \left[ \langle \nabla^\mu \phi, \nabla_\mu \phi \rangle^2 \right. \right.$$
$$\left. \left. - \langle \nabla^\mu \phi, \nabla^\nu \phi \rangle \langle \nabla_\mu \phi, \nabla_\nu \phi \rangle \right] \right\} d\mu_g, \tag{1.88}$$

with $\alpha$ being a constant having the dimension of length.

If we replace the Minkowski spacetime $\mathbb{R}^{3+1}$ by a general $n + 1$-dimensional Lorentzian manifold $M$ and we denote the matrix defined in (1.18) by

$$\sigma_{\mu\nu} \overset{\text{def}}{=} \langle \nabla_\mu \phi, \nabla_\nu \phi \rangle, \qquad (1.89)$$

then the Lagrangian of the wave map problem is simply

$$\mathbf{L}[\phi] = \int_M \left\{ \frac{1}{2} \text{tr}(\sigma) \right\} d\mu_g,$$

while the one for the modification proposed by Skyrme is given by

$$\begin{aligned}
\mathbf{L}_S[\phi] &= \int_M \left\{ \frac{1}{2} \text{tr}(\sigma) + \frac{\alpha^2}{4} \left( (\text{tr}(\sigma))^2 - \text{tr}(\sigma^2) \right) \right\} d\mu_g \\
&= \int_M \left\{ \frac{1}{2} \sigma_\mu^\mu + \frac{\alpha^2}{4} \left( \sigma_\mu^\mu \sigma_\nu^\nu - \sigma^{\mu\nu} \sigma_{\mu\nu} \right) \right\} d\mu_g.
\end{aligned} \qquad (1.90)$$

Now, we comment briefly on the connection of the Skyrme model to elasticity. Labeling the eigenvalues of the matrix $\sigma_\mu^\nu$ by $\{\lambda_\mu\}_{0 \le \mu \le n}$, we know that

$$\text{tr}(\sigma) = \sum_\mu \lambda_\mu, \qquad (\text{tr}(\sigma))^2 - \text{tr}(\sigma^2) = 2 \sum_{\mu \ne \nu} \lambda_\mu \lambda_\nu,$$

which implies

$$\mathbf{L}_S[\phi] = \int_M \left\{ \frac{1}{2} \sum_\mu \lambda_\mu + \frac{\alpha^2}{2} \sum_{\mu \ne \nu} \lambda_\mu \lambda_\nu \right\} d\mu_g.$$

This formulation should look familiar to people acquainted with nonlinear elasticity, where one writes Lagrangians in terms of the Newton symmetric polynomials. For more details and connected issues, we refer the reader to the book by Dafermos [34].

Next, we would like to derive the Euler-Lagrange equations for the Skyrme model and we want to do it in such a way that makes clear the relationship between this field theory and the wave map problem. For this purpose, let us introduce the matrix

$$L^{\mu\nu} \overset{\text{def}}{=} g^{\mu\nu} - \alpha^2 \left( \sigma^{\mu\nu} - g^{\mu\nu} \text{tr}(\sigma) \right). \qquad (1.91)$$

With the help of this definition and the short-hand notation $(L\nabla)^\mu \phi^a := L^{\mu\nu} \nabla_\nu \phi^a$, the Euler-Lagrange equations for the Skyrme model can be stated as

$$- \nabla_\mu \left\{ L^{\mu\nu} \nabla_\nu \phi^a \right\} + \Gamma^a_{bc} \langle \nabla \phi^b, L\nabla \phi^c \rangle = 0. \qquad (1.92)$$

One should compare these equations with the wave map system

$$-\nabla_\mu\{g^{\mu\nu}\nabla_\nu\phi^a\} + \Gamma^a_{bc}\langle\nabla\phi^b, \nabla\phi^c\rangle = 0.$$

Following this, we focus on obtaining the associated Hodge system to the Skyrme model, for which the approach is similar to the one used in the case of the wave map problem. We can place ourselves in a previously discussed framework for a 3-dimensional target manifold with conformal structure, as $\mathbb{S}^3$ is such a target. Thus, we rely on the definitions

$$\psi^a_\mu \overset{\text{def}}{=} \sqrt{h}\, S^a_\mu, \qquad u^b \overset{\text{def}}{=} \delta^{bc}\, \widehat{\nabla}_c\{\log(h)\}, \qquad A^a_\mu \overset{\text{def}}{=} \frac{1}{2}\widehat{\epsilon}\,^a_{bc}u^b S^c_\mu,$$

$$\nabla^A_\mu \overset{\text{def}}{=} \nabla_\mu + A_\mu\wedge, \qquad F_{\mu\nu} \overset{\text{def}}{=} \nabla_\mu A_\nu - \nabla_\nu A_\mu + A_\mu \wedge A_\nu, \qquad G_{\mu\nu} \overset{\text{def}}{=} \psi_\mu \wedge \psi_\nu.$$

Noticing that

$$\begin{aligned}
\big(\sigma^{\mu\nu} - g^{\mu\nu}\text{tr}(\sigma)\big)\nabla_\nu\phi^a &= \big(\psi^\mu \cdot \psi^\nu\big)S^a_\nu - \big(\psi^\nu \cdot S_\nu\big)\psi^{a,\mu} \\
&= \big((\psi^\mu \wedge S_\nu) \wedge \psi^\nu\big)^a,
\end{aligned}$$

we deduce that the Skyrme Hodge system takes the form

$$\nabla^A_\mu\{\psi^\mu - \alpha^2(\psi^\mu \wedge \psi^\nu) \wedge \psi_\nu\} = 0, \tag{1.93}$$

$$\nabla^A_\mu\psi_\nu = \nabla^A_\nu\psi_\mu, \tag{1.94}$$

$$F_{\mu\nu} = KG_{\mu\nu}. \tag{1.95}$$

When compared to the Hodge system (1.83) associated to the wave map problem, we find that only the first equation of the system has been modified for the Skyrme model.

The energy-momentum tensor for the Skyrme model is given by

$$T_{\mu\nu} \overset{\text{def}}{=} \psi_\mu\cdot\psi_\nu - \frac{1}{2}g_{\mu\nu}\big(\psi^\lambda\cdot\psi_\lambda\big) + \alpha^2\left\{G^\lambda_\mu \cdot G_{\nu\lambda} - \frac{1}{4}g_{\mu\nu}\big(G^{\lambda\rho}\cdot G_{\lambda\rho}\big)\right\} \tag{1.96}$$

and can be derived in the usual manner by considering the action of a Lie derivative on the Lagrangian (1.88). If we compare it to its counterpart (1.23) for the wave map problem, then we see that the Skyrme model introduces new terms similar to the ones coming from Maxwell's equations, with $G_{\mu\nu}$ playing the role of the electromagnetic field. Now, we would like to write the previous formula strictly in terms of the matrix $\sigma$ introduced by (1.89). For this goal, using the easily checked identity

$$\widehat{\epsilon}\,^a_{bc}\widehat{\epsilon}_{apq} = \delta_{bp}\delta_{cq} - \delta_{bq}\delta_{cp},$$

we argue that

$$G_\mu^\lambda \cdot G_{\nu\lambda} = (\psi_\mu \wedge \psi^\lambda) \cdot (\psi_\nu \wedge \psi_\lambda)$$
$$= \widehat{\epsilon}^{\,a}_{\;bc}\, \widehat{\epsilon}_{apq}\, \psi_\mu^b\, \psi^{c,\lambda}\, \psi_\nu^p\, \psi_\lambda^q$$
$$= (\delta_{bp}\delta_{cq} - \delta_{bq}\delta_{cp})\psi_\mu^b\, \psi^{c,\lambda}\, \psi_\nu^p\, \psi_\lambda^q$$
$$= \sigma_{\mu\nu}\, \mathrm{tr}(\sigma) - \sigma_{\mu\lambda}\, \sigma_\nu^\lambda,$$

from which it also follows that

$$G^{\lambda\rho} \cdot G_{\lambda\rho} = \left(\mathrm{tr}(\sigma)\right)^2 - \mathrm{tr}(\sigma^2).$$

Hence, we obtain

$$T_{\mu\nu} = \sigma_{\mu\nu} - \frac{1}{2}g_{\mu\nu}\, \mathrm{tr}(\sigma)$$
$$+ \alpha^2 \left\{ \sigma_{\mu\nu}\, \mathrm{tr}(\sigma) - \sigma_{\mu\lambda}\, \sigma_\nu^\lambda - \frac{1}{4}g_{\mu\nu}\left[\left(\mathrm{tr}(\sigma)\right)^2 - \mathrm{tr}(\sigma^2)\right]\right\}.$$

Next, we show that the energy-momentum tensor satisfies the structure equations

$$\nabla_\mu\{T_\nu^\mu\} = 0,$$

and we do this with the help of (1.93)-(1.94), the observation that

$$\nabla_\mu(X \cdot Y) = (\nabla_\mu^A X) \cdot Y + X \cdot (\nabla_\mu^A Y),$$

and the simple identity

$$(G^{\mu\lambda} \wedge \psi_\lambda) \cdot \psi_\nu = G^{\mu\lambda} \cdot (\psi_\lambda \wedge \psi_\nu)$$
$$= -G^{\mu\lambda} \cdot G_{\nu\lambda}.$$

Thus, we can write:

$$0 = \nabla_\mu^A\{\psi^\mu - \alpha^2(\psi^\mu \wedge \psi^\lambda) \wedge \psi_\lambda\} \cdot \psi_\nu$$
$$= \nabla_\mu^A\{\psi^\mu - \alpha^2 G^{\mu\lambda} \wedge \psi_\lambda\} \cdot \psi_\nu$$
$$= \nabla_\mu\{\psi^\mu \cdot \psi_\nu - \alpha^2(G^{\mu\lambda} \wedge \psi_\lambda) \cdot \psi_\nu\} - (\psi^\mu - \alpha^2 G^{\mu\lambda} \wedge \psi_\lambda) \cdot \nabla_\mu^A \psi_\nu$$
$$= \nabla_\mu\{\psi^\mu \cdot \psi_\nu + \alpha^2 G^{\mu\lambda} \cdot G_{\nu\lambda}\} - (\psi^\mu - \alpha^2 G^{\mu\lambda} \wedge \psi_\lambda) \cdot \nabla_\nu^A \psi_\mu$$
$$= \nabla_\mu\{\psi^\mu \cdot \psi_\nu + \alpha^2 G^{\mu\lambda} \cdot G_{\nu\lambda}\} - \frac{1}{2}\nabla_\nu(\psi^\mu \cdot \psi_\mu) - \frac{\alpha^2}{4}\nabla_\nu(G^{\mu\lambda} \cdot G_{\mu\lambda})$$
$$= \nabla_\mu\{T_\nu^\mu\}.$$

Finally, one can recognize relatively easily that the Skyrme Lagrangian can be stated in terms of $\psi_\mu$ and $G_{\mu\nu}$ as

$$\mathbf{L}_S[\phi] = \int_{\mathbb{R}^{3+1}} \left\{\frac{1}{2}\psi^\mu \cdot \psi_\mu + \frac{\alpha^2}{4}G^{\mu\nu} \cdot G_{\mu\nu}\right\} d\mu_g.$$

This comes in handy when discussing the energy associated to the Skyrme model, for which we also need to introduce the electric field $D$,

$$D_j^a \overset{\text{def}}{=} G_{0j}^a,$$

and the magnetic field $B$,

$$B_j^a \overset{\text{def}}{=} \frac{1}{2}\epsilon_j{}^{kl}G_{kl}^a.$$

Then, the conserved-in-time energy of the Skyrme model is the integral

$$\mathcal{E}(\phi) \overset{\text{def}}{=} \frac{1}{2}\int_{\{t\}\times\mathbb{R}^3}\left\{|\psi_0|^2 + \sum_{1\le j\le 3}|\psi_j|^2 + \alpha^2\big(|D|^2 + |B|^2\big)\right\}d\mathbf{x}. \quad (1.97)$$

Given that $G_{\mu\nu}$ is quadratic with respect to $\psi$, the portion of energy corresponding to the electric and magnetic fields is quartic in $\psi$, containing, however, only certain such combinations of $\psi$'s.

### 1.2.12 The Einstein vacuum equations: a symmetry reduction to a system containing wave map type problems

In this section, we would like to address an important connection between the wave map problem and a symmetry reduction for the Einstein vacuum equations. To put the issue in perspective, we recall that the Einstein vacuum equations admit a certain family of stationary solutions, which are either spherically symmetric or axisymmetric, with the latter generalizing the former. These are the celebrated Schwarzschild [140] and Kerr [82] (see also [23]) solutions, respectively. It was initially noticed by Bach-Weyl [5] that the Schwarzschild solutions, in the region outside the event horizon and with an appropriate choice of coordinates, can be thought of as solutions to the Laplace equation with prescribed singularities at the origin. This appears to be similar to what happens if one describes the Newton gravitation potential generated by an ideal point mass. The later discovery of Kerr solutions complicated matters and it was Ernst [41] who observed that the corresponding field equations can be reformulated as harmonic map equations with the target being the hyperbolic upper half-plane. Therefore, the Schwarzschild and Kerr solutions become, in fact, harmonic maps with prescribed singularities on the axis of symmetry.

Here, our goal is to show that if the Einstein vacuum equations admit a "cylindrical" symmetry (i.e., in the language of differential geometry, the metric possesses a Killing vector field with closed orbits), then they reduce to a system consisting of:

- the wave map system with the hyperbolic right half-plane (a model for $\mathbb{H}^2$) as its target;
- an equation stating that the Einstein tensor of the symmetry reduced metric equals the energy-momentum tensor of the previous wave map system.

In what follows, we present computations appearing in a series of papers by Weinstein [185, 186], who examined this symmetry reduction and, subsequently, the static problem. We focus here on the reduction that leads to a wave map system on a $2 + 1$-dimensional manifold. Throughout this section, we will rely heavily on basic material related to Lie derivatives and the curvature and Ricci tensors, for which we ask the reader to consult the appendix at the end of the book.

We start by considering a $3 + 1$-dimensional Lorentzian manifold $(M, g_{\mu\nu})$, which satisfies the Einstein vacuum equations,

$$R_{\mu\nu} = 0, \tag{1.98}$$

where the Ricci tensor $R_{\mu\nu}$ is the partial contraction of the curvature tensor in the $(2, 4)$ positions, i.e.,

$$R_{\mu\nu} \stackrel{\text{def}}{=} R_{\mu\lambda\nu}{}^{\lambda}.$$

We assume that the manifold $M$ has a symmetry described by a Killing vector field $K = K_{\mu}$, meaning

$$\mathcal{L}_K g_{\mu\nu} = \nabla_{\mu} K_{\nu} + \nabla_{\nu} K_{\mu} = 0, \tag{1.99}$$

and we introduce the scalar quantity

$$X \stackrel{\text{def}}{=} \langle K, K \rangle = K_{\nu} K^{\nu}, \tag{1.100}$$

where $\langle \cdot, \cdot \rangle$ denotes the total contraction of a tensor as in (1.19). Moving forward, it is helpful to recall the shorthand notation for anti-symmetrization,

$$T_{[\alpha\beta]} \stackrel{\text{def}}{=} \frac{1}{2} \left( T_{\alpha\beta} - T_{\beta\alpha} \right), \tag{1.101}$$

$$T_{[\alpha\beta\gamma]} \stackrel{\text{def}}{=} \frac{1}{6} \left( T_{\alpha\beta\gamma} + T_{\beta\gamma\alpha} + T_{\gamma\alpha\beta} - T_{\beta\alpha\gamma} - T_{\gamma\beta\alpha} - T_{\alpha\gamma\beta} \right), \tag{1.102}$$

and some of the basic symmetries of the curvature tensor,

$$R_{\mu\nu\alpha\beta} = R_{[\mu\nu]\alpha\beta} = R_{\mu\nu[\alpha\beta]} = R_{\alpha\beta\mu\nu}, \tag{1.103}$$

$$R_{[\mu\nu\alpha]\beta} = 0, \tag{1.104}$$

the last being the well-known Ricci identity. Later, we will also deal with contractions of the Levi-Civita tensor, for which the easily-derived formulas

$$\epsilon_{\alpha\mu\nu\beta}\,\epsilon^{\alpha\lambda\rho\sigma} = -6\delta^{\lambda}_{[\mu}\delta^{\rho}_{\nu}\delta^{\sigma}_{\beta]}, \tag{1.105}$$

$$\epsilon_{\mu\nu\alpha\beta}\,\epsilon^{\mu\nu\lambda\rho} = -2\delta^{\lambda}_{[\alpha}\delta^{\rho}_{\beta]}, \tag{1.106}$$

are important to have. Furthermore, we need certain equations for the second-order derivatives of the Killing vector field, which are included in the following lemma.

**Lemma 1.2.** *If $K = K_\sigma$ is a Killing vector field, then*

$$\nabla_\mu\nabla_\nu K_\sigma = K^\lambda R_{\lambda\mu\nu\sigma}. \tag{1.107}$$

*As a consequence[13],*

$$\Box_g K_\sigma = K^\lambda R_{\lambda\sigma} = 0 \tag{1.108}$$

*holds for metrics $g_{\mu\nu}$ satisfying the Einstein vacuum equations (1.98). Moreover, we have the commutation relation*

$$[\mathcal{L}_K, \nabla_\mu] = 0. \tag{1.109}$$

**Proof.** To see why (1.107) holds, we start by using the formula for the curvature tensor

$$\nabla_\mu\nabla_\nu K_\sigma - \nabla_\nu\nabla_\mu K_\sigma = R_{\mu\nu\sigma}{}^{\lambda}K_\lambda, \tag{1.110}$$

jointly with (1.99), to infer

$$\nabla_\mu\nabla_\nu K_\sigma + \nabla_\nu\nabla_\sigma K_\mu = R_{\mu\nu\sigma}{}^{\lambda}K_\lambda.$$

If one permutes above cyclically the indices $\mu$, $\nu$, and $\sigma$, and combines the resulting equations, then it follows that

$$2\nabla_\mu\nabla_\nu K_\sigma = \left(R_{\mu\nu\sigma}{}^{\lambda} - R_{\nu\sigma\mu}{}^{\lambda} + R_{\sigma\mu\nu}{}^{\lambda}\right)K_\lambda,$$

which yields (1.107) by applying (1.104).

As the derivation of (1.108) is immediate from (1.107) by contracting the $\mu$ and $\nu$ indices, we are left to prove (1.109), for which it is enough to check it for scalars and vectors. First, for a scalar $f$, we have

$$\mathcal{L}_K\nabla_\mu f = K^\alpha\nabla_\alpha\nabla_\mu f + \nabla_\mu K^\alpha\nabla_\alpha f,$$

$$\nabla_\mu\mathcal{L}_K f = \nabla_\mu\left(K^\alpha\nabla_\alpha f\right) = \nabla_\mu K^\alpha\nabla_\alpha f + K^\alpha\nabla_\mu\nabla_\alpha f,$$

---

[13]The d'Alembertian $\Box_g$ is the one defined in (1.20), i.e.,

$$\Box_g = \Box = -\nabla_\mu\nabla^\mu = -g^{\mu\nu}\nabla_\mu\nabla_\nu.$$

and (1.109) holds on the account of $\nabla_{[\mu}\nabla_{\alpha]}f = 0$. Finally, for a vector $V^\lambda$, we deduce

$$\mathcal{L}_K\nabla_\mu V^\lambda = K^\alpha\nabla_\alpha\nabla_\mu V^\lambda + \nabla_\alpha V^\lambda\nabla_\mu K^\alpha - \nabla_\mu V^\alpha\nabla_\alpha K^\lambda,$$

$$\nabla_\mu\mathcal{L}_K V^\lambda = \nabla_\mu\left(K^\alpha\nabla_\alpha V^\lambda - V^\alpha\nabla_\alpha K^\lambda\right)$$

$$= K^\alpha\nabla_\mu\nabla_\alpha V^\lambda + \nabla_\mu K^\alpha\nabla_\alpha V^\lambda - \nabla_\mu V^\alpha\nabla_\alpha K^\lambda - V^\alpha\nabla_\mu\nabla_\alpha K^\lambda.$$

If we subtract the two equations term by term and use (1.110) together with (1.107), then we derive

$$\left[\mathcal{L}_K, \nabla_\mu\right]V^\lambda = -K^\alpha R_{\alpha\mu\beta}{}^\lambda V^\beta + V^\alpha R_{\beta\mu\alpha}{}^\lambda K^\beta = 0,$$

which finishes the argument.                                        $\square$

Following this result, associated to the vector field $K$, we introduce the electromagnetic field $F_{\mu\nu}$ and its dual $F_{\mu\nu}^\dagger$ according to

$$F_{\mu\nu} \stackrel{\text{def}}{=} \nabla_\mu K_\nu - \nabla_\nu K_\mu = 2\nabla_\mu K_\nu, \tag{1.111}$$

$$F_{\mu\nu}^\dagger \stackrel{\text{def}}{=} \frac{1}{2}\epsilon_{\mu\nu\alpha\beta}F^{\alpha\beta}. \tag{1.112}$$

Furthermore, we define the one-form $\theta := \theta_\mu dx^\mu$, where

$$\theta_\mu \stackrel{\text{def}}{=} \epsilon_{\mu\alpha\beta\gamma}\left(\nabla^\alpha K^\beta\right) K^\gamma = \frac{1}{2}\epsilon_{\mu\alpha\beta\gamma}F^{\alpha\beta}K^\gamma = F_{\mu\gamma}^\dagger K^\gamma. \tag{1.113}$$

If we denote $k := K_\mu dx^\mu$ and $f := F_{\mu\nu}dx^\mu dx^\nu$, then it is easy to see that $dk = f$ and $\theta = \frac{1}{2}(k \wedge f)$. Hence,

$$2d\theta = dk \wedge dk + k \wedge d^2 k = 0,$$

which means that $\theta$ is exact and, in principle, we can find a scalar field $Y$, also named the Ernst potential, such that

$$\nabla_\mu Y = \theta_\mu. \tag{1.114}$$

Together with $X$ given by (1.100), $Y$ is one of the variables appearing in the reduced form of the Einstein vacuum equations that we seek out.

For both $X$ and $Y$, direct computations using the definitions (1.100), (1.111), (1.112), and (1.113) yield

$$\nabla_\mu X = F_{\mu\nu}K^\nu, \tag{1.115}$$

$$\nabla_\nu Y = F_{\mu\nu}^\dagger K^\nu, \tag{1.116}$$

and the fact that both $F_{\mu\nu}$ and $F_{\mu\nu}^\dagger$ are antisymmetric implies

$$K^\mu\nabla_\mu X = K^\mu\nabla_\mu Y = 0. \tag{1.117}$$

Relying on the previous lemma, we deduce

$$\nabla_\mu F^\mu_{\ \nu} = -2\Box_g K_\nu = 0,$$

$$\nabla_{[\mu} F_{\nu\lambda]} = 2\nabla_{[\mu}\nabla_\nu K_{\lambda]} = 2K^\alpha R_{\alpha[\mu\nu\lambda]} = 0,$$

with the second equation being, in fact, equivalent to $\nabla_\mu F^{\dagger\mu}_{\ \ \nu} = 0$. If we apply these two facts in the context of taking the divergence for both (1.115) and (1.116), then we can derive the system

$$\Box_g X + \frac{1}{2}\langle F, F\rangle = 0, \tag{1.118}$$

$$\Box_g Y + \frac{1}{2}\langle F^\dagger, F\rangle = 0, \tag{1.119}$$

where the notational convention $\langle F, F\rangle := F_{\mu\nu}F^{\mu\nu}$ indicates the full contraction of tensors.

Our next goal is to write the second terms in both of the above equations using only $X$ and $Y$ and thus obtain a true system in these variables. For this purpose, using (1.105), we can infer that

$$\begin{aligned}
\epsilon_{\mu\nu\alpha\beta}(\nabla^\alpha Y)K^\beta &= \frac{1}{2}\epsilon_{\alpha\mu\nu\beta}\,\epsilon^{\alpha\lambda\rho\sigma}F_{\lambda\rho}K_\sigma K^\beta \\
&= -3\delta_{[\mu}^{\ \lambda}\delta_\nu^{\ \rho}\delta_{\beta]}^{\ \sigma}F_{\lambda\rho}K_\sigma K^\beta \\
&= -\left(\delta_\mu^{\ \lambda}\delta_\nu^{\ \rho}\delta_\beta^{\ \sigma} + \delta_\nu^{\ \lambda}\delta_\beta^{\ \rho}\delta_\mu^{\ \sigma} + \delta_\beta^{\ \lambda}\delta_\mu^{\ \rho}\delta_\nu^{\ \sigma}\right)F_{\lambda\rho}K_\sigma K^\beta \\
&= -F_{\mu\nu}\langle K, K\rangle - K_\mu(\nabla_\nu X) + (\nabla_\mu X)K_\nu.
\end{aligned}$$

Hence,

$$XF_{\mu\nu} = (\nabla_\mu X)K_\nu - (\nabla_\nu X)K_\mu - \epsilon_{\mu\nu\alpha\beta}(\nabla^\alpha Y)K^\beta. \tag{1.120}$$

In a similar manner, but starting with $\epsilon_{\mu\nu\alpha\beta}(\nabla^\alpha X)K^\beta$ and noticing that

$$\nabla^\alpha X = -\frac{1}{2}\epsilon^{\alpha\lambda\rho\sigma}F^\dagger_{\lambda\rho}K_\sigma,$$

we deduce

$$XF^\dagger_{\mu\nu} = (\nabla_\mu Y)K_\nu - (\nabla_\nu Y)K_\mu + \epsilon_{\mu\nu\alpha\beta}(\nabla^\alpha X)K^\beta. \tag{1.121}$$

We follow this up by decomposing $F$ and $F^\dagger$ into "parallel" and "perpendicular" components given by

$$\overline{F}_{\mu\nu} \stackrel{\text{def}}{=} \frac{1}{X}\left((\nabla_\mu X)K_\nu - (\nabla_\nu X)K_\mu\right), \quad F^\perp_{\mu\nu} \stackrel{\text{def}}{=} \frac{1}{X}\left(-\epsilon_{\mu\nu\alpha\beta}(\nabla^\alpha Y)K^\beta\right),$$

$$\overline{F}^\dagger_{\mu\nu} \stackrel{\text{def}}{=} \frac{1}{X}\left((\nabla_\mu Y)K_\nu - (\nabla_\nu Y)K_\mu\right), \quad F^{\dagger\perp}_{\mu\nu} \stackrel{\text{def}}{=} \frac{1}{X}\left(\epsilon_{\mu\nu\alpha\beta}(\nabla^\alpha X)K^\beta\right).$$

To justify this terminology, it is easy to check that

$$F^\perp_{\mu\nu}K^\nu = F^{\dagger\perp}_{\mu\nu}K^\nu = 0 \tag{1.122}$$

and, furthermore,

$$\langle \overline{F}, F^{\perp} \rangle = \langle \overline{F}, F^{\dagger\perp} \rangle = \langle \overline{F}^{\dagger}, F^{\perp} \rangle = 0.$$

Thus, we derive

$$\langle F, F \rangle = \langle \overline{F}, \overline{F} \rangle + \langle F^{\perp}, F^{\perp} \rangle, \qquad \langle F, F^{\dagger} \rangle = \langle \overline{F}, \overline{F}^{\dagger} \rangle + \langle F^{\perp}, F^{\dagger\perp} \rangle.$$

Due to (1.117), we can compute the contractions

$$\langle \overline{F}, \overline{F} \rangle = \frac{2}{X} \langle \nabla X, \nabla X \rangle$$

and

$$\begin{aligned}
\langle F^{\perp}, F^{\perp} \rangle &= \frac{1}{X^2} \epsilon_{\mu\nu\alpha\beta} \epsilon^{\mu\nu\rho\sigma} \left( \nabla^{\alpha} Y \, K^{\beta} \, \nabla_{\rho} Y \, K_{\sigma} \right) \\
&= -\frac{2}{X^2} \delta_{[\alpha}^{\rho} \delta_{\beta]}^{\sigma} \left( \nabla^{\alpha} Y \, K^{\beta} \, \nabla_{\rho} Y \, K_{\sigma} \right) \\
&= -\frac{2}{X} \langle \nabla Y, \nabla Y \rangle.
\end{aligned}$$

Similarly, one obtains

$$\langle \overline{F}, \overline{F}^{\dagger} \rangle = \langle F^{\perp}, F^{\dagger\perp} \rangle = \frac{2}{X} \langle \nabla X, \nabla Y \rangle.$$

Therefore, the system formed by (1.118) and (1.119) becomes

$$\Box_g X + \frac{1}{X} \left\{ \langle \nabla X, \nabla X \rangle - \langle \nabla Y, \nabla Y \rangle \right\} = 0, \qquad (1.123)$$

$$\Box_g Y + \frac{2}{X} \langle \nabla X, \nabla Y \rangle = 0. \qquad (1.124)$$

When compared to the wave map system (1.21), both of the above equations seem to be of wave-map type. However, it is a well-known fact that the Einstein vacuum equations are of quasilinear nature, while the wave-map type equations have semilinear structure. The way to explain this apparent paradox is by recognizing that the system formed by (1.123) and (1.124) does not fully describe the Einstein vacuum equations and, as such, we need to discover the rest of the equations which complete the picture.

This is the moment in our computations when we bring about the "cylindrical" symmetry assumption mentioned at the start of the section, which we take it to mean that the orbits of the vector field $K$ are "circles". Additionally, we will ask for $K$ to be spacelike such that $X \geq 0$. If we mod-out the orbits generated by $K$, we are naturally led to the following reduction of the ambient manifold,

$$M^{3+1} \sim SO(2) \times N^{2+1},$$

where $N^{2+1}$ is a $2 + 1$-dimensional manifold inheriting a hyperbolic metric through $g_{\mu\nu}$.

To formalize this intuition, let us define the projection

$$h_\mu{}^\nu \overset{\text{def}}{=} \delta_\mu{}^\nu - \frac{K_\mu K^\nu}{X}, \tag{1.125}$$

and the "metric"

$$h_{\mu\nu} \overset{\text{def}}{=} h_\mu{}^\lambda g_{\lambda\nu} = g_{\mu\nu} - \frac{K_\mu K_\nu}{X}, \tag{1.126}$$

for which we easily note that

$$h_\mu{}^\nu K^\mu = h_\mu{}^\nu K_\nu = 0, \tag{1.127}$$

$$h_\mu{}^\lambda h_{\lambda\nu} = h_{\mu\nu}, \qquad h_\mu{}^\lambda h_\lambda{}^\nu = h_\mu{}^\nu, \tag{1.128}$$

and $h_{\mu\nu}$ becomes non-degenerate when restricted to vectors orthogonal to $K$. For vector fields which are both orthogonal to and Lie dragged by $K$, i.e.,

$$\langle K, V \rangle = 0, \qquad \mathcal{L}_K V = 0, \tag{1.129}$$

we introduce a derivation through

$$\widehat{\nabla}_\mu V^\nu \overset{\text{def}}{=} h_\mu{}^\alpha h_\beta{}^\nu \nabla_\alpha V^\beta, \tag{1.130}$$

and, for later use, the associated Ricci curvature coming from

$$(\widehat{\nabla}_\mu \widehat{\nabla}_\nu - \widehat{\nabla}_\nu \widehat{\nabla}_\mu) V^\mu = \widehat{R}_{\nu\mu} V^\mu. \tag{1.131}$$

Using the definitions (1.125)-(1.126) and (1.127), we deduce

$$\widehat{\nabla}_\mu h_{\nu\lambda} = h_\mu{}^\alpha h_\nu{}^\beta h_\lambda{}^\gamma \nabla_\alpha h_{\beta\gamma} = -h_\mu{}^\alpha h_\nu{}^\beta h_\lambda{}^\gamma \nabla_\alpha \left( \frac{K_\beta K_\gamma}{X} \right) = 0,$$

which tells us that the derivation of the "metric" tensor $h_{\mu\nu}$ is zero. Next, we define the d'Alembertian with respect to $h$,

$$\Box_h \overset{\text{def}}{=} -h_\mu{}^\beta \nabla_\beta \left( h^{\mu\alpha} \nabla_\alpha \right),$$

and, successively applying (1.127), (1.99), and (1.117), we derive

$$\begin{aligned}
h_\mu{}^\beta \nabla_\beta \left( h^{\mu\alpha} \right) &= -h_\mu{}^\beta \nabla_\beta \left( \frac{K^\mu K^\alpha}{X} \right) \\
&= -h_\mu{}^\beta \nabla_\beta K^\mu \frac{K^\alpha}{X} \\
&= \left\{ -\nabla_\mu K^\mu + \frac{K^\beta K_\mu \nabla_\beta K^\mu}{X} \right\} \frac{K^\alpha}{X} \\
&= \frac{1}{2} \frac{K^\beta \nabla_\beta X K^\alpha}{X^2} = 0.
\end{aligned}$$

Hence, if we also rely on (1.128), then we can infer that

$$\Box_h = -h^{\alpha\beta}\nabla_\alpha\nabla_\beta = \Box_g + \frac{K_\alpha K_\beta}{X}\nabla_\alpha\nabla_\beta.$$

Following these preliminaries, we proceed to compute $\Box_h X$ and $\Box_h Y$. For the former, we use consecutively (1.115), (1.111), (1.107), (1.103), again (1.115), and (1.123) to obtain

$$
\begin{aligned}
\Box_h X &= \Box_g X + \frac{K^\alpha K^\beta}{X}\nabla_\alpha\nabla_\beta X \\
&= \Box_g X + \frac{K^\alpha K^\beta}{X}\nabla_\alpha\left(F_{\beta\nu}K^\nu\right) \\
&= \Box_g X + \frac{K^\alpha K^\beta}{X}\left\{2\nabla_\alpha\nabla_\beta K_\nu\, K^\nu + \frac{1}{2}F_{\beta\nu}F_\alpha{}^\nu\right\} \\
&= \Box_g X + \frac{1}{X}\left\{2K^\alpha K^\beta K^\nu K^\lambda R_{\lambda\alpha\beta\nu} + \frac{1}{2}F^\nu{}_\alpha K^\alpha\, F_{\nu\beta}K^\beta\right\} \\
&= -\frac{1}{X}\left\{\langle\nabla X,\nabla X\rangle - \langle\nabla Y,\nabla Y\rangle\right\} + \frac{1}{2X}\langle\nabla X,\nabla X\rangle \\
&= -\frac{1}{X}\left\{\frac{1}{2}\langle\nabla X,\nabla X\rangle - \langle\nabla Y,\nabla Y\rangle\right\}.
\end{aligned}
$$

For $\Box_h Y$, we argue similarly, by relying in succession on (1.116), (1.111), (1.112), (1.107), (1.103), (1.115), again (1.116), and (1.124), to deduce

$$
\begin{aligned}
\Box_h Y &= \Box_g Y + \frac{K^\alpha K^\beta}{X}\nabla_\alpha\nabla_\beta Y \\
&= \Box_g Y + \frac{K^\alpha K^\beta}{X}\nabla_\alpha\left(F^\dagger_{\beta\nu}K^\nu\right) \\
&= \Box_g Y + \frac{K^\alpha K^\beta}{X}\left\{\nabla_\alpha\left(\frac{1}{2}\epsilon_{\beta\nu\rho\sigma}F^{\rho\sigma}\right)K^\nu + \frac{1}{2}F^\dagger_{\beta\nu}F_\alpha{}^\nu\right\} \\
&= \Box_g Y + \frac{1}{X}\left\{\epsilon_{\beta\nu}{}^{\rho\sigma}K^\alpha K^\beta K^\nu K^\lambda R_{\lambda\alpha\rho\sigma} + \frac{1}{2}F^\nu{}_\alpha K^\alpha\, F^\dagger_{\nu\beta}K^\beta\right\} \\
&= \Box_g Y + \frac{1}{2X}\langle\nabla X,\nabla Y\rangle = -\frac{3}{2X}\langle\nabla X,\nabla Y\rangle.
\end{aligned}
$$

In conclusion, the system formed by (1.123) and (1.124) can be rewritten with respect to $h_{\mu\nu}$ as

$$\Box_h X = -\frac{1}{X}\left\{\frac{1}{2}\langle\nabla X,\nabla X\rangle - \langle\nabla Y,\nabla Y\rangle\right\}, \tag{1.132}$$

$$\Box_h Y = -\frac{3}{2X}\langle\nabla X,\nabla Y\rangle. \tag{1.133}$$

Next, we would like to derive a formula for the Ricci curvature appearing in (1.131), in terms of $h$, $X$, and $Y$. For this purpose, we need to develop a few more prerequisites. First, we take advantage of the properties (1.129) for $V$ to obtain equations that allows us to replace derivatives of $V$ by derivatives of $K$:

$$\begin{aligned}
0 &= \nabla_\lambda \langle K, V \rangle = K^\rho \nabla_\lambda V_\rho + V_\rho \nabla_\lambda K^\rho, \\
0 &= \mathcal{L}_K V_\rho = K^\lambda \nabla_\lambda V_\rho + V_\lambda \nabla_\rho K^\lambda.
\end{aligned} \tag{1.134}$$

Secondly, if we use the decomposition $F = \overline{F} + F^\perp$ coupled with (1.117) and (1.122), then we can infer that

$$\begin{aligned}
h_\mu{}^\alpha F_{\alpha\lambda} &= \left( \delta_\mu{}^\alpha - \frac{K_\mu K^\alpha}{X} \right) \left\{ \frac{1}{X} \left[ (\nabla_\alpha X) K_\lambda - (\nabla_\lambda X) K_\alpha \right] + F_{\alpha\lambda}^\perp \right\} \\
&= F_{\mu\lambda} + \frac{1}{X} (\nabla_\lambda X) K_\mu \\
&= F_{\mu\lambda}^\perp + \frac{1}{X} (\nabla_\mu X) K_\lambda.
\end{aligned} \tag{1.135}$$

Multiplying this equation by $h_\sigma{}^\lambda$ and applying consecutively (1.127) and (1.122), it follows that

$$h_\sigma{}^\lambda h_\mu{}^\alpha F_{\alpha\lambda} = h_\sigma{}^\lambda F_{\mu\lambda}^\perp = F_{\mu\sigma}^\perp. \tag{1.136}$$

Thus, successively relying on (1.128), (1.127), (1.134), (1.135), and (1.136), we deduce

$$\begin{aligned}
&\widehat{\nabla}_\mu \widehat{\nabla}_\nu V_\sigma \\
&= \widehat{\nabla}_\mu \left( \widehat{\nabla}_\nu V_\sigma \right) = h_\mu{}^\alpha h_\nu{}^\beta h_\sigma{}^\gamma \nabla_\alpha \left( h_\beta{}^\lambda h_\gamma{}^\rho \nabla_\lambda V_\rho \right) \\
&= - h_\mu{}^\alpha h_\nu{}^\lambda h_\sigma{}^\gamma \nabla_\alpha \left( \frac{K_\gamma K^\rho}{X} \right) \nabla_\lambda V_\rho - h_\mu{}^\alpha h_\nu{}^\beta h_\sigma{}^\rho \nabla_\alpha \left( \frac{K_\beta K^\lambda}{X} \right) \nabla_\lambda V_\rho \\
&\quad + h_\mu{}^\alpha h_\nu{}^\beta h_\sigma{}^\gamma \nabla_\alpha \nabla_\beta V_\gamma \\
&= - h_\mu{}^\alpha h_\nu{}^\lambda h_\sigma{}^\gamma \frac{\nabla_\alpha K_\gamma}{X} \left( K^\rho \nabla_\lambda V_\rho \right) - h_\mu{}^\alpha h_\nu{}^\beta h_\sigma{}^\rho \frac{\nabla_\alpha K_\beta}{X} \left( K^\lambda \nabla_\lambda V_\rho \right) \\
&\quad + h_\mu{}^\alpha h_\nu{}^\beta h_\sigma{}^\gamma \nabla_\alpha \nabla_\beta V_\gamma \\
&= h_\mu{}^\alpha h_\nu{}^\lambda h_\sigma{}^\gamma \frac{F_{\alpha\gamma} F_\lambda{}^\rho V_\rho}{4X} + h_\mu{}^\alpha h_\nu{}^\beta h_\sigma{}^\rho \frac{F_{\alpha\beta} F_\rho{}^\lambda V_\lambda}{4X} \\
&\quad + h_\mu{}^\alpha h_\nu{}^\beta h_\sigma{}^\gamma \nabla_\alpha \nabla_\beta V_\gamma \\
&= \frac{1}{4X} \left\{ F_{\mu\sigma}^\perp F_\nu{}^{\perp\rho} V_\rho + F_{\mu\nu}^\perp F_\sigma{}^{\perp\lambda} V_\lambda \right\} + h_\mu{}^\alpha h_\nu{}^\beta h_\sigma{}^\gamma \nabla_\alpha \nabla_\beta V_\gamma.
\end{aligned}$$

If we antisymmetrize the indices $\mu$ and $\nu$ in the above equation, then we derive

$$2\widehat{\nabla}_{[\mu}\widehat{\nabla}_{\nu]}V_\sigma = \frac{1}{4X}\left\{2F^\perp_{[\mu|\sigma|}F^{\perp\lambda}_{\nu]} + 2F^\perp_{\mu\nu}F^{\perp\lambda}_\sigma\right\}V_\lambda + 2h_{[\mu}{}^\alpha h_{\nu]}{}^\beta h_\sigma{}^\gamma\nabla_\alpha\nabla_\beta V_\gamma$$

$$= \frac{1}{4X}\left\{2F^\perp_{[\mu|\sigma|}F^{\perp\lambda}_{\nu]} + 2F^\perp_{\mu\nu}F^{\perp\lambda}_\sigma\right\}V_\lambda + 2h_\mu{}^\alpha h_\nu{}^\beta h_\sigma{}^\gamma\nabla_{[\alpha}\nabla_{\beta]}V_\gamma,$$

which, by further contracting the indices $\sigma$ and $\mu$, yields

$$2\widehat{\nabla}_{[\mu}\widehat{\nabla}_{\nu]}V^\mu = \left\{\frac{3}{4X}F^\perp_{\mu\nu}F^{\perp\mu\lambda} + h_\nu{}^\beta h^{\alpha\gamma}R_{\alpha\beta\gamma}{}^\lambda\right\}V_\lambda.$$

Comparing this expression to (1.131), we arrive at the following formula for the Ricci tensor:

$$\widehat{R}_{\mu\nu} = \frac{3}{4X}F^\perp_{\lambda\mu}F^{\perp\lambda}{}_\nu + h_\nu{}^\beta h_\mu{}^\sigma h^{\alpha\gamma}R_{\alpha\beta\gamma\sigma}, \qquad (1.137)$$

where the right-hand side needs extra work to be written in terms of $h$, $X$, and $Y$. For the second term on that side, with the help of (1.126) and (1.98), we infer

$$h^{\alpha\gamma}R_{\alpha\beta\gamma\sigma} = \left(g^{\alpha\gamma} - \frac{K^\alpha K^\gamma}{X}\right)R_{\alpha\beta\gamma\sigma} = R_{\beta\sigma} - \frac{1}{X}K^\alpha K^\gamma R_{\alpha\beta\gamma\sigma}$$

$$= -\frac{1}{X}K^\alpha K^\gamma R_{\alpha\beta\gamma\sigma}.$$

On the other hand, differentiating twice (1.100) and using (1.107), we obtain

$$\nabla_\beta\nabla_\sigma X = 2\nabla_\beta\left\{(\nabla_\sigma K_\gamma)K^\gamma\right\}$$

$$= 2\left\{(\nabla_\beta\nabla_\sigma K_\gamma)K^\gamma + \nabla_\beta K^\gamma\nabla_\sigma K_\gamma\right\}$$

$$= -2K^\alpha K^\gamma R_{\alpha\beta\gamma\sigma} + \frac{1}{2}F_{\sigma\gamma}F_\beta{}^\gamma,$$

which, coupled with the previous equation, produces

$$h^{\alpha\gamma}R_{\alpha\beta\gamma\sigma} = \frac{1}{2X}\nabla_\beta\nabla_\sigma X - \frac{1}{4X}F_{\sigma\lambda}F_\beta{}^\lambda.$$

Hence, due to (1.135) and (1.122), we deduce that

$$h_\nu{}^\beta h_\mu{}^\sigma h^{\alpha\gamma}R_{\alpha\beta\gamma\sigma}$$

$$= \frac{1}{2X}\widehat{\nabla}_\nu\widehat{\nabla}_\mu X - \frac{1}{4X}\left(F^\perp_{\mu\lambda} + \frac{1}{X}(\nabla_\mu X)K_\lambda\right)\left(F^{\perp\lambda}_\nu + \frac{1}{X}(\nabla_\nu X)K^\lambda\right)$$

$$= \frac{1}{2X}\widehat{\nabla}_\nu\widehat{\nabla}_\mu X - \frac{1}{4X}F^\perp_{\mu\lambda}F^{\perp\lambda}_\nu - \frac{1}{4X^2}\widehat{\nabla}_\mu X\widehat{\nabla}_\nu X.$$

In order to finish the calculations for the Ricci tensor, all we are left to do is to rewrite in a favorable form the first term on the right-hand side of

(1.137). Relying on the original formula for $F^{\perp}$, (1.105), and (1.117), we derive

$$X^2 F_{\mu}^{\perp\lambda} F^{\perp\nu}{}_{\lambda} = \epsilon^{\lambda}{}_{\mu\alpha\beta} \epsilon_{\lambda}{}^{\nu\rho\sigma} (\nabla^{\alpha} Y) K^{\beta} (\nabla_{\rho} Y) K_{\sigma}$$
$$= -6 \delta_{[\mu}^{\nu} \delta_{\alpha}^{\rho} \delta_{\beta]}^{\sigma} (\nabla^{\alpha} Y) K^{\beta} (\nabla_{\rho} Y) K_{\sigma}$$
$$= -X \left\{ \langle \nabla Y, \nabla Y \rangle h_{\mu}^{\nu} - \nabla_{\mu} Y \nabla^{\nu} Y \right\}.$$

Therefore, if we combine this equation with (1.137) and the previous one, then we finally obtain:

$$\widehat{R}_{\mu\nu} = \frac{1}{2X} \widehat{\nabla}_{\nu} \widehat{\nabla}_{\mu} X + \frac{1}{2X^2} \widehat{\nabla}_{\mu} Y \widehat{\nabla}_{\nu} Y - \frac{1}{4X^2} \widehat{\nabla}_{\mu} X \widehat{\nabla}_{\nu} X$$
$$- \frac{1}{2X^2} \langle \widehat{\nabla} Y, \widehat{\nabla} Y \rangle h_{\mu\nu}. \tag{1.138}$$

We recall that our goal in this section is to reduce the Einstein vacuum equations, in the presence of a certain symmetry, to a system which contains wave-map type problems. So far, we have come up with a system formed by (1.132), (1.133), and (1.138). In order to exhibit a more transparent wave-map structure, we start by remembering that $h$ is, in fact, a metric on the $2+1$-dimensional manifold $N$ and, as such, can be denoted $h_{ab}$ for a set of coordinates $x^a$, with $0 \le a \le 2$. Next, we perform a conformal transformation and change the metric into a new one, given by

$$\widetilde{h}_{ab} \overset{\text{def}}{=} X h_{ab}.$$

Thus, one also has $\widetilde{h}^{ab} = h^{ab}/X$. The Christoffel symbols for the two metrics are related via

$$\widetilde{\Gamma}^c_{ab} = \widehat{\Gamma}^c_{ab} + S^c_{ab}, \tag{1.139}$$

where

$$S^c_{ab} \overset{\text{def}}{=} \frac{1}{2} \left( \phi_a \delta^c_b + \phi_b \delta^c_a - \phi^c h_{ab} \right), \qquad \phi_a \overset{\text{def}}{=} \frac{\partial_a X}{X}. \tag{1.140}$$

If we compute second-order derivatives, then we find that

$$\widetilde{\nabla}_a \widetilde{\nabla}^b = \widetilde{\nabla}_a \left( \frac{h^{bc}}{X} \partial_c \right) = \widehat{\nabla}_a \left( \frac{h^{bc}}{X} \partial_c \right) + S^b_{ac} \frac{h^{cd}}{X} \partial_d$$
$$= \frac{1}{X} \widehat{\nabla}_a \widehat{\nabla}^b - \frac{\partial_a X}{X^2} h^{bc} \partial_c$$
$$+ \frac{1}{2} \left( \frac{\partial_a X}{X^2} h^{bd} \partial_d + \delta_a^b h^{cd} \frac{\partial_c X}{X^2} \partial_d - h^{bc} \frac{\partial_c X}{X^2} \partial_a \right).$$

This implies

$$\Box_{\widetilde{h}} = \frac{1}{X} \left( \Box_h - \frac{1}{2} \frac{h^{ab}}{X} \partial_a X \partial_b \right). \tag{1.141}$$

Furthermore, using (1.139), we deduce that the curvature tensor for $\widetilde{h}$ has the formula

$$\widetilde{R}_{abc}{}^d = \widehat{R}_{abc}{}^d - 2\left(\widehat{\nabla}_{[a}S^d_{b]c} + S^d_{p[a}S^p_{b]c}\right)$$

and, as a consequence, the corresponding Ricci tensor satisfies

$$\widetilde{R}_{ac} = \widehat{R}_{ac} - 2\left(\widehat{\nabla}_{[a}S^b_{b]c} + S^b_{p[a}S^p_{b]c}\right).$$

By taking advantage now of (1.140), a tedious, yet straightforward, calculation yields

$$\widetilde{R}_{ab} = \widehat{R}_{ab} - \frac{1}{2X}\widehat{\nabla}_a\widehat{\nabla}_bX + \frac{3}{4X^2}\widehat{\nabla}_aX\widehat{\nabla}_bX$$
$$+ \frac{1}{2X}\left(\Box_h X + \frac{1}{2X}\langle\widehat{\nabla}X,\widehat{\nabla}X\rangle\right)h_{ab}. \tag{1.142}$$

With these prerequisites finished, we can first combine (1.132) and (1.133) with (1.141) to derive

$$\Box_{\widetilde{h}}X = \frac{1}{X}\left(\Box_h X - \frac{1}{2X}\langle\nabla X,\nabla X\rangle\right)$$
$$= -\frac{1}{X}\left(\langle\widetilde{\nabla}X,\widetilde{\nabla}X\rangle_{\widetilde{h}} - \langle\widetilde{\nabla}Y,\widetilde{\nabla}Y\rangle_{\widetilde{h}}\right) \tag{1.143}$$

and

$$\Box_{\widetilde{h}}Y = \frac{1}{X}\left(\Box_h Y - \frac{1}{2X}\langle\nabla X,\nabla Y\rangle\right) = -\frac{2}{X}\langle\widetilde{\nabla}X,\widetilde{\nabla}Y\rangle_{\widetilde{h}}. \tag{1.144}$$

Finally, if we put together (1.132), (1.138), and (1.142), then we obtain

$$\widetilde{R}_{ab} = \frac{1}{2X^2}\left(\widetilde{\nabla}_aX\widetilde{\nabla}_bX + \widetilde{\nabla}_aY\widetilde{\nabla}_bY\right). \tag{1.145}$$

The last three equations form a system, which is the desired reduced version for the Einstein vacuum equations. The first two (i.e., (1.143) and (1.144)) are the equations of a wave map, with the hyperbolic plane as its target, modeled in this case by the right half-plane $\{(X,Y) \mid X > 0\}$ endowed with the metric $(dX^2 + dY^2)/X^2$.

It is worth noticing that (1.145) implies

$$\widetilde{R} = \frac{1}{2X^2}\left(\widetilde{\nabla}_aX\widetilde{\nabla}^aX + \widetilde{\nabla}_aY\widetilde{\nabla}^aY\right),$$

where $\widetilde{R} = \operatorname{tr}(\widetilde{R}_{ab}) = \widetilde{h}^{ab}\widetilde{R}_{ab}$ is the Ricci scalar curvature. With this observation in mind, we discover that (1.143), (1.144), and (1.145) can be derived as the Euler-Lagrange equations for the functional

$$\mathbb{L}(\widetilde{h}_{ab}, X, Y) \stackrel{\text{def}}{=} \int_N \left\{\widetilde{h}^{ab}\widetilde{R}_{ab} - \frac{\widetilde{\nabla}_aX\widetilde{\nabla}^aX + \widetilde{\nabla}_aY\widetilde{\nabla}^aY}{2X^2}\right\}d\mu_{\widetilde{h}}.$$

One of the reasons for this fact is

$$\frac{\delta}{\delta \widetilde{h}_{ab}} \int_N \left\{ \widetilde{R} \right\} d\mu_{\widetilde{h}} = \widetilde{G}_{ab} \overset{\text{def}}{=} \widetilde{R}_{ab} - \frac{1}{2} \widetilde{h}_{ab} \widetilde{R},$$

where $\widetilde{G}_{ab}$ is the associated Einstein tensor. When performing this computation, it is important to be mindful of the divergence produced by the variation, which may contribute a boundary term. These details are outlined in the appendix.

We can see a connection of the previous system with the wave map problem when its target manifold is the unit disk by introducing

$$w \overset{\text{def}}{=} \frac{X - 1 + iY}{X + 1 + iY}$$

and thus transforming (1.143) and (1.144) into

$$\Box_{\widetilde{h}} w - \frac{2\overline{w}}{1 - |w|^2} \langle \widetilde{\nabla} w, \widetilde{\nabla} w \rangle_{\widetilde{h}} = 0.$$

However, this equation does not describe the entire system, given that the metric $\widetilde{h}$ is unknown and it has to be computed using (1.145). As a final remark, the study of the system in its full generality is still an open problem, although there is a very recent breakthrough by Anderson-Gudapati-Szeftel [3] for the radially symmetric case.

# Chapter 2

# Analytic tools

## 2.1 Preliminaries

We introduce first the space of Schwartz functions

$$\mathcal{S}(\mathbb{R}^n) = \left\{ f \in C^\infty(\mathbb{R}^n); \ \sup_x |x^\alpha \partial^\beta f(x)| < \infty, \ \forall \, \alpha, \beta \text{ multi-indices} \right\},$$

whose topology is given by the seminorms

$$\|f\|_{\alpha,\beta} = \sup_x |x^\alpha \partial^\beta f(x)|,$$

thus transforming it into a Fréchet space. Its dual is $\mathcal{S}'(\mathbb{R}^n)$, the space of tempered distributions.

For $f \in \mathcal{S}$, we define its Fourier transform by

$$\hat{f}(\xi) \stackrel{\text{def}}{=} \int_{\mathbb{R}^n} e^{-ix\cdot\xi} f(x)\, dx,$$

for which we have the following well-known result.

**Proposition 2.1.** *The map $f \mapsto \hat{f}$ is an isomorphism of $\mathcal{S}$ into $\mathcal{S}$, whose inverse is given by*

$$f(x) = \frac{1}{(2\pi)^n} \int_{\mathbb{R}^n} e^{ix\cdot\xi} \hat{f}(\xi)\, d\xi.$$

*Moreover, the above map is also an isomorphism of $\mathcal{S}'$ into $\mathcal{S}'$, where the Fourier transform of $u \in \mathcal{S}'$ is defined by*

$$\hat{u}(f) \stackrel{\text{def}}{=} u(\hat{f}), \qquad \forall f \in \mathcal{S}.$$

### 2.1.1   Littlewood-Paley theory

Next, we consider $\psi \in C_0^\infty(\mathbb{R}^n)$ to be a radially symmetric bump function satisfying

$$
\psi(\xi) = \begin{cases} 1, & |\xi| \leq 1, \\ 0, & |\xi| \geq 2. \end{cases}
$$

This allows us to define the sequence of functions $(\phi_\lambda)_{\lambda \in 2^{\mathbb{Z}}}$ by

$$
\phi_\lambda(\xi) \overset{\text{def}}{=} \psi\left(\frac{\xi}{\lambda}\right) - \psi\left(\frac{2\xi}{\lambda}\right), \tag{2.1}
$$

for which we can easily verify:

$$
\psi(\xi) + \sum_{\lambda \in 2^{\mathbb{Z}}, \lambda \geq 2} \phi_\lambda(\xi) = 1, \qquad \forall \xi \in \mathbb{R}^n, \tag{2.2}
$$

$$
\sum_{\lambda \in 2^{\mathbb{Z}}} \phi_\lambda(\xi) = 1, \qquad \forall \xi \in \mathbb{R}^n \backslash \{0\}. \tag{2.3}
$$

We now introduce the Littlewood-Paley projectors

$$
\widehat{S_\lambda f}(\xi) \overset{\text{def}}{=} \phi_\lambda(\xi)\, \hat{f}(\xi),
$$

$$
\widehat{S_{\leq \lambda} f}(\xi) \overset{\text{def}}{=} \psi\left(\frac{\xi}{\lambda}\right) \hat{f}(\xi),
$$

$$
\widehat{S_{> \lambda} f}(\xi) \overset{\text{def}}{=} \left(1 - \psi\left(\frac{\xi}{\lambda}\right)\right) \hat{f}(\xi),
$$

which localize the Fourier transform to the regions $\{\lambda/2 \leq |\xi| \leq 2\lambda\}$, $\{|\xi| \leq 2\lambda\}$, and $\{|\xi| \geq \lambda\}$, respectively, where $|\xi|^2 = \xi_1^2 + \xi_2^2 + \ldots + \xi_n^2$ if $\xi = (\xi_1, \xi_2, \ldots, \xi_n) \in \mathbb{R}^n$. It is an easy verification that all these projectors are bounded uniformly in $\lambda$ on any $L^p$ with $1 \leq p \leq \infty$.

To simplify the notation, when there is no ambiguity, we will use $f_\lambda$ for $S_\lambda f$, $f_{\leq \lambda}$ for $S_{\leq \lambda} f$, and $f_{> \lambda}$ for $S_{> \lambda} f$. The previous two identities, (2.2) and (2.3), can then be rewritten as

$$
f_{\leq 1} + \sum_{\lambda \in 2^{\mathbb{Z}}, \lambda \geq 2} f_\lambda = f, \qquad \forall f \in \mathcal{S}',
$$

$$
\sum_{\lambda \in 2^{\mathbb{Z}}} f_\lambda = f, \qquad \forall f \in \mathcal{S}',\ 0 \notin \operatorname{supp} \hat{f}.
$$

We summarize below some of the fundamental estimates related to the Littlewood-Paley theory.

**Proposition 2.2.** *a) (Bernstein's inequalities) For $s \geq 0$ and $1 \leq p \leq q \leq \infty$,*

$$\||D|^{\pm s} f_\lambda\|_{L^p} \simeq \lambda^{\pm s} \|f_\lambda\|_{L^p},$$
$$\||D|^{s} f_{\leq \lambda}\|_{L^p} \lesssim \lambda^{s} \|f_{\leq \lambda}\|_{L^p},$$
$$\||D|^{-s} f_{>\lambda}\|_{L^p} \lesssim \lambda^{-s} \|f_{>\lambda}\|_{L^p},$$

*with* $\widehat{|D|^s g}(\xi) := |\xi|^s \hat{g}(\xi)$, *and*

$$\|f_\lambda\|_{L^q(\mathbb{R}^n)} \lesssim \lambda^{n\left(\frac{1}{p}-\frac{1}{q}\right)} \|f_\lambda\|_{L^p(\mathbb{R}^n)},$$
$$\|f_{\leq\lambda}\|_{L^q(\mathbb{R}^n)} \lesssim \lambda^{n\left(\frac{1}{p}-\frac{1}{q}\right)} \|f_{\leq\lambda}\|_{L^p(\mathbb{R}^n)}.$$

*b) (Littlewood-Paley inequalities) For $1 < p < \infty$,*

$$\left\|\left(\sum_{\lambda \in 2^{\mathbb{Z}}} |f_\lambda|^2\right)^{\frac{1}{2}}\right\|_{L^p} \simeq \|f\|_{L^p}.$$

*As a consequence,*

$$\sum_{\lambda \in 2^{\mathbb{Z}}} \|f_\lambda\|_{L^p}^2 \lesssim \|f\|_{L^p}^2, \qquad 1 < p \leq 2,$$
$$\sum_{\lambda \in 2^{\mathbb{Z}}} \|f_\lambda\|_{L^p}^2 \gtrsim \|f\|_{L^p}^2, \qquad 2 \leq p < \infty.$$

### 2.1.2  Function spaces

We now have the background to introduce the classical Besov and Sobolev spaces.

**Definition 2.1.** Let $s \in \mathbb{R}$ and $1 \leq p, q \leq \infty$. Consider

$$\|f\|_{B^s_{p,q}} \stackrel{\text{def}}{=} \|f_{\leq 1}\|_{L^p} + \left(\sum_{\lambda \in 2^{\mathbb{Z}}, \lambda \geq 2} \lambda^{sq}\|f_\lambda\|_{L^p}^q\right)^{\frac{1}{q}},$$

$$\|f\|_{H^{s,p}} \stackrel{\text{def}}{=} \|\langle D\rangle^s f\|_{L^p},$$

where $\widehat{\langle D\rangle^s g}(\xi) := \langle\xi\rangle^s \hat{g}(\xi) = (1+|\xi|^2)^{\frac{s}{2}}\hat{g}(\xi)$. The Besov space $B^s_{p,q}$ and the Sobolev space $H^{s,p}$ are then defined by

$$B^s_{p,q} \stackrel{\text{def}}{=} \left\{f \in \mathcal{S}'; \|f\|_{B^s_{p,q}} < \infty\right\},$$
$$H^{s,p} \stackrel{\text{def}}{=} \left\{f \in \mathcal{S}'; \|f\|_{H^{s,p}} < \infty\right\},$$

with the obvious modification for the case $q = \infty$.

We will also work with the homogeneous counterparts of these spaces, for which we need to consider

$$\|f\|_{\dot{B}^s_{p,q}} \overset{\text{def}}{=} \left( \sum_{\lambda \in 2^{\mathbb{Z}}} \lambda^{sq} \|f_\lambda\|^q_{L^p} \right)^{\frac{1}{q}}, \tag{2.4}$$

$$\|f\|_{\dot{H}^{s,p}} \overset{\text{def}}{=} \left\| \sum_{\lambda \in 2^{\mathbb{Z}}} |D|^s f_\lambda \right\|_{L^p}. \tag{2.5}$$

The homogeneous Besov space $\dot{B}^s_{p,q}$ is the set of those $f \in \mathcal{S}'$ for which $\|f\|_{\dot{B}^s_{p,q}}$ is finite, whereas the homogeneous Sobolev space $\dot{H}^{s,p}$ represents those $f \in \mathcal{S}'$ for which $\sum_{\lambda \in 2^{\mathbb{Z}}} |D|^s f_\lambda$ converges in $\mathcal{S}'$ to a function in $L^p$. Both $\dot{B}^s_{p,q}$ and $\dot{H}^{s,p}$ are semi-normed spaces, with either $\|f\|_{\dot{B}^s_{p,q}} = 0$ or $\|f\|_{\dot{H}^{s,p}} = 0$ if and only if supp $\hat{f} = \{0\}$[1].

As was the case for Lebesgue spaces, the Littlewood-Paley projectors are bounded uniformly in $\lambda$ on any Sobolev space $H^{s,p}$ (or $\dot{H}^{s,p}$) with $1 < p < \infty$. For $s > 0$, we have $B^s_{p,q} = L^p \cap \dot{B}^s_{p,q}$ and $H^{s,p} = L^p \cap \dot{H}^{s,p}$. From here on, we use the notations $H^s = H^{s,2}$ and $\dot{H}^s = \dot{H}^{s,2}$. A relatively easy application of Plancherel's theorem yields

$$\|f\|_{H^s} \simeq \|f\|_{B^s_{2,2}} \quad \text{and} \quad \|f\|_{\dot{H}^s} \simeq \|f\|_{\dot{B}^s_{2,2}}.$$

We record below a collection of embeddings and interpolation relations that we will rely on extensively.

**Proposition 2.3.** *a) (Embeddings) If $s_1, s_2 \in \mathbb{R}$ satisfy*

$$s_1 - s_2 = n \left( \frac{1}{p_1} - \frac{1}{p_2} \right),$$

*then*

$$B^{s_1}_{p_1,q_1}(\mathbb{R}^n) \subset B^{s_2}_{p_2,q_2}(\mathbb{R}^n), \quad 1 \le p_1 \le p_2 \le \infty, 1 \le q_1 \le q_2, \tag{2.6}$$

$$H^{s_1,p_1}(\mathbb{R}^n) \subset H^{s_2,p_2}(\mathbb{R}^n), \quad 1 < p_1 \le p_2 < \infty. \tag{2.7}$$

*b) (Interpolation relations) If $s_1 \neq s_2$ are real, $1 \le p, p_1, p_2, q, q_1, q_2 \le \infty$, and*

$$s^* = (1-\theta)s_1 + \theta s_2, \quad \frac{1}{p^*} = \frac{1-\theta}{p_1} + \frac{\theta}{p_2}, \quad \frac{1}{q^*} = \frac{1-\theta}{q_1} + \frac{\theta}{q_2}, \quad 0 \le \theta \le 1,$$

*then*

$$\left(H^{s_1,p}, H^{s_2,p}\right)_{\theta,q} = B^{s^*}_{p,q}, \tag{2.8}$$

$$\left(B^{s_1}_{p,q_1}, B^{s_2}_{p,q_2}\right)_{\theta,q} = B^{s^*}_{p,q}, \tag{2.9}$$

$$\left(B^{s_1}_{p_1,q_1}, B^{s_2}_{p_2,q_2}\right)_{[\theta]} = B^{s^*}_{p^*,q^*}. \tag{2.10}$$

---

[1] This happens if and only if $f$ is a polynomial.

These results are also valid for homogeneous versions of the Besov and Sobolev spaces and, for more details and their full proofs, we refer the reader to the book by Bergh-Löfström [16] and references therein.

Furthermore, we include a couple of product estimates, which will be useful later on.

**Proposition 2.4.** *a) If $s \geq 0$, then*

$$\|fg\|_{H^s} \lesssim \|f\|_{H^s}\|g\|_{L^\infty} + \|f\|_{L^\infty}\|g\|_{H^s}. \qquad (2.11)$$

*b) If*

$$\frac{1}{p} = \frac{1}{p_1} + \frac{1}{q_1} = \frac{1}{p_2} + \frac{1}{q_2}, \quad 1 \leq p, p_1, p_2, q_1, q_2 \leq \infty, \quad s > 0,$$

*then*

$$\|fg\|_{\dot{B}^s_{p,1}} \lesssim \|f\|_{L^{p_1}}\|g\|_{\dot{B}^s_{q_1,1}} + \|g\|_{L^{p_2}}\|f\|_{\dot{B}^s_{q_2,1}}. \qquad (2.12)$$

**Proof.** a) For $s = 0$, the estimate is obvious. If $s > 0$,

$$\|fg\|^2_{H^s} \simeq \|S_{\leq 1}(fg)\|^2_{L^2} + \sum_{\lambda \geq 2} \lambda^{2s}\|S_\lambda(fg)\|^2_{L^2}$$

and the first term is bounded by

$$\|S_{\leq 1}(fg)\|_{L^2} \lesssim \|fg\|_{L^2} \leq \|f\|_{L^\infty}\|g\|_{L^2} \leq \|f\|_{L^\infty}\|g\|_{H^s}.$$

For the second one, we argue that

$$\sum_{\lambda \geq 2} \lambda^{2s}\|S_\lambda(fg)\|^2_{L^2} \lesssim \sum_{\lambda \geq 2} \lambda^{2s} \left[ \left( \sum_{\mu \geq \lambda} \|S_\lambda(f_\mu g)\|_{L^2} \right)^2 + \|S_\lambda(f_{<\lambda}g)\|^2_{L^2} \right],$$

and

$$\sum_{\lambda \geq 2} \lambda^{2s}\|S_\lambda(f_{<\lambda}g)\|^2_{L^2}$$

$$\lesssim \sum_{\lambda \geq 2} \lambda^{2s}\|S_\lambda(f_{<\lambda}\tilde{S}_\lambda g)\|^2_{L^2} \lesssim \sum_{\lambda \geq 2} \lambda^{2s}\|f_{<\lambda}\|^2_{L^\infty}\|\tilde{S}_\lambda g\|^2_{L^2}$$

$$\lesssim \|f\|^2_{L^\infty} \sum_{\lambda \geq 2} \lambda^{2s}\|\tilde{S}_\lambda g\|^2_{L^2} \lesssim \|f\|^2_{L^\infty}\|g\|^2_{H^s},$$

where $\tilde{S}_\lambda$ is also a Fourier multiplier localizing $|\xi| \simeq \lambda$, but with a slightly wider support than $S_\lambda$.

Finally, using $s > 0$ and the Cauchy-Schwarz inequality, we deduce

$$\sum_{\lambda \geq 2} \lambda^{2s} \left( \sum_{\mu \geq \lambda} \|S_\lambda(f_\mu g)\|_{L^2} \right)^2 \lesssim \|g\|_{L^\infty}^2 \sum_{\lambda \geq 2} \left( \sum_{\mu \geq \lambda} \lambda^s \|f_\mu\|_{L^2} \right)^2$$

$$\lesssim \|g\|_{L^\infty}^2 \sum_{\lambda \geq 2} \left( \sum_{\mu \geq \lambda} \frac{\lambda^s}{\mu^s} \|f_\mu\|_{H^s} \right)^2$$

$$\lesssim \|g\|_{L^\infty}^2 \sum_{\lambda \geq 2} \left( \sum_{\mu \geq \lambda} \frac{\lambda^s}{\mu^s} \right) \left( \sum_{\mu \geq \lambda} \frac{\lambda^s}{\mu^s} \|f_\mu\|_{H^s}^2 \right)$$

$$\lesssim \|g\|_{L^\infty}^2 \sum_{\mu \geq 2} \left( \sum_{\lambda \leq \mu} \frac{\lambda^s}{\mu^s} \right) \|f_\mu\|_{H^s}^2 \lesssim \|g\|_{L^\infty}^2 \|f\|_{H^s}^2,$$

which finishes the proof of (2.11).

b) The argument is direct, using only Bernstein's and Hölder's inequalities:

$$\|f\,g\|_{\dot{B}_{p,1}^s}$$

$$\lesssim \sum_\lambda \lambda^s \left( \|f_\lambda\,g_{<\lambda}\|_{L^p} + \|g_\lambda\,f_{<\lambda}\|_{L^p} + \sum_{\mu \gtrsim \lambda} \|f_\mu\,g_\mu\|_{L^p} \right)$$

$$\lesssim \sum_\lambda \lambda^s \left( \|f_\lambda\|_{L^{q_2}} \|g_{<\lambda}\|_{L^{p_2}} + \|g_\lambda\|_{L^{q_1}} \|f_{<\lambda}\|_{L^{p_1}} \right) + \sum_\mu \mu^s \|f_\mu\,g_\mu\|_{L^p}$$

$$\lesssim \|g\|_{L^{p_2}} \|f\|_{\dot{B}_{q_2,1}^s} + \|f\|_{L^{p_1}} \|g\|_{\dot{B}_{q_1,1}^s} + \sum_\mu \mu^s \|f_\mu\|_{L^{q_2}} \|g_\mu\|_{L^{p_2}}$$

$$\lesssim \|g\|_{L^{p_2}} \|f\|_{\dot{B}_{q_2,1}^s} + \|f\|_{L^{p_1}} \|g\|_{\dot{B}_{q_1,1}^s}.$$

$\square$

An immediate consequence of (2.11) is the following well-known result.

**Corollary 2.1.** *For $s > n/2$, the Sobolev space $H^s(\mathbb{R}^n)$ is an algebra.*

Next, we record decay estimates for radial functions which were obtained by Cho-Ozawa [26] as an extension of a well-known bound due to Strauss [162] [2].

**Proposition 2.5.** *i) Let $n \geq 2$ and assume that $1/2 < s < n/2$. For every radial function*

$$f \in L^{2n/(n-2s)} \cap \dot{H}^s(\mathbb{R}^n),$$

---

[2]The interested reader can also consult related work due to Sickel-Skrzypczak [149, 150] and Sickel-Skrzypczak-Vybiral [151]

*there exists a function $F$ continuous on $\mathbb{R}^n \setminus \{0\}$ such that it is equal to $f$ almost everywhere and*

$$\sup_{x \in \mathbb{R}^n \setminus \{0\}} |x|^{n/2-s} |F(x)| \lesssim \|f\|_{\dot{H}^s(\mathbb{R}^n)}. \tag{2.13}$$

*In the case when $s = 1/2$, the previous statement holds with $\dot{H}^s(\mathbb{R}^n)$ replaced by $\dot{B}_{2,1}^{1/2}(\mathbb{R}^n)$.*

*ii) Let $n \geq 2$ and assume that $1/2 \leq s < 1$. For every radial function $f \in H^1(\mathbb{R}^n)$, there exists a function $F$ continuous on $\mathbb{R}^n \setminus \{0\}$ such that it is equal to $f$ almost everywhere and*

$$\sup_{x \in \mathbb{R}^n \setminus \{0\}} |x|^{n/2-s} |F(x)| \lesssim \|f\|_{L^2(\mathbb{R}^n)}^{1-s} \|\nabla f\|_{L^2(\mathbb{R}^n)}^{s}. \tag{2.14}$$

Finally, we write down two versions of Hardy-type inequalities that can be found in the books by Opic-Kufner [127] and Bahouri-Chemin-Danchin [6]. Together with the previous decay estimates, these bounds will be used in the analysis of equivariant problems.

**Proposition 2.6.** *i) Consider $n \geq 1$ and assume that $1 \leq p,\, q < \infty$, $\alpha$, $\beta \in \mathbb{R}$, and $p - n < \beta < np - n$. The Hardy-type inequality*

$$\left[ \int_{\mathbb{R}^n} |f(x)|^q |x|^\alpha \, dx \right]^{1/q} \lesssim \left[ \int_{\mathbb{R}^n} |\nabla f(x)|^p |x|^\beta \, dx \right]^{1/p} \tag{2.15}$$

*holds for all functions $f$ satisfying*

$$\int_{\mathbb{R}^n} \left( |\nabla f(x)|^p + \frac{|f(x)|^p}{|x|^p} \right) |x|^\beta \, dx < \infty,$$

*if and only if the following conditions are true:*

$$1 \leq p \leq q < \infty, \qquad \frac{n}{q} - \frac{n}{p} + 1 \geq 0, \qquad \frac{\alpha}{q} - \frac{\beta}{p} + \frac{n}{q} - \frac{n}{p} + 1 = 0.$$

*ii) If $n \geq 1$ and $0 \leq s < n/2$, then*

$$\left[ \int_{\mathbb{R}^n} |f(x)|^2 |x|^{-2s} \, dx \right]^{1/2} \lesssim \|f\|_{\dot{H}^s(\mathbb{R}^n)} \tag{2.16}$$

*is valid for all $f \in \dot{H}^s(\mathbb{R}^n)$.*

## 2.2　Strichartz estimates

Our goal in this section is to prove Strichartz estimates for solutions of linear wave equations. The classical references[3] on this topic are the works of Strichartz [163], Ginibre-Velo [57], and Keel-Tao [78]. We use the notation $(t, x) = (t, x_1, x_2, \ldots, x_n)$ to denote an arbitrary point in the spacetime $\mathbb{R}^{n+1}$, and, for functions which depend both on time and spatial variables, we work with

$$\|u\|_{L^p L^q(I \times \mathbb{R}^n)} = \|u\|_{L^p_t L^q_x(I \times \mathbb{R}^n)} = \left( \int_I \left( \int_{\mathbb{R}^n} |u(t,x)|^q \, dx \right)^{\frac{p}{q}} dt \right)^{\frac{1}{p}},$$

where $I \subset \mathbb{R}$ is an arbitrary time interval. Next, we introduce the terminology used in formulating Strichartz inequalities.

**Definition 2.2.** Let $n \geq 2$ be an integer. The triple $(p, q, n)$ is called *wave-admissible* if

$$2 \leq p, q \leq \infty, \qquad \frac{2}{p} + \frac{n-1}{q} \leq \frac{n-1}{2},$$

and $(p, q, n) \neq (2, \infty, 3)$.

In this context, the main result we prove here is the dyadic version of what are known as *classical Strichartz estimates for the wave equation.*

**Theorem 2.1.** *For $n \geq 2$, $(p, q, n)$ is wave-admissible if and only if the inequality*

$$\|e^{\pm it|D|} f\|_{L^p L^q(\mathbb{R} \times \mathbb{R}^n)} \lesssim \|f\|_{L^2(\mathbb{R}^n)} \qquad (2.17)$$

*holds uniformly for $f \in L^2(\mathbb{R}^n)$ with $f = S_1 f$, where*

$$\widehat{e^{\pm it|D|} f}(\xi) \stackrel{\text{def}}{=} e^{\pm it|\xi|} \hat{f}(\xi).$$

The proof of this theorem relies upon a particular set of prerequisites, which are discussed next. First, we recall the well-known convolution estimates due to Young and Hardy-Littlewood-Sobolev and the Riesz-Thorin interpolation theorem.

**Lemma 2.1.** *a) (Young's inequality) If $1 \leq p, p_1, p_2 \leq \infty$ satisfy*

$$1 + \frac{1}{p} = \frac{1}{p_1} + \frac{1}{p_2},$$

---

[3]There exists a parallel literature on Strichartz estimates for the linear Schrödinger equation, for which we refer the reader to the book by Cazenave [24] and references therein.

*then*

$$\|f * g\|_{L^p} \lesssim \|f\|_{L^{p_1}} \|g\|_{L^{p_2}}. \tag{2.18}$$

b) *(Hardy-Littlewood-Sobolev inequality) If $0 < \gamma < n$ and $1 < p < q < \infty$ verify*

$$1 - \frac{\gamma}{n} = \frac{1}{p} - \frac{1}{q},$$

*then*

$$\||x|^{-\gamma} * f\|_{L^q(\mathbb{R}^n)} \lesssim \|f\|_{L^p(\mathbb{R}^n)}. \tag{2.19}$$

**Theorem 2.2.** *(Riesz-Thorin interpolation theorem) Let $U$ and $V$ be two measure spaces and let $T$ denote a linear mapping for which $T : L^{p_0}(U) \to L^{q_0}(V)$ has norm $M_0$ and $T : L^{p_1}(U) \to L^{q_1}(V)$ has norm $M_1$, where $1 \le p_0, p_1, q_0, q_1 \le \infty$, $p_0 \ne p_1$, and $q_0 \ne q_1$. Then, for all $0 < \theta < 1$, $T$ maps $L^p(U)$ into $L^q(V)$ with norm*

$$M \le M_0^\theta M_1^{1-\theta},$$

*where*

$$\frac{1}{p} = \frac{\theta}{p_0} + \frac{1-\theta}{p_1}, \quad \frac{1}{q} = \frac{\theta}{q_0} + \frac{1-\theta}{q_1}.$$

Secondly, we need the following result concerning translation-invariant operators, which is due to Hörmander [70].

**Lemma 2.2.** *If $1 \le p, q < \infty$ and $T : L^p(\mathbb{R}^n) \to L^q(\mathbb{R}^n)$ is a bounded linear operator which commutes with translations, i.e.,*

$$(Tf) \circ w_y = T(f \circ w_y), \quad \forall f \in L^p, \, y \in \mathbb{R}^n,$$

*where $w_y(x) = x + y$, then $p \le q$.*

**Proof.** We prove first that

$$\lim_{|y| \to \infty} \|f + f \circ w_y\|_{L^p} = 2^{1/p} \|f\|_{L^p}, \tag{2.20}$$

which can be reduced by density arguments to the case when $f \in C_0(\mathbb{R}^n)$. For this limit, we show that

$$\lim_{|y| \to \infty} \int |f + f \circ w_y|^p - |f \circ w_y|^p = \int |f|^p, \tag{2.21}$$

which is obviously a restatement of (2.20).

Let $R > 0$ such that $f(z) = 0$ for all $|z| \geq R$. It follows that $|y| \geq R + |x|$ implies

$$|f(x) + (f \circ w_y)(x)|^p - |(f \circ w_y)(x)|^p = |f(x)|^p.$$

Moreover,

$$\left| |f + f \circ w_y|^p - |f \circ w_y|^p \right| \lesssim \chi_{B(0,R)}$$

holds uniformly for $y \in \mathbb{R}^n$, where $\chi_{B(0,R)}$ is the characteristic function for the ball of radius $R$ centered at the origin. Based on these, if we now invoke the dominated convergence theorem, (2.21) follows immediately.

Using the boundedness and the translation invariance of $T$, we infer that

$$\|Tf + Tf \circ w_y\|_{L^q} \leq \|T\| \, \|f + f \circ w_y\|_{L^p},$$

which, based on (2.20), leads to

$$2^{1/q} \|Tf\|_{L^q} \leq 2^{1/p} \|T\| \, \|f\|_{L^p}.$$

The conclusion follows by optimizing the ratio $\|Tf\|_{L^q} / \|f\|_{L^p}$. $\qquad \square$

**Remark 2.1.** This result extends easily to operators between vector-valued $L^p$ spaces.

Thirdly, we discuss the so-called $TT^*$ framework, which will be used in demonstrating Theorem 2.1 and whose proof is a simple exercise in functional analysis. The main benefit of this method is that, in certain situations, it is easier to work with $TT^*$ than with either $T$ or $T^*$ individually. We present it formulated as follows.

**Lemma 2.3.** *Let $T : H \to B$ be a linear operator between a Hilbert space $H$ and a Banach space $B$. Consider also its adjoint $T^* : B^* \to H$, with $B^*$ being the dual of $B$, which satisfies*

$$\langle T^* u, v \rangle_H = \langle u, Tv \rangle_{B^*, B}, \qquad \forall \, (u, v) \in B^* \times H.$$

*The following statements are equivalent:*

- *$T$ is bounded;*
- *$T^*$ is bounded;*
- *$TT^* : B^* \to B$ is bounded;*
- *the bilinear form*

$$a : B^* \times B^* \to \mathbb{C}, \qquad a(u, w) \stackrel{\text{def}}{=} \langle T^* u, T^* w \rangle_H,$$

*is bounded.*

*Moreover, if any of these is true, then*

$$\|T\|^2 = \|T^\star\|^2 = \|TT^\star\|.$$

Finally, we include a classical result (e.g., Theorem 2 of Section 3.2 in Chapter 8 of Stein [158]) concerning decay estimates for the Fourier transform of surface-carried measures.

**Lemma 2.4.** *If $S \subset \mathbb{R}^n$ is a hypersurface with at least $k$ non-vanishing principal curvatures and $\phi \in C_0^\infty(S)$, then*

$$\widehat{\phi\,d\sigma}(\xi) = \int_S e^{-ix\cdot\xi}\,\phi(x)\,d\sigma(x)$$

*satisfies*

$$|\widehat{\phi\,d\sigma}(\xi)| \lesssim \frac{1}{(1+|\xi|)^{\frac{k}{2}}}.$$

In the particular case of the cone

$$\mathcal{C} = \{(t,x) \in \mathbb{R}^{n+1};\ t^2 = |x|^2\},$$

the principal curvatures at a point $(t_0, x_0) \neq (0,0)$ are the eigenvalues for the matrix

$$A = \left(\partial_{ij}^2 \psi(x_0)\right)_{1 \le i,j \le n},$$

where $\psi(x) = \pm|x|$. A direct computation leads to

$$A = \frac{1}{|x_0|}\left(I - \operatorname{proj}_{\hat{x}_0}\right),$$

where $I$ is the identity matrix and $\operatorname{proj}_{\hat{x}_0}$ denotes the projection along the unit vector $\hat{x}_0 = x_0/|x_0|$. If we construct an orthonormal basis of $\mathbb{R}^n$ which includes $\hat{x}_0$, we see immediately that $\hat{x}_0$ is an eigenvector for $A$ corresponding to an eigenvalue equal to 0, whereas the other $n-1$ vectors of the basis are all eigenvectors for $A$ corresponding to an eigenvalue equal to $1/|x_0|$. Hence, $\mathcal{C}$ has precisely $n-1$ non-vanishing principal curvatures at $(t_0, x_0)$. As a consequence of Lemma 2.4, we deduce that

$$\left|\int_{\mathbb{R}^n} e^{i(x\cdot\xi \pm t|\xi|)}\,\phi(\xi)\,d\xi\right| \lesssim \frac{1}{(1+|t|)^{\frac{n-1}{2}}}, \tag{2.22}$$

where

$$\phi \in C_0^\infty(\mathbb{R}^n), \qquad \operatorname{supp}\phi \subset \left\{\frac{1}{2} \le |\xi| \le 2\right\}.$$

We now have all the facts needed to prove the main result of this section.

**Proof of Theorem 2.1.** We work first on the forward direction. The statement to be proved (i.e., (2.17)) can be restated in the $TT^\star$ framework as

$$T : L^2(\mathbb{R}^n) \to L^p L^q(\mathbb{R} \times \mathbb{R}^n), \quad (Tf)(t, x) \stackrel{\text{def}}{=} \int e^{i(x \cdot \xi \pm t|\xi|)} \phi_1(\xi) \hat{f}(\xi) \, d\xi$$

is bounded, when $(p, q, n)$ is wave-admissible. Hence, according to Lemma 2.3, it will be enough to show the boundedness for

$$TT^\star : L^{p'} L^{q'}(\mathbb{R} \times \mathbb{R}^n) \to L^p L^q(\mathbb{R} \times \mathbb{R}^n), \quad \frac{1}{p} + \frac{1}{p'} = \frac{1}{q} + \frac{1}{q'} = 1.$$

A straightforward computation yields

$$\begin{aligned}
(TT^\star g)(t, x) &= \int e^{i((x-y) \cdot \xi \pm (t-s)|\xi|)} \, |\phi_1(\xi)|^2 \, g(s, y) \, d\xi \, ds \, dy \\
&= (K * g)(t, x),
\end{aligned} \tag{2.23}$$

with

$$K(t, x) = \int e^{i(x \cdot \xi \pm t|\xi|)} \, |\phi_1(\xi)|^2 \, d\xi.$$

For $h = h(x)$, we can argue that

$$\|K(t, \cdot) * h\|_{L^2} \simeq \|\hat{K}(t, \cdot) \hat{h}\|_{L^2} \simeq \|e^{\pm it|\xi|} \, |\phi_1(\xi)|^2 \, \hat{h}(\xi)\|_{L^2_\xi} \lesssim \|h\|_{L^2}.$$

On the other hand, based on (2.18) and (2.22), we deduce

$$\|K(t, \cdot) * h\|_{L^\infty} \lesssim \|K(t, \cdot)\|_{L^\infty} \|h\|_{L^1} \lesssim \frac{1}{(1 + |t|)^{\frac{n-1}{2}}} \|h\|_{L^1}.$$

Hence, if we rely on Theorem 2.2, it follows that

$$\|K(t, \cdot) * h\|_{L^q} \lesssim \frac{1}{(1 + |t|)^{\frac{n-1}{2}(1 - \frac{2}{q})}} \|h\|_{L^{q'}}, \quad \forall 2 \le q \le \infty.$$

Using this estimate in the context of (2.23), we infer that

$$\|(TT^\star g)(t, \cdot)\|_{L^q} \lesssim \int \frac{1}{(1 + |t-s|)^{\frac{n-1}{2}(1 - \frac{2}{q})}} \|g(s, \cdot)\|_{L^{q'}} \, ds,$$

for $q$ in the above range. This already gives us the desired outcome for $(p, q) = (\infty, 2)$.

For the case when

$$\frac{2}{p} + \frac{n-1}{q} < \frac{n-1}{2},$$

we can argue that

$$\frac{1}{(1 + |\cdot|)^{\frac{n-1}{2}(1 - \frac{2}{q})}} \in L^{\frac{p}{2}}(\mathbb{R})$$

and the conclusion follows as a result of (2.18).

Finally, when

$$\frac{2}{p} + \frac{n-1}{q} = \frac{n-1}{2} \quad \text{and} \quad (p,q) \neq (\infty, 2),$$

(2.18) cannot be used anymore. However, if we also assume that $p > 2$, then we can appeal to (2.19) to claim (2.17). For the remaining case when $n \geq 4$ and $(p,q) = (2, 2(n-1)/(n-3))$, the analysis is more intricate and we refer the reader to the argument in [78]. This finishes the discussion of the forward direction.

We assume now that the Strichartz bound (2.17) holds and show that $(p,q)$ is necessarily wave-admissible. We start with a construction commonly referred to as *Knapp's counterexample*, in which

$$\hat{f} = \chi_{A_\delta}, \qquad A_\delta = \left\{ \xi = (\xi_1, \xi') \in \mathbb{R}^n; \ |\xi_1 - 1| < \frac{1}{2}, |\xi'| < \delta \right\},$$

with $\delta \ll 1$, such that $f = S_1 f$. A simple calculation leads to $\|f\|_{L^2} \simeq \delta^{(n-1)/2}$.

On the other hand,

$$\left( e^{it|D|} f \right)(x_1, x') = e^{i(t+x_1)} \int_{A_\delta} e^{i[x'\cdot\xi' + t(|\xi|-\xi_1) + (t+x_1)(\xi_1-1)]} \, d\xi.$$

If we choose

$$B_\delta = \left\{ (t,x) = (t, x_1, x') \in \mathbb{R}^{n+1}; \ |t| < \frac{\delta^{-2}}{10}, |t+x_1| < \frac{1}{10}, |x'| < \frac{\delta^{-1}}{10} \right\},$$

due to

$$|\xi| - \xi_1 = \frac{|\xi'|^2}{|\xi| + \xi_1} < \delta^2,$$

we can infer that

$$|x' \cdot \xi' + t(|\xi| - \xi_1) + (t + x_1)(\xi_1 - 1)| < 1, \qquad \forall (t,x) \in B_\delta, \ \xi \in A_\delta.$$

This allows us to claim that

$$\left| \left( e^{it|D|} f \right)(x) \right| \gtrsim \int_{A_\delta} 1 \, d\xi \simeq \delta^{n-1}, \qquad \forall (t,x) \in B_\delta,$$

which further implies

$$\|e^{it|D|} f\|_{L^p L^q} \gtrsim \delta^{n-1} \|\chi_{B_\delta}\|_{L^p L^q} \simeq \delta^{n-1-\frac{2}{p}-\frac{n-1}{q}}.$$

Therefore, for this choice of $f$, we obtain

$$\frac{\|e^{it|D|} f\|_{L^p L^q}}{\|f\|_{L^2}} \gtrsim \delta^{\frac{n-1}{2}-\frac{2}{p}-\frac{n-1}{q}},$$

with the flexibility of choosing $\delta$ arbitrarily small. It follows that, in order for (2.17) to hold uniformly for $f = S_1 f$, we need to have

$$\frac{2}{p} + \frac{n-1}{q} \leq \frac{n-1}{2},$$

which also implies $q \geq 2$.

As argued in the proof of the forward direction, the Strichartz estimate (2.17) is equivalent to the boundedness for

$$TT^* : L^{p'} L^{q'} (\mathbb{R} \times \mathbb{R}^n) \to L^p L^q (\mathbb{R} \times \mathbb{R}^n),$$

when $(p,q)$ is wave-admissible. We can check easily that $TT^*$ is linear and translation-invariant. Hence, applying Lemma 2.2 and taking into account Remark 2.1, it follows that $p \geq p'$, which yields $p \geq 2$.

Finally, we show the failure of (2.17) for the case when $n = 3$ and $(p,q) = (2, \infty)$, which is a result due to Tao [167]. If this estimate would be true, using duality, we infer that

$$\left\| \int e^{it|D|} F(t, \cdot) \, dt \right\|_{L_x^2} \lesssim \|F\|_{L^2 L^1} \tag{2.24}$$

also holds for $F = F(t, x)$ satisfying $F(t, \cdot) = S_1 F(t, \cdot)$. Therefore, the proof of this theorem is complete if we are able to construct a function $F$ that violates the previous estimate.

We define first $b = b(t)$ to be a one-dimensional Brownian motion, for which $b(t) - b(s)$ is a Gaussian random variable for all $t \neq s$ and satisfies

$$E\left( e^{i(b(t)-b(s))\eta} \right) = \frac{1}{\sqrt{2\pi}} e^{-|t-s|\eta^2}, \qquad \forall \eta \in \mathbb{R}. \tag{2.25}$$

Then our choice for $F$ is given by

$$F(t, x) = \varphi(t/T) \, \Phi(x_1 - b(t), x_2, x_3 - t),$$

where $\varphi$ is a smooth, nonnegative, even function supported on $[-1, 1]$, $T$ is a positive constant, and $\Phi = \Phi(x)$ verifies $\hat{\Phi} = \phi_1$, which is defined in (2.1). We easily obtain that

$$\|F\|_{L^2 L^1} = T^{1/2} \|\varphi\|_{L^2} \|\Phi\|_{L^1}.$$

In what concerns the left-hand side of (2.24), a direct computation yields

$$\left\| \int e^{it|D|} F(t, \cdot) \, dt \right\|_{L_x^2} \simeq \left( \int \left| \int e^{i[t(|\xi|-\xi_3)-b(t)\xi_1]} \varphi(t/T) \, dt \right|^2 \phi_1^2(\xi) \, d\xi \right)^{1/2}.$$

Due to

$$\Omega = \left\{ \xi = (\xi_1, \xi_2, \xi_3) \in \mathbb{R}^3; \; \frac{1}{\sqrt{2}} \leq \xi_3 \leq 1, \; |\xi_2| \leq |\xi_1| \leq \frac{1}{4} \right\} \subset \text{supp } \phi_1,$$

it follows that (2.24) implies

$$\int_\Omega \int \int e^{i[(t-s)(|\xi|-\xi_3)-(b(t)-b(s))\xi_1]} \varphi(t/T)\, \varphi(s/T)\, dt\, ds\, d\xi \lesssim T.$$

Now we take advantage of (2.25) and deduce

$$\int_\Omega \int \int e^{i(t-s)(|\xi|-\xi_3)-|t-s|\xi_1^2} \varphi(t/T)\, \varphi(s/T)\, dt\, ds\, d\xi \lesssim T.$$

As $\varphi$ is even, if we make the change of variable $s \mapsto v = t - s$, we obtain

$$\int_\Omega \int e^{iv(|\xi|-\xi_3)-|v|\xi_1^2} (\varphi * \varphi)(v/T)\, dv\, d\xi \lesssim 1.$$

The dominated convergence theorem allows us then to take $T \to \infty$ and infer

$$\int_\Omega \int e^{iv(|\xi|-\xi_3)-|v|\xi_1^2} dv\, d\xi \lesssim 1,$$

which can be further rewritten as

$$\int_\Omega \frac{\xi_1^2}{(|\xi|-\xi_3)^2 + \xi_1^4}\, d\xi \lesssim 1,$$

after computing the inner integral. However, in $\Omega$:

$$|\xi| - \xi_3 = \frac{\xi_1^2 + \xi_2^2}{|\xi| + \xi_3} \simeq \xi_1^2.$$

Therefore, we conclude that

$$\int_\Omega \frac{1}{\xi_1^2}\, d\xi \lesssim 1,$$

which we can easily see to be false. This concludes the argument for this theorem. $\qquad\square$

If we assume now that $g \in L^2(\mathbb{R}^n)$ satisfies $g = S_\lambda g$, where $\lambda \in 2^{\mathbb{Z}}$ is an arbitrary dyadic scale, then $f = g(\cdot/\lambda)$ verifies $f = S_1 f$ and

$$\left(e^{\pm it|D|} f\right)(x) = \left(e^{\pm i\frac{t}{\lambda}|D|} g\right)\left(\frac{x}{\lambda}\right).$$

Therefore, based on Theorem 2.1, we infer that

$$\|e^{\pm it|D|} g\|_{L^p L^q(\mathbb{R} \times \mathbb{R}^n)} \lesssim \lambda^{\frac{n}{2}-\frac{1}{p}-\frac{n}{q}} \|g\|_{L^2(\mathbb{R}^n)}. \tag{2.26}$$

Using the Littlewood-Paley theory, we can now sum up these estimates and deduce the following result.

**Corollary 2.2.** *If $(p, q, n)$ is wave-admissible and*

$$s(p, q, n) = \frac{n}{2} - \frac{1}{p} - \frac{n}{q},$$

*then*

$$\|e^{\pm it|D|} h\|_{L^p L^q(\mathbb{R} \times \mathbb{R}^n)} \lesssim \|h\|_{\dot{B}^{s(p,q,n)}_{2,1}(\mathbb{R}^n)} \tag{2.27}$$

*holds uniformly for* $h \in \mathcal{S}'(\mathbb{R}^n)$. *If, in addition,* $q \neq \infty$, *then*

$$\|e^{\pm it|D|} h\|_{L^p L^q(\mathbb{R} \times \mathbb{R}^n)} \lesssim \|h\|_{\dot{H}^{s(p,q,n)}(\mathbb{R}^n)}. \tag{2.28}$$

**Remark 2.2.** If $(p, q, n)$ is wave-admissible, then

$$s(p, q, n) \geq 0 \qquad \text{and} \qquad s(p, q, n) = 0 \iff (p, q) = (\infty, 2).$$

As a consequence, (2.28) implies

$$\|e^{\pm it|D|} h\|_{L^p L^q(\mathbb{R} \times \mathbb{R}^n)} \lesssim \|h\|_{H^{s(p,q,n)}(\mathbb{R}^n)}, \tag{2.29}$$

for $q \neq \infty$. If $q = \infty$, we can use (2.26) to derive

$$
\begin{aligned}
\|e^{\pm it|D|} h\|_{L^p L^q(\mathbb{R} \times \mathbb{R}^n)} &\lesssim \sum_{\lambda \in 2^{\mathbb{Z}}} \lambda^{s(p,q,n)} \|h_\lambda\|_{L^2(\mathbb{R}^n)} \\
&\lesssim \sum_{\lambda \in 2^{\mathbb{Z}}, \lambda \leq 1} \lambda^{s(p,q,n)} \|h\|_{L^2(\mathbb{R}^n)} \\
&\quad + \left( \sum_{\lambda \in 2^{\mathbb{Z}}, \lambda \geq 1} \lambda^{2(s(p,q,n)-s)} \right)^{\frac{1}{2}} \|h\|_{\dot{H}^s(\mathbb{R}^n)} \\
&\lesssim \|h\|_{H^s(\mathbb{R}^n)},
\end{aligned}
\tag{2.30}
$$

for any $s > s(p, q, n)$.

### 2.2.1    *Homogeneous bounds*

A straightforward calculation, which relies on the Fourier transform, shows that the solution of the Cauchy problem for the homogeneous wave equation

$$
\begin{cases}
\Box u := -u_{tt} + \Delta u = 0, \qquad u = u(t, x) : \mathbb{R} \times \mathbb{R}^n \to \mathbb{R}, \\
u(0, x) = u_0(x), \qquad u_t(0, x) = u_1(x),
\end{cases}
\tag{2.31}
$$

is given by

$$u(t) = \frac{1}{2} e^{it|D|} \left( u_0 - i |D|^{-1} u_1 \right) + \frac{1}{2} e^{-it|D|} \left( u_0 + i |D|^{-1} u_1 \right). \tag{2.32}$$

Thus, a direct application of Corollary 2.2 yields:

**Theorem 2.3.** *Let* $u$ *be a solution for* (2.31). *If* $(p, q, n)$ *is wave-admissible, then*

$$\|u\|_{L^p L^q(\mathbb{R} \times \mathbb{R}^n)} \lesssim \|u_0\|_{\dot{B}^{s(p,q,n)}_{2,1}(\mathbb{R}^n)} + \|u_1\|_{\dot{B}^{s(p,q,n)-1}_{2,1}(\mathbb{R}^n)}. \tag{2.33}$$

*If, in addition,* $q \neq \infty$, *then*

$$\|u\|_{L^p L^q(\mathbb{R} \times \mathbb{R}^n)} \lesssim \|u_0\|_{\dot{H}^{s(p,q,n)}(\mathbb{R}^n)} + \|u_1\|_{\dot{H}^{s(p,q,n)-1}(\mathbb{R}^n)}. \tag{2.34}$$

**Remark 2.3.** In what concerns the validity of (2.34) when $q = \infty$, Fang-Wang [46] showed that this holds when

$$\max\left\{2, \frac{4}{n-1}\right\} < p < \infty.$$

They also proved that if (2.34) with $q = \infty$ is true for all $(u_0, u_1) \in \mathcal{S} \times \mathcal{S}$, then

$$\frac{4}{n-1} < p < \infty.$$

In applications, we are also interested in working with versions of (2.34) in which either the homogeneous Sobolev norms are replaced by inhomogeneous ones or $u$ is exchanged for its full gradient

$$\nabla u \overset{\text{def}}{=} (\partial_t u, \nabla_x u) = (\partial_t u, \partial_{x_1} u, \dots, \partial_{x_n} u).$$

We list them here in one of their most general forms and refer the reader for proofs and further details to [46]. Specifically, for the former:

**Theorem 2.4.** *Let $u$ be a solution for* (2.31).

*i) If $n \geq 3$, $(p, q, n)$ is wave-admissible, $(p, q) \neq (2, \infty)$, $(p, q) \neq (\infty, \infty)$, and $s(p, q, n) \geq 1$, then*

$$\|u\|_{L^p L^q(\mathbb{R} \times \mathbb{R}^n)} \lesssim \|u_0\|_{H^{s(p,q,n)}(\mathbb{R}^n)} + \|u_1\|_{H^{s(p,q,n)-1}(\mathbb{R}^n)}.$$

*Additionally, if this estimate is true for all $(u_0, u_1) \in \mathcal{S} \times \mathcal{S}$, then $n \geq 3$, $(p, q, n)$ is wave-admissible, $(p, q) \neq (\infty, \infty)$, and $s(p, q, n) \geq 1$.*

*ii) If $n \geq 3$, $(p, q, n)$ is wave-admissible, and $s(p, q, n) \geq 1$, then*

$$\|u\|_{L^p L^q(\mathbb{R} \times \mathbb{R}^n)} \lesssim \|u_0\|_{H^s(\mathbb{R}^n)} + \|u_1\|_{H^{s-1}(\mathbb{R}^n)}$$

*holds for any $s > s(p, q, n)$. Moreover, if this estimate is valid for all $(u_0, u_1) \in \mathcal{S} \times \mathcal{S}$, then $(p, q, n)$ is wave-admissible and $s(p, q, n) \geq 1$.*

For estimates on the gradient, we have the following result.

**Theorem 2.5.** *Let $u$ be a solution for* (2.31).

*i) If $(p, q, n)$ is wave-admissible, $(p, q) \neq (\max\{2, 4/(n-1)\}, \infty)$, and $(p, q) \neq (\infty, \infty)$, then*

$$\|\nabla u\|_{L^p L^q(\mathbb{R} \times \mathbb{R}^n)} \lesssim \|u_0\|_{H^{s(p,q,n)+1}(\mathbb{R}^n)} + \|u_1\|_{H^{s(p,q,n)}(\mathbb{R}^n)}. \tag{2.35}$$

*If this estimate is true for all $(u_0, u_1) \in \mathcal{S} \times \mathcal{S}$, then $(p, q, n)$ is wave-admissible and $(p, q) \neq (\infty, \infty)$.*

*ii) $(p, q, n)$ is wave-admissible if and only if*

$$\|\nabla u\|_{L^p L^q(\mathbb{R} \times \mathbb{R}^n)} \lesssim \|u_0\|_{H^{s+1}(\mathbb{R}^n)} + \|u_1\|_{H^s(\mathbb{R}^n)} \tag{2.36}$$

*holds for any $s > s(p, q, n)$.*

**Remark 2.4.** A particular case of (2.35) is the *energy estimate*

$$\|\nabla u\|_{L^\infty L^2(\mathbb{R}\times\mathbb{R}^n)} \lesssim \|u_0\|_{H^1(\mathbb{R}^n)} + \|u_1\|_{L^2(\mathbb{R}^n)}.$$

In fact, we know very well that if $u$ solves (2.31), then $\|\nabla u(t)\|_{L^2(\mathbb{R}^n)}$ is a conserved quantity. Therefore, we can improve the previous bound as

$$\|\nabla u\|_{L^\infty L^2(\mathbb{R}\times\mathbb{R}^n)} \lesssim \|\nabla_x u_0\|_{L^2(\mathbb{R}^n)} + \|u_1\|_{L^2(\mathbb{R}^n)}. \tag{2.37}$$

Given that both $|D|^s$ and $\langle D\rangle^s$ are Fourier multipliers which commute with the wave operator $\Box$, we also deduce that

$$\|\nabla u\|_{L^\infty \dot{H}_x^s(\mathbb{R}\times\mathbb{R}^n)} \lesssim \|\nabla_x u_0\|_{\dot{H}^s(\mathbb{R}^n)} + \|u_1\|_{\dot{H}^s(\mathbb{R}^n)}$$

and

$$\|\nabla u\|_{L^\infty H_x^s(\mathbb{R}\times\mathbb{R}^n)} \lesssim \|\nabla_x u_0\|_{H^s(\mathbb{R}^n)} + \|u_1\|_{H^s(\mathbb{R}^n)}$$

hold for any $s \in \mathbb{R}$.

In future chapters we will be dealing with functions exhibiting radial symmetry, i.e.,

$$u = u(t,x) = \tilde{u}(t,r), \qquad f = f(x) = \tilde{f}(r), \qquad r = |x|.$$

It will be important to know that, in this case, the range of wave-admissible triplets is more comprehensive than the one given by Definition 2.2. Such results appeared first in work by Sterbenz [159] for $n \geq 3$ and were later extended by Fang-Wong [46] to include the case $n = 2$.

**Theorem 2.6.** *Let* $n \geq 2$ *and* $2 \leq p, q \leq \infty$ *satisfy*

$$\frac{1}{p} + \frac{n-1}{q} < \frac{n-1}{2}.$$

*i) If* $(p,q) \neq (\infty, \infty)$, *then* (2.28) *holds uniformly for all radial functions* $h$.

*ii) If* $q < \infty$ *and* $u$ *solves* (2.31) *with both* $u_0$ *and* $u_1$ *being radial, then* (2.34) *is true.*

**Remark 2.5.** In the radial setting, the improvement in the availability of Strichartz triples can be traced to the fact that, for a radial function $f$ satisfying $f = S_1 f$, the wave $e^{\pm it|D|}f$ is highly localized in the physical space along the radial variable.

## 2.2.2  *Inhomogeneous bounds*

Next, we are concerned with Strichartz estimates for inhomogeneous wave equations. The Cauchy problem

$$\begin{cases} \Box u(t,x) = F(t,x), \\ u(0,x) = u_0(x), \qquad u_t(0,x) = u_1(x), \end{cases} \tag{2.38}$$

is solved by $u = v + w$, where

$$\Box v = 0, \qquad v(0) = u_0, \qquad v_t(0) = u_1, \tag{2.39}$$

and

$$\Box w = F, \qquad w(0) = w_t(0) = 0. \tag{2.40}$$

Given that we already have Strichartz estimates for $v$, we proceed further only with $w$. Duhamel's formula applied to the initial value problem for $w$ yields

$$w(t,x) = \int_0^t w^\tau(t-\tau,x)\,d\tau,$$

with $w^\tau = w^\tau(t,x)$ satisfying

$$\Box w^\tau(t,x) = 0, \qquad w^\tau(0,x) = 0, \qquad w_t^\tau(0,x) = F(\tau,x).$$

Hence, using (2.32), we infer that

$$\begin{aligned} w(t,x) &= -\int_0^t |D|^{-1} \frac{e^{i(t-\tau)|D|} - e^{-i(t-\tau)|D|}}{2i} F(\tau,x)\,d\tau \\ &= -\int_0^t \frac{\sin((t-\tau)|D|)}{|D|} F(\tau,x)\,d\tau. \end{aligned} \tag{2.41}$$

If we denote

$$K_s^\pm(t) = |D|^{-s}\,e^{\pm it|D|},$$

the Strichartz estimate (2.28) can be rewritten as

$$\|K_{s(p,q,n)}^\pm(t)\,h\|_{L^p L^q(\mathbb{R}\times\mathbb{R}^n)} \lesssim \|h\|_{L^2(\mathbb{R}^n)}. \tag{2.42}$$

Thus, $h \mapsto Th = K_{s(p,q,n)}^\pm(t)\,h$ is a bounded linear operator from $L^2$ to $L^p L^q$, for which we can apply Lemma 2.3 to deduce that $T^\star$ is bounded from $L^{p'} L^{q'}$ to $L^2$. A simple calculation shows that

$$T^\star F = \int K_{s(p,q,n)}^\mp(t)\,F(t,\cdot)\,dt,$$

which implies

$$\left\| \int K^{\pm}_{s(p,q,n)}(t)\, F(t,\cdot)\, dt \right\|_{L^2(\mathbb{R}^n)} \lesssim \|F\|_{L^{p'}L^{q'}(\mathbb{R}\times\mathbb{R}^n)}.$$

Combining this bound with (2.42), we infer that

$$\left\| \int K^{\pm}_{s(p,q,n)+s(\tilde{p},\tilde{q},n)}(t-\tau)\, F(\tau,\cdot)\, d\tau \right\|_{L^p L^q(\mathbb{R}\times\mathbb{R}^n)} \lesssim \|F\|_{L^{\tilde{p}'}L^{\tilde{q}'}(\mathbb{R}\times\mathbb{R}^n)}, \tag{2.43}$$

if both $(p,q,n)$ and $(\tilde{p},\tilde{q},n)$ are wave-admissible, and neither $q$ nor $\tilde{q}$ are equal to $\infty$. Therefore, if we enforce

$$s(p,q,n) + s(\tilde{p},\tilde{q},n) = 1 \iff \frac{1}{p} + \frac{n}{q} = \frac{1}{\tilde{p}'} + \frac{n}{\tilde{q}'} - 2, \tag{2.44}$$

it follows that

$$\left\| \int |D|^{-1} \frac{e^{i(t-\tau)|D|} - e^{-i(t-\tau)|D|}}{2i}\, F(\tau,\cdot)\, d\tau \right\|_{L^p L^q(\mathbb{R}\times\mathbb{R}^n)} \tag{2.45}$$

$$\lesssim \|F\|_{L^{\tilde{p}'}L^{\tilde{q}'}(\mathbb{R}\times\mathbb{R}^n)},$$

which can easily be seen to imply the same estimate, but with $\mathbb{R}$ replaced by either $\mathbb{R}_+$ or $\mathbb{R}_-$.

At this point, we would like to derive Strichartz bounds for $w$ based on the previous estimate and the formula (2.41). In order to achieve this, we need one more ingredient, which is a generalization of the Christ-Kiselev lemma [27]. We present it here as it appeared in works by Smith-Sogge [155] and Tao [168], and we include its proof for completeness.

**Lemma 2.5.** *Let $X$ and $Y$ be Banach spaces, let $I \subseteq \mathbb{R}$ be an arbitrary time interval, and consider $K : I \times I \to L(X,Y)$ to be a kernel taking values in the space of linear transformations from $X$ to $Y$. Assume that the linear operator $T$ defined by*

$$Tf(t) \overset{def}{=} \int_I K(t,\tau) f(\tau)\, d\tau$$

*satisfies*

$$\|Tf\|_{L^{p_2}(I,Y)} \le C \|f\|_{L^{p_1}(I,X)}, \tag{2.46}$$

*with $1 \le p_1 < p_2 \le \infty$. Then*

$$\tilde{T}f(t) \overset{def}{=} \int_{\tau \in I, \tau < t} K(t,\tau) f(\tau)\, d\tau$$

*verifies*

$$\|\tilde{T}f\|_{L^{p_2}(I,Y)} \le C \cdot C(p_1,p_2) \|f\|_{L^{p_1}(I,X)}, \tag{2.47}$$

*where $C(p_1,p_2)$ is a constant depending solely on $p_1$ and $p_2$.*

**Proof.** Without loss of generality, we start by working only with functions $f : I \to X$ which are continuous, nowhere equal to zero, and normalized, i.e., $\|f\|_{L^{p_1}(I,X)} = 1$. Therefore, if we define

$$t \in I \mapsto F(t) \overset{\text{def}}{=} \int_{\tau \in I, \tau < t} \|f(\tau)\|_X^{p_1} \, d\tau,$$

then $F : I \to [0, 1]$ is a bijection. For an interval $J \subseteq [0, 1]$, this implies:

$$|J| = \int_{F^{-1}(J)} \|f(\tau)\|_X^{p_1} \, d\tau = \left\| \chi_{F^{-1}(J)} f \right\|_{L^{p_1}(I,X)}^{p_1}. \tag{2.48}$$

Following this, we consider the set of all intervals coming from dyadic partitions of $[0, 1]$,

$$[0, 1] = \left[ 0, \frac{1}{2} \right] \cup \left[ \frac{1}{2}, 1 \right] = \left[ 0, \frac{1}{4} \right] \cup \left[ \frac{1}{4}, \frac{2}{4} \right] \cup \left[ \frac{2}{4}, \frac{3}{4} \right] \cup \left[ \frac{3}{4}, 1 \right] = \dots,$$

and define on this set the relationship $J_1 \sim J_2$ if

- $J_1$ and $J_2$ have the same length;
- $J_1$ lies to the left of $J_2$;
- $J_1$ and $J_2$ are not adjacent, but have adjacent parents (i.e., $J_1 \subset \tilde{J}_1$, $J_2 \subset \tilde{J}_2$, and $\tilde{J}_1$ and $\tilde{J}_2$ are dyadic intervals of twice the length).

For each $J_2$, there are at most two $J_1$ satisfying $J_1 \sim J_2$. The same statement holds with the roles of $J_1$ and $J_2$ reversed. Furthermore, for almost all $0 \le x < y \le 1$, there exists a unique pair $(J_1, J_2)$ with $x \in J_1$, $y \in J_2$, and $J_1 \sim J_2$.

Using these properties, we can infer that

$$\chi_{\{(t,\tau) \in I \times I, t > \tau\}}(t, \tau) = \chi_{\{(x,y) \in [0,1] \times [0,1], x > y\}}(F(t), F(\tau))$$

$$= \sum_{J_1 \sim J_2} \chi_{J_2}(F(t)) \chi_{J_1}(F(\tau))$$

$$= \sum_{J_1 \sim J_2} \chi_{F^{-1}(J_2)}(t) \chi_{F^{-1}(J_1)}(\tau)$$

holds almost all $(\tau, t) \in I^2$. Taking into account that, for fixed $t$ and $\tau$, $K(t, \tau)$ is linear, we deduce

$$\tilde{T} f(t) = \int_I \chi_{\{(t,\tau) \in I \times I, t > \tau\}}(t, \tau) \, K(t, \tau) f(\tau) \, d\tau$$

$$= \sum_{J_1 \sim J_2} \chi_{F^{-1}(J_2)}(t) \int_I K(t, \tau) \left( \chi_{F^{-1}(J_1)} f \right)(\tau) \, d\tau$$

$$= \sum_{J_1 \sim J_2} \chi_{F^{-1}(J_2)}(t) \, T \left( \chi_{F^{-1}(J_1)} f \right)(t),$$

which further implies

$$\|\tilde{T}f\|_{L^{p_2}(I,Y)}$$

$$\leq \sum_{k\geq 2} \left\| \sum_{|J_2|=2^{-k}} \chi_{F^{-1}(J_2)} \left( \sum_{J_1\sim J_2} T\left(\chi_{F^{-1}(J_1)}f\right) \right) \right\|_{L^{p_2}(I,Y)} . \qquad (2.49)$$

For fixed $k$, the supports of $\chi_{F^{-1}(J_2)}$ are mutually disjoint when $J_2$ runs through the set of dyadic intervals having length $2^{-k}$. Hence:

$$\left\| \sum_{|J_2|=2^{-k}} \chi_{F^{-1}(J_2)} \left( \sum_{J_1\sim J_2} T\left(\chi_{F^{-1}(J_1)}f\right) \right) \right\|_{L^{p_2}(I,Y)}$$

$$\leq \left( \sum_{|J_2|=2^{-k}} \left( \sum_{J_1\sim J_2} \left\|T\left(\chi_{F^{-1}(J_1)}f\right)\right\|_{L^{p_2}(I,Y)} \right)^{p_2} \right)^{\frac{1}{p_2}} \qquad (2.50)$$

for $p_2 \neq \infty$, and

$$\left\| \sum_{|J_2|=2^{-k}} \chi_{F^{-1}(J_2)} \left( \sum_{J_1\sim J_2} T\left(\chi_{F^{-1}(J_1)}f\right) \right) \right\|_{L^{p_2}(I,Y)}$$

$$\leq \sup_{|J_2|=2^{-k}} \sum_{J_1\sim J_2} \left\|T\left(\chi_{F^{-1}(J_1)}f\right)\right\|_{L^{\infty}(I,Y)} \qquad (2.51)$$

for $p_2 = \infty$.

Based on a previous observation, we know that there are at most two terms in $\sum_{J_1\sim J_2}$. Using also (2.46) and (2.48), we derive

$$\sum_{|J_2|=2^{-k}} \left( \sum_{J_1\sim J_2} \left\|T\left(\chi_{F^{-1}(J_1)}f\right)\right\|_{L^{p_2}(I,Y)} \right)^{p_2}$$

$$\leq 2^{p_2-1} \sum_{|J_2|=2^{-k}} \sum_{J_1\sim J_2} \left\|T\left(\chi_{F^{-1}(J_1)}f\right)\right\|_{L^{p_2}(I,Y)}^{p_2} \leq (2C)^{p_2} 2^{k\left(1-\frac{p_2}{p_1}\right)}$$

for $p_2 \neq \infty$, and

$$\sup_{|J_2|=2^{-k}} \sum_{J_1\sim J_2} \left\|T\left(\chi_{F^{-1}(J_1)}f\right)\right\|_{L^{\infty}(I,Y)}$$

$$\leq 2 \sup_{|J_1|=2^{-k}} \left\|T\left(\chi_{F^{-1}(J_1)}f\right)\right\|_{L^{\infty}(I,Y)} \leq C\, 2^{1-\frac{k}{p_1}},$$

for $p_2 = \infty$. If we combine these estimates with (2.49)-(2.51), we obtain (2.47), which finishes the argument.                                                            □

In our setting, we can apply this lemma by taking

$$X = L^q(\mathbb{R}^n), \quad Y = L^{\tilde{q}'}(\mathbb{R}^n), \quad I = \mathbb{R}_+$$

and

$$K(t,\tau) = |D|^{-1} \frac{e^{i(t-\tau)|D|} - e^{-i(t-\tau)|D|}}{2i},$$

with $p_1 = \tilde{p}'$ and $p_2 = p$. Based on our assumptions on $p$ and $\tilde{p}$, which include (2.44), the condition $1 \le p_1 < p_2 \le \infty$ is satisfied. The other hypothesis, (2.46), holds according to (2.45). Therefore, based on (2.41), we deduce

$$\|w\|_{L^p L^q(\mathbb{R} \times \mathbb{R}^n)} \lesssim \|F\|_{L^{\tilde{p}'} L^{\tilde{q}'}(\mathbb{R} \times \mathbb{R}^n)}.$$

If we combine this estimate with (2.34), which is applicable to (2.39), we obtain the following result.

**Theorem 2.7.** *Let $u$ be a solution to (2.38). If $(p,q,n)$ and $(\tilde{p}, \tilde{q}, n)$ are wave-admissible, neither $q$ nor $\tilde{q}$ are equal to $\infty$, and the scaling condition (2.44) is satisfied, then*

$$\|u\|_{L^p L^q(\mathbb{R} \times \mathbb{R}^n)} \lesssim \|u_0\|_{\dot{H}^{s(p,q,n)}(\mathbb{R}^n)} + \|u_1\|_{\dot{H}^{s(p,q,n)-1}(\mathbb{R}^n)} \tag{2.52}$$
$$+ \|F\|_{L^{\tilde{p}'} L^{\tilde{q}'}(\mathbb{R} \times \mathbb{R}^n)}.$$

Next, we intend to obtain Strichartz estimates for $\nabla u$, which generalize (2.36). Using (2.41), we derive

$$\partial_t w(t,x) = - \int_0^t \cos((t-\tau)|D|) F(\tau, x) \, d\tau, \tag{2.53}$$

$$\nabla_x w(t,x) = - \int_0^t \frac{iD}{|D|} \cdot \sin((t-\tau)|D|) F(\tau, x) \, d\tau. \tag{2.54}$$

A direct consequence of these formulas is the energy estimate

$$\|\nabla w\|_{L^\infty L^2(\mathbb{R} \times \mathbb{R}^n)} \lesssim \|F\|_{L^1 L^2(\mathbb{R} \times \mathbb{R}^n)}. \tag{2.55}$$

This complements (2.37), which was obtained for the homogeneous equation.

If we denote

$$\overline{K}_s^{\pm}(t) = \langle D \rangle^{-s} e^{\pm it|D|},$$

then (2.29) and (2.30) imply

$$\|\overline{K}_s^{\pm}(t) h\|_{L^p L^q(\mathbb{R} \times \mathbb{R}^n)} \lesssim \|h\|_{L^2(\mathbb{R}^n)},$$

when $(p, q, n)$ is wave-admissible and $s > s(p, q, n)$. An argument almost identical to the one that led from (2.42) to (2.43) yields, in this case,

$$\left\| \int \overline{K}_s^{\pm}(t - \tau) G(\tau, \cdot) \, d\tau \right\|_{L^p L^q (\mathbb{R} \times \mathbb{R}^n)} \lesssim \|G\|_{L^{\tilde{p}'} L^{\tilde{q}'} (\mathbb{R} \times \mathbb{R}^n)},$$

if $(p, q, n)$ and $(\tilde{p}, \tilde{q}, n)$ are both wave-admissible, and $s > s(p, q, n) + s(\tilde{p}, \tilde{q}, n)$.

Due to (2.53) and (2.54), we can write formally

$$\nabla w(t, x) \simeq \int_0^t \overline{K}_s^{\pm}(t - \tau) \langle D \rangle^s F(\tau, x) \, d\tau.$$

Hence, another application of Lemma 2.5 allows us to infer

$$\|\nabla w\|_{L^p L^q (\mathbb{R} \times \mathbb{R}^n)} \lesssim \|\langle D \rangle^s F\|_{L^{\tilde{p}'} L^{\tilde{q}'} (\mathbb{R} \times \mathbb{R}^n)},$$

if, in addition to the previous conditions, $(p, \tilde{p}) \neq (2, 2)$. Using this estimate in conjunction with (2.36), we obtain:

**Theorem 2.8.** *Let $u$ be a solution to (2.38). If $(p, q, n)$ and $(\tilde{p}, \tilde{q}, n)$ are wave-admissible, $(p, \tilde{p}) \neq (2, 2)$, $s_1 > s(p, q, n)$, and $s_2 > s(p, q, n) + s(\tilde{p}, \tilde{q}, n)$, then*

$$\|\nabla u\|_{L^p L^q (\mathbb{R} \times \mathbb{R}^n)} \lesssim \|u_0\|_{H^{s_1 + 1}(\mathbb{R}^n)} + \|u_1\|_{H^{s_1}(\mathbb{R}^n)} + \|\langle D \rangle^{s_2} F\|_{L^{\tilde{p}'} L^{\tilde{q}'} (\mathbb{R} \times \mathbb{R}^n)}.$$

An immediate consequence is as follows.

**Corollary 2.3.** *Let $u$ be a solution to (2.38). If $(p, q, n)$ is wave-admissible and $s > s(p, q, n)$, then*

$$\|\nabla u\|_{L^p L^q (\mathbb{R} \times \mathbb{R}^n)} \lesssim \|u_0\|_{H^{s+1}(\mathbb{R}^n)} + \|u_1\|_{H^s(\mathbb{R}^n)} + \|F\|_{L^1 H^s (\mathbb{R} \times \mathbb{R}^n)}. \quad (2.56)$$

## 2.3   Hyperbolic Sobolev spaces

Another set of norms in which it is natural to measure solutions of wave equations are the hyperbolic Sobolev ones, which were introduced formally by Klainerman-Machedon [86]. These are an extension of the norms originally used by Bourgain [20,21] in the context of Schrödinger and Korteweg-de Vries equations. Earlier accounts of similar ideas applied to semilinear wave equations appeared in works by Rauch-Reed [137] and Beals [13].

The most intuitive way to describe the function spaces corresponding to the above norms is that they are associated to the wave operator $\Box$ in the same way the classical Sobolev spaces are associated to the Laplacian $\Delta$. For a good introduction to this topic, we refer the reader to Selberg [141], Klainerman-Selberg [90], and D'Ancona-Georgiev [36].

**Definition 2.3.** Let $n \geq 1$, $s$, $\theta \in \mathbb{R}$, and consider

$$\|u\|_{X^{s,\theta}(\mathbb{R}^{n+1})}$$
$$\overset{\text{def}}{=} \| (1 + |\tau| + |\xi|)^s (1 + ||\tau| - |\xi||)^\theta \, \widehat{u}(\tau, \xi)\|_{L^2_{\tau,\xi}(\mathbb{R}^{n+1})}, \tag{2.57}$$

where

$$\widehat{u}(\tau, \xi) \overset{\text{def}}{=} \int_{\mathbb{R}^{n+1}} e^{-i(t\tau + x \cdot \xi)} u(t, x) \, dt \, dx$$

is the spacetime Fourier transform for $u = u(t, x) \in \mathcal{S}'(\mathbb{R}^{n+1})$. The hyperbolic Sobolev space $X^{s,\theta}(\mathbb{R}^{n+1})$ is defined by

$$X^{s,\theta}(\mathbb{R}^{n+1}) \overset{\text{def}}{=} \left\{ u \in \mathcal{S}'(\mathbb{R}^{n+1}); \; \|u\|_{X^{s,\theta}(\mathbb{R}^{n+1})} < \infty \right\}.$$

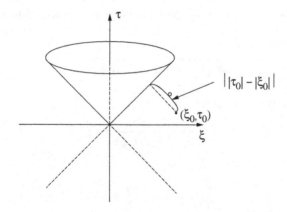

Fig. 2.1 The geometric representation of the weight $||\tau_0| - |\xi_0||$ in Fourier space.

**Remark 2.6.** The heuristic behind (2.57) is as follows: the hyperbolic Sobolev space $X^{s,\theta}(\mathbb{R}^{n+1})$ controls $s$ regular derivatives and $\theta$ *wave/null* derivatives of $u$ in $L^2(\mathbb{R}^{n+1})$.

**Remark 2.7.** In literature, other variations of $X^{s,\theta}$ have been used in investigating nonlinear wave equations:

$$\|u\|_{X_1^{s,\theta}} = \| (1 + |\xi|)^s (1 + ||\tau| - |\xi||)^\theta \, \widehat{u}(\tau, \xi)\|_{L^2(\mathbb{R}^{n+1})},$$
$$\|u\|_{X_2^{s,\theta}} = \|u\|_{X_1^{s,\theta}} + \|\partial_t u\|_{X_1^{s-1,\theta}}.$$

**Remark 2.8.** As $\widehat{\Box u}(\tau, \xi) = (\tau^2 - |\xi|^2)\,\widehat{u}(\tau, \xi)$, an immediate consequence of the above definition is that, for all $s$, $\theta \in \mathbb{R}$,

$$\|\Box u\|_{X^{s-1,\theta-1}(\mathbb{R}^{n+1})} \lesssim \|u\|_{X^{s,\theta}(\mathbb{R}^{n+1})}, \tag{2.58}$$

which shows the intimate relationship between the wave operator and the hyperbolic Sobolev spaces.

**Remark 2.9.** In order to simplify the notation, we will work from this point on with

$$w_\pm(\tau, \xi) \overset{\text{def}}{=} 1 + ||\tau| \pm |\xi||.$$

The first result we prove in this section concerns energy estimates for $X^{s,\theta}$ spaces.

**Proposition 2.7.** If $\theta > 1/2$ and $s + \theta > 3/2$, then $X^{s,\theta}(\mathbb{R}^{n+1})$ embeds continuously in $C(\mathbb{R}, H^s(\mathbb{R}^n)) \cap C^1(\mathbb{R}, H^{s-1}(\mathbb{R}^n))$, i.e.,

$$\|u\|_{L^\infty H^s(\mathbb{R} \times \mathbb{R}^n)} + \|\partial_t u\|_{L^\infty H^{s-1}(\mathbb{R} \times \mathbb{R}^n)} \lesssim \|u\|_{X^{s,\theta}(\mathbb{R}^{n+1})}. \tag{2.59}$$

**Proof.** A direct application of the Cauchy–Schwarz inequality yields

$$\|u(t)\|_{H^s}^2 + \|\partial_t u(t)\|_{H^{s-1}}^2$$

$$\lesssim \int_{\mathbb{R}^{n+1}} \left[ \int_{\mathbb{R}} \frac{\langle \xi \rangle^{2s} + \langle \xi \rangle^{2s-2}\tau^2}{w_+^{2s}(\tau, \xi)\, w_-^{2\theta}(\tau, \xi)}\, d\tau \right] w_+^{2s}(\tau_1, \xi)\, w_-^{2\theta}(\tau_1, \xi)\, |\widehat{u}(\tau_1, \xi)|^2\, d\tau_1\, d\xi.$$

Therefore, (2.59) follows if we show that

$$\int_{\mathbb{R}} \frac{\langle \xi \rangle^{2s} + \langle \xi \rangle^{2s-2}\tau^2}{w_+^{2s}(\tau, \xi)\, w_-^{2\theta}(\tau, \xi)}\, d\tau \lesssim 1$$

holds uniformly in $\xi$. If $|\xi| < 1/2$, then

$$\int_{\mathbb{R}} \frac{\langle \xi \rangle^{2s} + \langle \xi \rangle^{2s-2}\tau^2}{w_+^{2s}(\tau, \xi)\, w_-^{2\theta}(\tau, \xi)}\, d\tau \simeq \int_{\mathbb{R}} \langle \tau \rangle^{2(1-s-\theta)}\, d\tau \lesssim 1,$$

due to $s + \theta > 3/2$.

If $|\xi| \geq 1/2$, we separate the analysis into two cases: $|\tau| \leq 2|\xi|$ and $|\tau| > 2|\xi|$. For the former, we argue that

$$\int_{|\tau| \leq 2|\xi|} \frac{\langle \xi \rangle^{2s} + \langle \xi \rangle^{2s-2}\tau^2}{w_+^{2s}(\tau, \xi)\, w_-^{2\theta}(\tau, \xi)}\, d\tau \simeq \int_{|\tau| \leq 2|\xi|} \frac{1}{w_-^{2\theta}(\tau, \xi)}\, d\tau \simeq 1,$$

based on $\theta > 1/2$. For the latter, we deduce

$$\int_{|\tau| > 2|\xi|} \frac{\langle \xi \rangle^{2s} + \langle \xi \rangle^{2s-2}\tau^2}{w_+^{2s}(\tau, \xi)\, w_-^{2\theta}(\tau, \xi)}\, d\tau \simeq \int_{|\tau| > 2|\xi|} |\xi|^{2s-2}\, |\tau|^{2(1-s-\theta)}\, d\tau$$

$$\simeq |\xi|^{1-2\theta} \lesssim 1,$$

due to both $s + \theta > 3/2$ and $\theta > 1/2$. This concludes the argument. $\quad\square$

Next, we show that localization in time is a bounded operation on $X^{s,\theta}$ spaces.

**Proposition 2.8.** *If $s$, $\theta \in \mathbb{R}$ and $\varphi = \varphi(t) \in \mathcal{S}(\mathbb{R})$ is an arbitrary Schwartz function in time, then*

$$\|\varphi\, u\|_{X^{s,\theta}(\mathbb{R}^{n+1})} \lesssim \|u\|_{X^{s,\theta}(\mathbb{R}^{n+1})} \tag{2.60}$$

*holds uniformly for $u \in X^{s,\theta}$.*

**Proof.** Directly from (2.57), we derive

$$\|\varphi\, u\|_{X^{s,\theta}}^2 \simeq \int_{\mathbb{R}^{n+1}} w_+^{2s}(\tau,\xi)\, w_-^{2\theta}(\tau,\xi) \left| \int_{\mathbb{R}} \hat{\varphi}(\tau - \tau_1)\, \widehat{u}(\tau_1,\xi)\, d\tau_1 \right|^2 d\tau\, d\xi.$$

The simple bound

$$\langle \tau - \tau_1 \rangle^{-1} \le \frac{w_\pm(\tau,\xi)}{w_\pm(\tau_1,\xi)} \le \langle \tau - \tau_1 \rangle,$$

leads to

$$\|\varphi\, u\|_{X^{s,\theta}}^2 \lesssim \int_{\mathbb{R}^{n+1}} \left| \int_{\mathbb{R}} \langle \tau - \tau_1 \rangle^{|s|+|\theta|}\, |\hat{\varphi}(\tau - \tau_1)| \right.$$
$$\left. w_+^s(\tau_1,\xi)\, w_-^\theta(\tau_1,\xi)\, |\widehat{u}(\tau_1,\xi)|\, d\tau_1 \right|^2 d\tau\, d\xi.$$

Finally, we apply Young's inequality (2.18) for the integral in $\tau$ to infer

$$\|\varphi\, u\|_{X^{s,\theta}}^2 \lesssim \left( \int_{\mathbb{R}} \langle \tau \rangle^{|s|+|\theta|}\, |\hat{\varphi}(\tau)|\, d\tau \right)^2 \|u\|_{X^{s,\theta}}^2,$$

which implies (2.60) due to $\varphi \in \mathcal{S}$. $\qquad\square$

An identical reasoning gives a more precise estimate when $\varphi$ has compact support.

**Proposition 2.9.** *Let $s$, $\theta \in \mathbb{R}$ and consider $\varphi : \mathbb{R} \to \mathbb{R}_+$ to be an even, smooth cutoff function satisfying $\varphi \equiv 1$ on $[-1,1]$ and $\mathrm{supp}\ \varphi \subset [-2,2]$. If $T > 0$ is arbitrary and $\varphi_T : \mathbb{R} \to \mathbb{R}_+$ is defined by $\varphi_T(t) := \varphi(t/T)$, then*

$$\|\varphi_T\, u\|_{X^{s,\theta}(\mathbb{R}^{n+1})} \lesssim \left( 1 + T^{-(|s|+|\theta|)} \right) \|u\|_{X^{s,\theta}(\mathbb{R}^{n+1})} \tag{2.61}$$

*holds uniformly for $u \in X^{s,\theta}$.*

Following this, we focus on proving estimates in $X^{s,\theta}$ spaces for solutions of linear wave equations. As was the case in the previous section, we start by investigating the homogeneous Cauchy problem (2.31).

**Theorem 2.9.** *Let* $s, \theta \in \mathbb{R}$. *If* $\varphi \in C_0^\infty(\mathbb{R})$, $u_0 \in H^s(\mathbb{R}^n)$, $u_1 \in H^{s-1}(\mathbb{R}^n)$, *and* $u$ *solves* (2.31), *then*

$$\|\varphi \, u\|_{X^{s,\theta}(\mathbb{R}^{n+1})} \lesssim \|u_0\|_{H^s(\mathbb{R}^n)} + \|u_1\|_{H^{s-1}(\mathbb{R}^n)}. \qquad (2.62)$$

**Proof.** A routine computation based on the solution formula (2.32) yields

$$\widehat{\varphi u}(\tau, \xi) = \frac{(\widehat{\varphi}(\tau - |\xi|) + \widehat{\varphi}(\tau + |\xi|)) \, \widehat{u_0}(\xi)}{2}$$
$$+ \frac{(\widehat{\varphi}(\tau - |\xi|) - \widehat{\varphi}(\tau + |\xi|)) \, \widehat{u_1}(\xi)}{2i|\xi|}, \qquad (2.63)$$

and our approach will be to estimate separately the two terms on the right-hand side.

For the first one, we rely on $\varphi \in C_0^\infty(\mathbb{R})$ and $|\tau \pm |\xi|| \geq ||\tau| - |\xi||$ to infer that

$$\int_{\mathbb{R}^{n+1}} w_+^{2s}(\tau, \xi) \, w_-^{2\theta}(\tau, \xi) \, |\widehat{\varphi}(\tau - |\xi|) + \widehat{\varphi}(\tau + |\xi|)|^2 \, |\widehat{u_0}(\xi)|^2 \, d\tau \, d\xi$$
$$\lesssim \int_{\mathbb{R}^{n+1}} w_+^{2s}(\tau, \xi) \, w_-^{2(\theta - N)}(\tau, \xi) \, |\widehat{u_0}(\xi)|^2 \, d\tau \, d\xi$$

holds for arbitrarily large integers $N$. Next, we use the straightforward estimate

$$w_-^{-1}(\tau, \xi) \lesssim \frac{w_+(\tau, \xi)}{\langle \xi \rangle} \lesssim w_-(\tau, \xi), \qquad (2.64)$$

to deduce

$$\int_{\mathbb{R}^{n+1}} w_+^{2s}(\tau, \xi) \, w_-^{2(\theta - N)}(\tau, \xi) \, |\widehat{u_0}(\xi)|^2 \, d\tau \, d\xi$$
$$\lesssim \int_{\mathbb{R}^{n+1}} w_-^{2(|s| + \theta - N)}(\tau, \xi) \, \langle \xi \rangle^{2s} \, |\widehat{u_0}(\xi)|^2 \, d\tau \, d\xi \qquad (2.65)$$
$$\lesssim \|u_0\|_{H^s(\mathbb{R}^n)}^2,$$

if $N > |s| + \theta$.

For the term in (2.63) involving $u_1$, due to the singularity present in the denominator, we discuss individually the regimes $|\xi| \geq 1/2$ and $|\xi| < 1/2$.

When $|\xi| \geq 1/2$, we argue as in the analysis of the first term to derive

$$\int_{\{|\xi| \geq 1/2\} \times \mathbb{R}} \left\{ w_+^{2s}(\tau, \xi) \, w_-^{2\theta}(\tau, \xi) \right.$$

$$\left. |\widehat{\varphi}(\tau - |\xi|) - \widehat{\varphi}(\tau + |\xi|)|^2 \, \frac{|\widehat{u_1}(\xi)|^2}{|\xi|^2} \right\} d\tau \, d\xi \qquad (2.66)$$

$$\lesssim \int_{\{|\xi| \geq 1/2\} \times \mathbb{R}} w_-^{2(|s| + \theta - N)}(\tau, \xi) \, \langle \xi \rangle^{2s-2} \, |\widehat{u_1}(\xi)|^2 \, d\tau \, d\xi$$

$$\lesssim \|u_1\|_{H^{s-1}(\mathbb{R}^n)}^2,$$

under the same conditions on $N$ as above. When $|\xi| < 1/2$, we have

$$w_+(\tau, \xi) \simeq \langle \tau + \lambda|\xi| \rangle \simeq w_-(\tau, \xi), \qquad \forall \lambda \in [-1, 1],$$

which implies

$$|\widehat{\varphi}(\tau - |\xi|) - \widehat{\varphi}(\tau + |\xi|)| \lesssim |\xi| \sup_{\lambda \in [-1,1]} |\widehat{\varphi}'(\tau + \lambda|\xi|)|$$

$$\lesssim |\xi| \sup_{\lambda \in [-1,1]} \langle \tau + \lambda|\xi| \rangle^{-N} \lesssim |\xi| \, w_-^{-N}(\tau, \xi).$$

Hence, we can derive

$$\int_{\{|\xi| < 1/2\} \times \mathbb{R}} \left\{ w_+^{2s}(\tau, \xi) \, w_-^{2\theta}(\tau, \xi) \right.$$

$$\left. |\widehat{\varphi}(\tau - |\xi|) - \widehat{\varphi}(\tau + |\xi|)|^2 \, \frac{|\widehat{u_1}(\xi)|^2}{|\xi|^2} \right\} d\tau \, d\xi \qquad (2.67)$$

$$\lesssim \int_{\{|\xi| < 1/2\} \times \mathbb{R}} w_-^{2(s + \theta - N)}(\tau, \xi) \, |\widehat{u_1}(\xi)|^2 \, d\tau \, d\xi$$

$$\lesssim \int_{\{|\xi| < 1/2\} \times \mathbb{R}} w_-^{2(s + \theta - N)}(\tau, \xi) \, \langle \xi \rangle^{2s-2} \, |\widehat{u_1}(\xi)|^2 \, d\tau \, d\xi$$

$$\lesssim \|u_1\|_{H^{s-1}(\mathbb{R}^n)}^2,$$

if $N > s + \theta$.

The conclusion is then a consequence of (2.65), (2.66), and (2.67). $\square$

We work next on inhomogeneous linear wave equations, specifically (2.40). In this argument and onward, we adopt the notation $\widehat{w}(t, \xi)$ for the Fourier transform of $w = w(t, x)$ only with respect to the spatial variable, i.e.,

$$\widehat{w}(t, \xi) \stackrel{\text{def}}{=} \int_{\mathbb{R}^n} e^{-ix \cdot \xi} \, w(t, x) \, dx.$$

**Theorem 2.10.** *Let* $s$, $\theta \in \mathbb{R}$ *such that* $\theta > 1/2$ *and* $s + \theta > 3/2$. *If* $\varphi \in C_0^\infty(\mathbb{R})$, $F \in X^{s-1, \theta-1}(\mathbb{R}^{n+1})$, *and* $w$ *solves* (2.40), *then*

$$\|\varphi w\|_{X^{s, \theta}(\mathbb{R}^{n+1})} \lesssim \|F\|_{X^{s-1, \theta-1}(\mathbb{R}^{n+1})}. \qquad (2.68)$$

**Proof.** Using the solution formula (2.41), we can infer

$$
\begin{aligned}
\widehat{w}(t,\xi) &= -\int_0^t \frac{\sin((t-t')|\xi|)}{|\xi|}\, \widehat{F}(t',\xi)\, dt' \\
&= -\frac{1}{2\pi} \int_{[0,t]\times\mathbb{R}} \left\{ \left( e^{i(t'(\tau_1-|\xi|)+t|\xi|)} - e^{i(t'(\tau_1+|\xi|)-t|\xi|)} \right) \right. \\
&\qquad\qquad \left. \frac{\widehat{F}(\tau_1,\xi)}{2i|\xi|} \right\} d\tau_1\, dt' \\
&= -\frac{1}{2\pi} \int_{\mathbb{R}} \left( \frac{e^{it\tau_1}-e^{-it|\xi|}}{\tau_1+|\xi|} - \frac{e^{it\tau_1}-e^{it|\xi|}}{\tau_1-|\xi|} \right) \frac{\widehat{F}(\tau_1,\xi)}{2|\xi|}\, d\tau_1.
\end{aligned}
\tag{2.69}
$$

The strategy behind this argument is to decompose the above expression into integrals on smaller regions of the Fourier space, where it will be easy to obtain estimates for $\|\varphi\, w\|_{X^{s,\theta}(\mathbb{R}^{n+1})}$. First, we write the last line of (2.69) as:

$$
\begin{aligned}
\widehat{w}(t,\xi) &= \int_{\{\tau_1\in\mathbb{R};\,|\tau_1|+|\xi|<1/2\}} \cdots\, d\tau_1 + \int_{\{\tau_1\in\mathbb{R};\,|\tau_1|+|\xi|\geq 1/2\}} \cdots\, d\tau_1 \\
&= \widehat{w_1}(t,\xi) + \widehat{w_2}(t,\xi).
\end{aligned}
\tag{2.70}
$$

In particular, if $|\xi| \geq 1/2$, we have $\widehat{w_1}(t,\xi) = 0$.

A direct computation on $w_1$ yields

$$
\begin{aligned}
\widehat{\varphi w_1}(\tau,\xi) &= \int_{\mathbb{R}} e^{-it\tau}\, \varphi(t)\, \widehat{w_1}(t,\xi)\, dt \\
&= -\frac{1}{2\pi} \int_{|\tau_1|+|\xi|<1/2} \left( \frac{\widehat{\varphi}(\tau-\tau_1)-\widehat{\varphi}(\tau+|\xi|)}{\tau_1+|\xi|} \right. \\
&\qquad\qquad \left. - \frac{\widehat{\varphi}(\tau-\tau_1)-\widehat{\varphi}(\tau-|\xi|)}{\tau_1-|\xi|} \right) \frac{\widehat{F}(\tau_1,\xi)}{2|\xi|}\, d\tau_1.
\end{aligned}
$$

For the first factor of the integrand, we obtain

$$
\begin{aligned}
&\frac{\widehat{\varphi}(\tau-\tau_1)-\widehat{\varphi}(\tau+|\xi|)}{\tau_1+|\xi|} - \frac{\widehat{\varphi}(\tau-\tau_1)-\widehat{\varphi}(\tau-|\xi|)}{\tau_1-|\xi|} \\
&= \int_0^1 \widehat{\varphi}'(\tau-|\xi|-a(\tau_1-|\xi|)) - \widehat{\varphi}'(\tau+|\xi|-a(\tau_1+|\xi|))\, da \\
&= -|\xi| \int_0^1 \left( \int_{-1}^1 \widehat{\varphi}''(\tau-a\tau_1-b(1-a)|\xi|)\, db \right) (1-a)\, da.
\end{aligned}
$$

Thus, based on $\varphi \in C_0^\infty(\mathbb{R})$ and $|\tau_1| + |\xi| < 1/2$, we can estimate it as

$$
\left| \frac{\widehat{\varphi}(\tau-\tau_1)-\widehat{\varphi}(\tau+|\xi|)}{\tau_1+|\xi|} - \frac{\widehat{\varphi}(\tau-\tau_1)-\widehat{\varphi}(\tau-|\xi|)}{\tau_1-|\xi|} \right| \lesssim |\xi|\, \langle\tau\rangle^{-N},
$$

for an arbitrarily large integer $N$, which implies

$$|\widehat{\varphi w_1}(\tau, \xi)| \lesssim \langle \tau \rangle^{-N} \int_{|\tau_1| + |\xi| < 1/2} \left| \widehat{F}(\tau_1, \xi) \right| d\tau_1.$$

Moreover, due to $|\xi| < 1/2$, we have

$$w_+(\tau, \xi) \simeq w_-(\tau, \xi) \simeq \langle \tau \rangle.$$

Therefore, also using the Cauchy-Schwarz inequality, we deduce

$$\|\varphi \, w_1\|_{X^{s,\theta}}^2$$

$$\lesssim \int_{\{|\xi| < 1/2\} \times \mathbb{R}} \langle \tau \rangle^{2(s + \theta - N)} \left( \int_{|\tau_1| + |\xi| < 1/2} \left| \widehat{F}(\tau_1, \xi) \right| d\tau_1 \right)^2 d\tau \, d\xi$$

$$\lesssim \int_{\{(\tau_1, \xi) \in \mathbb{R}^{n+1}; |\tau_1| + |\xi| < 1/2\}} \left| \widehat{F}(\tau_1, \xi) \right|^2 d\tau_1 \, d\xi$$

$$\lesssim \|F\|_{X^{s-1, \theta - 1}}^2,$$

which finishes the analysis of $w_1$.

Next, we take $w_2$ from (2.70) and we further decompose it:

$$\widehat{w_2}(t, \xi) = \widehat{w_3}(t, \xi) + \widehat{w_4}(t, \xi), \tag{2.71}$$

with

$$\widehat{w_3}(t, \xi)$$
$$= -\frac{1}{2\pi} \int_{|\tau_1| + |\xi| \geq 1/2} e^{it\tau_1} \left( \frac{1 - \chi_{(-\infty, 0)}(\tau_1) \, a(|\tau_1| - |\xi|)}{\tau_1 + |\xi|} \right.$$
$$\left. - \frac{1 - \chi_{(0, \infty)}(\tau_1) \, a(|\tau_1| - |\xi|)}{\tau_1 - |\xi|} \right) \frac{\widehat{F}(\tau_1, \xi)}{2|\xi|} d\tau_1. \tag{2.72}$$

In this formula, $a \in C_0^\infty(\mathbb{R})$ satisfies $a \equiv 1$ on a small neighborhood of 0 and $a(\eta) = 0$ for $|\eta| \geq 1/4$. It follows that

$$\widehat{w_3}(\tau, \xi) = -\left( \frac{1 - \chi_{(-\infty, 0)}(\tau) \, a(|\tau| - |\xi|)}{\tau + |\xi|} \right.$$
$$\left. - \frac{1 - \chi_{(0, \infty)}(\tau) \, a(|\tau| - |\xi|)}{\tau - |\xi|} \right) \frac{\widehat{F}(\tau, \xi)}{2|\xi|} \tag{2.73}$$

if $|\tau| + |\xi| \geq 1/2$, and it is equal to 0 otherwise. Our goal is to show that

$$|\widehat{w_3}(\tau, \xi)| \lesssim w_+^{-1}(\tau, \xi) \, w_-^{-1}(\tau, \xi) \left| \widehat{F}(\tau, \xi) \right|, \tag{2.74}$$

which would then imply, based on (2.60),

$$\|\varphi \, w_3\|_{X^{s,\theta}} \lesssim \|w_3\|_{X^{s,\theta}} \lesssim \|F\|_{X^{s-1, \theta - 1}}.$$

In order to prove (2.74), we need to work only in the $|\tau| + |\xi| \geq 1/2$ regime, based on (2.73). Moreover, it is enough to discuss solely the $\tau > 0$ case. The reason is that, for $\xi$ fixed,

$$\tau \mapsto \frac{1 - \chi_{(-\infty,0)}(\tau)\, a(|\tau| - |\xi|)}{\tau + |\xi|} - \frac{1 - \chi_{(0,\infty)}(\tau)\, a(|\tau| - |\xi|)}{\tau - |\xi|}$$

is an even map. As $\tau > 0$, we deduce

$$\widehat{w_3}(\tau, \xi) = -\left( \frac{1}{\tau + |\xi|} - \frac{1 - a(|\tau| - |\xi|)}{\tau - |\xi|} \right) \frac{\widehat{F}(\tau, \xi)}{2|\xi|}.$$

If $||\tau| - |\xi|| \geq 1/4$, then $a(|\tau| - |\xi|) = 0$ and $w_\pm(\tau, \xi) \simeq ||\tau| \pm |\xi||$. Therefore, we obtain

$$|\widehat{w_3}(\tau, \xi)| = \frac{|\widehat{F}(\tau, \xi)|}{||\xi|^2 - \tau^2|} \simeq w_+^{-1}(\tau, \xi)\, w_-^{-1}(\tau, \xi) \left| \widehat{F}(\tau, \xi) \right|,$$

which implies (2.74).

If $||\tau| - |\xi|| < 1/4$, then, due to $|\tau| + |\xi| \geq 1/2$,

$$w_+(\tau, \xi) \simeq |\xi|, \qquad w_-(\tau, \xi) \simeq 1.$$

Moreover, based on the properties of $a$, we also have

$$|1 - a(|\tau| - |\xi|)| \lesssim ||\tau| - |\xi||. \tag{2.75}$$

Hence, we can infer

$$|\widehat{w_3}(\tau, \xi)| \lesssim \frac{|\widehat{F}(\tau, \xi)|}{|\xi|} \lesssim w_+^{-1}(\tau, \xi)\, w_-^{-1}(\tau, \xi) \left| \widehat{F}(\tau, \xi) \right|,$$

which finishes the proof for (2.74) and the analysis of $w_3$.

Now, we arrive at the most intricate part of the argument, which is estimating $w_4$. Using (2.70), (2.71), and (2.72), we derive

$$2\pi \cdot \widehat{w_4}(t, \xi) = \widehat{w_{41}}(t, \xi) + \widehat{w_{42}}(t, \xi) + \widehat{w_{43}}(t, \xi) + \widehat{w_{44}}(t, \xi), \tag{2.76}$$

where

$$\widehat{w_{41}}(t, \xi)$$
$$= -\int_{|\tau_1| + |\xi| \geq 1/2} \frac{e^{it|\xi|}}{\tau_1 - |\xi|} \left[ 1 - \chi_{(0,\infty)}(\tau_1)\, a(|\tau_1| - |\xi|) \right] \frac{\widehat{F}(\tau_1, \xi)}{2|\xi|}\, d\tau_1,$$

$$\widehat{w_{42}}(t, \xi)$$
$$= -\int_{|\tau_1| + |\xi| \geq 1/2} \frac{-e^{-it|\xi|}}{\tau_1 + |\xi|} \left[ 1 - \chi_{(-\infty,0)}(\tau_1)\, a(|\tau_1| - |\xi|) \right] \frac{\widehat{F}(\tau_1, \xi)}{2|\xi|}\, d\tau_1,$$

$$\widehat{w_{43}}(t,\xi)$$

$$= -\int_{|\tau_1|+|\xi|\geq 1/2} \frac{-e^{it\tau_1} + e^{it|\xi|}}{\tau_1 - |\xi|} \chi_{(0,\infty)}(\tau_1)\, a(|\tau_1| - |\xi|) \frac{\widehat{F}(\tau_1,\xi)}{2|\xi|}\, d\tau_1,$$

$$\widehat{w_{44}}(t,\xi)$$

$$= -\int_{|\tau_1|+|\xi|\geq 1/2} \frac{e^{it\tau_1} - e^{-it|\xi|}}{\tau_1 + |\xi|} \chi_{(-\infty,0)}(\tau_1)\, a(|\tau_1| - |\xi|) \frac{\widehat{F}(\tau_1,\xi)}{2|\xi|}\, d\tau_1.$$

We perform first the analysis for $w_{41}$ and $w_{42}$. Routine computations based on the previous formulae yield

$$\widehat{\varphi\, w_{41}}(\tau,\xi) = -\widehat{\varphi}(\tau - |\xi|)$$
$$\cdot \int_{|\tau_1|+|\xi|\geq 1/2} \frac{1 - \chi_{(0,\infty)}(\tau_1)\, a(|\tau_1| - |\xi|)}{\tau_1 - |\xi|} \frac{\widehat{F}(\tau_1,\xi)}{2|\xi|}\, d\tau_1, \qquad (2.77)$$

$$\widehat{\varphi\, w_{42}}(\tau,\xi) = \widehat{\varphi}(\tau + |\xi|)$$
$$\cdot \int_{|\tau_1|+|\xi|\geq 1/2} \frac{1 - \chi_{(-\infty,0)}(\tau_1)\, a(|\tau_1| - |\xi|)}{\tau_1 + |\xi|} \frac{\widehat{F}(\tau_1,\xi)}{2|\xi|}\, d\tau_1. \qquad (2.78)$$

If we denote

$$H(\tau_1,\xi) = \frac{1 - \chi_{(0,\infty)}(\tau_1)\, a(|\tau_1| - |\xi|)}{\tau_1 - |\xi|},$$

it follows that

$$H(-\tau_1,\xi) = -\frac{1 - \chi_{(-\infty,0)}(\tau_1)\, a(|\tau_1| - |\xi|)}{\tau_1 + |\xi|}.$$

Moreover, if we rely on the properties of $a$ and (2.75), we easily deduce

$$|H(\tau_1,\xi)| \lesssim \begin{cases} w_-^{-1}(\tau_1,\xi), & \text{for } \tau_1 > 0, \\ (|\tau_1| + |\xi|)^{-1}, & \text{for } \tau_1 \leq 0. \end{cases}$$

For $|\xi| \geq 1/8$, the previous bound implies

$$|\widehat{\varphi\, w_{41}}(\tau,\xi)| + |\widehat{\varphi\, w_{42}}(\tau,\xi)|$$
$$\lesssim \frac{|\widehat{\varphi}(\tau - |\xi|)| + |\widehat{\varphi}(\tau - |\xi|)|}{|\xi|} \int_{|\tau_1|+|\xi|\geq 1/2} w_-^{-1}(\tau_1,\xi)\, |\widehat{F}(\tau_1,\xi)|\, d\tau_1$$
$$\lesssim w_-^{-N}(\tau,\xi)\, |\xi|^{-1} \int_{|\tau_1|+|\xi|\geq 1/2} w_-^{-1}(\tau_1,\xi)\, |\widehat{F}(\tau_1,\xi)|\, d\tau_1,$$

for an arbitrarily large integer $N$. Therefore, using the Cauchy-Schwarz inequality, we obtain:

$$\int_{\{|\xi|\geq 1/8\}\times\mathbb{R}} w_+^{2s}(\tau,\xi)\, w_-^{2\theta}(\tau,\xi)\left(\left|\widehat{\varphi\, w_{41}}(\tau,\xi)\right|^2 + \left|\widehat{\varphi\, w_{42}}(\tau,\xi)\right|^2\right) d\tau\, d\xi$$

$$\lesssim \|F\|_{X^{s-1,\theta-1}}^2 \cdot \sup_{|\xi|\geq 1/8} |\xi|^{-2}\Bigg(\int_{\{|\tau_1|+|\xi|\geq 1/2\}\times\mathbb{R}} w_+^{2-2s}(\tau_1,\xi)$$

$$\cdot\, w_-^{-2\theta}(\tau_1,\xi)\, w_+^{2s}(\tau,\xi)\, w_-^{2(\theta-N)}(\tau,\xi)\, d\tau_1\, d\tau\Bigg).$$

For the above supremum, we argue that

$$w_-^{-1}(\tau_1,\xi) \lesssim \frac{w_+(\tau,\xi)}{w_+(\tau_1,\xi)} \lesssim w_-(\tau,\xi)$$

and, due to $|\xi|\geq 1/8$,

$$\frac{w_+(\tau,\xi)}{|\xi|} \lesssim w_-(\tau,\xi).$$

Based on $\theta > 1/2$ and $s+\theta > 3/2$, these estimates allow us to infer

$$|\xi|^{-2}\int_{\{|\tau_1|+|\xi|\geq 1/2\}\times\mathbb{R}}\left\{w_+^{2-2s}(\tau_1,\xi)\, w_-^{-2\theta}(\tau_1,\xi)\right.$$

$$\left. w_+^{2s}(\tau,\xi)\, w_-^{2(\theta-N)}(\tau,\xi)\right\}d\tau_1\, d\tau$$

$$\lesssim \int_{\{|\tau_1|+|\xi|\geq 1/2\}\times\mathbb{R}}\left\{w_-^{-2\theta}(\tau_1,\xi)\, w_-^{2(s+\theta-N)}(\tau,\xi)\right.$$

$$\left. +\, w_-^{2(1-s-\theta)}(\tau_1,\xi)\, w_-^{2(1+\theta-N)}(\tau,\xi)\right\}d\tau_1\, d\tau \lesssim 1,$$

if $N$ is sufficiently large. Hence,

$$\int_{\{|\xi|\geq 1/8\}\times\mathbb{R}} w_+^{2s}(\tau,\xi)\, w_-^{2\theta}(\tau,\xi)$$

$$\left(\left|\widehat{\varphi\, w_{41}}(\tau,\xi)\right|^2 + \left|\widehat{\varphi\, w_{42}}(\tau,\xi)\right|^2\right) d\tau\, d\xi \qquad (2.79)$$

$$\lesssim \|F\|_{X^{s-1,\theta-1}}^2.$$

Next, we discuss the case when $|\xi| < 1/8$. As $|\tau_1| + |\xi| \geq 1/2$, it follows that $|\tau_1| \geq 3/8$ and $||\tau_1| - |\xi|| > 1/4$. Therefore, in this setting, we have:

$$w_\pm(\tau,\xi) \simeq \langle\tau\rangle, \qquad w_\pm(\tau_1,\xi) \simeq |\tau_1|, \qquad a(|\tau_1|-|\xi|) = 0.$$

Combining (2.77) and (2.78), we derive

$$\widehat{\varphi\,(w_{41}+w_{42})}(\tau,\xi) = -\,\widehat{\varphi}(\tau-|\xi|)\int_{|\tau_1|+|\xi|\geq 1/2}\frac{\widehat{F}(\tau_1,\xi)}{\tau_1^2-|\xi|^2}\, d\tau_1$$

$$-\,\frac{\widehat{\varphi}(\tau-|\xi|)-\widehat{\varphi}(\tau+|\xi|)}{2|\xi|}\int_{|\tau_1|+|\xi|\geq 1/2}\frac{\widehat{F}(\tau_1,\xi)}{\tau_1+|\xi|}\, d\tau_1,$$

which implies

$$\left| \varphi\,\widehat{(w_{41} + w_{42})}(\tau, \xi) \right| \lesssim \langle \tau \rangle^{-N} \int_{|\tau_1| + |\xi| \geq 1/2} |\tau_1|^{-1} \left| \widehat{F}(\tau_1, \xi) \right| d\tau_1,$$

for an arbitrarily large integer $N$. Consequently, relying as before on $s + \theta > 3/2$ and the Cauchy-Schwarz inequality, we deduce

$$\int_{\{|\xi| < 1/8\} \times \mathbb{R}} w_+^{2s}(\tau, \xi)\, w_-^{2\theta}(\tau, \xi) \left| \varphi\,\widehat{(w_{41} + w_{42})}(\tau, \xi) \right|^2 d\tau\, d\xi$$

$$\lesssim \|F\|_{X^{s-1, \theta-1}}^2 \tag{2.80}$$

$$\cdot \sup_{|\xi| < 1/8} \int_{\{|\tau_1| + |\xi| \geq 1/2\} \times \mathbb{R}} |\tau_1|^{2(1-s-\theta)} \langle \tau \rangle^{2(s+\theta-N)} d\tau_1\, d\tau$$

$$\lesssim \|F\|_{X^{s-1, \theta-1}}^2,$$

if $N$ is sufficiently large.

In conclusion, based on (2.79) and (2.80), we obtain

$$\|\varphi\,(w_{41} + w_{42})\|_{X^{s, \theta}} \lesssim \|F\|_{X^{s-1, \theta-1}},$$

which finishes the analysis for $w_{41}$ and $w_{42}$.

All we have left to discuss is $w_{43}$ and $w_{44}$ from (2.76). Given that their formulae have very similar structures, we will only work on $w_{43}$ and we will leave the argument for $w_{44}$ to the interested reader. We start by rewriting (2.3) as

$$\widehat{w_{43}}(t, \xi)$$

$$= e^{it|\xi|} \int_{\{\tau_1 > 0; |\tau_1| + |\xi| \geq 1/2\}} \frac{e^{it(|\tau_1| - |\xi|)} - 1}{|\tau_1| - |\xi|}\, a(|\tau_1| - |\xi|)\, \frac{\widehat{F}(\tau_1, \xi)}{2|\xi|}\, d\tau_1$$

$$= e^{it|\xi|} \sum_{k=1}^{\infty} \frac{t^k}{k!}\, f_k(\xi),$$

where

$$f_k(\xi) = i^k \int_{\{\tau_1 > 0; |\tau_1| + |\xi| \geq 1/2\}} (|\tau_1| - |\xi|)^{k-1}\, a(|\tau_1| - |\xi|)\, \frac{\widehat{F}(\tau_1, \xi)}{2|\xi|}\, d\tau_1.$$

The commutation of the sum with the integral is motivated by the uniform convergence of the series, which is in turn due to the localization of $|\tau_1| - |\xi|$ by $a$ to $[-1/4, 1/4]$. Moreover, if we take into account that $\tau_1 > 0$ and $|\tau_1| + |\xi| \geq 1/2$, it follows that

$$\tau_1 = |\tau_1| \simeq |\xi| \gtrsim 1, \qquad w_+(\tau_1, \xi) \simeq |\xi|, \qquad w_-(\tau_1, \xi) \simeq 1.$$

As a consequence, we derive, by applying the Cauchy-Schwarz inequality, that

$$|f_k(\xi)| \lesssim \langle\xi\rangle^{-s} \left( \int_{\mathbb{R}} w_+^{2s-2}(\tau_1,\xi)\, w_-^{2\theta-2}(\tau_1,\xi) \left|\widehat{F}(\tau_1,\xi)\right| d\tau_1 \right)^{1/2}. \quad (2.81)$$

A simple computation yields

$$\widehat{\varphi\, w_{43}}(\tau,\xi) = \sum_{k=1}^{\infty} \frac{1}{k!}\, \widehat{\varphi_k}(\tau - |\xi|)\, f_k(\xi), \quad (2.82)$$

with $\varphi_k(t) = t^k\, \varphi(t)$. In order to obtain the desired estimate for $\varphi\, w_{43}$, we claim that it is enough to show that, for all $k \geq 1$,

$$|\widehat{\varphi_k}(\eta)| \lesssim \|\varphi\|_{H^{N,1}} k^N \langle R\rangle^k \langle\eta\rangle^{-N} \quad (2.83)$$

holds uniformly for $\eta \in \mathbb{R}$, where $N$ is an arbitrarily large integer and $R > 0$ satisfies $\operatorname{supp}\varphi \subset [-R, R]$. Indeed, using consequently (2.82), (2.64), (2.83), and (2.81), we deduce

$$\|\varphi\, w_{43}\|_{X^{s,\theta}}$$

$$\lesssim \sum_{k=1}^{\infty} \frac{1}{k!} \left( \int_{\mathbb{R}^{n+1}} w_+^{2s}(\tau,\xi)\, w_-^{2\theta}(\tau,\xi)\, |\widehat{\varphi_k}(\tau-|\xi|)\, f_k(\xi)|^2\, d\tau\, d\xi \right)^{1/2}$$

$$\lesssim \sum_{k=1}^{\infty} \frac{1}{k!} \left( \int_{\mathbb{R}^{n+1}} \langle\xi\rangle^{2s}\, w_-^{2(|s|+\theta)}(\tau,\xi)\, |\widehat{\varphi_k}(\tau-|\xi|)\, f_k(\xi)|^2\, d\tau\, d\xi \right)^{1/2}$$

$$\lesssim \|\varphi\|_{H^{N,1}} \sum_{k=1}^{\infty} \frac{k^N \langle R\rangle^k}{k!} \left( \int_{\mathbb{R}^{n+1}} \langle\xi\rangle^{2s}\, w_-^{2(|s|+\theta-N)}(\tau,\xi)\, |f_k(\xi)|^2\, d\tau\, d\xi \right)^{1/2}$$

$$\lesssim \sum_{k=1}^{\infty} \frac{k^N \langle R\rangle^k}{k!} \|\varphi\|_{H^{N,1}} \|w_-^{|s|+\theta-N}\|_{L_\xi^\infty L_\tau^2} \|F\|_{X^{s-1,\theta-1}} \lesssim \|F\|_{X^{s-1,\theta-1}},$$

if $N$ is sufficiently large. Therefore, to conclude the proof of this proposition, we just need to provide an argument for (2.83). However,

$$(i\eta)^N\, \widehat{\varphi_k}(\eta) = (-1)^N \int_{\mathbb{R}} \partial_t^N \left(e^{-it\eta}\right) t^k\, \varphi(t)\, dt = \int_{\mathbb{R}} e^{-it\eta}\, \partial_t^N \left(t^k\, \varphi(t)\right) dt$$

and the claim follows by applying the product rule and then estimating each resulting term. $\square$

Our final goal in this section is to prove product estimates in $X^{s,\theta}$ spaces, which will be useful later in establishing the local existence theory of wave maps. We follow the approach in D'Ancona-Georgiev [36] and, unlike in the other results obtained so far, we assume $n \geq 2$. This is needed in order to be able to work with the following integration result, whose detailed proof can also be found in [36].

**Lemma 2.6.** *Let $n \geq 2$ and $s > n/2$. If $\tau > |\xi|$, then*

$$\int_{|\xi_1|+|\xi-\xi_1|=\tau} \langle \xi_1 \rangle^{-2s} + \langle \xi - \xi_1 \rangle^{-2s} \, dS_{\xi_1} \lesssim w_-^{n-2s-1}(\tau, \xi). \tag{2.84}$$

*If $|\xi| > \tau > 0$, then*

$$\int_{|\xi-\xi_1|-|\xi_1|=\tau} \langle \xi_1 \rangle^{-2s} + \langle \xi - \xi_1 \rangle^{-2s} \, dS_{\xi_1} \lesssim w_-^{n-2s-1}(\tau, \xi). \tag{2.85}$$

A simple observation is that (2.85) is also true for $|\xi| > |\tau| > 0$. Indeed, for $\tau < 0$ we can make the change of variables $(\tau, \xi, \xi_1) \mapsto (-\tau, \xi, \xi - \tilde{\xi}_1)$ to see that the integral transforms into one that can be estimated by (2.85).

We will also rely on:

**Lemma 2.7.** *If $n \geq 1$ and $(\tau_1, \xi_1)$, $(\tau_2, \xi_2) \in \mathbb{R}^{n+1}$, then*

$$||\tau_1 + \tau_2| - |\xi_1 + \xi_2|| \leq ||\tau_1| - |\xi_1|| + ||\tau_2| - |\xi_2|| + r(\xi_1, \xi_2), \tag{2.86}$$

*where*

$$r(\xi_1, \xi_2)$$
$$\stackrel{\text{def}}{=} \begin{cases} |\xi_1| + |\xi_2| - |\xi_1 + \xi_2|, & \text{if } \tau_1 \tau_2 \geq 0, \\ |\xi_1 + \xi_2| + |\xi_2| - |\xi_1|, & \text{if } \tau_1 \tau_2 < 0, \ |\tau_1| > |\tau_2|, \\ |\xi_1 + \xi_2| + |\xi_1| - |\xi_2|, & \text{if } \tau_1 \tau_2 < 0, \ |\tau_1| \leq |\tau_2|. \end{cases} \tag{2.87}$$

**Proof.** The bound (2.86) is a direct consequence of the triangle inequality and of a case-by-case analysis in which $|\tau_1 + \tau_2|$ is expressed in terms of $|\tau_1|$ and $|\tau_2|$. □

Our first $X^{s,\theta}$ product result is the following one.

**Theorem 2.11.** *If $n \geq 2$, $s - n/2 \geq \theta - 1/2 > 0$, then the space $X^{s,\theta}$ is an algebra, i.e.,*

$$\|u_1 u_2\|_{X^{s,\theta}(\mathbb{R}^{n+1})} \lesssim \|u_1\|_{X^{s,\theta}(\mathbb{R}^{n+1})} \|u_2\|_{X^{s,\theta}(\mathbb{R}^{n+1})}. \tag{2.88}$$

**Proof.** First, we make the substitution

$$\widehat{U_i}(\tau, \xi) = w_+^s(\tau, \xi) \, w_-^\theta(\tau, \xi) \, \widehat{u}_i(\tau, \xi), \quad i = 1, 2,$$

and we use the notation

$$\tau_2 \stackrel{\text{def}}{=} \tau - \tau_1, \qquad \xi_2 \stackrel{\text{def}}{=} \xi - \xi_1,$$

for symmetry reasons, to rewrite (2.88) as

$$\left\| \int_{\mathbb{R}^{n+1}} \frac{w_+^s(\tau,\xi)\, w_-^\theta(\tau,\xi)}{w_+^s(\tau_1,\xi_1)\, w_-^\theta(\tau_1,\xi_1)\, w_+^s(\tau_2,\xi_2)\, w_-^\theta(\tau_2,\xi_2)} \right. \\ \left. \cdot \left|\widehat{U_1}(\tau_1,\xi_1)\right| \left|\widehat{U_2}(\tau_2,\xi_2)\right| d\tau_1\, d\xi_1 \right\|_{L^2_{\tau,\xi}} \tag{2.89}$$

$$\lesssim \|U_1\|_{L^2} \|U_2\|_{L^2}.$$

Next, we take advantage of the symmetry of the integrand in $(\tau_1, \xi_1)$ and $(\tau_2, \xi_2)$ to reduce the proof to the case when $|\tau_1| + |\xi_1| \geq |\tau_2| + |\xi_2|$. Due to $s \geq 0$, it follows by the triangle inequality that

$$w_+^s(\tau,\xi) \lesssim w_+^s(\tau_1,\xi_1).$$

Moreover, based on $\theta \geq 0$ and (2.86), we obtain

$$w_-^\theta(\tau,\xi) \leq \left(w_-(\tau_1,\xi_1) + w_-(\tau_2,\xi_2) + r(\xi_1,\xi_2)\right)^\theta \\ \lesssim w_-^\theta(\tau_1,\xi_1) + w_-^\theta(\tau_2,\xi_2) + r^\theta(\xi_1,\xi_2). \tag{2.90}$$

These considerations allow us to further reduce the argument for (2.89) to the proof of the following three estimates:

$$\left\| \int_{\mathbb{R}^{n+1}} w_+^{-s}(\tau_2,\xi_2) w_-^{-\theta}(\tau_2,\xi_2) \left|\widehat{U_1}(\tau_1,\xi_1)\right| \left|\widehat{U_2}(\tau_2,\xi_2)\right| d\tau_1 d\xi_1 \right\|_{L^2_{\tau,\xi}} \tag{2.91}$$

$$\lesssim \|U_1\|_{L^2} \|U_2\|_{L^2},$$

$$\left\| \int_{\mathbb{R}^{n+1}} w_-^{-\theta}(\tau_1,\xi_1) w_+^{-s}(\tau_2,\xi_2) \left|\widehat{U_1}(\tau_1,\xi_1)\right| \left|\widehat{U_2}(\tau_2,\xi_2)\right| d\tau_1 d\xi_1 \right\|_{L^2_{\tau,\xi}} \tag{2.92}$$

$$\lesssim \|U_1\|_{L^2} \|U_2\|_{L^2},$$

$$\left\| \int_{\mathbb{R}^{n+1}} \frac{r^\theta(\xi_1,\xi_2)}{w_-^\theta(\tau_1,\xi_1) w_+^s(\tau_2,\xi_2) w_-^\theta(\tau_2,\xi_2)} \right. \\ \left. \cdot \left|\widehat{U_1}(\tau_1,\xi_1)\right| \left|\widehat{U_2}(\tau_2,\xi_2)\right| d\tau_1 d\xi_1 \right\|_{L^2_{\tau,\xi}} \lesssim \|U_1\|_{L^2} \|U_2\|_{L^2}. \tag{2.93}$$

For the first inequality, we derive, based on (2.18) and the Cauchy-Schwarz inequality,

$$
\left\| \int_{\mathbb{R}^{n+1}} w_+^{-s}(\tau_2, \xi_2)\, w_-^{-\theta}(\tau_2, \xi_2)\, \left| \widehat{U_1}(\tau_1, \xi_1) \right|\, \left| \widehat{U_2}(\tau_2, \xi_2) \right|\, d\tau_1\, d\xi_1 \right\|_{L^2}
$$

$$
\lesssim \left\| \widehat{U_1} \right\|_{L^2} \left\| w_+^{-s} w_-^{-\theta}\, \widehat{U_2} \right\|_{L^1}
$$

$$
\lesssim \left\| \widehat{U_1} \right\|_{L^2} \left\| w_+^{-s} w_-^{-\theta} \right\|_{L^2} \left\| \widehat{U_2} \right\|_{L^2}
$$

$$
\lesssim \| U_1 \|_{L^2} \| U_2 \|_{L^2},
$$

where the last step is motivated by $s > n/2$ and $\theta > 1/2$.

For (2.92), a very similar approach leads to

$$
\left\| \int_{\mathbb{R}^{n+1}} w_-^{-\theta}(\tau_1, \xi_1)\, w_+^{-s}(\tau_2, \xi_2)\, \left| \widehat{U_1}(\tau_1, \xi_1) \right|\, \left| \widehat{U_2}(\tau_2, \xi_2) \right|\, d\tau_1\, d\xi_1 \right\|_{L^2_{\tau, \xi}}
$$

$$
\lesssim \left\| w_-^{-\theta}\, \widehat{U_1} \right\|_{L^2_\xi L^1_\tau} \left\| w_+^{-s}\, \widehat{U_2} \right\|_{L^1_\xi L^2_\tau}
$$

$$
\lesssim \left\| w_-^{-\theta} \right\|_{L^\infty_\xi L^2_\tau} \left\| \widehat{U_1} \right\|_{L^2} \left\| w_+^{-s} \right\|_{L^2_\xi L^\infty_\tau} \left\| \widehat{U_2} \right\|_{L^2}
$$

$$
\lesssim \| U_1 \|_{L^2} \| U_2 \|_{L^2}.
$$

Therefore, we are left to prove (2.93), which is the more intricate part of the argument. Using the positivity of the integrand and the Cauchy-Schwarz inequality, we infer

$$
\left\| \int_{\mathbb{R}^{n+1}} \frac{r^\theta(\xi_1, \xi_2)}{w_-^\theta(\tau_1, \xi_1) w_+^s(\tau_2, \xi_2)\, w_-^\theta(\tau_2, \xi_2)} \left| \widehat{U_1}(\tau_1, \xi_1) \right|\, \left| \widehat{U_2}(\tau_2, \xi_2) \right|\, d\tau_1\, d\xi_1 \right\|_{L^2_{\tau, \xi}}
$$

$$
\lesssim \left\| \left( \int_{\mathbb{R}^{n+1}} \frac{r^{2\theta}(\xi_1, \xi_2)}{w_-^{2\theta}(\tau_1, \xi_1) w_+^{2s}(\tau_2, \xi_2)\, w_-^{2\theta}(\tau_2, \xi_2)}\, d\tau_1\, d\xi_1 \right)^{1/2} \right.
$$

$$
\left. \cdot \left( \int_{\mathbb{R}^{n+1}} \left| \widehat{U_1}(\tau_1, \xi_1) \right|^2 \left| \widehat{U_2}(\tau_2, \xi_2) \right|^2 d\tau_1\, d\xi_1 \right)^{1/2} \right\|_{L^2_{\tau, \xi}}
$$

$$
\lesssim \left\| \int_{\mathbb{R}^{n+1}} \frac{r^{2\theta}(\xi_1, \xi_2)}{w_-^{2\theta}(\tau_1, \xi_1) w_+^{2s}(\tau_2, \xi_2)\, w_-^{2\theta}(\tau_2, \xi_2)}\, d\tau_1\, d\xi_1 \right\|_{L^\infty_{\tau, \xi}}^{1/2} \| U_1 \|_{L^2} \| U_2 \|_{L^2}.
$$

As a consequence, (2.93) follows if we show that

$$
\left\| \int_{\mathbb{R}^{n+1}} \frac{r^{2\theta}(\xi_1, \xi_2)}{w_-^{2\theta}(\tau_1, \xi_1) w_+^{2s}(\tau_2, \xi_2)\, w_-^{2\theta}(\tau_2, \xi_2)}\, d\tau_1\, d\xi_1 \right\|_{L^\infty_{\tau, \xi}} \lesssim 1.
$$

The strategy we take in proving the previous bound is to split the integral with respect to $\tau_1$ into four integrals whose domains are given by $\{\tau_1 > 0,\ \tau_2 > 0\}$, $\{\tau_1 < 0,\ \tau_2 < 0\}$, $\{\tau_1 > 0,\ \tau_2 < 0\}$, and $\{\tau_1 < 0,\ \tau_2 > 0\}$. This is motivated by the integrand being nonnegative and having the $(\tau_1, \tau_2) \mapsto (-\tau_1, -\tau_2)$ symmetry, which allows us to discuss only the $\{\tau_1 > 0,\ \tau_2 > 0\}$ and $\{\tau_1 > 0,\ \tau_2 < 0\}$ cases.

For the integral whose domain is $\{\tau_1 > 0,\ \tau_2 > 0\}$, we first make the change of variable $\tau_1 \mapsto \tilde{\tau}_1 = \tau_1 - |\xi_1|$ and then we foliate $\mathbb{R}^n$ by the ellipsoids $|\xi_1| + |\xi_2| = \eta$, with $\eta \geq |\xi|$. It implies that

$$r(\xi_1, \xi_2) = \eta - |\xi|, \quad w_-(\tau_1, \xi_1) \simeq \langle \tilde{\tau}_1 \rangle, \quad w_-(\tau_2, \xi_2) \simeq \langle \tau - \tilde{\tau}_1 - \eta \rangle.$$

Using also that $w_+(\tau_2, \xi_2) \geq \langle \xi_2 \rangle$, $s, \theta \geq 0$, we deduce

$$\int_{\mathbb{R}^n} \int_{\{\tau_1 > 0, \tau_2 > 0\}} \frac{r^{2\theta}(\xi_1, \xi_2)}{w_-^{2\theta}(\tau_1, \xi_1) w_+^{2s}(\tau_2, \xi_2) w_-^{2\theta}(\tau_2, \xi_2)} \, d\tau_1 \, d\xi_1$$
$$\lesssim \int_{|\xi|}^{\infty} \int_{|\xi_1| + |\xi_2| = \eta} \int_{\mathbb{R}} \frac{\langle \eta - |\xi| \rangle^{2\theta}}{\langle \tilde{\tau}_1 \rangle^{2\theta} \langle \xi_2 \rangle^{2s} \langle \tau - \tilde{\tau}_1 - \eta \rangle^{2\theta}} \, d\tilde{\tau}_1 \, dS_{\xi_1} \, d\eta.$$

For the innermost integral, we employ the straightforward bound

$$\langle \tilde{\tau}_1 \rangle^{2\theta} + \langle \tau - \tilde{\tau}_1 - \eta \rangle^{2\theta} \gtrsim \langle \tau - \eta \rangle^{2\theta}$$

and $\theta > 1/2$ to derive

$$\int_{\mathbb{R}} \langle \tilde{\tau}_1 \rangle^{-2\theta} \langle \tau - \tilde{\tau}_1 - \eta \rangle^{-2\theta} \, d\tilde{\tau}_1 \lesssim \langle \tau - \eta \rangle^{-2\theta}. \tag{2.94}$$

For the surface integral, we rely on $s > n/2$ and (2.84) to obtain

$$\int_{|\xi_1| + |\xi_2| = \eta} \langle \xi_2 \rangle^{-2s} \, dS_{\xi_1} \lesssim w_-^{n-2s-1}(\eta, \xi) \simeq \langle \eta - |\xi| \rangle^{n-2s-1}.$$

Therefore, we deduce

$$\int_{\mathbb{R}^n} \int_{\{\tau_1 > 0, \tau_2 > 0\}} \frac{r^{2\theta}(\xi_1, \xi_2)}{w_-^{2\theta}(\tau_1, \xi_1) w_+^{2s}(\tau_2, \xi_2) w_-^{2\theta}(\tau_2, \xi_2)} \, d\tau_1 \, d\xi_1$$
$$\lesssim \int_{|\xi|}^{\infty} \langle \eta - |\xi| \rangle^{n+2\theta-2s-1} \langle \tau - \eta \rangle^{-2\theta} \, d\eta.$$

Given that the hypothesis implies $n + 2\theta - 2s - 1 \leq 0$, we can easily estimate the last integrand by

$$\langle \eta - |\xi| \rangle^{n+2\theta-2s-1} \langle \tau - \eta \rangle^{-2\theta} \leq \langle \eta - |\xi| \rangle^{n-2s-1} + \langle \tau - \eta \rangle^{n-2s-1}.$$

This yields

$$\int_{\mathbb{R}^n} \int_{\{\tau_1 > 0, \tau_2 > 0\}} \frac{r^{2\theta}(\xi_1, \xi_2)}{w_-^{2\theta}(\tau_1, \xi_1) w_+^{2s}(\tau_2, \xi_2) w_-^{2\theta}(\tau_2, \xi_2)} \, d\tau_1 \, d\xi_1$$
$$\lesssim \int_{\mathbb{R}} \langle \eta - |\xi| \rangle^{n-2s-1} + \langle \tau - \eta \rangle^{n-2s-1} \, d\eta \lesssim 1,$$

due to $s > n/2$.

We discuss now the integral whose domain is $\{\tau_1 > 0,\ \tau_2 < 0\}$. As for the previous one, we make the change of variable $\tau_1 \mapsto \tilde{\tau}_1 = \tau_1 - |\xi_1|$, but then we foliate $\mathbb{R}^n$ by the hyperboloids $|\xi_1| - |\xi_2| = \eta$, with $|\eta| \leq |\xi|$. It follows that

$$r(\xi_1, \xi_2) = |\xi| - (\operatorname{sgn}\tau)\eta, \quad w_-(\tau_1, \xi_1) \simeq \langle\tilde{\tau}_1\rangle, \quad w_-(\tau_2, \xi_2) \simeq \langle\tau - \tilde{\tau}_1 - \eta\rangle.$$

Arguing as before, we deduce

$$\int_{\mathbb{R}^n}\int_{\{\tau_1>0,\tau_2<0\}} \frac{r^{2\theta}(\xi_1,\xi_2)}{w_-^{2\theta}(\tau_1,\xi_1)w_+^{2s}(\tau_2,\xi_2)\,w_-^{2\theta}(\tau_2,\xi_2)}\,d\tau_1\,d\xi_1$$

$$\lesssim \int_{-|\xi|}^{|\xi|}\int_{|\xi_1|-|\xi_2|=\eta}\int_{\mathbb{R}} \frac{(|\xi|-(\operatorname{sgn}\tau)\eta)^{2\theta}}{\langle\tilde{\tau}_1\rangle^{2\theta}\langle\xi_2\rangle^{2s}\langle\tau-\tilde{\tau}_1-\eta\rangle^{2\theta}}\,d\tilde{\tau}_1\,dS_{\xi_1}\,d\eta.$$

The innermost integral is handled by (2.94), whereas for the surface integral, we use (2.85) and the subsequent observation. This allows us to infer

$$\int_{|\xi_1|-|\xi_2|=\eta} \langle\xi_2\rangle^{-2s}\,dS_{\xi_1} \lesssim w_-^{n-2s-1}(\eta,\xi).$$

Hence,

$$\int_{\mathbb{R}^n}\int_{\{\tau_1>0,\tau_2<0\}} \frac{r^{2\theta}(\xi_1,\xi_2)}{w_-^{2\theta}(\tau_1,\xi_1)w_+^{2s}(\tau_2,\xi_2)\,w_-^{2\theta}(\tau_2,\xi_2)}\,d\tau_1\,d\xi_1$$

$$\lesssim \int_{-|\xi|}^{|\xi|} \frac{(|\xi|-(\operatorname{sgn}\tau)\eta)^{2\theta}}{w_-^{2s-n+1}(\eta,\xi)\,\langle\tau-\eta\rangle^{2\theta}}\,d\eta. \tag{2.95}$$

If either $\{\tau > 0$ and $\eta \in [-1/2, |\xi|]\}$ or $\{\tau < 0$ and $\eta \in [-|\xi|, 1/2]\}$, then

$$|\xi| - (\operatorname{sgn}\tau)\eta \leq w_-(\eta,\xi),$$

which implies

$$\frac{(|\xi|-(\operatorname{sgn}\tau)\eta)^{2\theta}}{w_-^{2s-n+1}(\eta,\xi)\,\langle\tau-\eta\rangle^{2\theta}} \leq w_-^{n+2\theta-2s-1}(\eta,\xi)\,\langle\tau-\eta\rangle^{-2\theta}$$

$$\leq w_-^{n-2s-1}(\eta,\xi) + \langle\tau-\eta\rangle^{n-2s-1}.$$

Therefore, unless $|\xi| \geq 1/2$,

$$\int_{-|\xi|}^{|\xi|} \frac{(|\xi|-(\operatorname{sgn}\tau)\eta)^{2\theta}}{w_-^{2s-n+1}(\eta,\xi)\,\langle\tau-\eta\rangle^{2\theta}}\,d\eta \lesssim 1, \tag{2.96}$$

due to $s > n/2$. If $|\xi| \geq 1/2$, we are left to estimate

$$\int_{-|\xi|}^{-1/2} \frac{(|\xi|-\eta)^{2\theta}}{w_-^{2s-n+1}(\eta,\xi)\,\langle\tau-\eta\rangle^{2\theta}}\,d\eta \qquad \text{for} \quad \tau > 0,$$

and

$$\int_{1/2}^{|\xi|} \frac{(|\xi| + \eta)^{2\theta}}{w_-^{2s-n+1}(\eta, \xi) \langle \tau - \eta \rangle^{2\theta}} \, d\eta \qquad \text{for} \quad \tau < 0.$$

However, if we make the change of variable $\eta \mapsto \tilde{\eta} = -\eta$ in the first integral, it transforms itself into the second one with $\tau$ replaced by $-\tau < 0$. This is why we analyze only the second integral. We notice first that

$$|\xi| + \eta \simeq |\xi|, \qquad w_-(\eta, \xi) = 1 + |\xi| - \eta, \qquad \langle \tau - \eta \rangle \simeq 1 + \eta + |\tau|.$$

If $\eta \geq (|\xi| - |\tau|)/2$, then $1 + \eta + |\tau| \gtrsim |\xi|$, which implies

$$\int_{\max\{1/2,(|\xi|-|\tau|)/2\}}^{|\xi|} \frac{(|\xi| + \eta)^{2\theta}}{w_-^{2s-n+1}(\eta, \xi) \langle \tau - \eta \rangle^{2\theta}} \, d\eta$$

$$\lesssim \int_{\max\{1/2,(|\xi|-|\tau|)/2\}}^{|\xi|} w_-^{n-2s-1}(\eta, \xi) \, d\eta \lesssim 1.$$

If $1/2 \leq \eta \leq \max\{1/2, (|\xi| - |\tau|)/2\}$, then $w_-(\eta, \xi) \simeq |\xi|$, and we derive

$$\int_{1/2}^{\max\{1/2,(|\xi|-|\tau|)/2\}} \frac{(|\xi| + \eta)^{2\theta}}{w_-^{2s-n+1}(\eta, \xi) \langle \tau - \eta \rangle^{2\theta}} \, d\eta$$

$$\lesssim |\xi|^{n+2\theta-2s-1} \int_{1/2}^{\max\{1/2,(|\xi|-|\tau|)/2\}} \langle \tau - \eta \rangle^{-2\theta} \, d\eta \lesssim 1,$$

based on $n + 2\theta - 2s - 1 < 0$ and $\theta > 1/2$. Together with (2.95) and (2.96), this analysis shows that

$$\int_{\mathbb{R}^n} \int_{\{\tau_1 > 0, \tau_2 < 0\}} \frac{r^{2\theta}(\xi_1, \xi_2)}{w_-^{2\theta}(\tau_1, \xi_1) w_+^{2s}(\tau_2, \xi_2) \, w_-^{2\theta}(\tau_2, \xi_2)} \, d\tau_1 \, d\xi_1 \lesssim 1,$$

which finishes the proof of (2.88). $\qquad \square$

Finally, we discuss a second product result in $X^{s,\theta}$ spaces, which will be particularly useful in estimating nonlinearities of the wave maps equation.

**Theorem 2.12.** *If $n \geq 2$ and $s - n/2 \geq \theta - 1/2 > 0$, then*

$$\|u_1 u_2\|_{X^{s-1,\theta-1}(\mathbb{R}^{n+1})} \lesssim \|u_1\|_{X^{s,\theta}(\mathbb{R}^{n+1})} \|u_2\|_{X^{s-1,\theta-1}(\mathbb{R}^{n+1})}. \qquad (2.97)$$

**Proof.** The argument has many similarities with the one for (2.88). In fact, some steps here will be motivated by facts proven in the previous theorem.

This is why we adopt the notational conventions used before, which allow us to recast (2.97) as

$$\left\| \int_{\mathbb{R}^{n+1}} \frac{w_+^{s-1}(\tau,\xi)\, w_-^{\theta-1}(\tau,\xi)}{w_+^s(\tau_1,\xi_1)\, w_-^\theta(\tau_1,\xi_1)\, w_+^{s-1}(\tau_2,\xi_2)\, w_-^{\theta-1}(\tau_2,\xi_2)} \right.$$
$$\left. \cdot \left|\widehat{U_1}(\tau_1,\xi_1)\right| \left|\widehat{U_2}(\tau_2,\xi_2)\right| d\tau_1\, d\xi_1 \right\|_{L^2_{\tau,\xi}} \lesssim \|U_1\|_{L^2} \|U_2\|_{L^2}. \tag{2.98}$$

Due to the lack of symmetry of the integrand with respect to $(\tau_1,\xi_1)$ and $(\tau_2,\xi_2)$, we have to treat the cases $|\tau_1| + |\xi_1| < |\tau_2| + |\xi_2|$ and $|\tau_1| + |\xi_1| \geq |\tau_2| + |\xi_2|$, respectively, separately.

For the former, as $s > n/2 \geq 1$ and $\theta > 1/2$, it follows that

$$w_+^{s-1}(\tau,\xi) \lesssim w_+^{s-1}(\tau_2,\xi_2), \tag{2.99}$$

$$\frac{w_-^{\theta-1}(\tau,\xi)}{w_-^{\theta-1}(\tau_2,\xi_2)} \leq \frac{w_-^\theta(\tau,\xi)}{w_-^\theta(\tau_2,\xi_2)} + \frac{w_-^\theta(\tau_2,\xi_2)}{w_-^\theta(\tau,\xi)}, \tag{2.100}$$

which reduces the proof of (2.98) to the one for

$$\left\| \int_{\mathbb{R}^{n+1}} \frac{w_-^\theta(\tau,\xi)}{w_+^s(\tau_1,\xi_1)\, w_-^\theta(\tau_1,\xi_1)\, w_-^\theta(\tau_2,\xi_2)} \right.$$
$$\left. \cdot \left|\widehat{U_1}(\tau_1,\xi_1)\right| \left|\widehat{U_2}(\tau_2,\xi_2)\right| d\tau_1\, d\xi_1 \right\|_{L^2_{\tau,\xi}} \lesssim \|U_1\|_{L^2} \|U_2\|_{L^2} \tag{2.101}$$

and

$$\left\| \int_{\mathbb{R}^{n+1}} \frac{w_-^\theta(\tau_2,\xi_2)}{w_+^s(\tau_1,\xi_1)\, w_-^\theta(\tau_1,\xi_1)\, w_-^\theta(\tau,\xi)} \right.$$
$$\left. \cdot \left|\widehat{U_1}(\tau_1,\xi_1)\right| \left|\widehat{U_2}(\tau_2,\xi_2)\right| d\tau_1\, d\xi_1 \right\|_{L^2_{\tau,\xi}} \lesssim \|U_1\|_{L^2} \|U_2\|_{L^2}. \tag{2.102}$$

Through the change of variables $(\tau_1,\xi_1) \mapsto (\tau-\tau_1, \xi-\xi_1)$, the first inequality becomes

$$\left\| \int_{\mathbb{R}^{n+1}} \frac{w_-^\theta(\tau,\xi)}{w_+^s(\tau_2,\xi_2)\, w_-^\theta(\tau_2,\xi_2)\, w_-^\theta(\tau_1,\xi_1)} \right.$$
$$\left. \cdot \left|\widehat{U_1}(\tau_2,\xi_2)\right| \left|\widehat{U_2}(\tau_1,\xi_1)\right| d\tau_1\, d\xi_1 \right\|_{L^2_{\tau,\xi}} \lesssim \|U_1\|_{L^2} \|U_2\|_{L^2},$$

which follows as a result of (2.90)-(2.93) by interchanging the roles of $U_1$ and $U_2$. Moreover, we show that the second inequality can be derived from the first one. For that, we use the self-duality of $L^2$ to claim that (2.102) is equivalent to

$$\left| \int_{\mathbb{R}^{n+1}} \int_{\mathbb{R}^{n+1}} \frac{w_-^\theta(\tau_2, \xi_2)}{w_+^s(\tau_1, \xi_1) \, w_-^\theta(\tau_1, \xi_1) \, w_-^\theta(\tau, \xi)} \right.$$

$$\left. \cdot |V_1(\tau_1, \xi_1)| \, |V_2(\tau_2, \xi_2)| \, |V(\tau, \xi)| \, d\tau_1 \, d\xi_1 \, d\tau \, d\xi \right| \lesssim \|V_1\|_{L^2} \|V_2\|_{L^2} \|V\|_{L^2}.$$

Next, making the change of variables $(\tau_1, \xi_1) \mapsto (-\tau_1, -\xi_1)^4$ in the inner integral and relying on $\tilde{V}_1(\tau_1, \xi_1) = V_1(-\tau_1, -\xi_1)$, we can rewrite the previous estimate as

$$\left| \int_{\mathbb{R}^{n+1}} \int_{\mathbb{R}^{n+1}} \frac{w_-^\theta(\tau + \tau_1, \xi + \xi_1)}{w_+^s(\tau_1, \xi_1) \, w_-^\theta(\tau_1, \xi_1) \, w_-^\theta(\tau, \xi)} \, \left| \tilde{V}_1(\tau_1, \xi_1) \right| \right.$$

$$\left. \cdot |V_2(\tau + \tau_1, \xi + \xi_1)| \, |V(\tau, \xi)| \, d\tau_1 \, d\xi_1 \, d\tau \, d\xi \right| \lesssim \|\tilde{V}_1\|_{L^2} \|V_2\|_{L^2} \|V\|_{L^2}.$$

Finally, we use Tonelli's theorem twice and, in between, perform the change of variables $(\tau_1, \xi_1, \tau, \xi) \mapsto (\tau_1, \xi_1, \tau + \tau_1, \xi + \xi_1)$ to further transform it into

$$\left| \int_{\mathbb{R}^{n+1}} \int_{\mathbb{R}^{n+1}} \frac{w_-^\theta(\tau, \xi)}{w_+^s(\tau_1, \xi_1) \, w_-^\theta(\tau_1, \xi_1) \, w_-^\theta(\tau - \tau_1, \xi - \xi_1)} \, \left| \tilde{V}_1(\tau_1, \xi_1) \right| \right.$$

$$\left. \cdot |V_2(\tau, \xi)| \, |V(\tau - \tau_1, \xi - \xi_1)| \, d\tau_1 \, d\xi_1 \, d\tau \, d\xi \right| \lesssim \|\tilde{V}_1\|_{L^2} \|V_2\|_{L^2} \|V\|_{L^2}.$$

However, this estimate follows by duality from (2.101), through choosing $\widehat{U}_1 = \tilde{V}_1$ and $\widehat{U}_2 = V$. This finishes the discussion of the $|\tau_1| + |\xi_1| < |\tau_2| + |\xi_2|$ case.

If $|\tau_1| + |\xi_1| \geq |\tau_2| + |\xi_2|$, then

$$w_+^{s-1}(\tau, \xi) \lesssim w_+^{s-1}(\tau_1, \xi_1).$$

Together with (2.100), it reduces the argument for (2.98) to proving

$$\left\| \int_{\mathbb{R}^{n+1}} \frac{w_-^\theta(\tau, \xi)}{w_+(\tau_1, \xi_1) \, w_-^\theta(\tau_1, \xi_1) \, w_+^{s-1}(\tau_2, \xi_2) \, w_-^\theta(\tau_2, \xi_2)} \right.$$

$$\left. \cdot \left| \widehat{U}_1(\tau_1, \xi_1) \right| \left| \widehat{U}_2(\tau_2, \xi_2) \right| \, d\tau_1 \, d\xi_1 \right\|_{L^2_{\tau, \xi}} \lesssim \|U_1\|_{L^2} \|U_2\|_{L^2} \tag{2.103}$$

---

[4]More precisely, we take first $(\tau_1, \xi_1) \mapsto (\tilde{\tau}_1, \tilde{\xi}_1) = (-\tau_1, -\xi_1)$ and then reverse $(\tilde{\tau}_1, \tilde{\xi}_1)$ to the old $(\tau_1, \xi_1)$ notation. To follow the argument more easily, this is the notation caveat we will adopt from this point on.

and
$$\left\|\int_{\mathbb{R}^{n+1}} \frac{w_-^\theta(\tau_2,\xi_2)}{w_+(\tau_1,\xi_1)\, w_-^\theta(\tau_1,\xi_1)\, w_+^{s-1}(\tau_2,\xi_2)\, w_-^\theta(\tau,\xi)} \right.$$

$$\left. \cdot \left|\widehat{U_1}(\tau_1,\xi_1)\right| \left|\widehat{U_2}(\tau_2,\xi_2)\right| d\tau_1\, d\xi_1 \right\|_{L^2_{\tau,\xi}} \lesssim \|U_1\|_{L^2}\, \|U_2\|_{L^2}. \tag{2.104}$$

Based on $w_+(\tau_1,\xi_1) \geq w_+(\tau_2,\xi_2)$, the first inequality would follow from
$$\left\|\int_{\mathbb{R}^{n+1}} \frac{w_-^\theta(\tau,\xi)}{w_-^\theta(\tau_1,\xi_1)\, w_+^s(\tau_2,\xi_2)\, w_-^\theta(\tau_2,\xi_2)} \right.$$

$$\left. \cdot \left|\widehat{U_1}(\tau_1,\xi_1)\right| \left|\widehat{U_2}(\tau_2,\xi_2)\right| d\tau_1\, d\xi_1 \right\|_{L^2_{\tau,\xi}} \lesssim \|U_1\|_{L^2}\, \|U_2\|_{L^2},$$

which, in turn, is the result of applying (2.90)-(2.93).

For (2.104), we invoke as before the self-duality of $L^2$ to rewrite it as
$$\left| \int_{\mathbb{R}^{n+1}} \int_{\mathbb{R}^{n+1}} \frac{w_-^\theta(\tau_2,\xi_2)}{w_+(\tau_1,\xi_1)\, w_-^\theta(\tau_1,\xi_1)\, w_+^{s-1}(\tau_2,\xi_2)\, w_-^\theta(\tau,\xi)} \right.$$

$$\left. \cdot |V_1(\tau_1,\xi_1)|\, |V_2(\tau_2,\xi_2)|\, |V(\tau,\xi)|\, d\tau_1\, d\xi_1\, d\tau\, d\xi \right| \lesssim \|V_1\|_{L^2}\, \|V_2\|_{L^2}\, \|V\|_{L^2}.$$

Using Tonelli's theorem twice and, in between, performing the change of variables $(\tau_1,\xi_1,\tau,\xi) \mapsto (-\tau_1,-\xi_1,\tau-\tau_1,\xi-\xi_1)$, the previous estimate becomes
$$\left| \int_{\mathbb{R}^{n+1}} \int_{\mathbb{R}^{n+1}} \frac{w_-^\theta(\tau,\xi)}{w_+(\tau_1,\xi_1)\, w_-^\theta(\tau_1,\xi_1)\, w_+^{s-1}(\tau,\xi)\, w_-^\theta(\tau_2,\xi_2)} \right.$$

$$\left. \cdot \left|\tilde{V}_1(\tau_1,\xi_1)\right|\, |V_2(\tau,\xi)|\, |V(\tau_2,\xi_2)|\, d\tau_1\, d\xi_1\, d\tau\, d\xi \right| \tag{2.105}$$

$$\lesssim \|\tilde{V}_1\|_{L^2}\, \|V_2\|_{L^2}\, \|V\|_{L^2},$$

where $\tilde{V}_1(\tau_1,\xi_1) = V_1(-\tau_1,-\xi_1)$. In analyzing this inequality, we rely on
$$w_+(\tau_1,\xi_1) \geq w_+(\tau,\xi),$$
which can be traced through the change of variables to the original assumption $|\tau_1| + |\xi_1| \geq |\tau_2| + |\xi_2|$, and (2.90). Hence, we can derive (2.105) from
$$\left| \int_{\mathbb{R}^{n+1}} \int_{\mathbb{R}^{n+1}} w_+^{-s}(\tau,\xi)\, w_-^{-\theta}(\tau_2,\xi_2) \right.$$

$$\left. \cdot \left|\tilde{V}_1(\tau_1,\xi_1)\right|\, |V_2(\tau,\xi)|\, |V(\tau_2,\xi_2)|\, d\tau_1\, d\xi_1\, d\tau\, d\xi \right| \tag{2.106}$$

$$\lesssim \|\tilde{V}_1\|_{L^2}\, \|V_2\|_{L^2}\, \|V\|_{L^2},$$

$$\left| \int_{\mathbb{R}^{n+1}} \int_{\mathbb{R}^{n+1}} w_-^{-\theta}(\tau_1, \xi_1)\, w_+^{-s}(\tau, \xi) \right.$$

$$\left. \cdot \left| \tilde{V}_1(\tau_1, \xi_1) \right| \, |V_2(\tau, \xi)| \, |V(\tau_2, \xi_2)| \, d\tau_1 \, d\xi_1 \, d\tau \, d\xi \right| \qquad (2.107)$$

$$\lesssim \|\tilde{V}_1\|_{L^2} \|V_2\|_{L^2} \|V\|_{L^2},$$

and

$$\left| \int_{\mathbb{R}^{n+1}} \int_{A(\tau, \xi)} \frac{r^\theta(\xi_1, \xi_2)}{w_+(\tau_1, \xi_1)\, w_-^\theta(\tau_1, \xi_1)\, w_+^{s-1}(\tau, \xi)\, w_-^\theta(\tau_2, \xi_2)} \right.$$

$$\left. \cdot \left| \tilde{V}_1(\tau_1, \xi_1) \right| \, |V_2(\tau, \xi)| \, |V(\tau_2, \xi_2)| \, d\tau_1 \, d\xi_1 \, d\tau \, d\xi \right| \qquad (2.108)$$

$$\lesssim \|\tilde{V}_1\|_{L^2} \|V_2\|_{L^2} \|V\|_{L^2},$$

where

$$A(\tau, \xi) = \left\{ (\tau_1, \xi_1) \in \mathbb{R}^{n+1};\ r(\xi_1, \xi_2) \geq 2(w_-(\tau_1, \xi_1) + w_-(\tau_2, \xi_2)) \right\}.$$

In what concerns (2.106) and (2.107), we argue using Hölder's and Young's (2.18) inequalities. For example, we deduce (2.106) as follows:

$$\left| \int_{\mathbb{R}^{n+1}} \int_{\mathbb{R}^{n+1}} w_+^{-s}(\tau, \xi)\, w_-^{-\theta}(\tau_2, \xi_2) \right.$$

$$\left. \cdot \left| \tilde{V}_1(\tau_1, \xi_1) \right| \, |V_2(\tau, \xi)| \, |V(\tau_2, \xi_2)| \, d\tau_1 \, d\xi_1 \, d\tau \, d\xi \right|$$

$$\lesssim \left\| \tilde{V}_1 * \left( w_-^{-\theta}\, V \right) \right\|_{L_\xi^\infty L_\tau^2} \left\| w_+^{-s}\, V_2 \right\|_{L_\xi^1 L_\tau^2}$$

$$\lesssim \left\| w_-^{-\theta} \right\|_{L_\xi^\infty L_\tau^2} \left\| w_+^{-s} \right\|_{L_\xi^2 L_\tau^\infty} \|\tilde{V}_1\|_{L^2} \|V_2\|_{L^2} \|V\|_{L^2}$$

$$\lesssim \|\tilde{V}_1\|_{L^2} \|V_2\|_{L^2} \|V\|_{L^2},$$

where the last line is due to $s > n/2$ and $\theta > 1/2$. An identical reasoning leads to (2.107). Thus, we are left to prove (2.108), which is the most intricate part of the whole argument.

If we take a look back at (2.87), we notice that we can write $r(\xi_1, \xi_2)$ in the more compact form

$$r(\xi_1, \xi_2) = \left| \|\xi_1| \pm |\xi_2|\| - |\xi| \right|,$$

where $\pm$ coincides with the sign of the product $\tau_1 \tau_2$. As a first step towards (2.108), we show that for $(\tau_1, \xi_1) \in A(\tau, \xi)$,

$$w_+(\tau, \xi) \gtrsim R(\xi_1, \xi_2) = 1 + \left| \|\xi_1| \pm |\xi_2|\| + |\xi|, \right. \qquad (2.109)$$

with $R(\xi_1, \xi_2)$ and $r(\xi_1, \xi_2)$ sharing the same $\pm$. When $\pm = -$, the result is immediate based on the triangle inequality. When $\pm = +$, we have

$$|\tau| = |\tau_1| + |\tau_2|, \quad r(\xi_1, \xi_2) = |\xi_1| + |\xi_2| - |\xi|, \quad R(\xi_1, \xi_2) \simeq 1 + |\xi_1| + |\xi_2|.$$

Hence, the inequality describing $A(\tau, \xi)$ implies

$$|\xi_1| + |\xi_2| - |\xi| \geq 2(2 + |\xi_1| + |\xi_2| - |\tau|),$$

which yields

$$|\tau| \gtrsim R(\xi_1, \xi_2).$$

This finishes the proof of (2.109). Together with $w_+(\tau_1, \xi_1) \geq |\xi_1|$ and $s > n/2 \geq 1$, (2.109) allows us to reduce the argument for (2.108) to simply showing

$$\left| \int_{\mathbb{R}^{n+1}} \int_{\mathbb{R}^{n+1}} \frac{r^\theta(\xi_1, \xi_2)}{R^{s-1}(\xi_1, \xi_2) \, |\xi_1| \, w_-^\theta(\tau_1, \xi_1) \, w_-^\theta(\tau_2, \xi_2)} \right.$$
$$\left. \cdot \left| \tilde{V}_1(\tau_1, \xi_1) \right| |V_2(\tau, \xi)| \, |V(\tau_2, \xi_2)| \, d\tau_1 \, d\xi_1 \, d\tau \, d\xi \right| \lesssim \|\tilde{V}_1\|_{L^2} \|V_2\|_{L^2} \|V\|_{L^2}.$$

Furthermore, given that $\tau \mapsto w_-(\tau, \xi)$ is an even function and $\|W^-\|_{L^2_{\tau, \xi}} = \|W\|_{L^2_{\tau, \xi}}$, where $W^-(\tau, \xi) = W(-\tau, \xi)$, we can restrict the domain of integration for $\tau_1$ in the above estimate to $\mathbb{R}_+$.

At this point, we claim that the desired estimate follows if we prove

$$\left\| \frac{||\eta| - |\xi||^\theta}{(1 + |\eta| + |\xi|)^{s-1}} \int_{|\xi - \zeta| \pm |\zeta| = \eta} \frac{W_1(\xi - \zeta) \, W_2(\zeta)}{|\zeta|} \, dS_\zeta \right\|_{L^2_{\eta, \xi}(\mathbb{R}^{n+1})} \quad (2.110)$$
$$\lesssim \|W_1\|_{L^2(\mathbb{R}^n)} \|W_2\|_{L^2(\mathbb{R}^n)},$$

with $s - n/2 \geq \theta - 1/2 > 0$. In order to see this, we first use Tonelli's theorem and the change of variables

$$(\tau_1, \tau) \mapsto (\mu_1, \mu) = (|\tau_1| - |\xi_1|, |\tau_2| - |\xi_2|),$$

to infer

$$\int_{\mathbb{R}^{n+1}} \int_{\mathbb{R}^n} \int_{\mathbb{R}_+} \frac{r^\theta(\xi_1, \xi_2)}{R^{s-1}(\xi_1, \xi_2) \, |\xi_1| \, w_-^\theta(\tau_1, \xi_1) \, w_-^\theta(\tau_2, \xi_2)}$$
$$\cdot \left| \tilde{V}_1(\tau_1, \xi_1) \right| |V_2(\tau, \xi)| \, |V(\tau_2, \xi_2)| \, d\tau_1 \, d\xi_1 \, d\tau \, d\xi$$
$$= \int_{\mathbb{R}} \int_{\mathbb{R}} \int_{\mathbb{R}^n} \int_{\mathbb{R}^n} \frac{r^\theta(\xi_1, \xi_2)}{R^{s-1}(\xi_1, \xi_2) \, |\xi_1| \, (1 + |\mu_1|)^\theta \, (1 + |\mu|)^\theta} \quad (2.111)$$
$$\cdot \left| \tilde{V}_1(\mu_1 + |\xi_1|, \xi_1) \right| |V_2(\mu + |\xi_1| \pm (\mu + |\xi_2|), \xi)|$$
$$\cdot |V(\pm(\mu + |\xi_2|), \xi_2)| \, d\xi_1 \, d\xi \, d\mu_1 \, d\mu,$$

where the choice of $\pm$ in the arguments of $V_2$ and $V$ coincides with the one corresponding to $r(\xi_1, \xi_2)$ (and $R(\xi_1, \xi_2)$). Next, we foliate the $\xi_1$-space by the ellipsoids/hyperboloids $|\xi_1| \pm |\xi_2| = \eta$ and rewrite the double integral in $(\xi_1, \xi)$ as

$$\int_{\mathbb{R}^{n+1}} \left\{ \int_{|\xi_1| \pm |\xi_2| = \eta} \frac{\left|\tilde{V}_1(\mu_1 + |\xi_1|, \xi_1)\right| \, |V(\pm(\mu + |\xi_2|), \xi_2)|}{|\xi_1|} \, dS_{\xi_1} \right\}$$
$$\cdot \frac{||\eta| - |\xi||^\theta}{(1 + |\eta| + |\xi|)^{s-1}} \, |V_2(\mu_1 \pm \mu + \eta, \xi)| \, d\eta \, d\xi.$$

We deduce, based on the Cauchy-Schwarz inequality with respect to the variables $(\eta, \xi)$ and (2.110), that

$$\int_{\mathbb{R}^{n+1}} \left\{ \int_{|\xi_1| \pm |\xi_2| = \eta} \frac{\left|\tilde{V}_1(\mu_1 + |\xi_1|, \xi_1)\right| \, |V(\pm(\mu + |\xi_2|), \xi_2)|}{|\xi_1|} \, dS_{\xi_1} \right\}$$
$$\cdot \frac{||\eta| - |\xi||^\theta}{(1 + |\eta| + |\xi|)^{s-1}} \, |V_2(\mu_1 \pm \mu + \eta, \xi)| \, d\eta \, d\xi$$
$$\lesssim \|\tilde{V}_1(\mu_1 + |\cdot|, \cdot)\|_{L^2(\mathbb{R}^n)} \|V(\pm(\mu + |\cdot|), \cdot)\|_{L^2(\mathbb{R}^n)} \|V_2\|_{L^2(\mathbb{R}^{n+1})}.$$

Hence, using (2.111), we obtain

$$\int_{\mathbb{R}^{n+1}} \int_{\mathbb{R}^n} \int_{\mathbb{R}_+} \frac{r^\theta(\xi_1, \xi_2)}{R^{s-1}(\xi_1, \xi_2) \, \langle \xi_1 \rangle \, w_-^\theta(\tau_1, \xi_1) \, w_-^\theta(\tau_2, \xi_2)} \left|\tilde{V}_1(\tau_1, \xi_1)\right|$$
$$\cdot |V_2(\tau, \xi)| \, |V(\tau_2, \xi_2)| \, d\tau_1 \, d\xi_1 \, d\tau \, d\xi$$
$$\lesssim \int_{\mathbb{R}} \int_{\mathbb{R}} \frac{\|\tilde{V}_1(\mu_1 + |\cdot|, \cdot)\|_{L^2(\mathbb{R}^n)} \|V(\pm(\mu + |\cdot|), \cdot)\|_{L^2(\mathbb{R}^n)}}{(1 + |\mu_1|)^\theta \, (1 + |\mu|)^\theta} \, d\mu_1 \, d\mu \cdot \|V_2\|_{L^2}$$
$$\lesssim \|\tilde{V}_1\|_{L^2} \|V\|_{L^2} \|V_2\|_{L^2},$$

where the last line is motivated by $\theta > 1/2$ and another application of the Cauchy-Schwarz inequality with respect to each of the variables $\mu_1$ and $\mu$.

Therefore, we are left to show that (2.110) holds. This follows if we prove

$$\left\| \frac{||\eta| - |\xi||^{2\theta}}{(1 + |\eta| + |\xi|)^{2s-2}} \int_{|\xi - \zeta| \pm |\zeta| = \eta} |\zeta|^{-2} \, dS_\zeta \right\|_{L^\infty_{\eta, \xi}(\mathbb{R}^{n+1})} \lesssim 1, \qquad (2.112)$$

for $s - n/2 \geq \theta - 1/2 > 0$. Indeed, relying on the Cauchy-Schwarz inequality with respect to $\zeta$, we derive

$$\left\| \frac{||\eta| - |\xi||^\theta}{(1 + |\eta| + |\xi|)^{s-1}} \int_{|\xi - \zeta| \pm |\zeta| = \eta} \frac{W_1(\xi - \zeta) W_2(\zeta)}{|\zeta|} dS_\zeta \right\|_{L^2_{\eta, \xi}(\mathbb{R}^{n+1})}$$

$$\lesssim \left\| \frac{||\eta| - |\xi||^\theta}{(1 + |\eta| + |\xi|)^{s-1}} \left( \int_{|\xi - \zeta| \pm |\zeta| = \eta} |\zeta|^{-2} dS_\zeta \right)^{1/2} \right.$$

$$\left. \cdot \left( \int_{|\xi - \zeta| \pm |\zeta| = \eta} W_1^2(\xi - \zeta) W_2^2(\zeta) dS_\zeta \right)^{1/2} \right\|_{L^2_{\eta, \xi}(\mathbb{R}^{n+1})}$$

$$\lesssim \left\| \frac{||\eta| - |\xi||^{2\theta}}{(1 + |\eta| + |\xi|)^{2s-2}} \int_{|\xi - \zeta| \pm |\zeta| = \eta} |\zeta|^{-2} dS_\zeta \right\|_{L^\infty_{\eta, \xi}(\mathbb{R}^{n+1})}^{1/2}$$

$$\cdot \left( \int_{\mathbb{R}^{n+1}} \int_{|\xi - \zeta| \pm |\zeta| = \eta} W_1^2(\xi - \zeta) W_2^2(\zeta) dS_\zeta \, d\eta \, d\xi \right)^{1/2}$$

$$\lesssim \|W_1\|_{L^2(\mathbb{R}^n)} \|W_2\|_{L^2(\mathbb{R}^n)}.$$

For proving (2.112), we rewrite the surface integral using polar coordinates,

$$\zeta = \rho\omega, \quad \rho \in \mathbb{R}_+, \quad \omega \in \mathbb{S}^{n-1}.$$

In the ellipsoid case, we have

$$\eta > |\xi| \quad \text{and} \quad \rho = \rho(\omega) = \frac{\eta^2 - |\xi|^2}{2(\eta - \xi \cdot \omega)}.$$

Hence,

$$\frac{||\eta| - |\xi||^{2\theta}}{(1 + |\eta| + |\xi|)^{2s-2}} \int_{|\xi - \zeta| + |\zeta| = \eta} |\zeta|^{-2} dS_\zeta$$

$$\simeq \frac{(|\eta| - |\xi|)^{2\theta}}{(1 + |\eta|)^{2s-2}(|\eta|^2 - |\xi|^2)} \int_{\mathbb{S}^{n-1}} \rho^{n-2}(\eta - \rho) dS_\omega$$

$$\lesssim \frac{(|\eta| - |\xi|)^{2\theta - 1}\eta^{n-2}}{(1 + |\eta|)^{2s-2}} \lesssim 1,$$

due to $\rho < \eta$, $n \geq 2$ and $s - n/2 \geq \theta - 1/2 > 0$.

For the hyperboloid, we have

$$|\eta| < |\xi| \quad \text{and} \quad \rho = \rho(\omega) = \frac{|\xi|^2 - \eta^2}{2(\eta + \xi \cdot \omega)},$$

which implies $\eta + \xi \cdot \omega > 0$ and $\rho > -\eta$. For $\rho \lesssim |\xi|$, it follows that

$$\frac{||\eta| - |\xi||^{2\theta}}{(1 + |\eta| + |\xi|)^{2s-2}} \int_{|\xi - \zeta| - |\zeta| = \eta} |\zeta|^{-2} \, dS_\zeta$$

$$\simeq \frac{(|\xi| - |\eta|)^{2\theta}}{(1 + |\xi|)^{2s-2}(|\xi|^2 - |\eta|^2)} \int_{\mathbb{S}^{n-1}} \rho^{n-2}(\eta + \rho) \, dS_\omega$$

$$\lesssim \frac{(|\xi| - |\eta|)^{2\theta - 1} |\xi|^{n-2}}{(1 + |\xi|)^{2s-2}} \lesssim 1,$$

based on the same facts listed in the ellipsoid case. The analysis for $\rho \gg |\xi|$ is slightly more involved and we refer the reader to a similar one in Section 4 of Klainerman-Selberg [89]. This concludes the argument for (2.112) and finishes the whole proof.                                                        □

## 2.4   Tataru's F-spaces

In this section, we discuss function spaces introduced by Tataru in his fundamental works [177] and [178], which are used in proving global regularity for wave maps with initial data in critical Besov spaces. Due to the overly technical nature of this material and the aim of our book, we chose to present in full detail only the high-dimensional case $n \geq 4$. We believe the reader can then make a clear transition to the considerably more involved $n = 2$ and $n = 3$ settings, for which we merely provide their definitions. The reasoning behind the construction of these spaces lies in building a norm which takes advantage of both Strichartz and $X^{s,\theta}$ estimates and is compatible with the critical regularity $s = n/2$.

We start by considering $\phi \in C_0^\infty(\mathbb{R})$ to be a smooth cutoff function satisfying

$$\text{supp } \phi \subset \left(\frac{1}{2}, 2\right), \qquad \sum_{\lambda \in 2^{\mathbb{Z}}} \phi\left(\frac{\eta}{\lambda}\right) = 1, \ \forall \eta \in \mathbb{R} \backslash \{0\}. \qquad (2.113)$$

For example, we can take $\phi$ to be the one-dimensional function $\phi_1$ defined in the context of the Littlewood-Paley theory. Next, we define the spacetime Fourier multipliers

$$A_\lambda(\nabla) \overset{\text{def}}{=} \mathcal{F}^{-1} \phi\left(\frac{|(\tau, \xi)|}{\lambda}\right) \mathcal{F}, \qquad (2.114)$$

$$B_\mu(\nabla) \overset{\text{def}}{=} \mathcal{F}^{-1} \phi\left(\frac{|\tau^2 - |\xi|^2|}{\mu \, |(\tau, \xi)|}\right) \mathcal{F}, \qquad \tilde{B}_\nu(\nabla) \overset{\text{def}}{=} \sum_{\nu \in 2^{\mathbb{Z}}, \, \nu \leq \mu} B_\nu(\nabla), \quad (2.115)$$

where $\lambda$, $\mu \in 2^{\mathbb{Z}}$, $\mathcal{F}$ is the spacetime Fourier transform in $\mathbb{R}^{n+1}$ ($n \geq 1$), and $|(\tau, \xi)| = \sqrt{\tau^2 + |\xi|^2}$. If $u \in \mathcal{S}'(\mathbb{R}^{n+1})$ has its spacetime Fourier transform supported at frequency $|(\tau, \xi)| \simeq \lambda$, then we introduce the norms

$$\|u\|_{X_\lambda^s(\mathbb{R}^{n+1})} \stackrel{\text{def}}{=} \sum_{\mu \in 2^{\mathbb{Z}}} \mu^s \|B_\mu(\nabla)u\|_{L^2(\mathbb{R}^{n+1})} \tag{2.116}$$

and

$$\|u\|_{Y_\lambda(\mathbb{R}^{n+1})} \stackrel{\text{def}}{=} \|u\|_{L^\infty L^2(\mathbb{R} \times \mathbb{R}^n)} + \lambda^{-1}\|\Box u\|_{L^1 L^2(\mathbb{R} \times \mathbb{R}^n)}. \tag{2.117}$$

**Remark 2.10.** In the definition of the $X_\lambda^s$ norm, we notice that, due to the frequency localization $|(\tau, \xi)| \simeq \lambda$ imposed by $A_\lambda(\nabla)$, the summation index $\mu$ is in fact restricted to the range $\mu \lesssim \lambda$, as the presence of $B_\mu(\nabla)$ forces $||\tau| - |\xi|| \simeq \mu$.

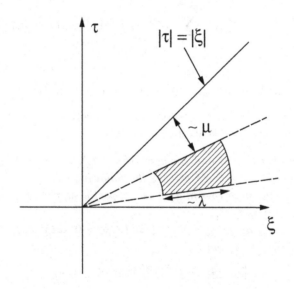

Fig. 2.2   The Fourier support for the multiplier $A_\lambda(\nabla)B_\mu(\nabla)$.

We are now ready to define Tataru's high-dimensional function spaces.

**Definition 2.4 ([177]).** Relying on (2.116) and (2.117), let us consider[5]

$$\|u\|_F \stackrel{\text{def}}{=} \sum_{\lambda \in 2^{\mathbb{Z}}} \lambda^{n/2} \|A_\lambda(\nabla)u\|_{F_\lambda}, \qquad F_\lambda = X_\lambda^{1/2} + Y_\lambda, \tag{2.118}$$

---

[5]In the construction of $F_\lambda$ and $\Box F_\lambda$, we use the classical definitions $\|a\|_{\alpha V} := |\alpha|^{-1}\|a\|_V$ and $\|a\|_{V+W} := \inf_{a=b+c}\{\|b\|_V + \|c\|_W\}$.

and

$$\|u\|_{\Box F} \overset{\text{def}}{=} \sum_{\lambda \in 2^{\mathbb{Z}}} \lambda^{n/2} \|A_\lambda(\nabla)u\|_{\Box F_\lambda}, \quad \Box F_\lambda = \lambda \left( X_\lambda^{-1/2} + (L^1 L^2)_\lambda \right). \quad (2.119)$$

The function spaces $F$ and $\Box F$ are then defined as the closures of $\mathcal{S}'(\mathbb{R}^{n+1})$ under their respective norms.

**Remark 2.11.** The intuition behind the $\Box F$ notation has to do with the estimate

$$\|\Box u\|_{\Box F} \lesssim \|u\|_F. \quad (2.120)$$

This is the result of

$$\|A_\lambda \Box u\|_{\lambda X_\lambda^{-1/2}} = \lambda^{-1} \sum_{\mu \in 2^{\mathbb{Z}}} \mu^{-1/2} \|A_\lambda B_\mu \Box u\|_{L^2}$$

$$\lesssim \lambda^{-1} \sum_{\mu \in 2^{\mathbb{Z}}} \mu^{-1/2} \lambda \mu \|A_\lambda B_\mu u\|_{L^2} = \|A_\lambda u\|_{X_\lambda^{1/2}}$$

and

$$\|A_\lambda \Box u\|_{\lambda (L^1 L^2)_\lambda} = \lambda^{-1} \|A_\lambda \Box u\|_{L^1 L^2} \leq \|A_\lambda u\|_{Y^\lambda}.$$

As a preamble for our discussion of these function spaces, we present a useful lemma concerning spacetime multipliers.

**Lemma 2.8.** *Let $C > 0$ and consider $M = M(\nabla)$ to be a Fourier multiplier. Its symbol $m = m(\tau, \xi)$ verifies the following conditions:*

*i) for all $\xi$, the map $\tau \mapsto m(\tau, \xi)$ is compactly supported on a set whose Lebesgue measure is $\lesssim C$;*

*ii) there exists an integer $N \geq 2$ and $C_N > 0$ such that $m$ admits $N$ partial derivatives in $\tau$ and*

$$\|m\|_{L^\infty_{\tau,\xi}} + C^N \|\partial_\tau^N m\|_{L^\infty_{\tau,\xi}} \leq C_N.$$

*Then, for all $1 \leq p \leq \infty$,*

$$\|M(\nabla)u\|_{L^p L^2} \lesssim C_N \|u\|_{L^p L^2}.$$

**Proof.** A direct computation shows that

$$\widehat{M(\nabla)u}(t, \xi) = \frac{1}{2\pi} \int_{\mathbb{R}^2} e^{i(t-s)\tau} m(\tau, \xi) \, \widehat{u}(s, \xi) \, d\tau \, ds$$

$$= \int_{\mathbb{R}} K(t - s, \xi) \, \widehat{u}(s, \xi) \, ds,$$

where

$$K(t,\xi) = \frac{1}{2\pi} \int_{\mathbb{R}} e^{it\tau} m(\tau,\xi) d\tau.$$

Then, an integration by parts yields

$$(it)^N K(t,\xi) = \frac{1}{2\pi} \int_{\mathbb{R}} \partial_\tau^N (e^{it\tau}) m(\tau,\xi) d\tau = \frac{(-1)^N}{2\pi} \int_{\mathbb{R}} e^{it\tau} \partial_\tau^N m(\tau,\xi) d\tau,$$

which allows us, based on hypotheses in the lemma, to infer

$$|K(t,\xi)| \lesssim \frac{C_N\, C}{(1 + C|t|)^N}.$$

As $N \geq 2$, we can apply Plancherel's theorem and Young's inequality (2.18) to finally derive

$$\|M(\nabla)u\|_{L^p L^2} \simeq \|\widehat{M(\nabla)}u\|_{L_t^p L_\xi^2} \lesssim \|K\|_{L_t^1 L_\xi^\infty} \|\widehat{u}\|_{L_t^p L_\xi^2} \lesssim C_N \|u\|_{L^p L^2}.$$

$$\square$$

Using this lemma, we are able to describe mapping properties for certain multipliers associated with Tataru's spaces.

**Corollary 2.4.** *For $1 \leq p \leq \infty$, the operators $A_\lambda(\nabla)$ are bounded on $L^p L^2$ uniformly in $\lambda$. The operators $\tilde{B}_\mu(\nabla)$ are bounded on $F_\lambda$ and on $\Box F_\lambda$ uniformly in $\mu \leq \lambda$. Furthermore, the operators $(1 - \tilde{B}_\mu)(\nabla) = \sum_{\nu \in 2^{\mathbb{Z}}, \nu > \mu} B_\nu(\nabla)$ are bounded from $Y_\lambda$ to $\mu^{-1} L^1 L^2$, also uniformly in $\mu \leq \lambda$.*

**Proof.** The first statement is a direct application of the lemma with $C = \lambda$, $M = A_\lambda$, $N = 2$, and $C_N = 1$. Based on (2.118), (2.119), and

$$\|\tilde{B}_\mu u\|_{X_\lambda^{1/2}} = \sum_{\nu \in 2^{\mathbb{Z}}} \nu^{1/2} \|B_\nu \tilde{B}_\mu u\|_{L^2} \lesssim \sum_{\nu \in 2^{\mathbb{Z}}, \nu \lesssim \mu} \nu^{1/2} \|B_\nu u\|_{L^2} \lesssim \|u\|_{X_\lambda^{1/2}},$$

it follows that the second claim reduces to proving that $\tilde{B}_\mu A_\lambda$ are bounded on $L^\infty L^2$ and on $L^1 L^2$, uniformly in $\mu \leq \lambda$. It is an immediate verification that $C = \mu$ and $M = \tilde{B}_\mu A_\lambda$ satisfy the hypotheses of the previous lemma with $N = 2$ and $C_N = 1$. Hence, we obtain the desired conclusion about the operators $\tilde{B}_\mu$.

For the mapping associated with $1 - \tilde{B}_\mu$, we prove in fact a stronger result, i.e.,

$$\|(1 - \tilde{B}_\mu)A_\lambda u\|_{L^1 L^2} \lesssim (\lambda\mu)^{-1} \|\Box A_\lambda u\|_{L^1 L^2},$$

which is equivalent to showing that the spacetime multiplier

$$\frac{\lambda\mu\, A_\lambda(1 - \tilde{B}_\mu)}{\Box}$$

is bounded on $L^1 L^2$, uniformly in $\mu \le \lambda$. If we denote its symbol

$$m_{\lambda,\mu}(\tau,\xi) = \frac{\lambda\mu\, a_\lambda(\tau,\xi)\,(1 - \tilde{b}_\mu(\tau,\xi))}{|\xi|^2 - \tau^2},$$

where $a_\lambda$ and $\tilde{b}_\mu$ are the symbols for $A_\lambda$ and $\tilde{B}_\mu$, respectively, then, arguing as in the previous lemma, the claim follows by proving

$$\left\| \int_{\mathbb{R}} e^{it\tau} m_{\lambda,\mu}(\tau,\xi)\, d\tau \right\|_{L^1_t L^\infty_\xi} \lesssim 1. \tag{2.121}$$

Due to the frequency localizations imposed by $A_\lambda$ and $B_\mu$, the symbol $m_{\lambda,\mu}$ is supported in the region

$$\{(\tau,\xi) \in \mathbb{R}^{n+1}; |\tau| + |\xi| \simeq \lambda, ||\tau| - |\xi|| \gtrsim \mu\}.$$

Moreover, simple calculations show that

$$||\tau| - |\xi||\,|m_{\lambda,\mu}(\tau,\xi)| + ||\tau| - |\xi||^3\,|\partial_\tau^2 m_{\lambda,\mu}(\tau,\xi)| \lesssim \mu.$$

These facts allow us to infer that

$$\left| \int_{\mathbb{R}} e^{it\tau} m_{\lambda,\mu}(\tau,\xi)\, d\tau \right| \lesssim \int_{\mu \lesssim ||\tau| - |\xi|| \lesssim \lambda} \frac{\mu}{||\tau| - |\xi||}\, d\tau \simeq \mu \ln \frac{\lambda}{\mu}$$

and

$$\left| t^2 \int_{\mathbb{R}} e^{it\tau} m_{\lambda,\mu}(\tau,\xi)\, d\tau \right| \lesssim \int_{||\tau| - |\xi|| \gtrsim \mu} \frac{\mu}{||\tau| - |\xi||^3}\, d\tau \simeq \mu^{-1}.$$

Therefore, we obtain

$$\int_{|t| \lesssim 1/\lambda} \left\| \int_{\mathbb{R}} e^{it\tau} m_{\lambda,\mu}(\tau,\xi)\, d\tau \right\|_{L^\infty_\xi}\, dt \lesssim \frac{\mu}{\lambda} \ln \frac{\lambda}{\mu} \lesssim 1 \tag{2.122}$$

and

$$\int_{|t| \gtrsim 1/\mu} \left\| \int_{\mathbb{R}} e^{it\tau} m_{\lambda,\mu}(\tau,\xi)\, d\tau \right\|_{L^\infty_\xi}\, dt \lesssim \int_{|t| \gtrsim 1/\mu} \frac{1}{\mu t^2}\, dt \simeq 1. \tag{2.123}$$

For $1/\lambda \ll |t| \ll 1/\mu$, we estimate as follows:

$$\left\| \int_{\mathbb{R}} e^{it\tau} m_{\lambda,\mu}(\tau,\xi)\, d\tau \right\|_{L^\infty_\xi}$$

$$\lesssim \int_{\mu \lesssim ||\tau| - |\xi|| \lesssim 1/t} \frac{\mu}{||\tau| - |\xi||}\, d\tau + \frac{1}{t^2} \int_{1/t \lesssim ||\tau| - |\xi|| \lesssim \lambda} \frac{\mu}{||\tau| - |\xi||^3}\, d\tau$$

$$\simeq \mu(1 - \ln(\mu|t|)),$$

which implies

$$\int_{1/\lambda \ll |t| \ll 1/\mu} \left\| \int_{\mathbb{R}} e^{it\tau} m_{\lambda,\mu}(\tau,\xi) \, d\tau \right\|_{L_\xi^\infty} dt \lesssim 1.$$

Together with (2.122) and (2.123), this estimate proves (2.121) and concludes the argument. $\square$

The first result we prove about the space $F$ is that it contains the solutions of the homogeneous wave equation (2.31) with initial data in $\dot{B}_{2,1}^{n/2} \times \dot{B}_{2,1}^{n/2-1}(\mathbb{R}^n)$.

**Theorem 2.13.** *If* $(u_0, u_1) \in \dot{B}_{2,1}^{n/2} \times \dot{B}_{2,1}^{n/2-1}(\mathbb{R}^n)$ *and* $u$ *solves* (2.31), *then*

$$\|u\|_F \lesssim \|u_0\|_{\dot{B}_{2,1}^{n/2}(\mathbb{R}^n)} + \|u_1\|_{\dot{B}_{2,1}^{n/2-1}(\mathbb{R}^n)}. \tag{2.124}$$

**Proof.** Using (2.118), we can infer that

$$\|u\|_F \leq \sum_{\lambda \in 2^{\mathbb{Z}}} \lambda^{n/2} \|A_\lambda u\|_{Y_\lambda} = \sum_{\lambda \in 2^{\mathbb{Z}}} \lambda^{n/2} \|A_\lambda u\|_{L^\infty L^2}$$

$$\simeq \sum_{\lambda \in 2^{\mathbb{Z}}} \lambda^{n/2} \|\widehat{A_\lambda u}(t,\xi)\|_{L_t^\infty L_\xi^2}.$$

An elementary computation relying on (2.32) leads to

$$\widehat{A_\lambda u}(t,\xi) \simeq a_\lambda(|\xi|,\xi) \left[ e^{it|\xi|} \left( \widehat{u_0}(\xi) - \frac{i\,\widehat{u_1}(\xi)}{|\xi|} \right) \right.$$

$$\left. + e^{-it|\xi|} \left( \widehat{u_0}(\xi) + \frac{i\,\widehat{u_1}(\xi)}{|\xi|} \right) \right].$$

Hence,

$$\sum_{\lambda \in 2^{\mathbb{Z}}} \lambda^{n/2} \|\widehat{A_\lambda u}(t,\xi)\|_{L_t^\infty L_\xi^2} \lesssim \sum_{\lambda \in 2^{\mathbb{Z}}} \lambda^{n/2} \left( \|a_\lambda(|\xi|,\xi)\,\widehat{u_0}(\xi)\|_{L_\xi^2} \right.$$

$$\left. + \||\xi|^{-1} a_\lambda(|\xi|,\xi)\,\widehat{u_1}(\xi)\|_{L_\xi^2} \right),$$

which implies (2.124), based on the properties of $\phi$ and the definition of homogeneous Besov norms (2.4). $\square$

Next, we address the inhomogeneous wave equation (2.40), in which we relabel the right-hand side by $H = H(t,x)$ to avoid confusion with the notation used for Tataru's space $F$. Together with (2.120), the following theorem reveals the intrinsic relation between the spaces $F$ and $\square F$.

**Theorem 2.14.** *If* $H \in \square F$ *and* $w$ *solves*

$$\square w = H, \qquad w(0) = w_t(0) = 0,$$

*then*

$$\|w\|_F \lesssim \|H\|_{\Box F}. \tag{2.125}$$

**Proof.** According to (2.118) and (2.119), the conclusion follows if we show that both

$$\|A_\lambda w\|_{X_\lambda^{1/2}} \lesssim \|A_\lambda H\|_{\lambda X_\lambda^{-1/2}} \tag{2.126}$$

and

$$\|A_\lambda w\|_{Y_\lambda} \lesssim \|A_\lambda H\|_{\lambda(L^1 L^2)_\lambda} \tag{2.127}$$

hold.

For the first estimate, we argue using Plancherel's theorem as follows:

$$\|A_\lambda w\|_{X_\lambda^{1/2}} = \sum_{\mu \in 2^{\mathbb{Z}}} \mu^{1/2} \|B_\mu A_\lambda w\|_{L^2}$$

$$\simeq \sum_{\mu \in 2^{\mathbb{Z}}} \frac{\mu^{1/2}}{\mu\lambda} \|B_\mu A_\lambda H\|_{L^2} = \lambda^{-1} \|A_\lambda H\|_{X_\lambda^{-1/2}}.$$

In what concerns (2.127), we have

$$\|A_\lambda w\|_{Y_\lambda} = \|A_\lambda w\|_{L^\infty L^2} + \lambda^{-1} \|A_\lambda H\|_{L^1 L^2}. \tag{2.128}$$

An application of Lemma 2.8 with $C = \lambda$,

$$\frac{m(\tau,\xi)}{a_\lambda(\tau,\xi)} \in \left\{ \frac{\lambda}{|\tau|+|\xi|}, \frac{\tau}{|\tau|}, \frac{\xi_i}{|\xi_i|}, \frac{|\xi|}{|\xi_1|+\ldots+|\xi_n|} \right\},$$

$N = 2$, $C_N = 1$, and $p = \infty$, yields

$$\|A_\lambda w\|_{L^\infty L^2} \lesssim \lambda^{-1} \|\nabla A_\lambda w\|_{L^\infty L^2},$$

whereas the energy estimate (2.55) implies

$$\|\nabla A_\lambda w\|_{L^\infty L^2} \lesssim \|A_\lambda H\|_{L^1 L^2}.$$

Hence, we obtain

$$\|A_\lambda w\|_{L^\infty L^2} \lesssim \lambda^{-1} \|A_\lambda H\|_{L^1 L^2},$$

which, based on (2.128), finishes the proof of (2.127) and the whole argument. $\qquad\square$

Following this, we prove Strichartz-type estimates in this functional setting, which are important in proving critical global regularity results for high-dimensional wave maps.

**Theorem 2.15.** *Let $n \geq 4$ and consider $(p,q,n)$ to be a wave-admissible triple. The following estimate holds uniformly in $\lambda$:*

$$\|A_\lambda u\|_{L^p L^q(\mathbb{R} \times \mathbb{R}^n)} \lesssim \lambda^{s(p,q,n)} \|A_\lambda u\|_{F_\lambda}. \tag{2.129}$$

**Proof.** We prove the claim by showing that both

$$\|A_\lambda u\|_{L^p L^q} \lesssim \lambda^{s(p,q,n)} \|A_\lambda u\|_{X_\lambda^{1/2}} \tag{2.130}$$

and

$$\|A_\lambda u\|_{L^p L^q} \lesssim \lambda^{s(p,q,n)} \|A_\lambda u\|_{Y_\lambda} \tag{2.131}$$

are true for all wave-admissible triples $(p, q, n)$.

The second inequality can be derived from the proof of (2.56) by switching from $u$ to $A_\lambda u$. This localization in frequency space allows for the inhomogeneous Sobolev norms in (2.56) to be replaced by homogeneous ones and for $s$ to be allowed to take the value $s(p, q, n)$.

In what concerns (2.130), we start by foliating the Fourier space using translates of the null cone

$$K = \{(\tau, \xi) \in \mathbb{R}^{n+1}; \tau^2 = |\xi|^2\}. \tag{2.132}$$

Accordingly,

$$
\begin{aligned}
\widehat{v}(\tau, \xi) &= \chi_{[0,\infty)}(\tau)\,\widehat{v}(\tau, \xi) + \chi_{(-\infty,0)}(\tau)\,\widehat{v}(\tau, \xi) \\
&= \chi_{[0,\infty)}(\tau) \int_{\mathbb{R}} \delta(\tau - |\xi| - a)\,\widehat{v}(|\xi| + a, \xi)\,da \\
&\quad + \chi_{(-\infty,0)}(\tau) \int_{\mathbb{R}} \delta(\tau + |\xi| - a)\,\widehat{v}(-|\xi| + a, \xi)\,da \\
&= \int_{\mathbb{R}} \widehat{v_a^+}(\tau, \xi) + \widehat{v_a^-}(\tau, \xi)\,da,
\end{aligned}
$$

with

$$\widehat{v_a^\pm}(\tau, \xi) \overset{\text{def}}{=} \chi_{[0,\infty)}(\pm\tau)\,\delta(\tau \mp |\xi| - a)\,\widehat{v}(\pm|\xi| + a, \xi).$$

Moreover, routine computations yield

$$
\begin{cases}
\Box(e^{-iat}\,v_a^\pm) &= 0, \\
e^{-iat}\,v_a^\pm(0, \xi) &= \chi_{[0,\infty)}(|\xi| \pm a)\,\widehat{v}(\pm|\xi| + a, \xi), \\
\partial_t\left(\widehat{e^{-iat}\,v_a^\pm}\right)(0, \xi) &= \pm|\xi|\,\chi_{[0,\infty)}(|\xi| \pm a)\,\widehat{v}(\pm|\xi| + a, \xi).
\end{cases}
$$

Using (2.34), the frequency localization imposed by $A_\lambda$, and that

$s(p,q,n) \geq 0$, we derive

$$\|A_\lambda u\|_{L^p L^q} \leq \int_{\mathbb{R}} \|(A_\lambda u)_a^+\|_{L^p L^q} + \|(A_\lambda u)_a^-\|_{L^p L^q} \, da$$

$$\lesssim \int_{\mathbb{R}} \left( \|\chi_{[0,\infty)}(|\xi| + a) \, \widehat{A_\lambda u}(|\xi| + a, \xi) \, |\xi|^{s(p,q,n)}\|_{L_\xi^2} \right.$$

$$\left. + \|\chi_{[0,\infty)}(|\xi| - a) \, \widehat{A_\lambda u}(-|\xi| + a, \xi) \, |\xi|^{s(p,q,n)}\|_{L_\xi^2} \right) da$$

$$\lesssim \lambda^{s(p,q,n)} \int_{\mathbb{R}} \left( \|\chi_{[0,\infty)}(|\xi| + a) \, \widehat{A_\lambda u}(|\xi| + a, \xi)\|_{L_\xi^2} \right.$$

$$\left. + \|\chi_{[0,\infty)}(|\xi| - a) \, \widehat{A_\lambda u}(-|\xi| + a, \xi)\|_{L_\xi^2} \right) da,$$

when $(p,q,n)$ is wave-admissible with $q \neq \infty$. If we denote the last integrand by $I(a)$ and we apply the Cauchy-Schwarz inequality, we deduce

$$\int_{\mathbb{R}} I(a) \, da = \sum_{\mu \in 2^{\mathbb{Z}}} \int_{\mu/2 \leq |a| \leq \mu} I(a) \, da \lesssim \sum_{\mu \in 2^{\mathbb{Z}}} \mu^{1/2} \|I(a)\|_{L^2(\mu/2 \leq |a| \leq \mu)}.$$

Applying now Tonelli's theorem, the change of variables $(\xi, a) \mapsto (\xi, \tau) = (\xi, \pm|\xi| + a)$, and taking advantage of the frequency localization imposed by $B_\mu$, we can infer that

$$\left\| \|\chi_{[0,\infty)}(|\xi| \pm a) \, \widehat{A_\lambda u}(\pm|\xi| + a, \xi)\|_{L_\xi^2} \right\|_{L^2(\mu/2 \leq |a| \leq \mu)}$$

$$\simeq \|\chi_{[0,\infty)}(\pm\tau) \, \chi_{[\mu/2,\mu]}(||\tau| - |\xi||) \, \widehat{A_\lambda u}(\tau, \xi)\|_{L_{\tau,\xi}^2}$$

$$\lesssim \|\chi_{[0,\infty)}(\pm\tau) \, \widehat{B_\mu A_\lambda u}(\tau, \xi)\|_{L_{\tau,\xi}^2}.$$

Hence, combining the previous three estimates, we conclude that

$$\|A_\lambda u\|_{L^p L^q} \lesssim \lambda^{s(p,q,n)} \sum_{\mu \in 2^{\mathbb{Z}}} \mu^{1/2} \|B_\mu A_\lambda u\|_{L^2} = \lambda^{s(p,q,n)} \|A_\lambda u\|_{X_\lambda^{1/2}}.$$

When $q = \infty$, (2.130) follows by using Bernstein's inequalities and the previous bound. This finishes the proof of this proposition. $\square$

Next, we prove the counterpart of Proposition 2.7 for Tataru's space $F$.

**Proposition 2.10.** *The function space $F$ embeds continuously in* $C(\mathbb{R}, \dot{B}_{2,1}^{n/2}(\mathbb{R}^n)) \cap C^1(\mathbb{R}, \dot{B}_{2,1}^{n/2-1}((\mathbb{R}^n))$, *i.e.*,

$$\|u\|_{L^\infty \dot{B}_{2,1}^{n/2}(\mathbb{R} \times \mathbb{R}^n)} + \|\partial_t u\|_{L^\infty \dot{B}_{2,1}^{n/2-1}(\mathbb{R} \times \mathbb{R}^n)} \lesssim \|u\|_F. \tag{2.133}$$

**Proof.** Based on the definition of $F$, (2.133) follows by proving

$$\|A_\lambda u\|_{L^\infty \dot{B}_{2,1}^{n/2}} + \|\partial_t A_\lambda u\|_{L^\infty \dot{B}_{2,1}^{n/2-1}} \lesssim \lambda^{n/2} \|A_\lambda u\|_{F_\lambda}.$$

However, applying Bernstein's inequalities in the context of the frequency localizations imposed by $A_\lambda$, we derive

$$\|A_\lambda u\|_{L^\infty \dot{B}_{2,1}^{n/2}} + \|\partial_t A_\lambda u\|_{L^\infty \dot{B}_{2,1}^{n/2-1}}$$

$$\lesssim \lambda^{n/2} \|A_\lambda u\|_{L^\infty L^2} + \lambda^{n/2-1} \|\partial_t A_\lambda u\|_{L^\infty L^2} \lesssim \lambda^{n/2} \|A_\lambda u\|_{L^\infty L^2},$$

and we can claim the desired conclusion on the account of (2.129). $\qquad\square$

After this result, we demonstrate product estimates in these function spaces. They mirror perfectly the ones obtained previously for $X^{s,\theta}$ spaces. First, we prove:

**Theorem 2.16.** *Let $n \geq 4$. The function space $F$ is an algebra, i.e.,*

$$\|uv\|_F \lesssim \|u\|_F \|v\|_F. \tag{2.134}$$

**Proof.** Given that the norm for $F$ is defined as an $l^1$ summation of its dyadic pieces, the proof of (2.134) reduces to the one of

$$\|A_\lambda u \cdot A_\nu v\|_F \lesssim \lambda^{n/2} \|A_\lambda u\|_{F_\lambda} \nu^{n/2} \|A_\nu v\|_{F_\nu}. \tag{2.135}$$

The symmetry of this bound with respect to $\lambda$ and $\nu$ allows us to assume $\lambda \geq \nu$. We discuss the cases $\lambda \simeq \nu$ and $\lambda \gg \nu$ separately.

For the former, we start by using the Cauchy-Schwarz inequality and the relative orthogonality of the multipliers $(B_\mu)_{\mu \in 2^{\mathbb{Z}}}$ to deduce

$$\|A_\kappa w\|_{X_\kappa^s} \simeq \sum_{\mu \in 2^{\mathbb{Z}}, \, \mu \lesssim \kappa} \mu^s \|B_\mu A_\kappa w\|_{L^2}$$

$$\lesssim \left( \sum_{\mu \in 2^{\mathbb{Z}}, \, \mu \lesssim \kappa} \mu^{2s} \right)^{1/2} \left( \sum_{\mu \in 2^{\mathbb{Z}}, \, \mu \lesssim \kappa} \|B_\mu A_\kappa w\|_{L^2}^2 \right)^{1/2}$$

$$\lesssim \kappa^s \|A_\kappa w\|_{L^2},$$

if $s > 0$. As the Fourier support of $A_\lambda u \cdot A_\nu v$ lies in the region $|\tau| + |\xi| \lesssim \lambda$, we can infer based on the previous bound, Bernstein's inequality, and (2.129) that

$$\|A_\lambda u \cdot A_\nu v\|_F \lesssim \sum_{\mu \in 2^{\mathbb{Z}}, \, \mu \lesssim \lambda} \mu^{n/2} \|A_\mu (A_\lambda u \cdot A_\nu v)\|_{X_\mu^{1/2}}$$

$$\lesssim \sum_{\mu \in 2^{\mathbb{Z}}, \, \mu \lesssim \lambda} \mu^{(n+1)/2} \|A_\mu (A_\lambda u \cdot A_\nu v)\|_{L^2}$$

$$\lesssim \lambda^{(n+1)/2} \|A_\lambda u \cdot A_\nu v\|_{L^2} \lesssim \lambda^{(n+1)/2} \|A_\lambda u\|_{L^2 L^\infty} \|A_\nu v\|_{L^\infty L^2}$$

$$\lesssim \lambda^n \|A_\lambda u\|_{F_\lambda} \|A_\nu v\|_{F_\nu} \simeq \lambda^{n/2} \|A_\lambda u\|_{F_\lambda} \nu^{n/2} \|A_\nu v\|_{F_\nu}.$$

For the $\lambda \gg \nu$ case, we notice first that the Fourier support of $A_\lambda u \cdot A_\nu v$ lies in the region $|\tau| + |\xi| \simeq \lambda$, which implies

$$\|A_\lambda u \cdot A_\nu v\|_F \simeq \lambda^{n/2} \|A_\lambda u \cdot A_\nu v\|_{F_\lambda}.$$

Hence, we are left to prove that

$$\|A_\lambda u \cdot A_\nu v\|_{F_\lambda} \lesssim \nu^{n/2} \|A_\lambda u\|_{F_\lambda} \|A_\nu v\|_{F_\nu}.$$

Performing the decomposition

$$A_\lambda u = \tilde{B}_\nu A_\lambda u + (1 - \tilde{B}_\nu) A_\lambda u,$$

the previous bound follows if we show

$$\|\tilde{B}_\nu A_\lambda u \cdot A_\nu v\|_{F_\lambda} \lesssim \nu^{n/2} \|A_\lambda u\|_{F_\lambda} \|A_\nu v\|_{F_\nu}, \qquad (2.136)$$

$$\|(1 - \tilde{B}_\nu) A_\lambda u \cdot A_\nu v\|_{X_\lambda^{1/2}} \lesssim \nu^{n/2} \|A_\lambda u\|_{X_\lambda^{1/2}} \|A_\nu v\|_{F_\nu}, \qquad (2.137)$$

and

$$\|(1 - \tilde{B}_\nu) A_\lambda u \cdot A_\nu v\|_{Y_\lambda} \lesssim \nu^{n/2} \|A_\lambda u\|_{Y_\lambda} \|A_\nu v\|_{F_\nu}. \qquad (2.138)$$

In what concerns (2.136), the Fourier support of $\tilde{B}_\nu A_\lambda u \cdot A_\nu v$ lies in the region $||\tau| - |\xi|| \lesssim \nu$. Hence, we can argue based on (2.129) and Corollary 2.4 to infer

$$\begin{aligned}
\|\tilde{B}_\nu A_\lambda u \cdot A_\nu v\|_{F_\lambda} &\lesssim \|\tilde{B}_\nu A_\lambda u \cdot A_\nu v\|_{X_\lambda^{1/2}} \\
&\lesssim \sum_{\mu \in 2^{\mathbb{Z}}, \, \mu \lesssim \nu} \mu^{1/2} \|B_\mu(\tilde{B}_\nu A_\lambda u \cdot A_\nu v)\|_{L^2} \\
&\lesssim \nu^{1/2} \|\tilde{B}_\nu A_\lambda u \cdot A_\nu v\|_{L^2} \lesssim \nu^{1/2} \|\tilde{B}_\nu A_\lambda u\|_{L^\infty L^2} \|A_\nu v\|_{L^2 L^\infty} \\
&\lesssim \nu^{n/2} \|\tilde{B}_\nu A_\lambda u\|_{F_\lambda} \|A_\nu v\|_{F_\nu} \lesssim \nu^{n/2} \|A_\lambda u\|_{F_\lambda} \|A_\nu v\|_{F_\nu}.
\end{aligned}$$

For (2.137), we remark that the Fourier support of $(1 - \tilde{B}_\nu) A_\lambda u$ lies in the region $||\tau| - |\xi|| \gg \nu$, whereas the one for $A_\nu v$ is contained in the region $|\tau| + |\xi| \simeq \nu$. Therefore, for any $\mu \in 2^{\mathbb{Z}}$, we have

$$B_\mu \left( (1 - \tilde{B}_\nu) A_\lambda u \cdot A_\nu v \right) \simeq B_\mu \left( \bar{B}_\mu (1 - \tilde{B}_\nu) A_\lambda u \cdot A_\nu v \right) \qquad (2.139)$$

where the multiplier $\bar{B}_\mu = \bar{B}_\mu(\nabla)$ imposes the same frequency localization as $B_\mu$ (i.e., $||\tau| - |\xi|| \simeq \mu$), but its symbol has a slightly larger support. It follows that

$$\begin{aligned}
\|(1 - \tilde{B}_\nu) A_\lambda u \cdot A_\nu v\|_{X_\lambda^{1/2}} &= \sum_{\mu \in 2^{\mathbb{Z}}} \mu^{1/2} \left\| B_\mu \left( (1 - \tilde{B}_\nu) A_\lambda u \cdot A_\nu v \right) \right\|_{L^2} \\
&\lesssim \sum_{\mu \in 2^{\mathbb{Z}}} \mu^{1/2} \left\| B_\mu \left( \bar{B}_\mu (1 - \tilde{B}_\nu) A_\lambda u \cdot A_\nu v \right) \right\|_{L^2} \\
&\lesssim \sum_{\mu \in 2^{\mathbb{Z}}} \mu^{1/2} \|\bar{B}_\mu (1 - \tilde{B}_\nu) A_\lambda u\|_{L^2} \|A_\nu v\|_{L^\infty} \\
&\lesssim \sum_{\mu \in 2^{\mathbb{Z}}} \mu^{1/2} \|\bar{B}_\mu A_\lambda u\|_{L^2} \cdot \nu^{n/2} \|A_\nu v\|_{F_\nu} \\
&\lesssim \nu^{n/2} \|A_\lambda u\|_{X_\lambda^{1/2}} \|A_\nu v\|_{F_\nu},
\end{aligned}$$

where we used, as above, (2.129).

In proving (2.138), we argue first on the basis of (2.129) and Corollary 2.4 to deduce

$$\|(1 - \tilde{B}_\nu)A_\lambda u \cdot A_\nu v\|_{L^\infty L^2} \lesssim \|(1 - \tilde{B}_\nu)A_\lambda u\|_{L^\infty L^2} \|A_\nu v\|_{L^\infty}$$
$$\lesssim \|A_\lambda u\|_{L^\infty L^2} \nu^{n/2} \|A_\nu v\|_{F_\nu} \qquad (2.140)$$
$$\lesssim \nu^{n/2} \|A_\lambda u\|_{Y_\lambda} \|A_\nu v\|_{F_\nu}.$$

For the $L^1 L^2$ component of $Y_\lambda$, we have

$$\|\Box((1 - \tilde{B}_\nu)A_\lambda u \cdot A_\nu v)\|_{L^1 L^2} \lesssim \|\Box\left((1 - \tilde{B}_\nu)A_\lambda u\right) \cdot A_\nu v\|_{L^1 L^2}$$
$$+ \|\nabla\left((1 - \tilde{B}_\nu)A_\lambda u\right) \cdot \nabla A_\nu v\|_{L^1 L^2} + \|(1 - \tilde{B}_\nu)A_\lambda u \cdot \Box A_\nu v\|_{L^1 L^2}$$

and we analyze each term separately. Using yet again (2.129) and Corollary 2.4, we derive

$$\|\Box\left((1 - \tilde{B}_\nu)A_\lambda u\right) \cdot A_\nu v\|_{L^1 L^2} \lesssim \|\Box\left((1 - \tilde{B}_\nu)A_\lambda u\right)\|_{L^1 L^2} \|A_\nu v\|_{L^\infty}$$
$$\lesssim \|\Box A_\lambda u\|_{L^1 L^2} \nu^{n/2} \|A_\nu v\|_{F_\nu}$$
$$\lesssim \lambda \nu^{n/2} \|A_\lambda u\|_{Y_\lambda} \|A_\nu v\|_{F_\nu}.$$

Together with Bernstein's inequality, the same reasons yield

$$\|\nabla\left((1 - \tilde{B}_\nu)A_\lambda u\right) \cdot \nabla A_\nu v\|_{L^1 L^2} \lesssim \|(1 - \tilde{B}_\nu)\nabla A_\lambda u\|_{L^1 L^2} \|\nabla A_\nu v\|_{L^\infty}$$
$$\lesssim \nu^{-1} \|\nabla A_\lambda u\|_{Y_\lambda} \nu \|A_\nu v\|_{L^\infty}$$
$$\lesssim \lambda \nu^{n/2} \|A_\lambda u\|_{Y_\lambda} \|A_\nu v\|_{F_\nu}$$

and

$$\|(1 - \tilde{B}_\nu)A_\lambda u \cdot \Box A_\nu v\|_{L^1 L^2} \lesssim \|(1 - \tilde{B}_\nu)A_\lambda u\|_{L^1 L^2} \|\Box A_\nu v\|_{L^\infty}$$
$$\lesssim \nu^{-1} \|A_\lambda u\|_{Y_\lambda} \nu^2 \|A_\nu v\|_{L^\infty}$$
$$\lesssim \nu^{n/2+1} \|A_\lambda u\|_{Y_\lambda} \|A_\nu v\|_{F_\nu}.$$

Thus, the last four estimates imply

$$\|\Box((1 - \tilde{B}_\nu)A_\lambda u \cdot A_\nu v)\|_{L^1 L^2} \lesssim \lambda \nu^{n/2} \|A_\lambda u\|_{Y_\lambda} \|A_\nu v\|_{F_\nu}.$$

By coupling this bound with (2.140), we obtain (2.138) and complete the proof of this theorem. $\qquad \Box$

The second product result concerning Tataru's spaces completes the picture about the interaction between $F$ and $\Box F$.

**Theorem 2.17.** *If $n \geq 4$, then the following estimate holds:*

$$\|uv\|_{\Box F} \lesssim \|u\|_{\Box F} \|v\|_F. \qquad (2.141)$$

**Proof.** This argument bears many similarities with the one in the previous theorem. The $l^1$ summation motivation used before reduces the proof of (2.141) to the one for

$$\|A_\lambda u \cdot A_\nu v\|_{\square F} \lesssim \lambda^{n/2} \|A_\lambda u\|_{\square F_\lambda} \nu^{n/2} \|A_\nu v\|_{F_\nu}.$$

This is analyzed separately in the complementary regimes $\lambda \lesssim \nu$ and $\lambda \gg \nu$. We start by noticing that

$$\square F_\lambda \subset \lambda^{3/2} L^2. \tag{2.142}$$

Indeed, relying on the frequency localizations imposed by $A_\lambda$ and $B_\mu$ and Bernstein's inequality, we obtain

$$\|A_\lambda u\|_{L^2} \lesssim \lambda^{1/2} \|A_\lambda u\|_{L^1 L^2}$$

and

$$\begin{aligned}
\|A_\lambda u\|_{L^2} &\lesssim \sum_{\mu \in 2^{\mathbb{Z}},\, \mu \lesssim \lambda} \|B_\mu A_\lambda u\|_{L^2} \\
&\lesssim \lambda^{1/2} \sum_{\mu \in 2^{\mathbb{Z}},\, \mu \lesssim \lambda} \mu^{-1/2} \|B_\mu A_\lambda u\|_{L^2} \\
&\lesssim \lambda^{1/2} \|A_\lambda u\|_{X_\lambda^{-1/2}},
\end{aligned}$$

which are sufficient to claim the above embedding.

If $\lambda \lesssim \nu$, the Fourier support of $A_\lambda u \cdot A_\nu v$ lies in the region $|\tau| + |\xi| \lesssim \nu$. Hence, an application of $n \geq 4$, Corollary 2.4, (2.129), and (2.142) leads to

$$\begin{aligned}
\|A_\lambda u \cdot A_\nu v\|_{\square F} &\lesssim \sum_{\mu \in 2^{\mathbb{Z}},\, \mu \lesssim \nu} \mu^{n/2} \|A_\mu (A_\lambda u \cdot A_\nu v)\|_{\square F_\mu} \\
&\lesssim \sum_{\mu \in 2^{\mathbb{Z}},\, \mu \lesssim \nu} \mu^{n/2-1} \|A_\mu (A_\lambda u \cdot A_\nu v)\|_{L^1 L^2} \\
&\lesssim \nu^{n/2-1} \|A_\lambda u \cdot A_\nu v\|_{L^1 L^2} \\
&\lesssim \nu^{n/2-1} \|A_\lambda u\|_{L^2 L^{2n/3}} \|A_\nu v\|_{L^2 L^{2n/(n-3)}} \\
&\lesssim \nu^{n/2-1} \lambda^{(n-3)/2} \|A_\lambda u\|_{L^2} \nu \|A_\nu v\|_{F_\nu} \\
&\lesssim \lambda^{n/2} \|A_\lambda u\|_{\square F_\lambda} \nu^{n/2} \|A_\nu v\|_{F_\nu}.
\end{aligned}$$

If $\lambda \gg \nu$, the Fourier support of $A_\lambda u \cdot A_\nu v$ lies in the region $|\tau| + |\xi| \simeq \lambda$, which implies

$$\|A_\lambda u \cdot A_\nu v\|_{\square F} \simeq \lambda^{n/2} \|A_\lambda u\, A_\nu v\|_{\square F_\lambda}.$$

Hence, we need to show that

$$\|A_\lambda u \cdot A_\nu v\|_{\square F_\lambda} \lesssim \nu^{n/2} \|A_\lambda u\|_{\square F_\lambda} \|A_\nu v\|_{F_\nu},$$

for which it is sufficient to prove

$$\|A_\lambda u \cdot A_\nu v\|_{\Box F_\lambda}$$
$$\lesssim \lambda^{-1} \nu^{n/2} \min \left\{ \|A_\lambda u\|_{X_\lambda^{-1/2}}, \|A_\lambda u\|_{L^1 L^2} \right\} \|A_\nu v\|_{F_\nu}. \qquad (2.143)$$

Using only (2.129), we infer

$$\|A_\lambda u \cdot A_\nu v\|_{\Box F_\lambda} \lesssim \lambda^{-1} \|A_\lambda u \cdot A_\nu v\|_{L^1 L^2}$$
$$\lesssim \lambda^{-1} \|A_\lambda u\|_{L^1 L^2} \|A_\nu v\|_{L^\infty}$$
$$\lesssim \lambda^{-1} \nu^{n/2} \|A_\lambda u\|_{L^1 L^2} \|A_\nu v\|_{F_\nu}.$$

In what concerns the $X_\lambda^{-1/2}$ inequality, it is sufficient to demonstrate

$$\|\tilde{B}_\nu A_\lambda u \cdot A_\nu v\|_{\Box F_\lambda} \lesssim \lambda^{-1} \nu^{n/2} \|\tilde{B}_\nu A_\lambda u\|_{X_\lambda^{-1/2}} \|A_\nu v\|_{F_\nu}$$

and

$$\|(1 - \tilde{B}_\nu) A_\lambda u \cdot A_\nu v\|_{\Box F_\lambda} \lesssim \lambda^{-1} \nu^{n/2} \|(1 - \tilde{B}_\nu) A_\lambda u\|_{X_\lambda^{-1/2}} \|A_\nu v\|_{F_\nu}.$$

For the first estimate, we remark that

$$\|\tilde{B}_\nu A_\lambda u\|_{X_\lambda^{-1/2}} \simeq \sum_{\mu \in 2^\mathbb{Z}, \, \mu \lesssim \nu} \mu^{-1/2} \|B_\mu \tilde{B}_\nu A_\lambda u\|_{L^2} \gtrsim \nu^{-1/2} \|\tilde{B}_\nu A_\lambda u\|_{L^2}.$$

Together with (2.129), this bound yields

$$\|\tilde{B}_\nu A_\lambda u \cdot A_\nu v\|_{\Box F_\lambda} \lesssim \lambda^{-1} \|\tilde{B}_\nu A_\lambda u \cdot A_\nu v\|_{L^1 L^2}$$
$$\lesssim \lambda^{-1} \|\tilde{B}_\nu A_\lambda u\|_{L^2} \|A_\nu v\|_{L^2 L^\infty}$$
$$\lesssim \lambda^{-1} \nu^{n/2} \|\tilde{B}_\nu A_\lambda u\|_{X_\lambda^{-1/2}} \|A_\nu v\|_{F_\nu}.$$

For the estimate involving $(1 - \tilde{B}_\nu) A_\lambda u \cdot A_\nu v$, we argue as in the proof of (2.137), relying on (2.139). We deduce:

$$\|(1 - \tilde{B}_\nu) A_\lambda u \cdot A_\nu v\|_{\Box F_\lambda} \lesssim \lambda^{-1} \|(1 - \tilde{B}_\nu) A_\lambda u \cdot A_\nu v\|_{X_\lambda^{-1/2}}$$
$$= \lambda^{-1} \sum_{\mu \in 2^\mathbb{Z}} \mu^{-1/2} \|B_\mu \left( (1 - \tilde{B}_\nu) A_\lambda u \cdot A_\nu v \right) \|_{L^2}$$
$$\lesssim \lambda^{-1} \sum_{\mu \in 2^\mathbb{Z}} \mu^{-1/2} \|B_\mu \left( \bar{B}_\mu (1 - \tilde{B}_\nu) A_\lambda u \cdot A_\nu v \right) \|_{L^2}$$
$$\lesssim \lambda^{-1} \sum_{\mu \in 2^\mathbb{Z}} \mu^{-1/2} \|\bar{B}_\mu (1 - \tilde{B}_\nu) A_\lambda u\|_{L^2} \|A_\nu v\|_{L^\infty}$$
$$\lesssim \lambda^{-1} \nu^{n/2} \|(1 - \tilde{B}_\nu) A_\lambda u\|_{X_\lambda^{-1/2}} \|A_\nu v\|_{F_\nu}.$$

This concludes the argument for (2.143) and finishes the proof of the theorem. $\qquad \Box$

To finish this section, we briefly describe the approach taken by Tataru [178] in defining $n = 2$ and $n = 3$ versions of these function spaces, for which he only modifies the $Y_\lambda$ and the $\lambda(L^1 L^2)_\lambda$ components of $F_\lambda$ and $\Box F_\lambda$, respectively. First, if $\alpha > 0$ is a certain fixed angle, Tataru introduces an overlapping partition of an $\alpha^2$ conical neighborhood of the characteristic cone $K$ (given by (2.132)) into sectors of angle $\alpha$:

$$\{(\tau, \xi) \in \mathbb{R}^{n+1}; \angle((\tau, \xi), K) \le \alpha^2\} = \bigcup_{j \in I_\alpha} C_\alpha^j.$$

Accordingly, each sector has angular dimension $\alpha^{n-1} \times \alpha^2$ and $|I_\alpha| \simeq \alpha^{1-n}$. In connection with this partition, we can then write

$$\sum_{\lambda \in 2^{\mathbb{Z}}, \lambda \le 1/2} \phi\left(\frac{\angle((\tau, \xi), K)}{\lambda \alpha^2}\right) = \sum_{j \in I_\alpha} r_\alpha^j(\tau, \xi), \qquad \forall (\tau, \xi) \in \mathbb{R}^{n+1},$$

where $\phi$ is given by (2.113) and, for each $j$, the function $r_\alpha^j = r_\alpha^j(\tau, \xi)$ is supported in $C_\alpha^j$. We denote by $R_\alpha^j = R_\alpha^j(\nabla)$ the spacetime Fourier multiplier whose symbol is $r_\alpha^j$. Furthermore, if $\omega \in \mathbb{S}^{n-1}$ is an arbitrary direction, then we consider its associated null vector,

$$\theta_\omega \stackrel{\text{def}}{=} \frac{1}{\sqrt{2}}(1, \omega) \in K, \tag{2.144}$$

and null coordinates,

$$t_\omega \stackrel{\text{def}}{=} (t, x) \cdot \theta_\omega, \qquad x_\omega \stackrel{\text{def}}{=} (t, x) - t_\omega \theta_\omega. \tag{2.145}$$

**Definition 2.5 ([178]).** Let $\lambda \in 2^{\mathbb{Z}}$ and $\alpha > 0$ be a fixed frequency and angle, respectively.

i) A function $u = u(t, x)$ is called a $Y_{\lambda, \alpha}$ *atom* if $u = A_\lambda \tilde{B}_{\frac{\alpha^2 \lambda}{20}} u$ and, for each $j \in I_\alpha$, there exists $\omega_j \in \mathbb{S}^{n-1}$ such that $\theta_{\omega_j}$ makes an angle $\alpha_j \ge \alpha$ with the Fourier support of $A_\lambda R_\alpha^j$ and

$$\sum_{j \in I_\alpha} (\lambda \alpha_j)^{-2} \|\Box R_\alpha^j u\|_{L^1_{t_{\omega_j}} L^2_{x_{\omega_j}}}^2 \le 1,$$

$$\sum_{j \in I_\alpha} \alpha_j^2 \|R_\alpha^j u\|_{L^\infty_{t_{\omega_j}} L^2_{x_{\omega_j}}}^2 \le 1.$$

ii) A function $f = f(t, x)$ is called a $\Box Y_{\lambda, \alpha}$ *atom* if $f = A_\lambda \tilde{B}_{\frac{\alpha^2 \lambda}{20}} f$ and, for each $j \in I_\alpha$, there exists $\omega_j \in \mathbb{S}^{n-1}$ such that $\theta_{\omega_j}$ makes an angle $\alpha_j \ge \alpha$ with the Fourier support of $A_\lambda R_\alpha^j$ and

$$\sum_{j \in I_\alpha} (\lambda \alpha_j)^{-2} \|R_\alpha^j f\|_{L^1_{t_{\omega_j}} L^2_{x_{\omega_j}}}^2 \le 1.$$

This definition allows us to introduce the atomic spaces $Y_{\lambda,\alpha}$ and $\Box Y_{\lambda,\alpha}$, which are given by

$$\|u\|_{Y_{\lambda,\alpha}} \overset{\text{def}}{=} \inf\left\{\sum_{k\geq1}|a_k|;\ u = \sum_{k\geq1}a_k u_k,\ u_k \text{ are } Y_{\lambda,\alpha} \text{ atoms}\right\}$$

and

$$\|f\|_{\Box Y_{\lambda,\alpha}} \overset{\text{def}}{=} \inf\left\{\sum_{k\geq1}|b_k|;\ f = \sum_{k\geq1}b_k f_k,\ f_k \text{ are } \Box Y_{\lambda,\alpha} \text{ atoms}\right\}.$$

Using these spaces, we can finally write down the dyadic pieces for the $n = 2$ and $n = 3$ versions of $F$ and $\Box F$ as

$$F_\lambda \overset{\text{def}}{=} X_\lambda^{1/2} + \sum_{\alpha\in2^{\mathbb{Z}},\,\alpha\leq1} Y_{\lambda,\alpha}$$

and

$$\Box F_\lambda \overset{\text{def}}{=} \lambda X_\lambda^{-1/2} + \sum_{\alpha\in2^{\mathbb{Z}},\,\alpha\leq1} \Box Y_{\lambda,\alpha}.$$

# Chapter 3

# Local and small data global well-posedness theory for wave maps

## 3.1 The Cauchy problem, the concept of well-posedness, and scaling heuristics

We recall that

$$\phi : (M, g) = (\mathbb{R}^{n+1}, \text{diag}(-1, 1, \dots, 1)) \to (N, h)$$

is a *wave map* from the Minkowski spacetime to a complete Riemannian manifold if it satisfies the equation

$$D^\mu \partial_\mu \phi = 0, \tag{3.1}$$

where $D$ stands for the covariant derivative induced by the metric $h$ on $N$ and the metric $g$ is used to raise and lower indices. As this is an evolution equation, one of the most natural things to study about it is the associated Cauchy problem, i.e.,

$$\begin{cases} D^\mu \partial_\mu \phi = 0, \\ \phi(0) = \phi_0 : \mathbb{R}^n \to N, \\ \partial_t \phi(0) = \phi_1 : \mathbb{R}^n \to T_{\phi_0} N. \end{cases} \tag{3.2}$$

In 1902, in reference to mathematical models of physical phenomena, Hadamard [66] introduced the concept of *well-posedness*, which, for Cauchy problems, means:

*Given a certain initial data, there exists a unique solution verifying the evolution equation, which is stable under small perturbations of the initial data.*

What does this mean for the wave maps equation? First, based on the conservation of energy

$$\mathcal{E}(\phi)(t) = \frac{1}{2} \int_{\mathbb{R}^n} |\nabla \phi(t)|_h^2 \, dx = \frac{1}{2} \int_{\mathbb{R}^n} |\phi_1|_h^2 + |\nabla_x \phi_0|_h^2 \, dx \tag{3.3}$$

proven in Chapter 1, which can be seen as a statement about the $\dot{H}^1$ norm of the solution, it makes sense to study (3.2) with initial data having Sobolev regularity. Therefore, we are led to consider $\phi_0 \in H^s(\mathbb{R}^n)$ and $\phi_1 \in H^{s-1}(\mathbb{R}^n)$ and look for a solution satisfying $\phi \in C([-T,T], H^s(\mathbb{R}^n))$ and $\partial_t \phi \in C([-T,T], H^{s-1}(\mathbb{R}^n))$, with $T > 0$. A more precise statement is as follows.

**Definition 3.1.** The Cauchy problem (3.2) is *locally well-posed* in $H^s \times H^{s-1}(\mathbb{R}^n)$ if, for all

$$(\phi_0, \phi_1) \in H^s \times H^{s-1}(\mathbb{R}^n),$$

there exist $T = T(\phi_0, \phi_1) > 0$ and

$$\phi \in C([-T,T], H^s(\mathbb{R}^n)) \cap C^1([-T,T], H^{s-1}(\mathbb{R}^n))$$

such that:

   i) $\phi$ verifies the equation in the sense of distributions;
   ii) $\phi(0) = \phi_0$ and $\partial_t \phi(0) = \phi_1$;
   iii) $\phi$ is unique in the above class;
   iv) $\phi$ depends continuously on $(\phi_0, \phi_1)$ in the following sense: if $\phi$ is defined up to time $T'$, then there exist two positive constants $C_1 = C_1(\phi_0, \phi_1)$ and $C_2 = C_2(\phi_0, \phi_1)$ such that, for all $(\bar{\phi}_0, \bar{\phi}_1) \in H^s \times H^{s-1}$ with

$$\|(\bar{\phi}_0 - \phi_0, \bar{\phi}_1 - \phi_1)\|_{H^s \times H^{s-1}} < C_1,$$

a solution $\bar{\phi}$ with $(\bar{\phi}_0, \bar{\phi}_1)$ as initial data is defined at least up to time $T'$ and

$$\|\bar{\phi} - \phi\|_{C([-T',T'], H^s) \cap C^1([-T',T'], H^{s-1})} \leq C_2 \|(\bar{\phi}_0 - \phi_0, \bar{\phi}_1 - \phi_1)\|_{H^s \times H^{s-1}}.$$

**Remark 3.1.** It is common in the literature to also include in the definition of well-posedness *the persistence of higher regularity*, i.e., if the initial data additionally satisfies $(\phi_0, \phi_1) \in H^\sigma \times H^{\sigma-1}$ with $\sigma > s$, then

$$\phi \in C([-T,T], H^\sigma) \cap C^1([-T,T], H^{\sigma-1}).$$

**Remark 3.2.** If the unique solution of (3.2) with $\phi_0 = \phi_1 = 0$ is given by $\phi = 0$, then the previous definition implies that, given $T > 0$, the solution $\phi$ is defined up to time $T$ provided $\|\phi_0\|_{H^s} + \|\phi_1\|_{H^{s-1}}$ is sufficiently small.

**Remark 3.3.** If one can choose $T = \infty$ in the definition of local well-posedness, then we say that the equation is *globally well-posed*. However, in this situation, we usually have restrictions on the size of initial data. As such, we define *small data global well-posedness* to be:

There exists $M > 0$ such that, for all initial data verifying $\|\phi_0\|_{H^s} + \|\phi_1\|_{H^{s-1}} < M$, a unique solution exists to the Cauchy problem (3.2) satisfying

$$\phi \in C(\mathbb{R}, H^s) \cap C^1(\mathbb{R}, H^{s-1})$$

and the solution map is continuous in the respective topologies.

In regards to Definition 3.1, one can naturally ask the following question: what are the values of $s$ for which (3.2) is locally well-posed and how small can they be? For low values of $s$, there are problems even defining the Sobolev regularity for maps with values in a manifold. If $s > n/2$, then the wave map is continuous and we can restrict it to a local chart. The regularity is then described using local coordinates. For $s \leq n/2$, complications arise and one needs to look at the global properties of $N$ (e.g., if it embeds isometrically in a Euclidean space).

As noticed in Chapter 1, the $n/2$ value is also related to the scaling invariance of (3.1). If $\phi$ is a solution on the time interval $[-T, T]$, then

$$\phi_\lambda = \phi_\lambda(t, x) = \phi(t/\lambda, x/\lambda), \qquad \lambda \in \mathbb{R},$$

satisfies (3.1) on the time interval $[-\lambda T, \lambda T]$. Moreover, if

$$\phi \in C([-T, T], \dot{H}^s(\mathbb{R}^n)) \cap C^1([-T, T], \dot{H}^{s-1}(\mathbb{R}^n)),$$

then

$$\phi_\lambda \in C([-\lambda T, \lambda T], \dot{H}^s(\mathbb{R}^n)) \cap C^1([-\lambda T, \lambda T], \dot{H}^{s-1}(\mathbb{R}^n)),$$

and

$$\|\phi_\lambda(0)\|_{\dot{H}^s(\mathbb{R}^n)} + \|\partial_t(\phi_\lambda)(0)\|_{\dot{H}^{s-1}(\mathbb{R}^n)}$$
$$= \lambda^{\frac{n}{2}-s}\left(\|\phi(0)\|_{\dot{H}^s(\mathbb{R}^n)} + \|\partial_t\phi(0)\|_{\dot{H}^{s-1}(\mathbb{R}^n)}\right).$$

By choosing $\lambda \gg 1$, we deduce the following three scenarios:

- if $s < n/2$, a small data short time result is equivalent to a large data long time result;
- if $s = n/2$, a short time result is equivalent to a long time result, without any restriction on the size of the initial data;
- if $s > n/2$, a large data short time result is equivalent to a small data long time result.

Hence, one expects the Cauchy problem to be

- *ill-posed*[1] for $s < n/2$;
- *globally well-posed for small data in* $\dot{H}^{n/2} \times \dot{H}^{n/2-1}(\mathbb{R}^n)$;
- *locally well-posed* for $s > n/2$.

## 3.2   Abstract existence theory

If we assume the wave map to be continuous[2], then we can use the local coordinates on the target manifold and write (3.1) in its intrinsic form,

$$\Box\phi^j + \Gamma^j_{ik}(\phi)\,\partial^\mu\phi^i\,\partial_\mu\phi^k = 0,$$

where $\Gamma^j_{ik}$ represent the Christoffel symbols associated to $h$. To simplify the exposition, we assume that $\Gamma^j_{ik}$ depend polynomially on $\phi$. This is not a significant constraint on $N$, as the most interesting target manifolds are uniformly analytic (e.g., the unit sphere $\mathbb{S}^m \subset \mathbb{R}^{m+1}$, the hyperbolic plane $\mathbb{H}^m \subset \mathbb{R}^{m+1}$, the Euclidean space $\mathbb{R}^m$). In this case, the Christoffel symbols can be expanded in power series, with uniform exponential bounds on the series coefficients, thus reducing the analysis to the one with polynomial dependence.

Schematically, the previous equation is of the type

$$\Box\phi = N(\phi) = \Gamma(\phi)\,Q(\phi,\phi), \tag{3.4}$$

with

$$Q(\phi,\psi) = \partial^\mu\phi\,\partial_\mu\psi = \frac{\Box(\phi\,\psi) - \phi\,\Box\psi - \phi\,\Box\psi}{2}. \tag{3.5}$$

Thus, we are dealing with a semilinear wave equation for which the usual approach is to treat the nonlinearity perturbatively and to absorb it through a contraction mapping argument. We present this framework in an abstract manner to stress its general applicability to evolution equations. We follow the exposition in Keel-Tao [79], as that was done also in the context of wave maps.

We start by considering two Banach spaces: $D$, the initial data space, and $X$, the solution or iteration space. $X$ is made up of functions defined

---

[1]For a more detailed discussion of the scaling heuristics in this case, we refer the reader to Subsection 9.1.3 of Selberg's lecture notes [142].

[2]This can be assured if we take the initial data in either $H^s \times H^{s-1}(\mathbb{R}^n)$, with $s > n/2$, or $\dot{B}^{n/2}_{2,1} \times \dot{B}^{n/2-1}_{2,1}(\mathbb{R}^n)$.

on $I \times \mathbb{R}^n$, where $I \subseteq \mathbb{R}$ is an arbitrary time interval, and $\mathcal{S}(I \times \mathbb{R}^n)$ is a dense set in $X$. We work with the Cauchy problem

$$\begin{cases} L\Phi = N(\Phi), \quad \Phi = \Phi(t,x), \quad (t,x) \in J \times \mathbb{R}^n, \\ \\ \Phi(0) = f, \quad f = (f_1, \ldots, f_d) \in D, \end{cases} \tag{3.6}$$

where $L$ is a linear evolution operator of order $d$, $N$ is a nonlinear operator satisfying $N(0) = 0$, and $0 \in J \subseteq I$ is another time interval.

We can recast this problem as

$$\Phi = \varphi\left(S(f) + L^{-1}N(\Phi)\right), \tag{3.7}$$

with $\varphi : I \to \mathbb{R}$ satisfying $\varphi \equiv 1$ on $J$, $f \mapsto S(f)$ solving

$$L(S(f)) = 0, \quad S(f)(0) = f,$$

and $F \mapsto L^{-1}F$ solving

$$L(L^{-1}F) = F, \quad L^{-1}F(0) = 0.$$

Moreover, we assume the map $f \in D \mapsto \varphi S(f) \in X$ is well-defined. As an application of the contraction mapping theorem, we have the following result.

**Theorem 3.1.** *Let $C > 0$ be a fixed number. Under the framework explained above, assume that the estimates*

$$\|\varphi S(f)\|_X \leq C \|f\|_D, \tag{3.8}$$

$$\sup_{t \in J} \|\Phi(t)\|_D \leq C \|\Phi\|_X, \tag{3.9}$$

*and*

$$\|\varphi L^{-1}(N(\Phi) - N(\Psi))\|_X \leq C \left(\|\Phi\|_X + \|\Psi\|_X\right) \|\Phi - \Psi\|_X \tag{3.10}$$

*are true for all $f \in D$ and all $\Phi, \Psi \in X$ with $\|\Phi\|_X$ and $\|\Psi\|_X$ sufficiently small. If $\|f\|_D$ is sufficiently small, depending only on $C$, then the Cauchy problem (3.6) has a unique solution $\Phi \in X \cap C(J, D)$ and the map $f \mapsto \Phi$ is Lipschitz continuous.*

**Proof.** We fix $f \in D$ with $\|f\|_D \leq \frac{3}{16C^2}$ and show that the map

$$T(\Phi) = \varphi\left(S(f) + L^{-1}N(\Phi)\right)$$

is a contraction on

$$B\left(0, \frac{1}{4C}\right) = \left\{\Omega \in X; \|\Omega\|_X \leq \frac{1}{4C}\right\}.$$

Indeed, if $\Phi \in B\left(0, \frac{1}{4C}\right)$, then

$$\|T(\Phi)\|_X \leq C(\|f\|_D + \|\Phi\|_X^2) \leq \frac{1}{4C},$$

according to (3.8) and (3.10). Furthermore, due to (3.10), we also have

$$\|T(\Phi) - T(\Psi)\|_X \leq C\left(\|\Phi\|_X + \|\Psi\|_X\right)\|\Phi - \Psi\|_X \leq \frac{1}{2}\|\Phi - \Psi\|_X.$$

Hence, the contraction mapping theorem yields a unique fixed point $\Phi \in B\left(0, \frac{1}{4C}\right)$ for the map $T$, which solves (3.7) and, as a consequence, (3.6). As $\Phi \in X$, (3.9) and a density argument using Schwartz functions show that $\Phi \in C(J, D)$.

In what concerns the Lipschitz continuity of the solution map, we can derive, using (3.8) and (3.10),

$$\|\Phi_1 - \Phi_2\|_X \leq C \left[\|f_1 - f_2\|_D + (\|\Phi_1\|_X + \|\Phi_2\|_X)\|\Phi_1 - \Phi_2\|_X\right]$$
$$\leq C \left[\|f_1 - f_2\|_D + \frac{1}{2C}\|\Phi_1 - \Phi_2\|_X\right],$$

where $\Phi_1$ and $\Phi_2$ are solutions to (3.6) with initial data $f_1$ and $f_2$, respectively. This implies

$$\|\Phi_1 - \Phi_2\|_X \leq 2C\|f_1 - f_2\|_D,$$

which finishes the proof. $\qquad\square$

**Remark 3.4.** The paper of Keel-Tao [79] also contains an abstract result which can be used to show persistence of regularity for the Cauchy problem (3.6).

## 3.3   Local well-posedness results

In this section, we plan to use the previously developed abstract framework to prove local well-posedness results for the Cauchy problem associated to (3.4). Theorem 3.1 reduces the well-posedness argument to the proof of a certain set of estimates in an appropriate functional setting. It is our intention to motivate in this way the emergence of energy estimates, Strichartz estimates, and hyperbolic Sobolev spaces as fundamental tools in studying the local regularity of wave maps.

### 3.3.1 *Energy arguments*

What is considered by now the "classical" approach is an argument based on energy estimates, which proves local well-posedness in the range $s > n/2+1$ for all $n \geq 1$. The main tool is the following result, whose proof we omit as it is standard and can be found, as stated here, in Section 5.1 of Selberg's lecture notes [142]. It generalizes the energy bounds discussed in Chapter 1.

**Theorem 3.2.** *Let $s \in \mathbb{R}$ and $T > 0$ be arbitrary. If $u_0 \in H^s(\mathbb{R}^n)$, $u_1 \in H^{s-1}(\mathbb{R}^n)$, and $F \in L^1 H^{s-1}([0,T] \times \mathbb{R}^n)$, then the Cauchy problem*

$$\begin{cases} \Box u = F, \\ \\ u(0) = u_0, \qquad u_t(0) = u_1, \end{cases} \tag{3.11}$$

*admits a unique solution $u$ on $[0,T] \times \mathbb{R}^n$ satisfying*

$$u \in C([0,T], H^s(\mathbb{R}^n)) \cap C^1([0,T], H^{s-1}(\mathbb{R}^n))$$

*and, for all $0 \leq t \leq T$,*

$$\begin{aligned} \|u(t)\|_{H^s} + \|\partial_t u(t)\|_{H^{s-1}} &\leq M(1+T) \Big( \|u_0\|_{H^s} + \|u_1\|_{H^{s-1}} \\ &\quad + \int_0^t \|F(t')\|_{H^{s-1}} \, dt' \Big), \end{aligned} \tag{3.12}$$

*where $M$ is a constant depending only on $s$.*

**Remark 3.5.** Using the time reversibility of the solution for (3.11), one can easily modify the statement of the previous theorem to switch from the time interval $[0,T]$ to $[-T,T]$.

We now start the "classical" approach by fixing $s > n/2+1$ and $T > 0$. We choose in the abstract framework

$$D = H^s \times H^{s-1}(\mathbb{R}^n), \quad \|(\phi_0, \phi_1)\|_D = \|\phi_0\|_{H^s} + \|\phi_1\|_{H^{s-1}}, \tag{3.13}$$

and

$$\begin{aligned} X &= C([-T,T], H^s(\mathbb{R}^n)) \cap C^1([-T,T], H^{s-1}(\mathbb{R}^n)), \\ \|\phi\|_X &= \|\phi\|_{L^\infty H^s} + \|\partial_t \phi\|_{L^\infty H^{s-1}}. \end{aligned} \tag{3.14}$$

We also set $I = J = [-T,T]$, $\varphi \equiv 1$, $L = \Box$, and the nonlinear operator $N$ to be given by the right-hand side of (3.4), which we decide to write it generically as

$$N(\phi) = \Gamma(\phi) \, \nabla\phi \, \nabla\phi.$$

The assumption concerning the map $f \in D \mapsto \varphi\, S(f) \in X$ and (3.8) are direct consequences of Theorem 3.2 and (3.12), with $C = M(1 + T)$. The estimate (3.9) is in fact an equality with $C = 1$, due to the definitions of $\|\cdot\|_D$ and $\|\cdot\|_X$. For (3.10), we use again (3.12) to infer

$$
\begin{aligned}
\|\varphi\, L^{-1}(N(\phi) - N(\psi))\|_X &= \|\Box^{-1}(N(\phi) - N(\psi))\|_X \\
&\le M(1 + T)\, \|N(\phi) - N(\psi)\|_{L^1 H^{s-1}([-T,T] \times \mathbb{R}^n)} \\
&\le 2M(T + T^2)\, \|N(\phi) - N(\psi)\|_{L^\infty H^{s-1}([-T,T] \times \mathbb{R}^n)}.
\end{aligned}
$$

As $s - 1 > n/2$, Corollary 2.1 implies that $L^\infty H^{s-1}([-T,T] \times \mathbb{R}^n)$ is an algebra. Taking into account the polynomial dependance of $\Gamma$ on $\phi$, we deduce

$$
\begin{aligned}
\|N(\phi) &- N(\psi)\|_{L^\infty H^{s-1}} \\
&\le \|(\Gamma(\phi) - \Gamma(\psi))\, \nabla\phi\, \nabla\phi\|_{L^\infty H^{s-1}} + \|\Gamma(\psi)\, \nabla(\phi - \psi)\, \nabla\phi\|_{L^\infty H^{s-1}} \\
&\quad + \|\Gamma(\psi)\, \nabla\psi\, \nabla(\phi - \psi)\|_{L^\infty H^{s-1}} \\
&\lesssim \|\Gamma(\phi) - \Gamma(\psi)\|_{L^\infty H^{s-1}}\, \|\phi\|_X^2 \\
&\quad + (1 + \|\Gamma(\psi) - \Gamma(0)\|_{L^\infty H^{s-1}})\, (\|\phi\|_X + \|\psi\|_X)\, \|\phi - \psi\|_X \\
&\lesssim (1 + \|\phi\|_X + \|\psi\|_X)^A\, \|\phi\|_X^2\, \|\phi - \psi\|_X \\
&\quad + (1 + \|\psi\|_X)^A\, (\|\phi\|_X + \|\psi\|_X)\, \|\phi - \psi\|_X,
\end{aligned}
$$

where $A$ is a nonnegative integer. Hence, (3.10) holds in this context for $\|\phi\|_X$ and $\|\psi\|_X$ sufficiently small, with a constant which is a multiple of $T + T^2$. This finishes the argument for the local well-posedness of (3.4) when $s > n/2 + 1$.

### 3.3.2  *Strichartz-type arguments*

Here, our goal is to show that, when $n \ge 2$, the Strichartz estimates proven in Chapter 2, combined with the energy bound (3.12), lead to an improvement over the energy argument in the range of $s$ for the local well-posedness of (3.4). Precisely, we obtain local regularity for

$$
s > \max\left\{\frac{n+1}{2}, \frac{n+5}{4}\right\}. \tag{3.15}
$$

Originally, for $n = 3$, this result is due to Ponce-Sideris [134].

As before, we fix $T > 0$ and let $D$ be given by (3.13), whereas

$$
\begin{aligned}
X = \{\phi \in\ &C([-T,T], H^s(\mathbb{R}^n)) \cap C^1([-T,T], H^{s-1}(\mathbb{R}^n)), \\
&\nabla\phi \in L^p L^\infty([-T,T] \times \mathbb{R}^n)\},
\end{aligned}
$$

$$\|\phi\|_X = \|\phi\|_{L^\infty H^s} + \|\partial_t \phi\|_{L^\infty H^{s-1}} + \|\nabla \phi\|_{L^p L^\infty},$$

with

$$\begin{cases} p = 4 & \text{if} \quad n = 2, \\ p > 2 & \text{if} \quad n = 3, \\ p = 2 & \text{if} \quad n \geq 4, \end{cases}$$

and

$$s > \frac{n+2}{2} - \frac{1}{p}.$$

We can see immediately that this numerology allows us to obtain any $s$ verifying (3.15) by appropriately choosing $p$. Moreover, it tells us that $(p, \infty, n)$ is wave-admissible and $s > s(p, \infty, n) + 1$. The choices for $I$, $J$, $\varphi$, $L$, and $N$ do not change from the energy argument.

The map $f \in D \mapsto \varphi S(f) \in X$ is well-defined according to Corollary 2.3 and Theorem 3.2. The homogeneous estimate (3.8) is a direct consequence of (2.56) and (3.12). The consistency bound (3.9) is trivial, due to

$$\sup_{t \in J} \|\phi(t)\|_D = \|\phi\|_{L^\infty H^s} + \|\partial_t \phi\|_{L^\infty H^{s-1}}.$$

The most intricate part of the argument is verifying (3.10). Relying on (2.56) and (3.12), we first derive

$$\|\varphi L^{-1}(N(\phi) - N(\psi))\|_X \lesssim \|N(\phi) - N(\psi)\|_{L^1 H^{s-1}}. \tag{3.16}$$

Next, we decompose $N(\phi) - N(\psi)$ according to

$$N(\phi) - N(\psi)$$
$$= (\Gamma(\phi) - \Gamma(\psi)) \nabla \phi \nabla \phi + \Gamma(\psi) \nabla(\phi - \psi) \nabla \phi + \Gamma(\psi) \nabla \psi \nabla(\phi - \psi),$$

in order to see that it contains three types of generic terms:

$$\phi^A \psi^B (\phi - \psi) \nabla \phi \nabla \phi, \qquad \psi^A \nabla(\phi - \psi) \nabla \phi, \qquad \psi^B \nabla(\phi - \psi) \nabla \psi,$$

where $A$ and $B$ are nonnegative integers. In this setting, $L^\infty H^{s-1}$ is no longer an algebra. However, as $s > 1$, we can control products of functions in $L^1 H^{s-1}$ using (2.11) and Hölder's inequality. Thus, only one of the functions in the product is measured in the $H_x^{s-1}$ norm, with the remaining ones being measured in the $L_x^\infty$. The guidelines in applying Hölder's inequality (in the time variable) afterwards are as follows:

- if a gradient term is measured in the $H_x^{s-1}$ norm, then it will be estimated in $L^\infty H^{s-1}$;

- if a gradient term is measured in the $L_x^\infty$ norm, then it will be estimated in $L^p L^\infty$.

Moreover, due to $s > n/2$, we notice that

$$\|\phi\|_{L_{t,x}^\infty} \lesssim \|\phi\|_{L^\infty H^s} \le \|\phi\|_X.$$

We present below some of the most representative estimates in the analysis and leave the remaining ones as an exercise to the reader. For the first type of generic terms, we have

$$\left\| \|\phi^A \psi^B (\phi - \psi) \nabla\phi\|_{L_x^\infty} \cdot \|\nabla\phi\|_{H_x^{s-1}} \right\|_{L_t^1}$$
$$\lesssim T^{1-1/p} \|\phi\|_{L_{t,x}^\infty}^A \|\psi\|_{L_{t,x}^\infty}^B \|\phi - \psi\|_{L_{t,x}^\infty} \|\nabla\phi\|_{L^p L^\infty} \|\nabla\phi\|_{L^\infty H^{s-1}}$$
$$\lesssim T^{1-1/p} \|\phi\|_X^{A+2} \|\psi\|_X^B \|\phi - \psi\|_X$$

and

$$\left\| \|\phi^A \psi^B \nabla\phi \nabla\phi\|_{L_x^\infty} \cdot \|\phi - \psi\|_{H_x^{s-1}} \right\|_{L_t^1}$$
$$\lesssim T^{1-2/p} \|\phi\|_{L_{t,x}^\infty}^A \|\psi\|_{L_{t,x}^\infty}^B \|\nabla\phi\|_{L^p L^\infty}^2 \|\phi - \psi\|_{L^\infty H^{s-1}}$$
$$\lesssim T^{1-2/p} \|\phi\|_X^{A+2} \|\psi\|_X^B \|\phi - \psi\|_X.$$

For the second and third types of terms, we argue that

$$\left\| \|\psi^A \nabla(\phi - \psi)\|_{L_x^\infty} \cdot \|\nabla\phi\|_{H_x^{s-1}} \right\|_{L_t^1}$$
$$\lesssim T^{1-1/p} \|\psi\|_{L_{t,x}^\infty}^A \|\nabla(\phi - \psi)\|_{L^p L^\infty} \|\nabla\phi\|_{L^\infty H^{s-1}}$$
$$\lesssim T^{1-1/p} \|\psi\|_X^A \|\phi\|_X \|\phi - \psi\|_X$$

and

$$\left\| \|\psi^{A-1} \nabla(\phi - \psi) \nabla\phi\|_{L_x^\infty} \cdot \|\psi\|_{H_x^{s-1}} \right\|_{L_t^1}$$
$$\lesssim T^{1-2/p} \|\psi\|_{L_{t,x}^\infty}^{A-1} \|\nabla(\phi - \psi)\|_{L^p L^\infty} \|\nabla\phi\|_{L^p L^\infty} \|\psi\|_{L^\infty H^{s-1}}$$
$$\lesssim T^{1-2/p} \|\psi\|_X^A \|\phi\|_X \|\phi - \psi\|_X.$$

Together with (3.16), these bounds lead to (3.10), which finishes the argument.

### 3.3.3  *Hyperbolic Sobolev spaces approach*

Finally, using $X^{s,\theta}$ technology, we are able to prove the sharp local result: the equation (3.4) is locally well-posed in $H^s \times H^{s-1}(\mathbb{R}^n)$ for any $s > n/2$ and $n \ge 2$. This is due to Klainerman-Selberg [89], with the $n = 3$ result appearing prior to that in Klainerman-Machedon [87].

There are two important remarks to be made about this argument. The first one is that $N(\phi)$ will no longer be treated as a generic quadratic nonlinearity. Instead, the *null structure* of $Q(\phi, \phi)$, which is on display in (3.5), plays a crucial role in the analysis. The second remark pertains to the abstract framework, which allows us to work with functions that are defined globally in spacetime in order to prove a local-in-time regularity result. Thus, we can involve in the analysis the spacetime Fourier transform of the functions in question.

In this case, we fix $T > 0$, $\theta > 1/2$, and $s = (n - 1)/2 + \theta$. We work with $D$ as in (3.13), $I = \mathbb{R}$, and $X = X^{s,\theta}(\mathbb{R}^{n+1})$ with the norm given by (2.57). Furthermore, we let $J = [-T, T]$ and let $\varphi$ be an even, smooth cutoff function satisfying $\varphi \equiv 1$ on $J$. The operators $L$ and $N$ are as before.

The fact that the map $f \in D \mapsto \varphi S(f) \in X$ is well-defined and continuous follows as a result of (2.62). The consistency estimate (3.9) is given by (2.59), due to

$$\theta > 1/2 \qquad \text{and} \qquad s + \theta = \frac{n-1}{2} + 2\theta > \frac{3}{2}.$$

In what concerns the inhomogeneous bound (3.10), we rely first on (2.68) to infer

$$\|\varphi L^{-1}(N(\phi) - N(\psi))\|_X \lesssim \|N(\phi) - N(\psi)\|_{X^{s-1,\theta-1}}.$$

Using (3.5), we deduce

$$
\begin{aligned}
N(\phi) - N(\psi) &= \frac{\Gamma(\phi)\Box\phi^2 - \Gamma(\psi)\Box\psi^2}{2} + (\Gamma(\psi)\psi\Box\psi - \Gamma(\phi)\phi\Box\phi) \\
&= \frac{(\Gamma(\phi) - \Gamma(\psi))\Box\phi^2 + \Gamma(\psi)\Box(\phi^2 - \psi^2)}{2} \\
&\quad + (\Gamma(\psi) - \Gamma(\phi))\psi\Box\psi + \Gamma(\phi)\psi\Box(\psi - \phi) + \Gamma(\phi)(\psi - \phi)\Box\phi,
\end{aligned}
$$

which contains five types of generic terms:

$$\phi^A\psi^B(\phi - \psi)\Box\phi^2, \qquad \psi^A\Box(\phi^2 - \psi^2), \qquad \phi^A\psi^{B+1}(\psi - \phi)\Box\psi,$$
$$\phi^A\psi\Box(\psi - \phi), \qquad \phi^A(\psi - \phi)\Box\phi,$$

where $A$ and $B$ are nonnegative integers.

These multilinear terms are estimated in $X^{s-1,\theta-1}$ by applying (2.97), (2.88), and (2.58), as

$$n \geq 2 \qquad \text{and} \qquad s - \frac{n}{2} = \theta - \frac{1}{2}.$$

More precisely, for the first two types of generic terms we obtain

$$
\begin{aligned}
\|\phi^A\psi^B(\phi - \psi)\Box\phi^2\|_{X^{s-1,\theta-1}} &\lesssim \|\phi^A\psi^B(\phi - \psi)\|_{X^{s,\theta}}\|\Box\phi^2\|_{X^{s-1,\theta-1}} \\
&\lesssim \|\phi\|_{X^{s,\theta}}^A\|\psi\|_{X^{s,\theta}}^B\|\phi - \psi\|_{X^{s,\theta}}\|\phi^2\|_{X^{s,\theta}} \\
&\lesssim \|\phi\|_X^{A+2}\|\psi\|_X^B\|\phi - \psi\|_X
\end{aligned}
$$

and

$$\|\psi^A \Box(\phi^2 - \psi^2)\|_{X^{s-1,\theta-1}} \lesssim \|\psi^A\|_{X^{s,\theta}} \|\Box(\phi^2 - \psi^2)\|_{X^{s-1,\theta-1}}$$
$$\lesssim \|\psi\|_{X^{s,\theta}}^A \|\phi^2 - \psi^2\|_{X^{s,\theta}}$$
$$\lesssim \|\psi\|_{X^{s,\theta}}^A \|\phi + \psi\|_{X^{s,\theta}} \|\phi - \psi\|_{X^{s,\theta}}$$
$$\lesssim \|\psi\|_X^A (\|\phi\|_X + \|\psi\|_X) \|\phi - \psi\|_X,$$

with the last three types of generic terms being bounded very similarly. Hence, we conclude that (3.10) is verified, which finishes the argument.

**Remark 3.6.** For the $1 + 1$-dimensional problem, after a partially successful attempt by Keel-Tao [79], Machihara-Nakanishi-Tsugawa [111] proved the sharp local well-posedness result by rewriting the equation in the null coordinates

$$\alpha = t + x, \qquad \beta = t - x,$$

and relying again on $X^{s,\theta}$-type iteration spaces.

## 3.4 Small data global well-posedness results

Our focus shifts now to global well-posedness results for the initial value problem (3.2) and, based on the scaling heuristics mentioned before, we should consider initial data that have $\dot{H}^{n/2} \times \dot{H}^{n/2-1}(\mathbb{R}^n)$ regularity. Here, we discuss results for small data, with the more challenging large data problem to be addressed in the next chapter.

In describing the intuition behind small data results, we follow the presentation in Tataru [179]. One of the methods used in proving global well-posedness for certain nonlinear wave equations is to demonstrate a local regularity result in which the lifespan of the solution is controlled by a conserved quantity. However, in the case of the wave maps equation (3.1), the only such quantity is the energy (3.3), which is at $\dot{H}^1(\mathbb{R}^n)$ regularity level. Hence, this approach cannot be used in proving global results for wave maps when $n > 2$.

### 3.4.1 *Semilinear approach*

Instead, we could first rely on the argument used in showing sharp local well-posedness and choose, in the abstract framework, $I = J = \mathbb{R}$ and $\varphi \equiv 1$. Moreover, due to scaling, we should consider working with homogeneous versions of $X^{s,\theta}$ spaces for $s = n/2$ and $\theta = 1/2$:

$$\dot{X}^{\frac{n}{2},\frac{1}{2}}(\mathbb{R}^{n+1}) = \left\{ u \in \mathcal{S}'(\mathbb{R}^{n+1}); (|\tau| + |\xi|)^{\frac{n}{2}} ||\tau| - |\xi||^{\frac{1}{2}} \widehat{u}(\tau,\xi) \in L^2_{\tau,\xi} \right\}$$

as the iteration space and

$$\dot{X}^{\frac{n}{2}-1,-\frac{1}{2}}(\mathbb{R}^{n+1}) = \left\{ u \in \mathcal{S}'(\mathbb{R}^{n+1}); (|\tau| + |\xi|)^{\frac{n}{2}-1} ||\tau| - |\xi||^{-\frac{1}{2}} \widehat{u}(\tau,\xi) \in L^2_{\tau,\xi} \right\}$$

as the space in which we measure the nonlinearity (i.e., the counterpart of $X^{s-1,\theta-1}$ in this setting).

Unfortunately, the first space is not well-defined as a space of distributions, whereas the second one does not contain all test functions. This can be easily explained as follows. Choose a test function $u$ such that

$$\widehat{u} \equiv 1 \quad \text{on} \quad \{(\tau,\xi) \in \mathbb{R}^{n+1}; |\tau| + |\xi| \leq 1\}$$

and assume

$$u \in \dot{X}^{s,\theta}(\mathbb{R}^{n+1}) = \left\{ w \in \mathcal{S}'(\mathbb{R}^{n+1}); (|\tau| + |\xi|)^s ||\tau| - |\xi||^{\theta} \widehat{w}(\tau,\xi) \in L^2_{\tau,\xi} \right\}.$$

Then

$$\int_{|\tau|\leq 1} |\tau|^{2s+2\theta+n}\, d\tau \simeq \int_{\{|\tau|+|\xi|\leq 1,|\xi|\ll|\tau|\}} (|\tau| + |\xi|)^{2s} ||\tau| - |\xi||^{2\theta}\, d\tau\, d\xi < \infty,$$

which implies $s + \theta > -(n + 1)/2$. A similar argument shows that if $\dot{X}^{s,\theta}(\mathbb{R}^{n+1})$ contains all test functions, then $\theta > -1/2$. On the other hand, if we suppose that $\dot{X}^{s,\theta}(\mathbb{R}^{n+1})$ is well-defined as a space of distributions, then, by duality, $\dot{X}^{-s,-\theta}(\mathbb{R}^{n+1})$ should contain all test functions. By the previous discussion, we deduce that $s + \theta < (n + 1)/2$ and $\theta < 1/2$.

As was seen from the proof of local results, the main challenge in implementing the abstract approach is the inhomogeneous estimate (3.10). For a global-in-time argument, it reads

$$\|\Box^{-1}(N(\phi) - N(\psi))\|_X \lesssim (\|\phi\|_X + \|\psi\|_X) \|\phi - \psi\|_X,$$

where $X$ is the iteration space. One scheme to prove this would be to also introduce a function space $Y$ for the nonlinearity $N = N(\phi)$ and show

$$\Box^{-1} : Y \to X, \quad X \cdot X \to X, \quad X \cdot Y \to Y, \quad \partial^\mu X \cdot \partial_\mu X \to Y. \quad (3.17)$$

This can be further streamlined by using a Littlewood-Paley decomposition associated to the family of spacetime Fourier multipliers $(A_\lambda)_{\lambda \in 2^{\mathbb{Z}}}$ defined by (2.114). Accordingly, we denote by $X_\lambda$ and $Y_\lambda$ the subsets of $X$ and $Y$, respectively, containing only those functions whose spacetime Fourier transform is supported in the region

$$\{(\tau,\xi); \lambda/2 \leq |\tau| + |\xi| \leq 2\lambda\}.$$

Due to the $l^2$ summability embedded in the structure of $\dot{H}^{n/2} \times \dot{H}^{n/2-1}(\mathbb{R}^n)$, we should first expect that

$$\|\phi\|_X^2 \simeq \sum_{\lambda \in 2^{\mathbb{Z}}} \|A_\lambda \phi\|_{X_\lambda}^2, \qquad \|\phi\|_Y^2 \simeq \sum_{\lambda \in 2^{\mathbb{Z}}} \|A_\lambda \phi\|_{Y_\lambda}^2. \qquad (3.18)$$

Secondly, dyadic versions of (3.17) should read as follows:

• for $\lambda \in 2^{\mathbb{Z}}$,

$$\Box^{-1} : Y_\lambda \to X_\lambda; \qquad (3.19)$$

• for $\nu \leq \lambda \in 2^{\mathbb{Z}}$,

$$A_\nu(X_\lambda \cdot X_\lambda) \to X_\nu, \qquad (3.20)$$
$$A_\nu(X_\lambda \cdot Y_\lambda) \to Y_\nu, \qquad (3.21)$$
$$A_\nu(\partial^\mu X_\lambda \cdot \partial_\mu X_\lambda) \to Y_\nu; \qquad (3.22)$$

• for $\nu < \lambda \in 2^{\mathbb{Z}}$,

$$X_\nu \cdot X_\lambda \to X_\lambda, \qquad (3.23)$$
$$X_\nu \cdot Y_\lambda \to Y_\lambda, \qquad X_\lambda \cdot Y_\nu \to Y_\lambda, \qquad (3.24)$$
$$\partial^\mu X_\nu \cdot \partial_\mu X_\lambda \to Y_\lambda. \qquad (3.25)$$

The estimates (3.20)-(3.22) describe the interaction of two comparable frequencies, while (3.23)-(3.25) pertain to the interaction between a high frequency and a low one.

In constructing function spaces suitable for this approach, the first obstacle, commonly referred to as *"the division problem"*, is dealing with the operator $\Box^{-1}$. In Fourier space, it corresponds roughly to a multiplication by $(\tau^2 - |\xi|^2)^{-1}$. Unfortunately, this function is not locally integrable and this can lead to nontrivial divergences in the above estimates.

Even if one solves this issue, the next difficulty appears in summing up (3.20)-(3.22) and (3.23)-(3.25) to derive the bilinear mappings of (3.17), as, a priori, a logarithmic divergence in $\lambda$ is present from the summation. This is usually called in literature *"the summation problem"*.

*"The division problem"* was solved by Tataru for $n \geq 4$ in [177] and for $n = 2$ or 3 in [178], who constructed as $X$ and $Y$ the function spaces $F$ and $\Box F$, respectively, which were discussed in Chapter 2. Using these spaces, Tataru proved the following global well-posedness result.

**Theorem 3.3 ([177], [178]).** *There exist $C_1$, $C_2 > 0$ such that for any initial data satisfying*

$$\|\phi_0\|_{\dot{B}_{2,1}^{n/2}(\mathbb{R}^n)} + \|\phi_1\|_{\dot{B}_{2,1}^{n/2-1}(\mathbb{R}^n)} < C_1,$$

*there exists a global solution $\phi$ to* (3.2) *verifying*

$$\|\phi\|_{L^\infty \dot{B}_{2,1}^{n/2}(\mathbb{R} \times \mathbb{R}^n)} + \|\partial_t \phi\|_{L^\infty \dot{B}_{2,1}^{n/2-1}(\mathbb{R} \times \mathbb{R}^n)} \leq C_2,$$

*which is the unique limit of smooth solutions. Furthermore, the solution map is Lipschitz continuous.*

*If, in addition, $(\phi_0, \phi_1) \in \dot{H}^s \times \dot{H}^{s-1}(\mathbb{R}^n)$ with $s > n/2$, then*

$$\|\phi\|_{L^\infty \dot{H}^s(\mathbb{R} \times \mathbb{R}^n)} + \|\partial_t \phi\|_{L^\infty \dot{H}^{s-1}(\mathbb{R} \times \mathbb{R}^n)} \lesssim \|\phi_0\|_{\dot{H}^s(\mathbb{R}^n)} + \|\phi_1\|_{\dot{H}^{s-1}(\mathbb{R}^n)}.$$

The benefit of working with Besov spaces is twofold. First, it allows one to bypass *"the summation problem"* by replacing the $l^2$ summation in (3.18) with an $l^1$ one. Secondly, due to the embedding $\dot{B}_{2,1}^{n/2}(\mathbb{R}^n) \subset L^\infty(\mathbb{R}^n)$, the problem becomes local with respect to $N$, thus rendering the geometry of $N$ irrelevant.

In proving the previous theorem, we can use the abstract framework by taking

$$D = \dot{B}_{2,1}^{n/2} \times \dot{B}_{2,1}^{n/2-1}(\mathbb{R} \times \mathbb{R}^n) \qquad \text{and} \qquad X = F$$

with their respective norms, $I = J = \mathbb{R}$, and $\varphi \equiv 1$, with $L$ and $N$ as in the section of local results. It follows that (3.8) holds due to (2.124), while (3.9) is true according to (2.133). For (3.10), we first apply (2.125) to infer

$$\|\Box^{-1}(N(\phi) - N(\psi))\|_F \lesssim \|N(\phi) - N(\psi)\|_{\Box F}.$$

Based on (2.134) and (2.141), we can then write an identical argument with the one in the $X^{s,\theta}$ approach to derive

$$\|N(\phi) - N(\psi)\|_{\Box F} \lesssim (\|\phi\|_F + \|\psi\|_F) \|\phi - \psi\|_F,$$

which finishes the proof.

A natural question is whether one can modify Tataru's F-spaces in order to also solve *"the summation problem"*. The answer is negative, as it was shown by D'Ancona-Georgiev [35] that the solution map is not uniformly continuous for data with $\dot{H}^{n/2} \times \dot{H}^{n/2-1}(\mathbb{R}^n)$ regularity. This tells us that, for pursuing global results in critical Sobolev spaces, we have to treat (3.1) as a fully nonlinear equation.

### 3.4.2 *Fully nonlinear approach*

Based on the previous discussion, we would also need to work with a weaker notion of global well-posedness, which is defined as follows.

**Definition 3.2.** The Cauchy problem (3.2) is *weakly globally well-posed for small data* in $\dot{H}^s \times \dot{H}^{s-1}(\mathbb{R}^n)$ if there exists $M > 0$ such that, for all

$$(\phi_0, \phi_1) \in H^\sigma \times H^{\sigma-1}(\mathbb{R}^n), \qquad \sigma > s,$$

with $\|\phi_0\|_{\dot{H}^s} + \|\phi_1\|_{\dot{H}^{s-1}} < M$, there exists a unique global solution

$$\phi \in C(\mathbb{R}, H^\sigma(\mathbb{R}^n)) \cap C^1(\mathbb{R}, H^{\sigma-1}(\mathbb{R}^n)).$$

In this context, Tao was the first one to solve *"the summation problem"* for the case when the target manifold is the unit sphere $\mathbb{S}^{m-1} \subset \mathbb{R}^m$ in [169] ($n \geq 5$) and [170] ($n \geq 2$). For $n \geq 4$, these results were further generalized to include more general target manifolds by Klainerman-Rodnianski [88], Shatah-Struwe [146], and Nahmod-Stefanov-Uhlenbeck [123]. The more challenging $n = 2$ and $n = 3$ problems were settled by Krieger [92, 93] and Tataru [180] for targets which include the hyperbolic plane $\mathbb{H}^{m-1} \subset \mathbb{R}^m$ and manifolds that are uniformly isometrically embeddable into Euclidean spaces, respectively. The most comprehensive of these results, which, apart from weak global well-posedness, includes also weak stability estimates and continuous dependence for the solution map, is contained in [180].

**Theorem 3.4 ([180]).** *Let $n \geq 2$ and consider $N$ to be a uniformly isometrically embeddable manifold in a Euclidean space. Then, for sufficiently small data*

$$(\phi_0, \phi_1) \in \dot{H}^{n/2} \times \dot{H}^{n/2-1}(\mathbb{R}^n),$$

*the following statements hold:*

*i) if the data is also smooth, then there exists a global smooth solution $\phi$ to (3.2) which, for $s \geq n/2$, satisfies*

$$\|\phi\|_{L^\infty \dot{H}^s(\mathbb{R}\times\mathbb{R}^n)} + \|\partial_t \phi\|_{L^\infty \dot{H}^{s-1}(\mathbb{R}\times\mathbb{R}^n)} < \infty$$

*and*

$$\|\phi\|_{L^\infty \dot{H}^s(\mathbb{R}\times\mathbb{R}^n)} + \|\partial_t \phi\|_{L^\infty \dot{H}^{s-1}(\mathbb{R}\times\mathbb{R}^n)} \lesssim \|\phi_0\|_{\dot{H}^s(\mathbb{R}^n)} + \|\phi_1\|_{\dot{H}^{s-1}(\mathbb{R}^n)},$$

*whenever the right-hand side is finite;*

*ii) there exists a global solution $\phi$ to (3.2) verifying*

$$\|\phi\|_{L^\infty \dot{H}^{n/2}(\mathbb{R}\times\mathbb{R}^n)} + \|\partial_t \phi\|_{L^\infty \dot{H}^{n/2-1}(\mathbb{R}\times\mathbb{R}^n)} \lesssim \|\phi_0\|_{\dot{H}^{n/2}(\mathbb{R}^n)} + \|\phi_1\|_{\dot{H}^{n/2-1}(\mathbb{R}^n)},$$

*which is the limit of smooth solutions in the $L^\infty_{loc}(\dot{H}^{n/2} \times \dot{H}^{n/2-1})$ topology;*

*iii) for two smooth solutions, $\phi^1$ and $\phi^2$, and $-\epsilon < s - n/2 < 0$, with $\epsilon$ sufficiently small, one has*

$$\|\phi^1 - \phi^2\|_{L^\infty \dot{H}^s(\mathbb{R}\times\mathbb{R}^n)} + \|\partial_t\phi^1 - \partial_t\phi^2\|_{L^\infty \dot{H}^{s-1}(\mathbb{R}\times\mathbb{R}^n)}$$
$$\lesssim \|\phi^1(0) - \phi^2(0)\|_{\dot{H}^s(\mathbb{R}^n)} + \|\partial_t\phi^1(0) - \partial_t\phi^2(0)\|_{\dot{H}^{s-1}(\mathbb{R}^n)};$$

*iv) if $-\epsilon < s - n/2 < 0$, with $\epsilon$ sufficiently small, and*

$$\lim_{n\to\infty} \|\phi_n(0) - \phi_0\|_{\dot{H}^{n/2} \cap \dot{H}^s(\mathbb{R}^n)} + \|\partial_t\phi_n(0) - \phi_1\|_{\dot{H}^{n/2-1} \cap \dot{H}^{s-1}(\mathbb{R}^n)} = 0,$$

*then for the sequence of corresponding solutions one has*

$$\lim_{n\to\infty} \|\phi_n - \phi\|_{L^\infty(\dot{H}^{n/2} \cap \dot{H}^s)(\mathbb{R}\times\mathbb{R}^n)} + \|\partial_t\phi_n - \partial_t\phi\|_{L^\infty(\dot{H}^{s-1} \cap \dot{H}^{n/2-1})(\mathbb{R}\times\mathbb{R}^n)} = 0.$$

**Remark 3.7.** Examples of uniformly isometrically embeddable manifolds include the compact manifolds, which is one of the celebrated results of Nash [124].

**Remark 3.8.** What we call smooth data in i) is data which is constant outside of a compact set and has $C^\infty$ regularity.

The solutions obtained above in ii) are in fact part of a function space $S$ verifying[3]

$$S \subseteq C(\mathbb{R}, \dot{H}^{n/2}(\mathbb{R}^n)) \cap \dot{C}^1(\mathbb{R}, \dot{H}^{n/2-1}(\mathbb{R}^n))$$

and, when $n = 2$, these solutions are called *strong finite energy wave maps* and satisfy the stronger estimate

$$\|\phi\|_S \lesssim \|\phi_0\|_{\dot{H}^1(\mathbb{R}^n)} + \|\phi_1\|_{L^2(\mathbb{R}^n)}.$$

For completeness purposes, we chose to briefly describe this solution space, whose origins can be traced back to Tataru's $F$-spaces and whose first variant appeared in work by Tao [170]. After a slight variation proposed by Krieger [93], the version we present here is due to Tataru [180]. It has the same dyadic pieces as in [170], but they are assembled differently.

In [180], one works with the closure of $\mathcal{S}'(\mathbb{R}^{n+1})$ under the norm

$$\|\phi\|_S \simeq \left(\sum_{\lambda \in 2^{\mathbb{Z}}} \|\phi\|_{S[\lambda]}^2\right)^{1/2}, \tag{3.26}$$

---

[3]The notation $\phi \in \dot{C}^1(\mathbb{R}, \dot{H}^{n/2-1}(\mathbb{R}^n))$ stands for $\partial_t\phi \in C(\mathbb{R}, \dot{H}^{n/2-1}(\mathbb{R}^n))$.

where, for a fixed $\lambda \in 2^{\mathbb{Z}}$,

$$\|\phi\|_{S[\lambda]} \overset{\text{def}}{=} \|\nabla S_\lambda \phi\|_{L^\infty \dot{H}^{n/2-1}(\mathbb{R} \times \mathbb{R}^n)}$$
$$+ \lambda^{n/2-1} \sup_{\mu \in 2^{\mathbb{Z}}} \mu^{1/2} \|\nabla B_\mu S_\lambda \phi\|_{L^2(\mathbb{R}^{n+1})} \tag{3.27}$$
$$+ \sup_{\mu \in 2^{\mathbb{Z}}, \, \mu/\lambda < 2^{-20}} \|\phi\|_{S[\lambda,\mu]}.$$

In the above, $S_\lambda = S_\lambda(\nabla_x)$ is the Littlewood-Paley projector localizing the spatial frequency to the region $\{\xi; \lambda/2 \leq |\xi| \leq 2\lambda\}$, $B_\mu = B_\mu(\nabla)$ is the Fourier multiplier given by (2.115), and $\|\cdot\|_{S[\lambda,\mu]}$ is usually referred to as a *modulational Strichartz* norm.

To describe the latter, we first introduce the concept of spherical cap $\mathcal{K} \subset \mathbb{S}^{n-1}$,

$$\mathcal{K} = \{\omega \in \mathbb{S}^{n-1}; \, |\omega - \omega_k| < r_k\},$$

with $\omega_k \in \mathbb{S}^{n-1}$ and $0 < r_k < 2$ being the center and radius of $\mathcal{K}$, respectively. For the area of $\mathcal{K}$ we use the notation $|\mathcal{K}| \simeq r_k^{n-1}$. Moreover, $C\mathcal{K}$, with $C > 0$, denotes a cap with center $\omega_k$ and radius $Cr_k$, whereas $-\mathcal{K}$ designates a cap with center $-\omega_k$ and radius $r_k$. Next, for $\nu > 2^{10}$, we consider $K_\nu$ to be a finitely overlapping cover of $\mathbb{S}^{n-1}$ into spherical caps of radius $\simeq \nu^{-1}$, which induces the decomposition

$$S_\lambda(\nabla_x) = \sum_{\mathcal{K} \in K_\nu} S_{\lambda,\mathcal{K}}(\nabla_x),$$

i.e., on the Fourier support of $S_{\lambda,\mathcal{K}}$, one has

$$|\xi| \simeq \lambda, \qquad \frac{\xi}{|\xi|} \in \mathcal{K}.$$

Lastly, for a fixed $\lambda \in 2^{\mathbb{Z}}$ and $\mathcal{K} \in K_\nu$, we define the norm

$$\|\psi\|_{S[\lambda,\mathcal{K}]} \overset{\text{def}}{=} \lambda^{n/2} \sup_{\omega \in \mathbb{S}^{n-1}, \, \omega \notin 2\mathcal{K}} d(\omega, \mathcal{K}) \|\psi\|_{L_{t_\omega}^\infty L_{x_\omega}^2} + \lambda^{n/2} \|\psi\|_{L_t^\infty L_x^2}$$
$$+ |\mathcal{K}|^{-1/2} \lambda^{1/2} \|\psi\|_{S[\mathcal{K}]},$$

where $d(\omega, \mathcal{K})$ stands for the usual distance between $\omega$ and $\mathcal{K}$, $(t_\omega, x_\omega)$ are the null coordinates associated to $\omega$ defined in (2.145), and $\|\cdot\|_{S[\mathcal{K}]}$ is an atomic norm whose atoms are functions $\psi_\omega = \psi_\omega(t,x)$ satisfying

$$\|\psi\|_{L_{t_\omega}^2 L_{x_\omega}^\infty} \leq 1, \qquad \text{for some } \omega \in \mathcal{K}.$$

Now, we have all the prerequisites to define the $\|\cdot\|_{S[\lambda,\mu]}$ norm for $\mu/\lambda < 2^{-20}$, which is as follows:

$$\|\phi\|_{S[\lambda,\mu]} \overset{\text{def}}{=} \sup_{\pm} \left( \sum_{\mathcal{K} \in K_{\sqrt{\lambda/\mu}}} \|B_\mu^\pm S_{\lambda,\pm\mathcal{K}} \phi\|_{S[\lambda,\mathcal{K}]}^2 \right)^{1/2}$$

with $B_\mu^\pm = B_\mu^\pm(\nabla)$ being the Fourier multiplier whose symbol is the restriction of the symbol for $B_\mu$ to the region $\{\pm\tau > 0\}$. This concludes the discussion of the function space $S$, which also has local versions adapted to arbitrary time intervals $I \subset \mathbb{R}$,

$$\|\phi\|_{S(I\times\mathbb{R}^n)} \overset{\text{def}}{=} \inf\{\|\tilde\phi\|_S; \ \tilde\phi \text{ is an extension of } \phi\}.$$

Using the finite speed of propagation property, Tataru [180] also derived a local version of Theorem 3.4.

**Theorem 3.5 ([180]).** *For $R > 0$, let*

$$B_R(x_0) \overset{\text{def}}{=} \{|x - x_0| \le R\}, \qquad Q_R(t_0, x_0) \overset{\text{def}}{=} \{|t - t_0| + |x - x_0| \le R\},$$

*and consider initial data* $\phi[t_0] := (\phi(t_0), \partial_t\phi(t_0))$ *in* $B_R(x_0)$, *which has small* $\dot{H}^{n/2} \times \dot{H}^{n/2-1}(B_R(x_0))$ *norm. Then:*

*i) if the data is also smooth, then there exists a smooth solution $\phi$ to (3.1) in $Q_R(t_0, x_0)$ which, for $s \ge n/2$, satisfies*

$$\|\phi\|_{L^\infty \dot{H}^s(Q_R(t_0,x_0))} + \|\partial_t\phi\|_{L^\infty \dot{H}^{s-1}(Q_R(t_0,x_0))} \lesssim \|\phi[t_0]\|_{\dot{H}^s\times\dot{H}^{s-1}(B_R(x_0))};$$

*ii) there exists a solution $\phi$ to (3.1) in $Q_R(t_0, x_0)$ verifying*

$$\|\phi\|_{L^\infty \dot{H}^{n/2}(Q_R(t_0,x_0))} + \|\partial_t\phi\|_{L^\infty \dot{H}^{n/2-1}(Q_R(t_0,x_0))}$$
$$\lesssim \|\phi[t_0]\|_{\dot{H}^{n/2}\times\dot{H}^{n/2-1}(B_R(x_0))},$$

*which is the unique limit of smooth solutions in the $L^\infty(\dot{H}^{n/2} \times \dot{H}^{n/2-1})$ topology;*

*iii) if*

$$\lim_{n\to\infty} \|\phi_n[t_0] - \phi[t_0]\|_{\dot{H}^{n/2}\times\dot{H}^{n/2-1}(B_R(x_0))} = 0,$$

*then for the sequence of corresponding solutions one has*

$$\lim_{n\to\infty} \|\phi_n - \phi\|_{L^\infty \dot{H}^{n/2}(Q_R(t_0,x_0))} + \|\partial_t\phi_n - \partial_t\phi\|_{L^\infty \dot{H}^{n/2-1}(Q_R(t_0,x_0))} = 0.$$

In line with the aim of this book, we chose to discuss in the remainder of this section the high-dimensional results of Tao [169] and Shatah-Struwe [146]. Our choice is motivated on one hand by the increased level of intricacy for the analysis of the low-dimensional problem, which sometimes obscures the fundamental ideas of these works. Nevertheless, we believe the avid reader, after following our exposition, can then make a smooth transition to this more involved setting. The other motivation for our selection has to do with:

- the novel concepts of *renormalization* and *frequency envelope*, which appeared first in [169];
- the entirely physical space approach used in [146].

### 3.4.3   Tao's result

This is a small data weak global well-posedness result for the $n \geq 5$ case of (3.2), with $N = \mathbb{S}^{m-1} \subset \mathbb{R}^m$. In this setting, as was derived in Chapter 1, (3.1) takes the form

$$\Box \phi = -\langle \partial^\mu \phi, \partial_\mu \phi \rangle \, \phi, \qquad (3.28)$$

where $\phi$ is a column vector and $\langle \cdot, \cdot \rangle$ designates the inner product in $\mathbb{R}^m$. The wave map $\phi$ is required to have $C^1_t L^2_x \cap C^0_t H^1_x$ regularity in order to be a distributional solution for the previous equation. Tao also works with *classical wave maps*, which are smooth wave maps that are constant outside of a finite union of light cones (see also Remark 3.8). This apparent restriction has to do with the inability of defining Sobolev regularity for $x \mapsto \phi(t,x)$ when $\phi$ takes values on the sphere. In this context, the compromise is to allow for constant functions to be elements of $H^s$ with zero norm and then to claim $\phi(t,\cdot) \in H^s$ when, in fact, $\phi(t,\cdot) - \tilde{C} \in H^s$ for some specific constant $\tilde{C} \in \mathbb{R}^m$. Under these conventions, we can state Tao's main result.

**Theorem 3.6 ([169]).** *Let $n \geq 5$. The Cauchy problem associated to (3.28) is weakly globally well-posed for small data in $\dot{H}^{n/2} \times \dot{H}^{n/2-1}(\mathbb{R}^n)$, i.e., for $s > n/2$, initial data in $H^s \times H^{s-1}(\mathbb{R}^n)$ with sufficiently small $\dot{H}^{n/2} \times \dot{H}^{n/2-1}(\mathbb{R}^n)$ norm leads to global-in-time $H^s \times H^{s-1}$ solutions for (3.28). Hence, smooth solutions remain smooth when the initial data has small $\dot{H}^{n/2} \times \dot{H}^{n/2-1}(\mathbb{R}^n)$ norm. For these solutions, if $|s - n/2| < 1/2$, then*

$$\|\phi\|_{L^\infty \dot{H}^s(\mathbb{R} \times \mathbb{R}^n)} + \|\partial_t \phi\|_{L^\infty \dot{H}^{s-1}(\mathbb{R} \times \mathbb{R}^n)} \lesssim \|\phi_0\|_{\dot{H}^s(\mathbb{R}^n)} + \|\phi_1\|_{\dot{H}^{s-1}(\mathbb{R}^n)}.$$

In proving this theorem, Tao mainly works with:

- microlocalization techniques combined with Strichartz estimates;
- the geometry of the sphere;
- a *renormalization* procedure which constructs a coordinate frame that simplifies the equation under investigation.

We present first an informal discussion of the argument, which is followed by a more detailed description of its key steps.

### 3.4.3.1 *Informal proof*

If we use a Littlewood-Paley decomposition with respect to the spatial variable and project (3.28) to a certain frequency, we will be dealing with the following product trichotomy for $\langle \partial^\mu \phi, \partial_\mu \phi \rangle$:

$$S_\lambda(\langle \partial^\mu \phi, \partial_\mu \phi \rangle) \simeq \sum_{\lambda_1 \simeq \lambda_2 \gtrsim \lambda} S_\lambda(\langle \partial^\mu \phi_{\lambda_1}, \partial_\mu \phi_{\lambda_2} \rangle)$$

$$+ \sum_{\lambda_1 \simeq \lambda \gg \lambda_2} S_\lambda(\langle \partial^\mu \phi_{\lambda_1}, \partial_\mu \phi_{\lambda_2} \rangle) + \sum_{\lambda_2 \simeq \lambda \gg \lambda_1} S_\lambda(\langle \partial^\mu \phi_{\lambda_1}, \partial_\mu \phi_{\lambda_2} \rangle).$$

Tao claims that the first sum is manageable as one estimates the low frequency output of an interaction between two high frequencies. This can be done, for example, by using the null structure (3.5) of the nonlinearity. Applying this heuristic to the entire right-hand side of (3.28), the generic trilinear term $\langle \partial^\mu \phi_{\lambda_1}, \partial_\mu \phi_{\lambda_2} \rangle \phi_{\lambda_3}$ can be written as

$$\langle \partial^\mu \phi_{\lambda_1}, \partial_\mu \phi_{\lambda_2} \rangle \phi_{\lambda_3} \simeq \frac{\lambda_2}{\lambda_3} \langle \partial^\mu \phi_{\lambda_1}, \partial_\mu \phi_{\lambda_3} \rangle \phi_{\lambda_2}.$$

Hence, for $\lambda_2 > \lambda_3$,

$$\langle \partial^\mu \phi_{\lambda_1}, \partial_\mu \phi_{\lambda_3} \rangle \phi_{\lambda_2}$$

is small in size when compared to

$$\langle \partial^\mu \phi_{\lambda_1}, \partial_\mu \phi_{\lambda_2} \rangle \phi_{\lambda_3}$$

and it should be, in principle, more manageable. Tao labels these terms as *derivative falls on low frequency* terms. In conclusion, a microlocalization analysis which trims these manageable terms from (3.28) yields

$$\Box \phi_\lambda = -2 \langle \partial^\mu \phi_\lambda, \partial_\mu \phi_{\ll \lambda} \rangle \phi_{\ll \lambda}, \tag{3.29}$$

which is the equation satisfied by the dyadic piece $\phi_\lambda$, with $\phi_{\ll \lambda}$ denoting the projection of $\phi$ to frequencies that are much smaller than $\lambda$ (e.g., $\phi_{\ll \lambda} = \phi_{\leq 2^{-10}\lambda}$).

Even though this equation is linear, it does not have a structure that would allow us to treat it through an iterative scheme. This is the point in the argument where Tao uses the geometry of $\mathbb{S}^{m-1}$ in the form of the identities

$$\langle \phi, \phi \rangle = 1 \quad \text{and} \quad \langle \phi, \partial_\mu \phi \rangle = 0.$$

If we project these identities using $S_\lambda$ and $S_{\ll \lambda}$ and, as before, neglect *high-high* interactions and *derivative falls on low frequency* terms, then it follows that

$$\langle \phi_\lambda, \phi_{\ll \lambda} \rangle \simeq 0, \qquad \langle \phi_{\ll \lambda}, \partial_\mu \phi_\lambda \rangle \simeq 0,$$

and

$$\langle \phi_{\ll\lambda}, \phi_{\ll\lambda} \rangle \simeq 1, \qquad \langle \phi_{\ll\lambda}, \partial_\mu \phi_{\ll\lambda} \rangle \simeq 0,$$

respectively. Tao interprets these formulas as $\phi_{\ll\lambda}$ always lying on the sphere and $\phi_\lambda$ lying in its tangent space. This argument allows us to rewrite (3.29) as

$$\begin{aligned} \Box \phi_\lambda &= -2 \left( \langle \partial^\mu \phi_\lambda, \partial_\mu \phi_{\ll\lambda} \rangle \phi_{\ll\lambda} - \langle \partial^\mu \phi_\lambda, \phi_{\ll\lambda} \rangle \partial_\mu \phi_{\ll\lambda} \right) \\ &= -2 A_\mu \, \partial^\mu \phi_\lambda, \end{aligned} \qquad (3.30)$$

where

$$A_\mu \stackrel{\text{def}}{=} \phi_{\ll\lambda} \left( \partial_\mu \phi_{\ll\lambda} \right)^t - \partial_\mu \phi_{\ll\lambda} \left( \phi_{\ll\lambda} \right)^t \qquad (3.31)$$

is an antisymmetric $m \times m$ matrix. This is the reason why (3.30) is better to work with than (3.29), as it has a more canceling structure. Ignoring some more *derivative falls on low frequency* terms, one can further recast (3.30) as the covariant wave equation

$$D_\mu D^\mu \phi_\lambda = 0, \qquad D_\mu = \partial_\mu + A_\mu. \qquad (3.32)$$

Covariant equations are typically hard to analyze and one usually tries to transform them into free equations which are more tractable. This is what Tao calls the *renormalization* procedure, at the heart of which lies the invariance or gauge freedom of $D_\mu D^\mu \psi = 0$ under the transformation

$$\psi \mapsto U\psi, \qquad A_\mu \mapsto U A_\mu U^t - \partial_\mu U U^t, \qquad U \in SO(m).$$

The measure of how different the covariant equation is from the free one is given by the curvature associated to $A_\mu$, which is defined as

$$F_{\mu\nu} \stackrel{\text{def}}{=} D_\mu D_\nu - D_\nu D_\mu = \partial_\mu A_\nu - \partial_\nu A_\mu + [A_\mu, A_\nu].$$

A direct computation with $A_\mu$ given by (3.31) shows that the curvature is quadratic in derivatives of $\phi_{\ll\lambda}$, thus being manageable. This allows one to construct a gauge transformation $U$ such that both $U A_\mu U^t - \partial_\mu U U^t$ and its curvature are small, thus transforming (3.32) into a free wave problem which does not have a high-low interaction and is solvable by standard methods. In the article under discussion, $U$ is constructed by iteration, through a dyadic scheme, relying again on discarding manageable terms. This finishes the informal part of the presentation.

### 3.4.3.2 *Preliminaries*

We proceed next to explain in more detail important steps of the argument. We start by introducing the function spaces used in the proof and the concept of *frequency envelope*, which are central to Tao's analysis. To facilitate an easy passage for the reader from our presentation to the article and back, we adopt the $2^k$, $k \in \mathbb{Z}$, (or $2^{k'}$) notation for spatial dyadic frequencies, instead of the $\lambda \in 2^{\mathbb{Z}}$ (or $\nu$) one used so far.

**Definition 3.3.** Let $n \geq 2$ and $k \in \mathbb{Z}$ be fixed. The $\dot{H}^{n/2}$-normalized Strichartz space at frequency $2^k$, denoted by $X_k(I \times \mathbb{R}^n)$, is the space of functions $\phi = \phi(t, x)$ defined on $I \times \mathbb{R}^n$, where $I \subset \mathbb{R}$ is an arbitrary time interval, for which

$$\|\phi\|_{X_k(I\times\mathbb{R}^n)} \overset{\text{def}}{=} \sup_{\substack{(p,q,n) \\ \text{wave admissible}}} 2^{(1/p+n/q)k}\big(\|\phi\|_{L^pL^q(I\times\mathbb{R}^n)} \tag{3.33}$$
$$+ 2^{-k}\|\partial_t\phi\|_{L^pL^q(I\times\mathbb{R}^n)}\big)$$

is well-defined and finite.

**Remark 3.9.** The actual Strichartz pairs used by Tao are

$$(p,q) = \left(2, \frac{2(n-1)}{(n-3)}\right), (2,4), (2,n-1), (2,\infty), (4,2(n-1)), (\infty,2), (\infty,\infty).$$

The restriction $n \geq 5$ in Theorem 3.6 comes from the need of having the values $(2,4)$ and $(2, n-1)$ available. Moreover, Tao notices that one can work without using the endpoint $(2, 2(n-1)/(n-3))$ when $n \geq 6$.

Based on (2.26), an easy adaptation of the arguments in Sections 2.2.1 and 2.2.2, yields the following result.

**Proposition 3.1.** *In the context of the previous definition, if $0 \in I$, then any function $\phi$ localized at spatial frequency $2^k$ (i.e., $\phi \simeq S_{2^k}\phi$) satisfies*

$$\|\phi\|_{X_k(I\times\mathbb{R}^n)} \lesssim \|\phi(0)\|_{\dot{H}^{n/2}(\mathbb{R}^n)} + \|\partial_t\phi(0)\|_{\dot{H}^{n/2-1}(\mathbb{R}^n)} \tag{3.34}$$
$$+ 2^{(n/2-1)k}\|\Box\phi\|_{L^1L^2(I\times\mathbb{R}^n)}.$$

Next, we define the *frequency envelope* concept.

**Definition 3.4.** Let $0 < \sigma < 1/2$ and $0 < \epsilon \ll 1$ be two fixed constants. A sequence $c = (c_k)_{k\in\mathbb{Z}}$ of positive reals is called a *frequency envelope* if

$$\|c\|_{l^2} \lesssim \epsilon \quad \text{and} \quad 2^{-\sigma|k-k'|} \lesssim \frac{c_k}{c_{k'}} \lesssim 2^{\sigma|k-k'|}, \ \forall\, k, k' \in \mathbb{Z}. \tag{3.35}$$

Furthermore, for a pair of functions $(f, g)$ defined on $\mathbb{R}^n$, we say that $(f, g)$ lies underneath the frequency envelope $c$ and write $(f, g) \prec c$ if

$$\|S_{2^k} f\|_{\dot{H}^{n/2}(\mathbb{R}^n)} + \|S_{2^k} g\|_{\dot{H}^{n/2-1}(\mathbb{R}^n)} \leq c_k, \ \forall k \in \mathbb{Z}. \tag{3.36}$$

**Remark 3.10.** An immediate consequence of this definition is that if $(f, g) \prec c$, then $\|f\|_{\dot{H}^{n/2}} + \|g\|_{\dot{H}^{n/2-1}} \lesssim \epsilon$. On the other hand, if $\|f\|_{\dot{H}^{n/2}} + \|g\|_{\dot{H}^{n/2-1}} \lesssim \epsilon$, then we can easily verify that, for an arbitrary $0 < \tilde{\sigma} < 1/2$, the sequence $c = (c_k)_{k \in \mathbb{Z}}$ given by

$$c_k \overset{\text{def}}{=} \sum_{k' \in \mathbb{Z}} 2^{-\tilde{\sigma}|k-k'|} \left( \|S_{2^{k'}} f\|_{\dot{H}^{n/2}} + \|S_{2^{k'}} g\|_{\dot{H}^{n/2-1}} \right), \ \forall k \in \mathbb{Z}, \tag{3.37}$$

is a frequency envelope satisfying $(f, g) \prec c$.

### 3.4.3.3 *Initial step*

The initial step in Tao's argument is reducing the proof of Theorem 3.6 to the one for the following proposition.

**Proposition 3.2.** *Let $n \geq 5$ and consider $0 < T < \infty$, $c$ a frequency envelope with $\|c\|_{l^2} \lesssim \epsilon$, and $\phi$ a classical wave map defined on $[0, T] \times \mathbb{R}^n$ verifying $(\phi(0), \partial_t \phi(0)) \prec c$. If $\epsilon$ is sufficiently small, then there exists a constant $C_0 = C_0(n, m)$ such that*

$$\|S_{2^k} \phi\|_{X_k([0,T] \times \mathbb{R}^n)} \leq C_0 c_k, \ \forall k \in \mathbb{Z}.$$

*As a consequence of this estimate, one has*

$$(\phi(t), \partial_t \phi(t)) \prec C_0 c, \ \forall t \in [0, T].$$

To achieve this reduction we first downgrade Theorem 3.6 to

**Theorem 3.7.** *Let $n \geq 5$ and $0 < \sigma < 1/2$. If $n/2 < s < n/2 + \sigma$, then compactly supported smooth initial data in $H^s \times H^{s-1}(\mathbb{R}^n)$ with sufficiently small $\dot{H}^{n/2} \times \dot{H}^{n/2-1}(\mathbb{R}^n)$ norm leads to global in time classical wave maps for (3.28).*

The cutback in the range of $s$ is achieved by invoking the persistence of higher regularity, which is part of the local well-posedness of the problem for $s > n/2$. In terms of considering only compactly supported smooth initial data, we argue that this is true based on the density of such data in the above Sobolev spaces and on the continuity of the solution map which is inherited from the local well-posedness theory.

The initial step will be completed if we show that Proposition 3.2 implies Theorem 3.7. For initial data $(f, g)$ and $\sigma$ as in the hypothesis of Theorem 3.7, we consider the frequency envelope $c$ given by (3.37) with $\tilde{\sigma} = \sigma$. Due to $(f, g) \in H^s \times H^{s-1}(\mathbb{R}^n)$ $(s > n/2)$, the local theory assures us of the existence of a classical wave map $\phi$, which is defined on $[0, T] \times \mathbb{R}^n$ $(T > 0)$ and satisfies $(\phi(0), \partial_t \phi(0)) = (f, g)$. Therefore, if $\|f\|_{\dot{H}^{n/2}} + \|g\|_{\dot{H}^{n/2-1}}$ is sufficiently small, we can apply Proposition 3.2 to deduce

$$\|S_{2^k}\phi\|_{L^\infty \dot{H}^s([0,T]\times\mathbb{R}^n)} + \|S_{2^k}\partial_t\phi\|_{L^\infty \dot{H}^{s-1}([0,T]\times\mathbb{R}^n)}$$
$$\lesssim 2^{(s-n/2)k}\|S_{2^k}\phi\|_{X_k([0,T]\times\mathbb{R}^n)} \leq C_0\, 2^{(s-n/2)k}c_k$$
$$= C_0\, 2^{(s-n/2)k} \sum_{k'\in\mathbb{Z}} 2^{-\sigma|k-k'|} \left( \|S_{2^{k'}}f\|_{\dot{H}^{n/2}(\mathbb{R}^n)} + \|S_{2^{k'}}g\|_{\dot{H}^{n/2-1}(\mathbb{R}^n)} \right)$$
$$\lesssim C_0 \sum_{k'\in\mathbb{Z}} 2^{(|s-n/2|-\sigma)|k-k'|} \left( \|S_{2^{k'}}f\|_{\dot{H}^s(\mathbb{R}^n)} + \|S_{2^{k'}}g\|_{\dot{H}^{s-1}(\mathbb{R}^n)} \right),$$

for all $k \in \mathbb{Z}$. Since $|s - n/2| < \sigma$, we obtain, based on Young's inequality (2.18) for $l^2$, that

$$\|\phi\|_{L^\infty \dot{H}^s([0,T]\times\mathbb{R}^n)} + \|\partial_t\phi\|_{L^\infty \dot{H}^{s-1}([0,T]\times\mathbb{R}^n)} \lesssim \|f\|_{\dot{H}^s(\mathbb{R}^n)} + \|g\|_{\dot{H}^{s-1}(\mathbb{R}^n)}.$$

This estimate allows us to say that $\phi$ can be extended globally in time, thus finishing the discussion of the initial step.

### 3.4.3.4 *Second step*

The second step in the proof consists in further reducing Proposition 3.2 to a bootstrap argument. To attain this, we first show that one can assume $c$ to behave like $(2^{-\sigma|k|})_{k\in\mathbb{Z}}$, where $\sigma$ is associated to $c$ through (3.35). We already know from (3.35) that

$$c_k \gtrsim c_0\, 2^{-\sigma|k|}, \quad \forall\, k \in \mathbb{Z}.$$

On the other hand, as $\phi$ is a classical wave map on $[0, T] \times \mathbb{R}^n$, we can rely on controlling quantities like $\|\nabla\phi\|_{L^\infty \dot{H}^s([0,T]\times\mathbb{R}^n)}$, for arbitrary values of $s \in \mathbb{R}$. As a consequence, we derive that

$$2^{|k'|} \left( \|S_{2^{k'}}\phi(0)\|_{\dot{H}^{n/2}(\mathbb{R}^n)} + \|S_{2^{k'}}\partial_t\phi(0)\|_{\dot{H}^{n/2-1}(\mathbb{R}^n)} \right)$$
$$\lesssim \|\nabla\phi\|_{L^\infty(\dot{H}^{n/2}\cap\dot{H}^{n/2-2})([0,T]\times\mathbb{R}^n)} < \infty \tag{3.38}$$

is true for all $k' \in \mathbb{Z}$. This implies that the frequency envelope $c^\phi$, which is defined by (3.37) with $(f, g) = (\phi(0), \partial_t\phi(0))$, satisfies

$$c_k^\phi \overset{\text{def}}{=} \sum_{k'\in\mathbb{Z}} 2^{-\sigma|k-k'|} \left( \|S_{2^{k'}}\phi(0)\|_{\dot{H}^{n/2}(\mathbb{R}^n)} + \|S_{2^{k'}}\partial_t\phi(0)\|_{\dot{H}^{n/2-1}(\mathbb{R}^n)} \right)$$
$$\lesssim \sum_{k'\in\mathbb{Z}} 2^{-\sigma|k-k'|} 2^{-|k'|} \lesssim 2^{-\sigma|k|} \sum_{k'\in\mathbb{Z}} 2^{(\sigma-1)|k'|} \lesssim 2^{-\sigma|k|}, \quad \forall\, k \in \mathbb{Z},$$

where the last estimate is due to $\sigma < 1/2$. Thus, eventually replacing $c$ by

$$\tilde{c} = (\tilde{c}_k)_{k \in \mathbb{Z}} \overset{\text{def}}{=} (\min\{c_k, c_k^\phi\})_{k \in \mathbb{Z}},$$

which still verifies the hypotheses of Proposition 3.2, we can assume $c_k \lesssim 2^{-\sigma|k|}$. Hence, we arrive at the desired behavior for the frequency envelope $c$.

Following this, using the notation of Proposition 3.2, we define the set

$$A \overset{\text{def}}{=} \{0 \leq T' \leq T; \, \|S_{2^k}\phi\|_{X_k([0,T'] \times \mathbb{R}^n)} \leq C_0 \, c_k, \, \forall k \in \mathbb{Z}\}.$$

The proposition in question would follow if we show that there exists a constant $C_0 = C_0(n, m) > 0$ such that $A = [0, T]$. This can be achieved by proving that $A$ is nonempty, closed, and open. Based on Bernstein's inequalities, we infer that

$$\|S_{2^k}\phi\|_{X_k(\{0\} \times \mathbb{R}^n)} = \sup_{q \geq 2} 2^{nk/q} \left( \|S_{2^k}\phi(0)\|_{L^q(\mathbb{R}^n)} + 2^{-k}\|S_{2^k}\partial_t\phi(0)\|_{L^q(\mathbb{R}^n)} \right)$$

$$\lesssim 2^{nk/2} \left( \|S_{2^k}\phi(0)\|_{L^2(\mathbb{R}^n)} + 2^{-k}\|S_{2^k}\partial_t\phi(0)\|_{L^2(\mathbb{R}^n)} \right)$$

$$\simeq \|S_{2^k}\phi(0)\|_{\dot{H}^{n/2}(\mathbb{R}^n)} + \|S_{2^k}\partial_t\phi(0)\|_{\dot{H}^{n/2-1}(\mathbb{R}^n)} \leq c_k,$$

holds for all $k \in \mathbb{Z}$. This implies that $0 \in A$ if $C_0$ is sufficiently large.

For showing that $A$ is both closed and open, we will need the following preliminary result.

**Lemma 3.1.** *If $0 < T < \infty$ and $\phi$ is a classical wave map defined on $[0, T] \times \mathbb{R}^n$, then, for any fixed $k \in \mathbb{Z}$ and wave admissible triplet $(p, q, n)$, the function*

$$t \mapsto F_{k,p,q,n}(t) \overset{\text{def}}{=} 2^{(1/p+n/q)k} \big( \|S_{2^k}\phi\|_{L^p L^q([0,t] \times \mathbb{R}^n)}$$
$$+ 2^{-k}\|S_{2^k}\partial_t\phi\|_{L^p L^q([0,t] \times \mathbb{R}^n)} \big)$$

*is increasing and continuous on $[0, T]$ and satisfies*

$$F_{k,p,q,n}(t) \lesssim 2^{-|k|} \max\{1, t^{1/2}\} \|\nabla\phi\|_{L^\infty(\dot{H}^{(n+1)/2} \cap \dot{H}^{n/2-2})([0,t] \times \mathbb{R}^n)}. \quad (3.39)$$

**Proof.** The fact that $F_{k,p,q,n}$ is increasing is immediate by its definition, whereas Bernstein's inequalities imply

$$F_{k,p,q,n}(t) \lesssim \|S_{2^k}\nabla\phi\|_{L^p \dot{H}^{n/2+1/p-1}([0,t] \times \mathbb{R}^n)}$$

$$\lesssim 2^{-|k|}\|S_{2^k}\nabla\phi\|_{L^p(\dot{H}^{n/2+1/p} \cap \dot{H}^{n/2+1/p-2})([0,t] \times \mathbb{R}^n)}$$

$$\lesssim 2^{-|k|} \max\{1, t^{1/2}\} \|\nabla\phi\|_{L^\infty(\dot{H}^{(n+1)/2} \cap \dot{H}^{n/2-2})([0,t] \times \mathbb{R}^n)}.$$

A similar argument yields

$$0 \leq F_{k,p,q,n}(t_2) - F_{k,p,q,n}(t_1)$$

$$\lesssim \|S_{2^k}\nabla\phi\|_{L^p\dot{H}^{n/2+1/p-1}([t_1,t_2]\times\mathbb{R}^n)}$$

$$\lesssim (t_2 - t_1)^{1/p}\|\nabla\phi\|_{L^\infty(\dot{H}^{(n-1)/2}\cap\dot{H}^{n/2-1})([0,T]\times\mathbb{R}^n)},$$

when $0 \leq t_1 \leq t_2 \leq T$ and $p < \infty$. Thus, all which is left to argue is the continuity of the function in the case $p = \infty$.

To simplify the computations, we will work with

$$t \mapsto 2^{(n/q-1)k}\|S_{2^k}\nabla\phi\|_{L^\infty L^q([0,t]\times\mathbb{R}^n)} \tag{3.40}$$

instead of $F_{k,\infty,q,n}(t)$. If $q < \infty$, the smoothness of $\phi$ combined with the fundamental theorem of calculus and Hölder's inequality implies

$$\left|\|S_{2^k}\nabla\phi(t_2)\|_{L^q(\mathbb{R}^n)}^q - \|S_{2^k}\nabla\phi(t_1)\|_{L^q(\mathbb{R}^n)}^q\right|$$

$$\lesssim C(q)|t_2 - t_1|\|S_{2^k}\nabla\phi\|_{L^\infty L^q([0,T]\times\mathbb{R}^n)}^{q-1}\|S_{2^k}\partial_t\nabla\phi\|_{L^\infty L^q([0,T]\times\mathbb{R}^n)},$$

where $C(q)$ is a constant depending strictly on $q$. On the right-hand side, Bernstein's inequalities give us control over

$$\|S_{2^k}\nabla\phi\|_{L^\infty L^q} \quad \text{and} \quad \|S_{2^k}\partial_t\nabla_x\phi\|_{L^\infty L^q}$$

by using $\|\nabla\phi\|_{L^\infty\dot{H}^s}$-type norms, for various $s$. The only term that cannot be estimated like this is

$$\|S_{2^k}\partial_{tt}\phi\|_{L^\infty L^q},$$

for which one needs to use the main equation satisfied by $\phi$ (i.e., (3.28)), together with $\|\phi\|_{L^\infty_{t,x}} = 1$ and Sobolev embeddings. This proves that

$$t \mapsto \|S_{2^k}\nabla\phi(t)\|_{L^q(\mathbb{R}^n)}^q$$

is Lipschitz continuous on $[0,T]$. It is now an elementary real analysis exercise to deduce that (3.40) is continuous. A similar approach settles the last remaining case, $p = q = \infty$. $\qquad\square$

One recognizes from (3.33) that

$$\|S_{2^k}\phi\|_{X_k([0,T']\times\mathbb{R}^n)} = \sup_{\substack{p,q,n \\ \text{wave admissible}}} F_{k,p,q,n}(T'), \quad \forall 0 \leq T' \leq T. \tag{3.41}$$

Using the continuity in $T'$ of $F_{k,p,q,n}(T')$ proven above, it follows that the left-hand side is lower semicontinuous in $T'$, thus showing that $A$ is closed.

In order to prove that $A$ is open, Tao considers also the set

$$B \stackrel{\text{def}}{=} \{0 \leq T' \leq T; \|S_{2^k}\phi\|_{X_k([0,T']\times\mathbb{R}^n)} \leq 2C_0\,c_k, \forall k \in \mathbb{Z}\}$$

and argues that $A \subseteq \mathrm{int}\, B$. This can be seen as follows. Using (3.39) and (3.41), we derive

$$\|S_{2^k}\phi\|_{X_k([0,T']\times\mathbb{R}^n)} \lesssim 2^{-|k|},$$

which holds uniformly for $0 \leq T' \leq T$ and $k \in \mathbb{Z}$. Therefore, as $c_k \simeq 2^{-\sigma|k|}$, we infer that for $T' \in A$, with $T' < T$, there exists $T' < T'' \leq T$ such that

$$\|S_{2^k}\phi\|_{X_k([0,T'']\times\mathbb{R}^n)} \leq 2\,C_0\,c_k$$

holds if $|k|$ is sufficiently large. It remains then to show that for $T' \in A$, with $T' < T$, and $K$ a fixed positive integer, we can find $T' < T''' \leq T$ such that

$$\|S_{2^k}\phi\|_{X_k([0,T''']\times\mathbb{R}^n)} \leq 2\,C_0\,c_k, \qquad \forall\, k \in \mathbb{Z},\ |k| \leq K.$$

This would follow if one proves that

$$t \mapsto \|S_{2^k}\phi\|_{X_k([0,t]\times\mathbb{R}^n)}$$

is not only lower semicontinuous, but continuous on $[0,T]$. We leave this as an exercise for the avid reader.

Based on $A \subseteq \mathrm{int}\, B$, a sufficient condition for $A$ to be open is the inclusion $B \subseteq A$, which is in fact the bootstrap argument. Thus, all which is left to prove is:

**Proposition 3.3.** *Let $n \geq 5$ and consider $0 < T < \infty$, $c$ a frequency envelope with $\|c\|_{l^2} \lesssim \epsilon$, and $\phi$ a classical wave map defined on $[0,T] \times \mathbb{R}^n$ verifying $(\phi(0), \partial_t\phi(0)) \prec c$ and*

$$\|S_{2^k}\phi\|_{X_k([0,T]\times\mathbb{R}^n)} \leq 2\,C_0\,c_k,\ \forall\, k \in \mathbb{Z}, \tag{3.42}$$

*with $C_0$ being an absolute constant. Then*

$$\|S_{2^k}\phi\|_{X_k([0,T]\times\mathbb{R}^n)} \leq C_0\,c_k,\ \forall\, k \in \mathbb{Z}, \tag{3.43}$$

*provided $\epsilon$ is sufficiently small and $C_0$ is sufficiently large.*

Due to a scaling argument, **it is enough to prove** (3.43) **just for $k = 0$.** Indeed, if $T$, $c$, and $\phi$ are as above and $k \in \mathbb{Z}$ is fixed, let us define

$$\bar{\phi}(t,x) \overset{\text{def}}{=} \phi(t/2^k, x/2^k), \qquad \forall\, (t,x) \in [0, 2^k T] \times \mathbb{R}^n,$$

and

$$\bar{c} = (\bar{c}_{k'})_{k'\in\mathbb{Z}} \overset{\text{def}}{=} (c_{k+k'})_{k'\in\mathbb{Z}}.$$

Simple computations yield $\|\bar{c}\|_{l^2} = \|c\|_{l^2}$, $(\bar{\phi}(0), \partial_t\bar{\phi}(0)) \prec \bar{c}$, and

$$\|S_{2^{k'}}\bar{\phi}\|_{X_{k'}([0,2^k T]\times\mathbb{R}^n)} = \|S_{2^{k+k'}}\phi\|_{X_{k+k'}([0,T]\times\mathbb{R}^n)}$$
$$\leq 2\,C_0\,c_{k+k'} = 2\,C_0\,\bar{c}_{k'},\ \forall\, k' \in \mathbb{Z}.$$

Hence, $2^k T$, $\bar{c}$, and $\bar{\phi}$ verify the hypotheses of Proposition 3.3, from which it would follow that

$$\|S_{2^k}\phi\|_{X_k([0,T]\times\mathbb{R}^n)} = \|S_{2^0}\bar{\phi}\|_{X_0([0,2^k T]\times\mathbb{R}^n)} \leq C_0\bar{c}_0 = C_0\,c_k.$$

### 3.4.3.5 *Third step*

The next step in the main argument corresponds to the heuristics that led us from (3.28) to (3.29). As he needs to prove that

$$\|S_{2^0}\phi\|_{X_0([0,T]\times\mathbb{R}^n)} \leq C_0 c_0,$$

Tao projects (3.28) using $S_{2^0}$ and shows that

$$\Box S_{2^0}\phi = -2\langle\partial^\mu S_{2^0}\phi, \partial_\mu S_{\leq 2^{-10}}\phi\rangle S_{\leq 2^{-10}}\phi + Error, \tag{3.44}$$

where *Error* is defined as any function $F = F(t,x)$ on $[0,T]\times\mathbb{R}^n$ verifying

$$\|F\|_{L^1 L^2([0,T]\times\mathbb{R}^n)} \lesssim C_0^3\,\epsilon\, c_0. \tag{3.45}$$

To see this, we start by relying on paradifferential calculus to deduce

$$\phi_1\phi_2\phi_3 = \sum_{(k_1,k_2,k_3)\in\mathbb{Z}^3} S_{2^{k_1}}\phi_1\, S_{2^{k_2}}\phi_2\, S_{2^{k_3}}\phi_3$$

$$= \sum_{\max\{k_2,k_3\}\geq 10} + \sum_{\max\{k_2,k_3\}<10}$$

$$= \sum_{\substack{\max\{k_2,k_3\}\geq 10\\|k_2-k_3|\leq 5}} + \sum_{\substack{\max\{k_2,k_3\}\geq 10\\|k_2-k_3|>5}} + \sum_{\substack{\max\{k_2,k_3\}<10\\k_1>-10}} + \sum_{\substack{\max\{k_2,k_3\}<10\\k_1\leq -10}}$$

and, furthermore,

$$\sum_{\substack{\max\{k_2,k_3\}<10\\k_1\leq -10}}$$

$$= \sum_{\substack{-10<k_2,k_3<10\\k_1\leq -10}} + \sum_{\substack{-10<k_2<10\\\max\{k_1,k_3\}\leq -10}} + \sum_{\substack{-10<k_3<10\\\max\{k_1,k_2\}\leq -10}} + \sum_{\max\{k_1,k_2,k_3\}<-10}.$$

Therefore, based on (3.28), we can write

$$\Box S_{2^0}\phi = -S_{2^0}(\langle\partial^\mu\phi, \partial_\mu\phi\rangle\,\phi)$$

$$= -S_{2^0}\left(\sum_{(k_1,k_2,k_3)\in\mathbb{Z}^3}\langle\partial^\mu S_{2^{k_2}}\phi, \partial_\mu S_{2^{k_3}}\phi\rangle\, S_{2^{k_1}}\phi\right)$$

$$= I + II + III + IV + V + VI + VII,$$

where

$$I = -S_{2^0}\left(\sum_{\substack{\max\{k_2,k_3\}\geq 10 \\ |k_2-k_3|\leq 5}} \langle \partial^\mu S_{2^{k_2}}\phi, \partial_\mu S_{2^{k_3}}\phi\rangle\, S_{2^{k_1}}\phi\right),$$

$$II = -S_{2^0}\left(\sum_{\substack{\max\{k_2,k_3\}\geq 10 \\ |k_2-k_3|>5}} \langle \partial^\mu S_{2^{k_2}}\phi, \partial_\mu S_{2^{k_3}}\phi\rangle\, S_{2^{k_1}}\phi\right),$$

$$III = -S_{2^0}\left(\sum_{\substack{\max\{k_2,k_3\}<10 \\ k_1>-10}} \langle \partial^\mu S_{2^{k_2}}\phi, \partial_\mu S_{2^{k_3}}\phi\rangle\, S_{2^{k_1}}\phi\right),$$

$$IV = -S_{2^0}\left(\sum_{\substack{-10<k_2,k_3<10 \\ k_1\leq-10}} \langle \partial^\mu S_{2^{k_2}}\phi, \partial_\mu S_{2^{k_3}}\phi\rangle\, S_{2^{k_1}}\phi\right),$$

$$V = -S_{2^0}\left(\sum_{\substack{-10<k_2<10 \\ \max\{k_1,k_3\}\leq-10}} \langle \partial^\mu S_{2^{k_2}}\phi, \partial_\mu S_{2^{k_3}}\phi\rangle\, S_{2^{k_1}}\phi\right),$$

$$VI = -S_{2^0}\left(\sum_{\substack{-10<k_3<10 \\ \max\{k_1,k_2\}\leq-10}} \langle \partial^\mu S_{2^{k_2}}\phi, \partial_\mu S_{2^{k_3}}\phi\rangle\, S_{2^{k_1}}\phi\right),$$

$$VII = -S_{2^0}\left(\sum_{\max\{k_1,k_2,k_3\}<-10} \langle \partial^\mu S_{2^{k_2}}\phi, \partial_\mu S_{2^{k_3}}\phi\rangle\, S_{2^{k_1}}\phi\right).$$

From here on, to simplify the notation, we work with

$$\psi = S_{2^0}\phi \quad \text{and} \quad \tilde{\phi} = S_{\leq 2^{-10}}\phi.$$

In regards to $V$ and $VI$, we notice easily that

$$V = VI = -S_{2^0}\left(\langle \partial^\mu\left(\sum_{-10<k<10} S_{2^k}\phi\right), \partial_\mu\tilde{\phi}\rangle\tilde{\phi}\right).$$

An immediate frequency analysis reveals that $VII$ vanishes. Next, we show that $I - IV$ are Error-type terms, for which we use (3.33), (3.35), Bernstein and Cauchy-Schwarz inequalities, $\|\phi\|_{L^\infty_{t,x}([0,T]\times\mathbb{R}^n)} = 1$ (as $\phi : [0,T]\times\mathbb{R}^n \to \mathbb{S}^{m-1}$), $\sigma + 1/2 < 1 < n/4$, and (3.42). For $I$, we argue as follows:

$$\|I\|_{L^1L^2([0,T]\times\mathbb{R}^n)} \lesssim \sum_{\substack{\max\{k_2,k_3\}\geq 10 \\ |k_2-k_3|\leq 5}} \|S_{2^0}\left(\langle \partial^\mu S_{2^{k_2}}\phi, \partial_\mu S_{2^{k_3}}\phi\rangle \phi\right)\|_{L^1L^2}$$

$$\lesssim \sum_{\substack{\max\{k_2,k_3\}\geq 10 \\ |k_2-k_3|\leq 5}} \|\nabla S_{2^{k_2}}\phi\|_{L^2L^4}\|\nabla S_{2^{k_3}}\phi\|_{L^2L^4}\|\phi\|_{L^\infty_{t,x}}$$

$$\lesssim \sum_{\substack{\max\{k_2,k_3\}\geq 10 \\ |k_2-k_3|\leq 5}} 2^{(1/2-n/4)k_2}\|S_{2^{k_2}}\phi\|_{X_{k_2}} 2^{(1/2-n/4)k_3}\|S_{2^{k_3}}\phi\|_{X_{k_3}}$$

$$\lesssim C_0^2 \sum_{\substack{\max\{k_2,k_3\}\geq 10 \\ |k_2-k_3|\leq 5}} 2^{(1/2-n/4)(k_2+k_3)} c_{k_2} c_{k_3}$$

$$\lesssim C_0^2 c_0^2 \sum_{k\geq 10} 2^{2(\sigma+1/2-n/4)k}$$

$$\lesssim C_0^2 c_0^2.$$

For $II$, we can choose by symmetry that $k_2 > k_3 + 5$, which implies through a direct frequency analysis that $II$ vanishes unless $|k_1 - k_2| \leq 5$. As a result, we obtain

$$\|II\|_{L^1L^2([0,T]\times\mathbb{R}^n)} \lesssim \sum_{\substack{k_2>10 \\ |k_1-k_2|\leq 5}} \|S_{2^0}\left(\langle \partial^\mu S_{2^{k_2}}\phi, \partial_\mu S_{<2^{k_2-5}}\phi\rangle S_{2^{k_1}}\phi\right)\|_{L^1L^2}$$

$$\lesssim \sum_{\substack{k_2>10 \\ |k_1-k_2|\leq 5}} \|\nabla S_{2^{k_2}}\phi\|_{L^2L^4}\|\nabla S_{<2^{k_2-5}}\phi\|_{L^\infty_{t,x}}\|S_{2^{k_1}}\phi\|_{L^2L^4}$$

$$\lesssim \sum_{\substack{k_2>10 \\ |k_1-k_2|\leq 5}} 2^{(1/2-n/4)k_2}\|S_{2^{k_2}}\phi\|_{X_{k_2}} 2^{k_2} 2^{(-1/2-n/4)k_1}\|S_{2^{k_1}}\phi\|_{X_{k_1}}$$

$$\lesssim C_0^2 \sum_{\substack{k_2>10 \\ |k_1-k_2|\leq 5}} 2^{(3/2-n/4)k_2} 2^{(-1/2-n/4)k_1} c_{k_2} c_{k_1}$$

$$\lesssim C_0^2 c_0^2 \sum_{k\geq 10} 2^{2(\sigma+1/2-n/4)k}$$

$$\lesssim C_0^2 c_0^2.$$

In what concerns $III$, we argue that this term is relevant only if $k_1 < 15$, which can be seen again by a frequency analysis. Therefore, we have

$$\|III\|_{L^1L^2([0,T]\times\mathbb{R}^n)} \lesssim \sum_{-10<k_1<15} \|S_{2^0}\left(\langle\partial^\mu S_{<2^{10}}\phi, \partial_\mu S_{<2^{10}}\phi\rangle\, S_{2^{k_1}}\phi\right)\|_{L^1L^2}$$

$$\lesssim \|\nabla S_{<2^{10}}\phi\|^2_{L^2L^\infty} \sum_{-10<k_1<15} \|S_{2^{k_1}}\phi\|_{L^\infty L^2}$$

$$\lesssim \left(\sum_{k<10} 2^{k/2}\|S_{2^k}\phi\|_{X_k}\right)^2 \sum_{-10<k_1<15} 2^{-nk_1/2}\|S_{2^{k_1}}\phi\|_{X_{k_1}}$$

$$\lesssim C_0^3 \left(\sum_{k<10} 2^{k/2} c_k\right)^2 \sum_{-10<k_1<15} 2^{-nk_1/2} c_{k_1}$$

$$\lesssim C_0^3 c_0 \sum_{k<10} 2^k \sum_{k<10} c_k^2 \sum_{-10<k_1<15} 2^{\sigma|k_1|-nk_1/2}$$

$$\lesssim C_0^3 \epsilon^2 c_0.$$

Finally, for $IV$, we can estimate directly as follows:

$$\|IV\|_{L^1L^2([0,T]\times\mathbb{R}^n)} \lesssim \sum_{-10<k_2,k_3<10} \|S_{2^0}\left(\langle\partial^\mu S_{2^{k_2}}\phi, \partial_\mu S_{2^{k_3}}\phi\rangle\, \tilde{\phi}\right)\|_{L^1L^2}$$

$$\lesssim \sum_{-10<k_2,k_3<10} \|\nabla S_{2^{k_2}}\phi\|_{L^2L^4}\|\nabla S_{2^{k_3}}\phi\|_{L^2L^4}\|\tilde{\phi}\|_{L^\infty_{t,x}}$$

$$\lesssim \sum_{-10<k_2,k_3<10} 2^{(1/2-n/4)k_2}\|S_{2^{k_2}}\phi\|_{X_{k_2}}\, 2^{(1/2-n/4)k_2}\|S_{2^{k_3}}\phi\|_{X_{k_3}}$$

$$\lesssim C_0^2 \sum_{-10<k_2,k_3<10} 2^{(1/2-n/4)(k_2+k_3)} c_{k_2} c_{k_3}$$

$$\lesssim C_0^2 c_0^2 \sum_{-10<k_2,k_3<10} 2^{\sigma(|k_2|+|k_3|)+(1/2-n/4)(k_2+k_3)}$$

$$\lesssim C_0^2 c_0^2.$$

Therefore, at this point, the previous considerations yield

$$\Box\psi = -2\, S_{2^0}\left(\langle\partial^\mu\left(\sum_{-10<k<10} S_{2^k}\phi\right), \partial_\mu\tilde{\phi}\rangle\tilde{\phi}\right) + Error.$$

In order to claim (3.44) from the previous equation, Tao uses the commutator estimate

$$\|[S_{2^0}, f]\, g\|_{L^r(\mathbb{R}^n)} \lesssim \|\nabla_x f\|_{L^p(\mathbb{R}^n)}\|g\|_{L^q(\mathbb{R}^n)},$$
$$1 \le r,p,q \le \infty, \qquad 1/r = 1/p + 1/q, \tag{3.46}$$

whose proof is a simple exercise in integration. Indeed, as

$$S_{2^0}\left(\sum_{-10<k<10} S_{2^k}\phi\right) = \psi,$$

we only need to show that

$$[S_{2^0}, \langle\cdot,\partial_\mu\tilde\phi\rangle\tilde\phi]\partial^\mu\left(\sum_{-10<k<10} S_{2^k}\phi\right) \tag{3.47}$$

is an Error-type term. Based on (3.46) with $(r,p,q) = (2, n-1, 2(n-1)/(n-3))$, we infer

$$\left\|[S_{2^0}, \langle\cdot,\partial_\mu\tilde\phi\rangle\tilde\phi]\partial^\mu\left(\sum_{-10<k<10} S_{2^k}\phi\right)\right\|_{L^1L^2([0,T]\times\mathbb{R}^n)}$$

$$\lesssim \|\nabla_x(\nabla\tilde\phi\,\tilde\phi)\|_{L^2L^{n-1}}\left\|\sum_{-10<k<10} \nabla S_{2^k}\phi\right\|_{L^2L^{2(n-1)/(n-3)}}$$

$$\lesssim \left(\|\nabla\tilde\phi\|^2_{L^4L^{2(n-1)}} + \|\nabla_x\nabla\tilde\phi\|_{L^2L^{n-1}}\|\tilde\phi\|_{L^\infty_{t,x}}\right)$$
$$\cdot \sum_{-10<k<10}\|\nabla S_{2^k}\phi\|_{L^2L^{2(n-1)/(n-3)}}.$$

Arguing as in the analysis of $IV$, we derive

$$\sum_{-10<k<10}\|\nabla S_{2^k}\phi\|_{L^2L^{2(n-1)/(n-3)}}$$

$$\lesssim \sum_{-10<k<10} 2^{(1/2-n(n-3)/2(n-1))k}\|S_{2^k}\phi\|_{X_k}$$

$$\lesssim C_0 \sum_{-10<k<10} 2^{(1/2-n(n-3)/2(n-1))k}\,c_k \tag{3.48}$$

$$\lesssim C_0 c_0 \sum_{-10<k<10} 2^{\sigma|k|+(1/2-n(n-3)/2(n-1))k}$$

$$\lesssim C_0 c_0.$$

For the terms involving $\tilde\phi$, we have $\|\tilde\phi\|_{L^\infty_{t,x}} \lesssim 1$,

$$\|\nabla\tilde\phi\|_{L^4L^{2(n-1)}} \lesssim \sum_{k\le-10} 2^{(3/4-n/2(n-1))k}\|S_{2^k}\phi\|_{X_k}$$

$$\lesssim C_0 \sum_{k\le-10} 2^{(3/4-n/2(n-1))k}\,c_k$$

$$\lesssim C_0 \left(\sum_{k\le-10} 2^{(3/2-n/(n-1))k}\right)^{1/2}\left(\sum_{k\le-10} c_k^2\right)^{1/2} \tag{3.49}$$

$$\lesssim C_0\,\epsilon,$$

and

$$\|\nabla_x \nabla \tilde{\phi}\|_{L^2 L^{n-1}} \lesssim \sum_{k \leq -10} 2^{(3/2 - n/(n-1))k} \|S_{2^k} \phi\|_{X_k}$$

$$\lesssim C_0 \sum_{k \leq -10} 2^{(3/2 - n/(n-1))k} c_k$$

$$\lesssim C_0 \left( \sum_{k \leq -10} 2^{(3 - 2n/(n-1))k} \right)^{1/2} \left( \sum_{k \leq -10} c_k^2 \right)^{1/2} \quad (3.50)$$

$$\lesssim C_0 \epsilon,$$

due to $3/2 - n/(n-1) > 0$ for $n \geq 5$. Hence, the last four estimates show that (3.47) is an Error-type term, which finishes the argument for (3.44).

### 3.4.3.6 *Fourth step*

The fourth step in Tao's proof consists in motivating the intuition behind (3.30), which can be recast as

$$\Box \psi = -2 A_\mu \partial^\mu \psi + Error, \quad (3.51)$$

with

$$A_\mu \stackrel{\text{def}}{=} \tilde{\phi} (\partial_\mu \tilde{\phi})^t - \partial_\mu \tilde{\phi} \tilde{\phi}^t \quad (3.52)$$

being an $m \times m$ antisymmetric matrix. This new equation for $\psi$ can be deduced from (3.44) if we show that $\langle \partial^\mu \psi, \tilde{\phi} \rangle \partial_\mu \tilde{\phi}$ is of Error type. An identical argument with the one producing (3.49) yields $\|\nabla \tilde{\phi}\|_{L^2 L^\infty} \lesssim C_0 \epsilon$. Therefore, the previous claim follows if we prove

$$\|\langle \tilde{\phi}, \partial^\mu \psi \rangle\|_{L^2_{t,x}} \lesssim C_0^2 \epsilon c_0.$$

However,

$$\|\langle \partial^\mu \tilde{\phi}, \psi \rangle\|_{L^2_{t,x}} \lesssim \|\nabla \tilde{\phi}\|_{L^2 L^\infty} \|\psi\|_{L^\infty L^2} \lesssim C_0 \epsilon \|S_{2^0} \phi\|_{X_0} \lesssim C_0^2 \epsilon c_0.$$

Hence, this step is reduced to showing

$$\|\partial^\mu \langle \tilde{\phi}, \psi \rangle\|_{L^2_{t,x}} \lesssim C_0^2 \epsilon c_0.$$

This is the second time in the main argument when Tao uses the geometry of the problem, in the form of $\langle \phi, \phi \rangle = 1$, to rewrite the previous estimate as

$$\|\partial^\mu (S_{2^0}(\langle \phi, \phi \rangle) - 2 \langle \tilde{\phi}, \psi \rangle)\|_{L^2_{t,x}} \lesssim C_0^2 \epsilon c_0. \quad (3.53)$$

Through a simple frequency analysis, we obtain

$$\partial^\mu(S_{2^0}(\langle\phi,\phi\rangle)) - 2\langle\tilde\phi,\psi\rangle) = \sum_{\substack{k_1,k_2>-10 \\ |k_1-k_2|<20}} \partial^\mu S_{2^0}\left(\langle S_{2^{k_1}}\phi, S_{2^{k_2}}\phi\rangle\right)$$

$$+ 2\partial^\mu\left([S_{2^0},\langle\tilde\phi,\cdot\rangle]\sum_{-10<k<10} S_{2^k}\phi\right).$$

For the first term, we derive

$$\left\|\sum_{\substack{k_1,k_2>-10 \\ |k_1-k_2|<20}} \partial^\mu S_{2^0}\left(\langle S_{2^{k_1}}\phi, S_{2^{k_2}}\phi\rangle\right)\right\|_{L^2_{t,x}}$$

$$\lesssim \sum_{\substack{k_1,k_2>-10 \\ |k_1-k_2|<20}} \|\nabla S_{2^{k_1}}\phi\|_{L^\infty L^2}\|S_{2^{k_2}}\phi\|_{L^2 L^\infty} + \|S_{2^{k_1}}\phi\|_{L^2 L^\infty}\|\nabla S_{2^{k_2}}\phi\|_{L^\infty L^2}$$

$$\lesssim \sum_{\substack{k_1,k_2>-10 \\ |k_1-k_2|<20}} (2^{(-n/2+1)k_1-k_2/2} + 2^{(-n/2+1)k_2-k_1/2})\|S_{2^{k_1}}\phi\|_{X_{k_1}}\|S_{2^{k_2}}\phi\|_{X_{k_2}}$$

$$\lesssim C_0^2 \sum_{k>-10} 2^{-(n-1)k/2} c_k^2$$

$$\lesssim C_0^2 c_0 \sum_{k>-10} 2^{\sigma|k|-(n-1)k/2} c_k$$

$$\lesssim C_0^2 c_0 \left(\sum_{k>-10} 2^{2(\sigma|k|-(n-1)k/2)}\right)^{1/2}\left(\sum_{k>-10} c_k^2\right)^{1/2}$$

$$\lesssim C_0^2 \epsilon c_0,$$

where we also relied on $\sigma - (n-1)/2 < 0$ for $\sigma < 1/2$ and $n \geq 5$. For the second term, we notice first that its spatial Fourier transform is supported in the region $\{|\xi| \lesssim 1\}$. Therefore, if $\mu$ is a spatial index, then reasoning similar to the one for (3.48), which also uses (3.46), implies

$$\left\|\partial^\mu\left([S_{2^0},\langle\tilde\phi,\cdot\rangle]\sum_{-10<k<10} S_{2^k}\phi\right)\right\|_{L^2_{t,x}}$$

$$\lesssim \left\|[S_{2^0},\langle\tilde\phi,\cdot\rangle]\sum_{-10<k<10} S_{2^k}\phi\right\|_{L^2_{t,x}}$$

$$\lesssim \|\nabla\tilde\phi\|_{L^2 L^\infty}\left\|\sum_{-10<k<10} S_{2^k}\phi\right\|_{L^\infty L^2}$$

$$\lesssim C_0^2 \epsilon c_0.$$

If $\partial^\mu = -\partial_t$, then

$$\left\| \partial_t \left( [S_{2^0}, \langle \tilde{\phi}, \cdot \rangle] \sum_{-10 < k < 10} S_{2^k} \phi \right) \right\|_{L^2_{t,x}}$$

$$\lesssim \left\| \left( [S_{2^0}, \langle \partial_t \tilde{\phi}, \cdot \rangle] \sum_{-10 < k < 10} S_{2^k} \phi \right) \right\|_{L^2_{t,x}}$$

$$+ \left\| \left( [S_{2^0}, \langle \tilde{\phi}, \cdot \rangle] \sum_{-10 < k < 10} \partial_t S_{2^k} \phi \right) \right\|_{L^2_{t,x}}$$

$$\lesssim \|\nabla_x \partial_t \tilde{\phi}\|_{L^2 L^\infty} \left\| \sum_{-10 < k < 10} S_{2^k} \phi \right\|_{L^\infty L^2}$$

$$+ \|\nabla_x \tilde{\phi}\|_{L^2 L^\infty} \left\| \sum_{-10 < k < 10} \partial_t S_{2^k} \phi \right\|_{L^\infty L^2}$$

$$\lesssim \|\nabla \tilde{\phi}\|_{L^2 L^\infty} \left( \left\| \sum_{-10 < k < 10} S_{2^k} \phi \right\|_{L^\infty L^2} + \left\| \sum_{-10 < k < 10} \partial_t S_{2^k} \phi \right\|_{L^\infty L^2} \right)$$

$$\lesssim C_0 \, \epsilon \sum_{-10 < k < 10} (1 + 2^k) \, 2^{-nk/2} \|S_{2^k} \phi\|_{X_k}$$

$$\lesssim C_0^2 \, \epsilon \sum_{-10 < k < 10} (1 + 2^k) \, 2^{-nk/2} c_k$$

$$\lesssim C_0^2 \, \epsilon \, c_0.$$

This concludes the argument for (3.53) and finishes the discussion of the fourth step.

### 3.4.3.7  Fifth step

The fifth step is the one that deals with the *renormalization* procedure and it is the most intricate part of the proof. The goal is to construct an almost orthogonal invertible $m \times m$ matrix $U$ such that the substitution $\psi = Uw$ transforms (3.51) into

$$\Box w \; = \; Error. \tag{3.54}$$

A straightforward computation shows that

$$\begin{aligned}
\Box w \; = \; & - 2 \, U^{-1} (\partial_\mu U + A_\mu U) \, U^{-1} (\partial^\mu \psi - \partial^\mu U \, U^{-1} \psi) \\
& - U^{-1} (2 \, A_\mu \, \partial^\mu U + \Box U) \, U^{-1} \psi + U^{-1} Error.
\end{aligned} \tag{3.55}$$

Thus, in order for the renormalization to be successful, we should, in principle, have good control on $U$, $U^{-1}$, $\nabla U$, $\Box U$, and $\partial_\mu U + A_\mu U$.

The matrix $U$ is constructed inductively through the following dyadic scheme. We start by choosing a large integer $M$ depending on $T$ and we let $U_{-M} := I$, where $I$ is the $m \times m$ identity matrix. For integer values of $k > -M$, we define by induction

$$U_k \stackrel{\text{def}}{=} \left(S_{2^k}\phi\left(S_{<2^k}\phi\right)^t - S_{<2^k}\phi\left(S_{2^k}\phi\right)^t\right)U_{<k},$$

$$U_{<k} \stackrel{\text{def}}{=} \sum_{-M \leq k' < k} U_{k'}. \tag{3.56}$$

As an example, if $k = -M + 1$, then $U_{<-M+1} = U_{-M} = I$ and

$$U_{-M+1} = S_{2^{-M+1}}\phi\left(S_{<2^{-M+1}}\phi\right)^t - S_{<2^{-M+1}}\phi\left(S_{2^{-M+1}}\phi\right)^t.$$

The matrix used in the renormalization procedure is given by $U = U_{<-9}$.

**Remark 3.11.** First, using frequency analysis, one can easily show that the spatial Fourier transform of $U_{<k}$ is supported in the region $\{|\xi| \leq 2^{k+2}\}$, for all $-M \leq k \leq -9$. Secondly, a direct computation based on (3.56) yields

$$U_k^t U_{<k} + U_{<k}^t U_k = 0,$$

which implies

$$
\begin{aligned}
U_{<k}^t U_{<k} &= \sum_{-M \leq k_1, k_2 < k} U_{k_1}^t U_{k_2} \\
&= \sum_{-M \leq k_1, k_2 < k} U_{k_1}^t U_{k_2} + U_{k_2}^t U_{k_1} + \sum_{-M \leq k_1 < k} U_{k_1}^t U_{k_1} \\
&= \sum_{-M < k_2 < k} U_{<k_2}^t U_{k_2} + U_{k_2}^t U_{<k_2} + \sum_{-M \leq k_1 < k} U_{k_1}^t U_{k_1} \\
&= \sum_{-M \leq k_1 < k} U_{k_1}^t U_{k_1}.
\end{aligned}
\tag{3.57}
$$

In this context, Tao is able to prove the following result.

**Proposition 3.4.** *Under the hypotheses of Proposition 3.3, let $\epsilon$ be sufficiently small depending on $C_0$ and let $M$ in the above dyadic scheme be sufficiently large, depending on $T$, $C_0$, and $\epsilon$. Then the matrix $U$ is invertible and satisfies*

$$\|U\|_{L^\infty_{t,x}([0,T]\times\mathbb{R}^n)} + \|U^{-1}\|_{L^\infty_{t,x}([0,T]\times\mathbb{R}^n)} \lesssim 1 \tag{3.58}$$

*and*

$$\|U^t U - I\|_{L^\infty_{t,x}([0,T]\times\mathbb{R}^n)} + \|\partial_t(U^t U - I)\|_{L^\infty_{t,x}([0,T]\times\mathbb{R}^n)}$$

$$+ \|\partial_\mu U + A_\mu U\|_{L^1 L^\infty([0,T]\times\mathbb{R}^n)} + \|\partial_\mu U\|_{L^\infty_{t,x}([0,T]\times\mathbb{R}^n)} \qquad (3.59)$$

$$+ \|\partial_\mu U\|_{L^2 L^\infty([0,T]\times\mathbb{R}^n)} + \|\Box U\|_{L^2 L^{n-1}([0,T]\times\mathbb{R}^n)} \lesssim C_0^2\,\epsilon,$$

*for all $\mu$.*

**Proof.** In the proof of this proposition, we will rely extensively on

$$\|S_{2^k}\phi\|_{L^\infty_{t,x}} + 2^{-k}\|\nabla S_{2^k}\phi\|_{L^\infty_{t,x}} + 2^{k/2}\|S_{2^k}\phi\|_{L^2 L^\infty}$$

$$+ 2^{-k/2}\|\nabla S_{2^k}\phi\|_{L^2 L^\infty} \lesssim C_0\,c_k \qquad (3.60)$$

and

$$\|S_{<2^k}\phi\|_{L^\infty_{t,x}} \lesssim 1,$$

$$2^{-k}\|\nabla S_{<2^k}\phi\|_{L^\infty_{t,x}} + 2^{-k/2}\|\nabla S_{<2^k}\phi\|_{L^2 L^\infty} \lesssim C_0\,c_k, \qquad (3.61)$$

which can be deduced immediately using (3.33), (3.35), and (3.42) (similar arguments have also appeared in the third step).

First, we argue by induction that

$$\|U_{<k}\|_{L^\infty_{t,x}} \le 2, \qquad \forall -M < k \le -9. \qquad (3.62)$$

In particular, for $k = -9$, this gives the desired control of $\|U\|_{L^\infty_{t,x}}$ in (3.58). The base case $k = -M+1$ is trivially satisfied, as $U_{<-M+1} = I$. We assume that the same is true for all $k < N \le -9$, where $N$ is a fixed integer. Then, based on (3.56), (3.60), and (3.61), we derive

$$\|U_k\|_{L^\infty_{t,x}} \lesssim \|S_{2^k}\phi\|_{L^\infty_{t,x}}\|S_{<2^k}\phi\|_{L^\infty_{t,x}}\|U_{<k}\|_{L^\infty_{t,x}} \lesssim C_0\,c_k, \qquad \forall -M < k < N.$$

Combining this estimate with (3.57), we further deduce

$$\|U^t_{<N}U_{<N} - I\|_{L^\infty_{t,x}} \le \sum_{-M<k<N} \|U^t_k U_k\|_{L^\infty_{t,x}}$$

$$\lesssim \sum_{-M<k<N} \|U_k\|^2_{L^\infty_{t,x}} \qquad (3.63)$$

$$\lesssim C_0^2\,\epsilon^2.$$

Finally, a simple matrix computation shows that

$$\|U_{<N}\|^2_{L^\infty_{t,x}} \le \|U^t_{<N}U_{<N}\|_{L^\infty_{t,x}} \le 1 + \|U^t_{<N}U_{<N} - I\|_{L^\infty_{t,x}},$$

which implies $\|U_{<N}\|_{L^\infty_{t,x}} \le 2$, if $\epsilon$ is sufficiently small depending on $C_0$. This finishes the proof of (3.62).

Moreover, as a byproduct of this argument, the bound (3.63), which was obtained for $U_{<N}$, can be similarly derived for $U_{<-9}$ once we have (3.62). This is even better than the estimate we are seeking for $\|U^t U - I\|_{L^\infty_{t,x}}$, i.e.,

$$\|U^t U - I\|_{L^\infty_{t,x}} \lesssim C_0^2 \,\epsilon.$$

It follows that, for sufficiently small $\epsilon$ depending on $C_0$, $U^t U$ is invertible and, thus, $U$ is invertible too. Relying on the identity

$$U^{-1} = (I - U^t U)\, U^{-1} + U^t,$$

we derive

$$
\begin{aligned}
\|U^{-1}\|_{L^\infty_{t,x}} &\lesssim \|I - U^t U\|_{L^\infty_{t,x}} \|U^{-1}\|_{L^\infty_{t,x}} + \|U^t\|_{L^\infty_{t,x}} \\
&\lesssim C_0^2\, \epsilon \|U^{-1}\|_{L^\infty_{t,x}} + \|U\|_{L^\infty_{t,x}} \\
&\lesssim C_0^2\, \epsilon \|U^{-1}\|_{L^\infty_{t,x}} + 1,
\end{aligned}
$$

which implies, based on the smallness of $\epsilon$, that $\|U^{-1}\|_{L^\infty_{t,x}} \lesssim 1$. This concludes the argument for (3.58).

Next, we prove that

$$\|\partial_\mu U\|_{L^\infty_{t,x}} + \|\partial_\mu U\|_{L^2 L^\infty} \lesssim C_0^2\, \epsilon, \tag{3.64}$$

for all $\mu$. We notice that the desired estimate for $\|\partial_t(U^t U - I)\|_{L^\infty_{t,x}}$ follows immediately by combining the bounds for $\|U\|_{L^\infty_{t,x}}$ and $\|\partial_\mu U\|_{L^\infty_{t,x}}$. In proving (3.64), we argue by induction and show that

$$\|\partial_\mu U_{<k}\|_{L^\infty_{t,x}} \leq C_1 C_0^2\, 2^k c_k \tag{3.65}$$

and

$$\|\partial_\mu U_{<k}\|_{L^2 L^\infty} \leq C_1 C_0^2\, 2^{k/2} c_k \tag{3.66}$$

hold for all $-M < k \leq -9$, where $C_1$ is a sufficiently large absolute constant. The base case $k = -M + 1$ is immediately verified and we assume that both claims are true for $k \leq N < -9$. We use (3.56), (3.60), (3.61), (3.62), and the induction hypothesis to infer

$$
\begin{aligned}
\|\partial_\mu U_N\|_{L^\infty_{t,x}} &\lesssim \|\nabla S_{2^N}\phi\|_{L^\infty_{t,x}} \|S_{<2^N}\phi\|_{L^\infty_{t,x}} \|U_{<N}\|_{L^\infty_{t,x}} \\
&\quad + \|S_{2^N}\phi\|_{L^\infty_{t,x}} \|\nabla S_{<2^N}\phi\|_{L^\infty_{t,x}} \|U_{<N}\|_{L^\infty_{t,x}} \\
&\quad + \|S_{2^N}\phi\|_{L^\infty_{t,x}} \|S_{<2^N}\phi\|_{L^\infty_{t,x}} \|\partial_\mu U_{<N}\|_{L^\infty_{t,x}} \\
&\lesssim C_0\, 2^N c_N + C_0^2\, 2^N c_N^2 + C_1 C_0^3\, 2^N c_N^2 \\
&\lesssim C_0\, 2^N c_N (1 + C_1 C_0^2\, c_N)
\end{aligned}
$$

and

$$\|\partial_\mu U_N\|_{L^2 L^\infty} \lesssim \|\nabla S_{2^N}\phi\|_{L^2 L^\infty} \|S_{<2^N}\phi\|_{L^\infty_{t,x}} \|U_{<N}\|_{L^\infty_{t,x}}$$
$$+ \|S_{2^N}\phi\|_{L^\infty_{t,x}} \|\nabla S_{<2^N}\phi\|_{L^2 L^\infty} \|U_{<N}\|_{L^\infty_{t,x}}$$
$$+ \|S_{2^N}\phi\|_{L^\infty_{t,x}} \|S_{<2^N}\phi\|_{L^\infty_{t,x}} \|\partial_\mu U_{<N}\|_{L^2 L^\infty}$$
$$\lesssim C_0\, 2^{N/2} c_N + C_0^2\, 2^{N/2} c_N^2 + C_1 C_0^3\, 2^{N/2} c_N^2$$
$$\lesssim C_0\, 2^{N/2} c_N (1 + C_1 C_0^2\, c_N).$$

If we also take advantage of (3.35), it follows that

$$\frac{\|\partial_\mu U_N\|_{L^\infty_{t,x}}}{C_1 C_0^2 2^{N+1} c_{N+1}} \lesssim 2^{\sigma-1}(C_1^{-1} C_0^{-1} + C_0\, \epsilon)$$

and

$$\frac{\|\partial_\mu U_N\|_{L^2 L^\infty}}{C_1 C_0^2 2^{(N+1)/2} c_{N+1}} \lesssim 2^{\sigma-1/2}(C_1^{-1} C_0^{-1} + C_0\, \epsilon),$$

which can both be made arbitrarily small on account of choosing $C_1 C_0 \gg 1$ and $C_0 \epsilon \ll 1$. Hence, we can prove that (3.65) and (3.66) are true for $k = N + 1$ if

$$\frac{c_N}{2\, c_{N+1}} \qquad \text{and} \qquad \frac{c_N}{2^{1/2}\, c_{N+1}}$$

admit uniform (i.e., in terms of $N$) upper bounds, which are both less than 1. The only control we have over these quantities is through (3.35), which yields

$$\frac{c_N}{2\, c_{N+1}} \lesssim 2^{\sigma-1} \qquad \text{and} \qquad \frac{c_N}{2^{1/2}\, c_{N+1}} \lesssim 2^{\sigma-1/2}.$$

We know that $0 < \sigma = \sigma(n) < 1/2$. Therefore, one more adjustment in the value of $\sigma$ might be needed to close the induction and claim (3.64).

Following this, we estimate $\|\partial_\mu U + A_\mu U\|_{L^1 L^\infty}$. We work with the identity

$$\sum_{-M < k < -9} [\partial_\mu U_k + (A_{\mu, <k+1} U_{<k+1} - A_{\mu, <k} U_{<k})]$$
$$= \partial_\mu (U_{<-9} - U_{<-M+1}) \qquad\qquad (3.67)$$
$$+ (A_{\mu, <-9} U_{<-9} - A_{\mu, <-M+1} U_{<-M+1})$$
$$= \partial_\mu U + A_\mu U - A_{\mu, <-M+1},$$

where we used the notation[4]

$$A_{\mu, <k} \overset{\text{def}}{=} S_{<2^k}\phi\, (\partial_\mu S_{<2^k}\phi)^t - \partial_\mu S_{<2^k}\phi\, (S_{<2^k}\phi)^t, \qquad (3.68)$$

_____
[4]This is in accord with (3.52), as $A_\mu = A_{\mu, <-9}$.

for all $-M < k \le -9$. We deduce

$$\|\partial_\mu U + A_\mu U\|_{L^1 L^\infty}$$
$$\lesssim \sum_{-M < k < -9} \|\partial_\mu U_k + (A_{\mu,<k+1} U_{<k+1} - A_{\mu,<k} U_{<k})\|_{L^1 L^\infty}$$
$$+ \|A_{\mu,<-M+1}\|_{L^1 L^\infty} .$$

For the last term, we have

$$\|A_{\mu,<-M+1}\|_{L^1 L^\infty} \lesssim T \|A_{\mu,<-M+1}\|_{L^\infty_{t,x}}$$
$$\lesssim T \|S_{<2^{-M+1}}\phi\|_{L^\infty_{t,x}} \|\nabla S_{<2^{-M+1}}\phi\|_{L^\infty_{t,x}} \qquad (3.69)$$
$$\lesssim C_0 \epsilon 2^{-M} T,$$

due to (3.61). This is the point in the argument where we enforce controlling $M$ in terms of $C_0$ and $T$, in order to obtain the desired bound for $\partial_\mu U + A_\mu U$, i.e.,

$$\|\partial_\mu U + A_\mu U\|_{L^1 L^\infty} \lesssim C_0^2 \epsilon. \qquad (3.70)$$

For that, we impose $2^{-M} < C_0/T$.

In what concerns the generic term of the sum in (3.67), a careful computation based on (3.56) and (3.68) shows that

$$\partial_\mu U_k + (A_{\mu,<k+1} U_{<k+1} - A_{\mu,<k} U_{<k})$$
$$= A_{\mu,<k+1} U_k + \left((S_{2^k}\phi \,(S_{<2^k}\phi)^t - S_{<2^k}\phi \,(S_{2^k}\phi)^t)\right) \partial_\mu U_{<k}$$
$$+ \big[2 \left(S_{2^k}\phi \,(\partial_\mu S_{<2^k}\phi)^t - \partial_\mu S_{<2^k}\phi \,(S_{2^k}\phi)^t\right)$$
$$+ \left(S_{2^k}\phi \,(\partial_\mu S_{2^k}\phi)^t - \partial_\mu S_{2^k}\phi \,(S_{2^k}\phi)^t\right)\big] U_{<k}.$$

We notice immediately the absence from this expression of any dangerous *derivative falls on high frequency* terms coming from *high-low* interactions, which is the main contribution of the dyadic scheme producing $U$. Our goal here is to prove

$$\|\partial_\mu U_k + (A_{\mu,<k+1} U_{<k+1} - A_{\mu,<k} U_{<k})\|_{L^1 L^\infty} \lesssim C_1 C_0^3 c_k^2, \qquad (3.71)$$

for all $-M < k < -9$. Together with (3.69), this bound is enough to claim (3.70) by eventually readjusting the value of $\epsilon$. Using the definitions (3.56) and (3.68) and the bounds (3.60), (3.61), (3.62), and (3.66), we derive:

$$\|A_{\mu,<k+1} U_k\|_{L^1 L^\infty}$$
$$\lesssim \|A_{\mu,<k+1}\|_{L^2 L^\infty} \|U_k\|_{L^2 L^\infty}$$
$$\lesssim \|\nabla S_{<2^{k+1}}\phi\|_{L^2 L^\infty} \|S_{<2^{k+1}}\phi\|_{L^\infty_{t,x}} \|S_{2^k}\phi\|_{L^2 L^\infty} \|S_{<2^k}\phi\|_{L^\infty_{t,x}} \|U_{<k}\|_{L^\infty_{t,x}}$$
$$\lesssim C_0^2 c_k^2$$
$$\lesssim C_1 C_0^3 c_k^2,$$

$$\|\big((S_{2^k}\phi\,(S_{<2^k}\phi)^t - S_{<2^k}\phi\,(S_{2^k}\phi)^t)\big)\partial_\mu U_{<k}\|_{L^1 L^\infty}$$
$$\lesssim \|S_{2^k}\phi\|_{L^2 L^\infty}\|S_{<2^k}\phi\|_{L^\infty_{t,x}}\|\partial_\mu U_{<k}\|_{L^2 L^\infty}$$
$$\lesssim C_1 C_0^3 \, c_k^2,$$

$$\|\big(S_{2^k}\phi\,(\partial_\mu S_{<2^k}\phi)^t - \partial_\mu S_{<2^k}\phi\,(S_{2^k}\phi)^t\big)U_{<k}\|_{L^1 L^\infty}$$
$$\lesssim \|S_{2^k}\phi\|_{L^2 L^\infty}\|\nabla S_{<2^k}\phi\|_{L^2 L^\infty}\|U_{<k}\|_{L^\infty_{t,x}}$$
$$\lesssim C_0^2 \, c_k^2$$
$$\lesssim C_1 C_0^3 \, c_k^2,$$

$$\|\big(S_{2^k}\phi\,(\partial_\mu S_{2^k}\phi)^t - \partial_\mu S_{2^k}\phi\,(S_{2^k}\phi)^t\big)U_{<k}\|_{L^1 L^\infty}$$
$$\lesssim \|S_{2^k}\phi\|_{L^2 L^\infty}\|\nabla S_{2^k}\phi\|_{L^2 L^\infty}\|U_{<k}\|_{L^\infty_{t,x}}$$
$$\lesssim C_0^2 \, c_k^2$$
$$\lesssim C_1 C_0^3 \, c_k^2.$$

Combining all these estimates, we deduce (3.71), which finishes the analysis of $\partial_\mu U + A_\mu U$.

All which is left to prove is

$$\|\Box U\|_{L^2 L^{n-1}} \lesssim C_0^2 \, \epsilon.$$

The argument is similar to the one for (3.64), as one proceeds by induction to show that

$$\|\Box U_{<k}\|_{L^2 L^{n-1}} \leq C_2 C_0^2 \, \epsilon \, 2^{(3/2-n/(n-1))k}, \qquad \forall - M < k \leq -9,$$

with $C_2$ being a suitable, arbitrarily large absolute constant. For this approach to work, we need one new ingredient by comparison to the proofs of (3.65) and (3.66). This is the bound

$$\|\Box S_{2^k}\phi\|_{L^2 L^{n-1}} \lesssim C_0^3 \, 2^{(3/2-n/(n-1))k} \, c_k, \qquad \forall\, k \in \mathbb{Z}. \qquad (3.72)$$

Using (3.33) and (3.42), we infer

$$\|\nabla_x \nabla S_{2^k}\phi\|_{L^2 L^{n-1}} \lesssim C_0 \, 2^{(3/2-n/(n-1))k} \, c_k, \qquad \forall\, k \in \mathbb{Z}.$$

Hence, the challenge in estimating $\Box S_{2^k}\phi$ is how to control $\partial_{tt} S_{2^k}\phi$. The most natural way to do this is through the main equation (3.28).

Based on a scaling argument, we argue first that it is enough to prove (3.72) only for the case when $k = 0$, i.e.,

$$\|\Box\psi\|_{L^2 L^{n-1}} \lesssim C_0^3 \, c_0. \qquad (3.73)$$

Relying on (3.28) and paradifferential calculus, we deduce

$$\Box\psi = -S_{2^0}(\langle \partial^\mu\phi, \partial_\mu\phi\rangle\,\phi)$$

$$= -S_{2^0}\left(\sum_{(k_1,k_2)\in\mathbb{Z}^2} \langle \partial^\mu S_{2^{k_1}}\phi, \partial_\mu S_{2^{k_2}}\phi\rangle\,\phi\right)$$

$$= I + II + III + IV,$$

with

$$
I = -S_{2^0}\left(\sum_{\substack{k_1>5\\k_2>10}} \langle \partial^\mu S_{2^{k_1}}\phi, \partial_\mu S_{2^{k_2}}\phi\rangle\,\phi\right),
$$

$$
II = -S_{2^0}\left(\sum_{k_2>10} \langle \partial^\mu S_{\le 2^5}\phi, \partial_\mu S_{2^{k_2}}\phi\rangle\,\phi\right),
$$

$$
III = -S_{2^0}\left(\sum_{k_1>5} \langle \partial^\mu S_{2^{k_1}}\phi, \partial_\mu S_{\le 2^{10}}\phi\rangle\,\phi\right),
$$

$$
IV = -S_{2^0}\left(\langle \partial^\mu S_{\le 2^5}\phi, \partial_\mu S_{\le 2^{10}}\phi\rangle\,\phi\right).
$$

In estimating these terms, as in the analysis for the **third step**, we use (3.33), (3.35), Bernstein and Cauchy-Schwarz inequalities, $\|\phi\|_{L^\infty_{t,x}} = 1$, $\sigma < 1/2$, and (3.42).

For $I$, we derive

$$
\begin{aligned}
\|I\|_{L^2 L^{n-1}} &\lesssim \|I\|_{L^2_{t,x}}\\
&\lesssim \sum_{\substack{k_1>5\\k_2>10}} \|\nabla S_{2^{k_1}}\phi\|_{L^4_{t,x}}\|\nabla S_{2^{k_2}}\phi\|_{L^4_{t,x}}\|\phi\|_{L^\infty_{t,x}}\\
&\lesssim \left(\sum_{k>5}\|\nabla S_{2^k}\phi\|_{L^4_{t,x}}\right)^2\\
&\lesssim C_0^2\left(\sum_{k>5} 2^{(3-n)k/4}c_k\right)^2\\
&\lesssim C_0^2 c_0^2\left(\sum_{k>5} 2^{((3-n)/4+\sigma)k}\right)^2\\
&\lesssim C_0^2 c_0^2,
\end{aligned}
\tag{3.74}
$$

as $(3-n)/4 + \sigma < 0$. For $II$, a frequency analysis shows that

$$
II = -S_{2^0}\left(\sum_{\substack{k_2>10\\k_3>5}} \langle \partial^\mu S_{\le 2^5}\phi, \partial_\mu S_{2^{k_2}}\phi\rangle\,S_{2^{k_3}}\phi\right).
$$

Hence, we obtain

$$
\begin{aligned}
\|II\|_{L^2 L^{n-1}} &\lesssim \|II\|_{L^2_{t,x}} \\
&\lesssim \sum_{\substack{k_2>10 \\ k_3>5}} \|\nabla S_{\leq 2^5}\phi\|_{L^\infty_{t,x}} \|\nabla S_{2^{k_2}}\phi\|_{L^4_{t,x}} \|S_{2^{k_3}}\phi\|_{L^4_{t,x}} \\
&\lesssim \left(\sum_{k>5} \|\nabla S_{2^k}\phi\|_{L^4_{t,x}}\right)\left(\sum_{k>5} \|S_{2^k}\phi\|_{L^4_{t,x}}\right) \\
&\lesssim C_0^2 \left(\sum_{k>5} 2^{(3-n)k/4} c_k\right)\left(\sum_{k>5} 2^{-(n+1)k/4} c_k\right) \\
&\lesssim C_0^2 c_0^2 \left(\sum_{k>5} 2^{((3-n)/4+\sigma)k}\right)^2 \\
&\lesssim C_0^2 c_0^2.
\end{aligned}
\tag{3.75}
$$

For $III$, we argue that its profile is identical with the one of $II$ and, as a consequence, we can skip its discussion. Finally, in what concerns $IV$, we conduct a frequency analysis to infer

$$
\begin{aligned}
IV = -S_{2^0}\Big( &\sum_{-5 \leq k_3 \leq 15} \langle \partial^\mu S_{\leq 2^5}\phi, \partial_\mu S_{\leq 2^{10}}\phi\rangle\, S_{2^{k_3}}\phi \\
&+ \sum_{-5 \leq k_1 \leq 5} \langle \partial^\mu S_{2^{k_1}}\phi, \partial_\mu S_{\leq 2^{10}}\phi\rangle\, S_{<2^{-5}}\phi \\
&+ \sum_{-5 \leq k_2 \leq 10} \langle \partial^\mu S_{<2^5}\phi, \partial_\mu S_{2^{k_2}}\phi\rangle\, S_{<2^{-5}}\phi\Big),
\end{aligned}
$$

and we estimate each of these terms individually:

$$
\begin{aligned}
\left\| S_{2^0}\left(\sum_{-5 \leq k_3 \leq 15} \langle \partial^\mu S_{\leq 2^5}\phi, \partial_\mu S_{\leq 2^{10}}\phi\rangle\, S_{2^{k_3}}\phi\right)\right\|_{L^2 L^{n-1}} & \\
\lesssim \sum_{-5 \leq k_3 \leq 15} \|\nabla S_{\leq 2^5}\phi\|_{L^\infty_{t,x}} \|\nabla S_{\leq 2^{10}}\phi\|_{L^\infty_{t,x}} \|S_{2^{k_3}}\phi\|_{L^2 L^{n-1}} & \\
\lesssim C_0^3 c_0^2 \sum_{-5 \leq k_3 \leq 15} 2^{-(1/2+n/(n-1))k_3} c_{k_3} & \\
\lesssim C_0^3 c_0^3 \sum_{-5 \leq k \leq 15} 2^{\sigma|k|-(1/2+n/(n-1))k} & \\
\lesssim C_0^3 c_0^3, &
\end{aligned}
$$

$$\left\| S_{2^0} \left( \sum_{-5 \leq k_1 \leq 5} \langle \partial^\mu S_{2^{k_1}} \phi, \partial_\mu S_{\leq 2^{10}} \phi \rangle \, S_{<2^{-5}} \phi \right) \right\|_{L^2 L^{n-1}}$$

$$\lesssim \sum_{-5 \leq k_1 \leq 5} \| \nabla S_{2^{k_1}} \phi \|_{L^2 L^{n-1}} \| \nabla S_{\leq 2^{10}} \phi \|_{L^\infty_{t,x}} \| S_{<2^{-5}} \phi \|_{L^\infty_{t,x}}$$

$$\lesssim C_0^2 \, c_0 \sum_{-5 \leq k_1 \leq 5} 2^{(1/2-n/(n-1))k_1} \, c_{k_1}$$

$$\lesssim C_0^2 \, c_0^2 \sum_{-5 \leq k \leq 5} 2^{\sigma|k|+(1/2-n/(n-1))k}$$

$$\lesssim C_0^2 \, c_0^2,$$

$$\left\| S_{2^0} \left( \sum_{-5 \leq k_2 \leq 10} \langle \partial^\mu S_{<2^5} \phi, \partial_\mu S_{2^{k_2}} \phi \rangle \, S_{<2^{-5}} \phi \right) \right\|_{L^2 L^{n-1}}$$

$$\lesssim \sum_{-5 \leq k_2 \leq 10} \| \nabla S_{\leq 2^5} \phi \|_{L^\infty_{t,x}} \| \nabla S_{2^{k_2}} \phi \|_{L^2 L^{n-1}} \| S_{<2^{-5}} \phi \|_{L^\infty_{t,x}}$$

$$\lesssim C_0^2 \, c_0 \sum_{-5 \leq k_2 \leq 10} 2^{(1/2-n/(n-1))k_2} \, c_{k_2}$$

$$\lesssim C_0^2 \, c_0^2 \sum_{-5 \leq k \leq 10} 2^{\sigma|k|+(1/2-n/(n-1))k}$$

$$\lesssim C_0^2 \, c_0^2.$$

Using these three estimates, we deduce

$$\| IV \|_{L^2 L^{n-1}} \lesssim C_0^2 \, c_0^2, \tag{3.76}$$

based on $C_0 \, c_0 \lesssim C_0 \, \epsilon \ll 1$. Together with (3.74), (3.75), and the considerations on $III$, this bound implies (3.73), thus finishing the proof of this proposition. □

At this point, let us remember that what is left to prove in order to finish the whole argument is Proposition 3.3. More precisely, we need to show that (3.43) is true for $k = 0$, i.e.,

$$\| \psi \|_{X_0} \leq C_0 \, c_0. \tag{3.77}$$

### 3.4.3.8  *Final step*

Tao's final step consists in using the previously constructed matrix $U$ and Proposition 3.4 to achieve two goals. First, he motivates the transformation of (3.51) into (3.54). Secondly, relying on (3.54), he obtains estimates for

$w$ which are enough to claim (3.77). Throughout this step, Tao adjusts one more time the values of $C_0$ and $\epsilon$ to satisfy $C_0^2 \epsilon \ll 1$.

For the former objective, we know that the substitution $\psi = Uw$ transforms (3.51) into (3.55). Therefore, we need to argue that the right-hand side of (3.55) is of Error type. Based on (3.33), (3.42), (3.58), (3.59), (3.60), and (3.61), we derive

$$\|U^{-1}(\partial_\mu U + A_\mu U)\,U^{-1}(\partial^\mu \psi - \partial^\mu U\,U^{-1}\psi)\|_{L^1 L^2}$$
$$\lesssim \|U^{-1}\|_{L^\infty_{t,x}}^2 \|\partial_\mu U + A_\mu U)\|_{L^1 L^\infty} \big(\|\nabla \psi\|_{L^\infty L^2}$$
$$+ \|\partial_\mu U\|_{L^\infty_{t,x}} \|U^{-1}\|_{L^\infty_{t,x}} \|\psi\|_{L^\infty L^2}\big)$$
$$\lesssim C_0^2 \epsilon (C_0 c_0 + C_0^3 \epsilon c_0)$$
$$\lesssim C_0^3 \epsilon c_0,$$

$$\|U^{-1}(2\,A_\mu\,\partial^\mu U + \Box U)\,U^{-1}\psi\|_{L^1 L^2}$$
$$\lesssim \|U^{-1}\|_{L^\infty_{t,x}}^2 \big(\|A_\mu\|_{L^2 L^\infty} \|\nabla_\mu U\|_{L^2 L^\infty} \|\psi\|_{L^\infty L^2}$$
$$+ \|\Box U\|_{L^2 L^{n-1}} \|\psi\|_{L^2 L^{2(n-1)/(n-3)}}\big)$$
$$\lesssim C_0^4 \epsilon c_0^2 + C_0^3 \epsilon c_0$$
$$\lesssim C_0^3 \epsilon c_0,$$

and

$$\|U^{-1} Error\|_{L^1 L^2} \lesssim \|U^{-1}\|_{L^\infty_{t,x}} \|Error\|_{L^1 L^2} \lesssim C_0^3 \epsilon c_0,$$

which prove (3.54).

In what concerns the second goal, we notice that

$$\|\psi\|_{X_0} \lesssim \|w\|_{X_0},$$

which can be easily deduced by using (3.33), (3.58), and (3.59). Thus, in order to prove (3.77), it is enough to show

$$\|w\|_{X_0} \ll C_0 c_0. \tag{3.78}$$

For this, we prove first that we have a favorable upper bound for

$$\|w(0)\|_{L^2} + \|\partial_t w(0)\|_{L^2}.$$

Direct computations yield

$$w(0) = U^{-1}(0)\,\psi(0), \quad \partial_t w(0) = U^{-1}(0)\,\partial_t \psi(0) - U^{-1}(0)\,\partial_t U(0)\,U^{-1}(0)\,\psi(0).$$

Hence, we can infer with the help of (3.58) and (3.59) that

$$\|w(0)\|_{L^2} + \|\partial_t w(0)\|_{L^2} \lesssim \|\psi(0)\|_{L^2} + \|\partial_t \psi(0)\|_{L^2}$$
$$\lesssim \|S_{2^0}\phi(0)\|_{\dot{H}^{n/2}} + \|S_{2^0}\partial_t \phi(0)\|_{\dot{H}^{n/2-1}}$$
$$\lesssim c_0.$$

Together with (3.34), this bound will allow us to derive

$$\left\| \sum_{-10<k<10} S_{2^k} w \right\|_{X_0} \ll C_0 \, c_0. \tag{3.79}$$

Indeed, applying (3.34) and Bernstein's inequality, we obtain

$$\| S_{2^k} w \|_{X_k}$$
$$\lesssim \| S_{2^k} w(0) \|_{\dot{H}^{n/2}} + \| S_{2^k} \partial_t w(0) \|_{\dot{H}^{n/2-1}} + 2^{(n/2-1)k} \| \Box S_{2^k} w \|_{L^1 L^2}$$
$$\lesssim 2^{nk/2} \| w(0) \|_{L^2} + 2^{(n/2-1)k} \| \partial_t w(0) \|_{L^2} + 2^{(n/2-1)k} \| \Box w \|_{L^1 L^2},$$

which, coupled with the trivial bound

$$\| G \|_{X_0} \leq 2^{10} \| G \|_{X_k}, \qquad \forall \, -10 < k < 10,$$

leads to

$$\left\| \sum_{-10<k<10} S_{2^k} w \right\|_{X_0} \lesssim \sum_{-10<k<10} \| S_{2^k} w \|_{X_k}$$
$$\lesssim \sum_{-10<k<10} \left( 2^{nk/2} + 2^{(n/2-1)k} \right) c_0 + 2^{(n/2-1)k} C_0^3 \, \epsilon \, c_0$$
$$\lesssim c_0 + C_0^3 \, \epsilon \, c_0.$$

This estimate surely implies (3.79) on the account of $C_0$ being sufficiently large, depending on $n$ and $m$, and $C_0^2 \epsilon \ll 1$. As a result, all we are left to prove in order to claim (3.78) and finish the argument is

$$\left\| \left( \mathrm{Id} - \sum_{-10<k<10} S_{2^k} \right) w \right\|_{X_0} \ll C_0 \, c_0, \tag{3.80}$$

where Id is the identity operator. According to Remark 3.11, the spatial Fourier transform of $U$ is supported at frequency $|\xi| \leq 2^{-7}$. As the spatial Fourier transform of $\psi$ is supported in the region $\{1/2 \leq |\xi| \leq 2\}$, it follows that

$$\left( \mathrm{Id} - \sum_{-10<k<10} S_{2^k} \right) U^t \psi = 0.$$

Due to $w = U\psi$, we have

$$w = U^t \psi - (U^t U - I) \, w.$$

Therefore, we deduce

$$\left( \mathrm{Id} - \sum_{-10<k<10} S_{2^k} \right) w = - \left( \mathrm{Id} - \sum_{-10<k<10} S_{2^k} \right) (U^t U - I) \, w.$$

Using one more time Bernstein's inequality and (3.59), we derive

$$\left\| \left( \mathrm{Id} - \sum_{-10 < k < 10} S_{2^k} \right) w \right\|_{X_0} \lesssim \left\| \left( U^t U - I \right) w \right\|_{X_0}$$

$$\lesssim \left( \| U^t U - I \|_{L^\infty_{t,x}} + \| \partial_t (U^t U - I) \|_{L^\infty_{t,x}} \right) \| w \|_{X_0}$$

$$\lesssim C_0^2 \, \epsilon \| w \|_{X_0},$$

which implies

$$\left\| \left( \mathrm{Id} - \sum_{-10 < k < 10} S_{2^k} \right) w \right\|_{X_0} \lesssim C_0^2 \, \epsilon \left\| \sum_{-10 < k < 10} S_{2^k} w \right\|_{X_0}.$$

Based on (3.79), it follows that (3.80) holds and Tao's proof is complete.

### 3.4.4   Shatah-Struwe's result

In [146], Shatah and Struwe developed an alternative approach to Tao's for obtaining small data global regularity for wave maps in $n \geq 4$ dimensions. There, they tackle the associated Hodge system (1.83) instead of the wave maps equation. The Hodge system has built-in gauge invariance, which one can exploit (by choosing the most convenient gauge fixing) in order to prove existence results. In this section, we would like to present Shatah-Struwe's argument for the $4 + 1$-dimensional case. As it turns out, this is the hardest case in which one has to employ endpoint Strichartz estimates as well as real interpolation techniques and Lorentz spaces. Given these basic ingredients, however, the proof is surprisingly simple, which makes the approach of Shatah-Struwe particularly appealing.

The work under discussion establishes global existence assuming that the critical norm of the initial data is small, for any space dimension $n \geq 4$ and for quite general manifolds. The only assumption required is that the manifold is parallelizable and its curvature is bounded. For pedagogical reasons, we will present the argument in the simpler setting, i.e., assuming that the space dimension is $n = 4$ and the target manifold is three dimensional with curvature $K = \pm 1$. Considering such a simple case, we can pinpoint the essential ingredients in the proof and avoid distracting generalities. One can see easily how the method generalizes (e.g., for higher dimensions, more general manifolds, extra regularity, etc.) and we leave this for the interested reader.

### 3.4.4.1  *Preliminary observations*

The subject of investigation is the wave maps equation with the base manifold being the $4+1$-dimensional Minkowski spacetime and the target being either the three dimensional unit sphere $\mathbb{S}^3$ or the three dimensional "unit" hyperboloid $\mathbb{H}^3$. This means that we study maps

$$\phi^a\left(x^\mu\right) : \mathbb{R}^{4+1} \mapsto \mathbb{S}^3 \quad \text{or} \quad \mathbb{H}^3, \qquad a = 1, 2, 3,$$

which satisfy the equation

$$\Box \phi^a + \Gamma^a_{bc}(\phi)\langle \nabla\phi^b, \nabla\phi^c \rangle = 0\,.$$

The idea however is to study the associated Hodge system for the unknown quantities $\left(\psi^a_\mu, A^a_\mu\right)$ ($A$ being the gauge field) where $\mu = 0, 1, 2, 3, 4$ and $a = 1, 2, 3$. We will use $A$-covariant derivatives defined via

$$\nabla^A_\mu \psi^a_\nu \stackrel{\text{def}}{=} \nabla_\mu \psi^a_\mu + \left(A_\mu \wedge \psi_\nu\right)^a,$$

where the wedge product is given by $(A_\mu \wedge \psi_\nu)^a := \epsilon^a_{\ bc} A^b_\mu \psi^c_\nu$, with $\epsilon_{abc}$ being the Levi-Civita (or totally antisymmetric) tensor. The Hodge system is the first order system:

$$\nabla^A_\mu \psi^{a,\mu} = 0, \qquad \nabla^A_{[\mu} \psi^a_{\nu]} = 0, \qquad F^a_{\mu\nu} = K G^a_{\mu\nu}\,, \qquad (3.81)$$

where $\psi^a_\mu$ is a vector field, $K$ is the curvature (we assume for simplicity $K = \pm 1$), and

$$F^a_{\mu\nu} \stackrel{\text{def}}{=} \nabla_\mu A^a_\nu - \nabla_\nu A^a_\mu + \left(A_\mu \wedge A_\nu\right)^a,$$

$$G^a_{\mu\nu} \stackrel{\text{def}}{=} \left(\psi_\nu \wedge \psi_\mu\right)^a.$$

Notice that $A$-covariant derivatives do not commute, but rather satisfy

$$\left[\nabla^A_\mu, \nabla^A_\nu\right]^a = F^a_{\mu\nu} \wedge\ .$$

The observation is that the critical (scale-invariant) norm requires $\phi \in H^2$ and, since $\psi$ are derivatives of $\phi$, this translates to $\psi \in H^1$. Thus, the critical norm requirement asks for existence of solutions for the above Hodge system (3.81) with data $\psi(0) \in \dot{H}^1$ and $\partial_t \psi(0) \in L^2$.

As we have seen earlier in the introduction, we can readily derive from (3.81) the system

$$\Box^A \psi^a_\nu + \left(F_{\mu\nu} \wedge \psi^\mu\right)^a = 0, \qquad F^a_{\mu\nu} = K G^a_{\mu\nu}, \qquad K = \pm 1, \quad (3.82)$$

where $\Box^A := -(\nabla_\mu + A_\mu \wedge)(\nabla^\mu + A^\mu \wedge)$ and the gauge field $A_\mu$ is computed from the equation $F_{\mu\nu} = K G_{\mu\nu}$ modulo a gauge fixing of our choice. Now, the idea is to study the previous system instead of the original Hodge system, the reason being that for the equations in (3.82) we can apply endpoint Strichartz estimates. There is however a question of whether the system (3.82) is equivalent to the Hodge system (3.81) and, for this purpose, we give a brief explanation which we will need later in the construction of solutions. The situation is analogous to that of Maxwell's equations, where the electric and magnetic fields satisfy the wave equation, but we need to impose compatibility conditions which must be preserved by the flow. The computations are in the same spirit as the ones presented earlier in the introduction[5].

We form the quantities

$$\Lambda^a_{\mu\nu\rho} \overset{\text{def}}{=} \epsilon_{\mu\nu\rho}{}^{\alpha\beta}\nabla^A_\alpha\psi^a_\beta, \qquad \tau^a \overset{\text{def}}{=} \nabla^A_\mu\psi^{a,\mu}. \tag{3.83}$$

For the Hodge system to hold, these must be identically zero, so we would like to see if, assuming that they are initially zero, the evolution (3.82) guarantees that they remain zero at later times. In what follows, we will write $\Lambda_{\mu\nu\rho}$ and $\tau$ for simplicity (omitting the dependence on $a = 1, 2, 3$). In addition, we construct the following 3-tensors,

$$E^a_{\mu\nu\rho} \overset{\text{def}}{=} \frac{1}{2}\epsilon_{\mu\nu\rho}{}^{\alpha\beta}F^a_{\alpha\beta}, \qquad H^a_{\mu\nu\rho} \overset{\text{def}}{=} \frac{1}{2}\epsilon_{\mu\nu\rho}{}^{\alpha\beta}G^a_{\alpha\beta},$$

for which we know, of course, that $E = KH$. We would like to take the derivative of $\Lambda_{\mu\nu\rho}$, namely the 1-form

$$\left(d^A \wedge \Lambda\right)_\mu \overset{\text{def}}{=} \epsilon_{\mu\alpha}{}^{\beta\gamma\rho}\nabla^{A,\alpha}\Lambda_{\beta\gamma\rho},$$

which turns out to satisfy the identity

$$\left(d^A \wedge \Lambda\right)_\mu = \Box^A\psi_\mu + \left(F_{\alpha\mu} \wedge \psi^\alpha\right) + \nabla^A_\mu\tau.$$

Thus, if (3.82) is satisfied, we conclude that

$$\left(d^A \wedge \Lambda\right)_\mu = \nabla^A_\mu\tau. \tag{3.84}$$

Let us first notice that $\Lambda$ satisfies

$$\nabla^A_\mu\Lambda^\mu{}_{\nu\rho} = \frac{K}{2}\epsilon_{\mu\nu\rho}{}^{\alpha\beta}\left(\psi^\mu \wedge \psi_\alpha\right) \wedge \psi_\beta = 0.$$

Taking derivatives of the identity (3.84), we discover the pair of equations,

$$\Box^A\tau = \langle F, \Lambda \rangle, \tag{3.85}$$

$$\Box^A\Lambda_{\mu\nu\rho} + 3!F^\alpha_{[\mu} \wedge \Lambda_{|\alpha|\nu\rho]} = 2E \wedge \tau. \tag{3.86}$$

---

[5]The computations are carried out in 4+1 dimensions, but they can easily be generalized to the case of $n + 1$ dimensions.

For the former one we use the short-hand notation

$$\langle F, \Lambda \rangle \overset{\text{def}}{=} \epsilon^{\mu\nu\alpha\beta\gamma} F_{\mu\nu} \wedge \Lambda_{\alpha\beta\gamma},$$

while for second term on the left-hand side of the latter equation we anti-symmetrize the $(\mu, \nu, \rho)$ indices.

We would like to derive energy-type estimates for this system of equations, our ultimate goal being to employ Grönwall's inequality. For this purpose, we take (3.85) and form its energy-momentum tensor

$$T_{\mu\nu} \overset{\text{def}}{=} \langle \nabla^A_\mu \tau, \nabla^A_\nu \tau \rangle - \frac{1}{2} g_{\mu\nu} \langle \nabla^A \tau, \nabla^A \tau \rangle,$$

which satisfies the identity

$$\nabla_\mu T^\mu_{\ \nu} = \langle \nabla^A_\nu \tau, F^\mu_{\ \nu} \wedge \tau \rangle - \left\langle \langle F, \Lambda \rangle, \nabla^A_\nu \tau \right\rangle.$$

The construction is similar for (3.86), where we treat each component of $\Lambda$ individually (since $\mathbb{R}^{4+1}$ is flat). With the notation convention $(t, \mathbf{x}) \in \mathbb{R}^{4+1}$ for the time and space coordinates, the overall energy is given by

$$\mathcal{E}_t(\tau, \Lambda) \overset{\text{def}}{=} \frac{1}{2} \int_{\{t\} \times \mathbb{R}^4} \left\{ \left| \nabla^A_t \tau \right|^2 + \left| \nabla^A_\mathbf{x} \tau \right|^2 + \sum_{\lambda, \mu, \rho} \left| \nabla^A_t \Lambda_{\lambda\mu\rho} \right|^2 \right.$$
$$\left. + \left| \nabla^A_\mathbf{x} \Lambda_{\lambda\mu\rho} \right|^2 \right\} d\mathbf{x}$$

and standard energy estimates lead to the following differential inequality,

$$\frac{d}{dt} \mathcal{E}_t(\tau, \Lambda) \leq \|F\|_{L^4_\mathbf{x}} \left( \|(\tau, \Lambda)\|_{L^4_\mathbf{x}} \right) \sqrt{\mathcal{E}_t(\tau, \Lambda)}. \tag{3.87}$$

We can relate $A$-derivatives to ordinary derivatives if we observe, for example,

$$|\nabla^A \tau|^2 = |\nabla \tau|^2 + |A \wedge \tau|^2 + 2\langle A \wedge \tau, \nabla \tau \rangle \geq \frac{1}{2} |\nabla \tau|^2 - |A \wedge \tau|^2,$$

which implies

$$\int |\nabla \tau|^2 d\mathbf{x} \leq 2 \int |\nabla^A \tau|^2 d\mathbf{x} + 2 \int |A \wedge \tau|^2 d\mathbf{x}.$$

As Sobolev inequality yields

$$\|A \wedge \tau\|_{L^2_\mathbf{x}} \leq \|A\|_{L^4_\mathbf{x}} \|\tau\|_{L^4_\mathbf{x}} \leq C \|A\|_{L^4_\mathbf{x}} \|\nabla \tau\|_{L^2_\mathbf{x}},$$

we conclude that there exists some $\epsilon > 0$ such that, if $\|A\|_{L^4_\mathbf{x}} < \epsilon$, then

$$\|\nabla \tau\|_{L^2_\mathbf{x}} \leq C \|\nabla^A \tau\|_{L^2_\mathbf{x}}.$$

Similar considerations for $\Lambda$, in which one relies again on Sobolev inequality, show that

$$\|(\tau, \Lambda)\|_{L_x^4} \le C\sqrt{\mathcal{E}_t}.$$

Hence, our differential inequality (3.87) becomes

$$\frac{d}{dt}\mathcal{E}_t(\tau, \Lambda) \le C\|F\|_{L_x^4}\mathcal{E}_t(\tau, \Lambda) . \tag{3.88}$$

Finally, we are in a position to state a lemma.

**Lemma 3.2.** *For some $\epsilon$ sufficiently small, assume that[6] $\|A\|_{L_t^\infty L_x^4} \le \epsilon$. Moreover, for a time interval $[0, T]$, assume that*

$$\|F\|_{L_t^1 L_x^4([0,T] \times \mathbb{R}^4)} < \infty .$$

*Then we have the inequality*

$$\mathcal{E}_t(\tau, \Lambda) \le \exp\left(C\|F\|_{L_t^1 L_x^4}\right)\mathcal{E}_0(\tau, \Lambda), \qquad \forall t \in [0, T].$$

*Thus, if $\mathcal{E}_0 = 0$, then $\mathcal{E}_t = 0$ and, consequently, $\tau = \Lambda = 0$ in $[0, T] \times \mathbb{R}^4$.*

**Proof.** The proof is straightforward. After integrating (3.88), we conclude that $\mathcal{E}_t = 0$ if $\mathcal{E}_0 = 0$. The smallness of $\|A\|_{L_t^\infty L_x^4}$ guarantees that $\mathcal{E}_t$ controls the $\dot{H}^1$ norm of $(\tau, \Lambda)$, hence they are constant equal to zero. The lemma demonstrates that solutions of (3.82) satisfy the Hodge system (3.81) provided that $\|F\|_{L_t^1 L_x^4}$ is finite. Since $F = KG$ and $G \sim \psi^2$, the previous norm for $F$ is finite if $\|\psi\|_{L_t^2 L_x^8} < \infty$. This, in turn, follows by applying endpoint Strichartz estimates.  □

### 3.4.4.2   *Statement of the global existence result*

After these preliminary observations, we can now state a small data global existence theorem, whose proof will be discussed later.

**Theorem 3.8.** *Consider solutions of the Hodge system*

$$\nabla_\mu^A \psi^{a,\mu} = 0, \qquad \nabla_{[\mu}^A \psi_{\nu]}^a = 0, \qquad F_{\mu\nu}^a = KG_{\mu\nu}^a , \tag{3.89}$$

*with initial data prescribed at time $t = 0$,*

$$\psi(0) := \psi_0, \qquad \partial_t \psi(0) := \psi_1 . \tag{3.90}$$

*Moreover, the gauge $A$ is constructed from $F_{\mu\nu} = KG_{\mu\nu}$ and the Coulomb gauge fixing $\nabla_j A^j = 0$. We assume that $\psi_0$ and $\psi_1$ satisfy the compatibility conditions $\tau(0) = \Lambda(0) = 0$ (see (3.83)).*

---

[6]The smallness assumption on $\|A\|_{L_t^\infty L_x^4}$ will be a consequence of the construction of solutions in Theorem 3.8.

*Under these assumptions, there exists a sufficiently small $\epsilon > 0$ such that, if*

$$\|\psi_0\|_{\dot{H}^1} + \|\psi_1\|_{L^2} \leq \epsilon, \tag{3.91}$$

*then the system (3.89) admits a unique global solution satisfying*

$$\|\psi\|_{L_t^2 L_x^{(8,2)}} + \|\psi\|_{L_t^\infty \dot{H}_x^1} + \|\partial_t \psi\|_{L_t^\infty L_x^2} \leq C\epsilon. \tag{3.92}$$

Let us make some comments regarding this theorem. We write $\psi$ to denote collectively $\psi_\nu$, where $\nu = 0, 1, 2, 3, 4$. For convenience, we separate the time and space components and write $\psi = (\psi_0, \psi_j)$, using $j$ to indicate spatial indices. The initial data are prescribed at $t = 0$, but the compatibility conditions imply that they are not in fact independent. First, recall the structure of the antisymmetric tensor $G_{\mu\nu} = \psi_\nu \wedge \psi_\mu$. We will see shortly that the gauge field is constructed from the following set of elliptic equations:

$$\Delta A_k + 2\nabla^j \left( A_j \wedge A_k \right) = K\nabla_j G^j{}_k,$$
$$\Delta A_0 + 2\nabla^j \left( A_j \wedge A_0 \right) = K\nabla_j G^j{}_0.$$

Thus, in prescribing $\partial_t \psi_\nu(0)$ for $\nu = 0, 1, 2, 3, 4$ compatible with (3.89), we actually have at $t = 0$ that

$$\partial_t \psi_0 = A_0 \wedge \psi_0 + \nabla_j^A \psi^j, \qquad \partial_t \psi_j = -A_0 \wedge \psi_j + \nabla_j^A \psi^j.$$

Hence, in view of these observations, it is enough to prescribe $\psi_\nu(0)$, solve for $A$ at $t = 0$ (as we will also do in Lemma 3.3), and then compute $\partial_t \psi_\nu$ from the equations above.

The $L^{(8,2)}$ norm appearing in the last estimate of the above theorem is the $L^{(p,q)}$ Lorentz space norm. These function spaces are refinements of the standard $L^p$ spaces and their norms are defined as follows:

$$\|f\|_{L^{(p,q)}} \stackrel{\text{def}}{=} \left( p \int_0^\infty \left| \{ |f| > \lambda \} \right|^{q/p} \lambda^{q-1} d\lambda \right)^{1/q}. \tag{3.93}$$

We will use Lorentz norms and elliptic estimates in order to control $\|A\|_{L_t^1 L_x^\infty}$, which will be crucial in the construction of solutions for (3.89). For more information on Lorentz spaces, we refer the reader to the works of O'Neil [125] and Hunt [75].

The construction of solutions proceeds in the following manner. We take the $\nabla^{A,\mu}$ derivative on the equation $\nabla_\mu^A \psi_\nu = \nabla_\nu^A \psi_\mu$, commute the derivatives, and use the fact $\nabla^{A,\mu} \psi_\mu = 0$ to discover the associated system,

$$\Box^A \psi_\nu + F_{\mu\nu} \wedge \psi^\mu = 0, \qquad F_{\mu\nu} = KG_{\mu\nu}, \tag{3.94}$$

where

$$F_{\mu\nu} = \nabla_\mu^A A_\nu - \nabla_\nu^A A_\mu, \qquad G_{\mu\nu} = \psi_\nu \wedge \psi_\mu. \qquad (3.95)$$

The first equation can be written in the form $\Box \psi_\nu = f_\nu$, where $f_\nu = f_\nu(A, \psi)$ is a nonlinearity (to be given explicitly later) containing $\psi$ as well as the gauge field $A$. For this equation we would like to apply Strichartz-type estimates with the hope that we can balance our inequalities. A quick calculation of $f_\nu$ reveals that it contains terms of the form $A(\nabla\psi)$, $(\nabla A)\psi$, etc. Thus, we need to obtain estimates for the gauge field $A$, which will be our first task.

### 3.4.4.3 *The Coulomb gauge*

We reserve the indices $j$ and $k$ to label spatial variables and use the index 0 for the time coordinate. Hence, we use the notation $(t, \mathbf{x}) := (x^0, x^1, x^2, x^3, x^4)$. The first ingredient in the proof is to use the Coulomb gauge, i.e., to fix the gauge field $A$ by imposing the condition $\nabla_j A^j = 0$, where we sum over repeated spatial indices, the metric on $\mathbb{R}^4$ is the Euclidean one, and $j = 1, 2, 3, 4$. Now, the idea is to take advantage of elliptic estimates, the crucial one being an $L^\infty$ estimate stated below.

**Lemma 3.3.** *(Estimates on the Coulomb gauge) Assume that for some $\epsilon$ sufficiently small we have*

$$\|\psi\|_{L_t^\infty \dot{H}_{\mathbf{x}}^1} \leq \epsilon, \qquad (3.96)$$

*and moreover, the gauge field $A := (A_0, A_j)$ with $j = 1, 2, 3, 4$ satisfies the Coulomb gauge condition, namely, $\nabla_j A^j = 0$. Then we have the following a priori estimates:*

$$\|\nabla_j A\|_{L_{\mathbf{x}}^2} \leq C\|\psi\|_{L_{\mathbf{x}}^4}^2 \leq C\|\psi\|_{\dot{H}_{\mathbf{x}}^1}^2, \qquad (3.97)$$

$$\|\nabla_0 A\|_{L_{\mathbf{x}}^2} \leq C\|\nabla\psi\|_{L_{\mathbf{x}}^2}\|\psi\|_{L_{\mathbf{x}}^4} \leq C\|\psi\|_{\dot{H}_{\mathbf{x}}^1}^2, \qquad (3.98)$$

$$\|A\|_{L_{\mathbf{x}}^\infty} \leq C\|\psi\|_{L_{\mathbf{x}}^{(8,2)}}^2, \qquad (3.99)$$

$$\|A\|_{L_{\mathbf{x}}^8} \leq \|\nabla A\|_{L_{\mathbf{x}}^{8/3}} \leq C\|\nabla\psi\|_{L_{\mathbf{x}}^2}\|\psi\|_{L_{\mathbf{x}}^8}. \qquad (3.100)$$

**Remark 3.12.** The estimates in the above lemma are fixed-time estimates. Integrating over time in an appropriate manner, we deduce, for example,

$$\|A\|_{L_t^1 L_{\mathbf{x}}^\infty} \leq C\|\psi\|_{L_t^2 L_{\mathbf{x}}^{(8,2)}}^2, \qquad (3.101)$$

$$\|\nabla A\|_{L_t^\infty L_{\mathbf{x}}^2} \leq C\|\psi\|_{L_t^\infty \dot{H}_{\mathbf{x}}^1}^2, \qquad (3.102)$$

and similarly other space-time estimates, which will be later used in the argument.

**Proof.** We will use $C$, $C_1$, $C_2$, etc., to denote various constants that may change from one line to the next one, but which are independent of the functions involved. The equation $F = KG$ in (3.94) can be written in components,

$$\nabla_j A_k - \nabla_k A_j + 2A_j \wedge A_k = KG_{jk},$$
$$\nabla_j A_0 - \nabla_0 A_j + 2A_j \wedge A_0 = KG_{j0},$$

and the Coulomb gauge implies that we have to solve the elliptic system

$$\Delta A_k + 2\nabla^j \left( A_j \wedge A_k \right) = K\nabla_j G^j{}_k,$$
$$\Delta A_0 + 2\nabla^j \left( A_j \wedge A_0 \right) = K\nabla_j G^j{}_0.$$

The solution has the form

$$A_k = R^j * \left( 2A_j \wedge A_k - KG_{jk} \right), \tag{3.103}$$
$$A_0 = R^j * \left( 2A_j \wedge A_0 - KG_{j0} \right), \tag{3.104}$$

where $R^j$ is the derivative of the Green's function of the Laplacian,

$$R^j(\mathbf{x}) \overset{\text{def}}{=} c\frac{-x^j}{|\mathbf{x}|^4}. \tag{3.105}$$

Unfortunately, there is a minor complication in the sense that the equations for $A$ are nonlinear. However, we are interested in small solutions, i.e., we may assume that $\psi$ is small in some appropriate norm. We can solve (3.103) and (3.104) by an iteration scheme starting from $A^{(-1)} = 0$ and using the iteration step, written schematically,

$$-\Delta A^{(n+1)} = \nabla \left( 2A^{(n)} \wedge A^{(n)} - KG(\psi) \right).$$

Thus, $A^{(0)}$ satisfies $\Delta A^{(0)} = K\nabla G(\psi)$. Since $G = \psi \wedge \psi \sim \psi^2$, we can use standard elliptic estimates to obtain

$$\|\nabla_j A^{(0)}\|_{L^2_{\mathbf{x}}} \le \|\psi\|^2_{L^4_{\mathbf{x}}} \le C\|\psi\|^2_{\dot{H}^1_{\mathbf{x}}} \le C\epsilon^2,$$

due also to Sobolev inequality and the smallness assumption (3.96).

By similar reasoning, the equation for $A^{(1)}$ implies the elliptic estimate

$$\|\nabla_j A^{(1)}\|_{L^2_{\mathbf{x}}} \le C_1\|A^{(0)}\|^2_{L^4_{\mathbf{x}}} + C_2\|\psi\|^2_{L^2_{\mathbf{x}}} \le C_1\epsilon^4 + C\epsilon^2.$$

Iterating, we realize that we add powers of $\epsilon$ to the elliptic estimate and, for $\epsilon$ sufficiently small, the series converges. It is an easy exercise, which we leave to the interested reader, to show that for $\epsilon$ sufficiently small the sequence $\{A^{(n)}\}$ satisfies

$$\|\nabla_j A^{(n)}\|_{L_x^2} \leq C\epsilon^2,$$

$$\|\nabla_j(A^{(n+1)} - A^{(n)})\|_{L_x^2} \leq \frac{1}{2}\|\nabla_j(A^{(n)} - A^{(n-1)})\|_{L_x^2},$$

for all $n$. We conclude now that $A^{(n)} \to A$ in the $\dot{H}^1$ norm, with $A$ being a solution of (3.103) and (3.104) that satisfies the estimate

$$\|\nabla_j A\|_{L_x^2} \leq C_1\|A\|_{L_x^4}^2 + C_2\|\psi\|_{L_x^4}^2.$$

The Sobolev inequality $\|A\|_{L_x^4} \leq C\|\nabla A\|_{L_x^2}$ and the fact that $\|\nabla_j A\|_{L_x^2} \leq C\epsilon^2$ imply (3.97).

Time derivatives can be treated in the following manner. Observe first that taking time derivative of $A$, we obtain the equation

$$\nabla_0 A = R^j * (2\nabla_0 A_j \wedge A + 2A_j \wedge \nabla_0 A - K(\nabla_0 \psi_j \wedge \psi + \psi_j \wedge \nabla_0 \psi)).$$

We employ the Sobolev inequality

$$\|f\|_{L_x^2} \leq C\|f\|_{\dot{H}_x^{1,4/3}}$$

to obtain from the above equation for $\nabla_0 A$ that

$$\|\nabla_0 A\|_{L_x^2} \leq C_1\|A\nabla_0 A\|_{L_x^{4/3}} + C_2\|\psi \nabla_0 \psi\|_{L_x^{4/3}}$$

$$\leq C_1\|A\|_{L_x^4}\|\nabla_0 A\|_{L_x^2} + C_2\|\psi\|_{L_x^4}\|\nabla_0 \psi\|_{L_x^2}.$$

Since $\|A\|_{L_x^4} \leq C\epsilon^2$, we obtain the estimate (3.98).

Next, notice that from (3.103) and (3.104), we can express the derivative of $A$ in the schematic form

$$\nabla A = R^j * (4A \wedge \nabla A - 2K\psi \wedge \nabla\psi).$$

Employing the Sobolev inequality

$$\|f\|_{L_x^{8/3}} \leq C\|f\|_{\dot{H}_x^{1,8/5}},$$

we obtain the estimates

$$\|\nabla A\|_{L_x^{8/3}} \leq C_1\|A\nabla A\|_{L_x^{8/5}} + C_2\|\psi \nabla\psi\|_{L_x^{8/5}}$$

$$\leq C_1\|A\|_{L_x^4}\|\nabla A\|_{L_x^{8/3}} + C_2\|\psi\|_{L_x^8}\|\nabla\psi\|_{L_x^2}.$$

The fact that $\|A\|_{L_x^4} \leq C\epsilon^2$ implies the bound (3.100).

Finally, we come to the crucial ingredient, which is the $L^\infty$ estimate (3.99). The fundamental observation and the reason one uses Lorentz spaces is the fact that

$$R^j \in L_x^{(4/3,\infty)}, \qquad \left(L_x^{(4/3,\infty)}\right)^* = L_x^{(4,1)}.$$

Hence, from (3.103) and (3.104), we have the Lorentz space estimates

$$\|A\|_{L_x^\infty} \le C_1 \|A \wedge A\|_{L_x^{(4,1)}} + C_2 \|\psi \wedge \psi\|_{L_x^{(4,1)}}$$
$$\le C_1 \|A\|^2_{L_x^{(8,2)}} + C_2 \|\psi\|^2_{L_x^{(8,2)}} . \tag{3.106}$$

In order to handle the term $\|A\|^2_{L_x^{(8,2)}}$, we first define

$$E_\lambda \stackrel{\text{def}}{=} \{|A| > \lambda\}$$

so that the Lorentz norm is, based on (3.106) and writing $\lambda = \lambda^{1/4+3/4}$,

$$\|A\|^2_{L_x^{(8,2)}} = 8 \int_0^{\|A\|_{L_x^\infty}} |E_\lambda|^{1/4} \lambda \, d\lambda$$
$$\le 8 \left( \int_0^{\|A\|_{L_x^\infty}} |E_\lambda| \lambda^3 d\lambda \right)^{1/4} \left( \int_0^{\|A\|_{L_x^\infty}} \lambda^{1/3} d\lambda \right)^{3/4}$$
$$\le 8 \|A\|_{L_x^4} \|A\|_{L_x^\infty}$$
$$\le C \|\psi\|^2_{L_x^4} \|A\|_{L_x^\infty} .$$

Since, by our assumption $\|\psi\|_{L_x^4} \le C\epsilon$, we can absorb this term on the right-hand side of the inequality and therefore prove (3.99). This concludes the proof of the lemma. $\qquad \square$

### 3.4.4.4 *The proof of the global existence result*

The estimates for the gauge field contained in the previous lemma are the last prerequisites needed in order to start proving Theorem 3.8. The main idea in Shatah-Struwe's argument for this result is to derive a wave equation that $\psi_\nu$ satisfies and then apply endpoint Strichartz estimates.

**Proof of Theorem 3.8.** We start the construction of solutions claimed in the theorem by untangling the pure wave equation in (3.82). As the $A$-covariant d'Alembertian is given by $\Box^A = -(\nabla_\mu + A_\mu \wedge)(\nabla^\mu + A^\mu \wedge)$, we deduce

$$\Box \psi_\nu = f_\nu, \tag{3.107}$$

where

$$f_\nu \stackrel{\text{def}}{=} K(\psi_\mu \wedge \psi_\nu) \wedge \psi^\mu + 2A_\mu \wedge (\nabla^\mu \psi_\nu)$$
$$+ (\nabla_\mu A^\mu) \wedge \psi_\nu + A_\mu \wedge (A^\mu \wedge \psi_\nu) . \tag{3.108}$$

Thus, $\psi_\nu$ is a solution of the inhomogeneous wave equation with $f_\nu$ as its right-hand side.

This is the point in the argument where we state the Strichartz estimates used in the analysis. According to Definition 2.2, the triple

$$(p, q, n) = \left(2, \frac{2(n-1)}{n-3}, n\right), \qquad n \geq 4,$$

is wave-admissible. The corresponding Strichartz bounds are also called endpoint Strichartz estimates due to $(1/2, (n-3)/2(n-1))$ being one of the corners of the wave-admissibility region in the $(1/p, 1/q)$ plane, for a fixed $n \geq 4$. These were obtained in the seminal work of Keel-Tao [78]. For $n = 4$, as a consequence of (2.52) and (2.56), one has the Strichartz estimates

$$\|\psi\|_{L_t^2 L_x^6} + \|\psi\|_{L_t^\infty \dot{H}_x^{5/6}} + \|\partial_t \psi\|_{L_t^\infty \dot{H}_x^{-1/6}}$$
$$\leq C \left( \|\psi(0)\|_{\dot{H}_x^{5/6}} + \|\partial_t \psi(0)\|_{\dot{H}_x^{-1/6}} + \|\Box\psi\|_{L^1 \dot{H}_x^{-1/6}} \right) \tag{3.109}$$

and

$$\|\psi\|_{L_t^2 \dot{H}_x^{1,6}} + \|\psi\|_{L_t^\infty \dot{H}_x^{11/6}} + \|\partial_t \psi\|_{L_t^\infty \dot{H}_x^{5/6}}$$
$$\leq C \left( \|\psi(0)\|_{\dot{H}_x^{11/6}} + \|\partial_t \psi(0)\|_{\dot{H}_x^{5/6}} + \|\Box\psi\|_{L^1 \dot{H}_x^{5/6}} \right). \tag{3.110}$$

Next, we use (2.8) to perform real interpolations in the spatial variables and derive

$$\left( L_x^6, \dot{H}_x^{1,6} \right)_{1/6,2} = \dot{B}_{6,2}^{5/6},$$

$$\left( \dot{H}_x^{5/6}, \dot{H}_x^{11/6} \right)_{1/6,2} = \dot{H}_x^1,$$

$$\left( \dot{H}_x^{-1/6}, \dot{H}_x^{5/6} \right)_{1/6,2} = L_x^2.$$

Moreover, if one relies again on interpolation relations, it follows that

$$\dot{B}_{6,2}^{5/6}(\mathbb{R}^4) \subset L^{(8,2)}(\mathbb{R}^4).$$

Hence, combining the previous two facts with (3.109) and (3.110), we obtain

$$\|\psi\|_{L_t^2 L_x^{(8,2)}} + \|\psi\|_{L_t^\infty \dot{H}_x^1} + \|\partial_t \psi\|_{L_t^\infty L_x^2}$$
$$\leq C \left( \|\psi(0)\|_{\dot{H}_x^1} + \|\partial_t \psi(0)\|_{L_x^2} + \|\Box\psi\|_{L_t^1 L_x^2.} \right), \tag{3.111}$$

which is the estimate to be applied to (3.107). The fact that this bound is enough to control the $L_x^\infty$ norm of the gauge field under the Coulomb gauge condition is the crucial observation in Shatah-Struwe's work.

For convenience, we can define a norm,

$$\|\psi\|_N \overset{\text{def}}{=} \|\partial_t \psi\|_{L_t^\infty L_x^2} + \|\psi\|_{L_t^\infty \dot{H}_x^1} + \|\psi\|_{L_t^2 L_x^{(8,2)}}, \tag{3.112}$$

and write (3.111) in the concise form

$$\|\psi\|_N \leq C \left( \|\psi_0\|_{\dot{H}^1} + \|\psi_1\|_{L^2} + \|f\|_{L^1_t L^2_x} \right), \qquad (3.113)$$

where

$$\Box \psi = f, \qquad \psi(0) = \psi_0, \qquad \partial_t \psi(0) = \psi_1.$$

Therefore, in applying the previous inequality for the wave equation $\Box \psi_\nu = f_\nu$, with $f_\nu$ given by (3.108), we would like to estimate all the terms appearing in $f_\nu$ in terms of the $\|\psi\|_N$ norm. For this, we will make use of the $L^\infty$ estimate (3.101) for the gauge field $A$ and the Sobolev inequality

$$\|\psi(t)\|_{L^4_x} \leq C\|\psi(t)\|_{\dot{H}^1_x}. \qquad (3.114)$$

The analysis for the four terms in $f_\nu$ proceeds as follows:

- the first term can be thought of as $\psi^3$, due to

$$(\psi_\mu \wedge \psi_\nu) \wedge \psi^\mu = \langle \psi_\mu, \psi^\mu \rangle \psi_\nu - \langle \psi_\mu, \psi_\nu \rangle \psi^\mu \sim \psi^3,$$

and thus can be estimated by

$$\int_0^T \left( \int_{\mathbb{R}^4} |\psi|^6 dx \right)^{1/2} dt \leq \|\psi\|_{L^\infty_t L^4_x} \|\psi\|^2_{L^2_t L^8_x} \qquad (3.115)$$

$$\leq C\|\psi\|_{L^\infty_t \dot{H}^1_x} \|\psi\|^2_{L^2_t L^{(8,2)}_x};$$

- the second term in $f_\nu$ looks like $A\nabla\psi$ and, with the help of (3.101), is bounded by

$$\int_0^T \left( \int_{\mathbb{R}^4} |A_\mu \wedge \nabla^\mu \psi|^2 dx \right)^{1/2} dt \leq \|A\|_{L^1_t L^\infty_x} \|\psi\|_{L^\infty_t \dot{H}^1_x} \qquad (3.116)$$

$$\leq C\|\psi\|^2_{L^2_t L^{(8,2)}_x} \|\psi\|_{L^\infty_t \dot{H}^1_x};$$

- the third term has the form $(\nabla A)\psi$ and, based on (3.100) and the Sobolev inequality, is estimated by

$$\int_0^T \left( \int_{\mathbb{R}^4} |\nabla A|^2 |\psi|^2 dx \right)^{1/2} dt \leq \|\nabla A\|_{L^2_t L^{8/3}_x} \|\psi\|_{L^2_t L^8_x} \qquad (3.117)$$

$$\leq C\|\psi\|^2_{L^2_t L^8_x} \|\nabla\psi\|_{L^\infty_t L^2_x}$$

$$\leq C\|\psi\|^2_{L^2_t L^{(8,2)}_x} \|\psi\|_{L^\infty_t \dot{H}^1_x};$$

- the last term can be thought of as $A^2\psi$ and, relying on (3.101), (3.102), and the Sobolev inequality, is controlled by

$$\int_0^T \left( \int_{\mathbb{R}^4} |A|^4 |\psi|^2 dx \right)^{1/2} dt \leq \|A\|_{L_t^1 L_x^\infty}^{1/2} \|A\|_{L_t^\infty L_x^4}^{3/2} \|\psi\|_{L_t^2 L_x^8}$$

$$\leq C\|\psi\|_{L_t^2 L_x^{(8,2)}}^2 \|\nabla A\|_{L_t^\infty L_x^2}^{3/2}$$

$$\leq C\|\psi\|_{L_t^2 L_x^{(8,2)}}^2 \|\psi\|_{L_t^\infty \dot{H}_x^1}^3 .$$

$$(3.118)$$

Finally, combining the Strichartz estimate (3.113) with the last four bounds (3.115)-(3.118), we obtain an inequality of the form

$$\|\psi\|_N \leq C \left( \|\psi_0\|_{\dot{H}^1} + \|\psi_1\|_{L^2} + \|\psi\|_N^3 + \|\psi\|_N^5 \right) . \qquad (3.119)$$

We are now in the position to construct solutions assuming that

$$\|\psi_0\|_{\dot{H}^1} + \|\psi_1\|_{L^2} \leq C\epsilon,$$

with $\epsilon$ being sufficiently small. The idea is to write $f_\nu = f_\nu(A, \psi)$ and solve iteratively,

$$\Box \psi^{(n+1)} + f\big(A^{(n)}, \psi^{(n)}\big) = 0, \qquad \psi^{(n+1)}(0) = \psi_0, \qquad \partial_t \psi^{(n+1)}(0) = \psi_1 .$$

For a given $\psi^{(n)}$, we compute $A^{(n)}$ via the set of equations

$$F\big(A^{(n)}\big) = KG\big(\psi^{(n)}\big), \qquad\qquad \nabla_j A^{j(n)} = 0 .$$

Let us start with $\psi^{(-1)} = 0$, which forces $A^{(-1)} = 0$, and hence $\psi^{(0)}$ solves

$$\Box \psi^{(0)} = 0, \qquad \psi^{(0)}(0) = \psi_0, \qquad \partial_t \psi^{(0)}(0) = \psi_1 .$$

Applying (3.113), we deduce

$$\|\psi^{(0)}\|_N \leq C \left( \|\psi_0\|_{\dot{H}^1} + \|\psi_1\|_{L^2} \right) \leq C\epsilon .$$

Next, as $\Box \psi^{(1)} + f\big(A^{(0)}, \psi^{(0)}\big) = 0$ has the same initial data, an application of (3.119) yields

$$\|\psi^{(1)}\|_N \leq C\epsilon + C_1 \epsilon^3 + C_2 \epsilon^5 .$$

Proceeding in this way, we obtain for a sufficiently small $\epsilon$ (such that the associated power series converges) that

$$\|\psi^{(n)}\|_N \leq C\epsilon .$$

Moreover, with some extra but straightforward work, one can check that, eventually choosing $\epsilon$ even smaller than before, the iteration is in fact a contraction, i.e.,

$$\|\psi^{(n+1)} - \psi^{(n)}\|_N \leq \frac{1}{2}\|\psi^{(n)} - \psi^{(n-1)}\|_N, \qquad \forall\, n \geq 0.$$

This implies that $\psi^{(n)} \to \psi$ in the $N$-norm with $\|\psi\|_N \leq C\epsilon$. It is then easy to verify that $\psi$ is the desired solution.

Finally, we can employ Lemma 3.2 to conclude, based on $\|F\|_{L^1_t L^4_x} < \infty$, $\|A\|_{L^4_x} \leq C\epsilon^2$, and $\tau(0) = \Lambda(0) = 0$, that $\tau$ and $\Lambda$ are in fact equal to zero for all times. The fact that they vanish implies that the Hodge system (3.81) is indeed satisfied and concludes the proof of the theorem. $\qquad\square$

**Remark 3.13.** In order to check that the iteration scheme for $\{\psi^{(n)}\}$ is indeed a contraction, one needs to verify that $f(\psi, A)$ is a smooth function of $\psi$. This is a tedious, but not very rewarding work. To see why this is the case, without going into too many details, we first observe that $G \sim \psi^2$ and that the equations for $A$ imply essentially

$$A \sim \nabla\Delta^{-1}\psi^2\,.$$

Consequently, we can express $f$ in the schematic form

$$f \sim \psi^3 + \left(\nabla\Delta^{-1}\psi^2\right)\nabla\psi + \left(\nabla^2\Delta^{-1}\psi^2\right)\psi + \left(\nabla\Delta^{-1}\psi^2\right)^2\psi\,.$$

Thus, $f$ consists of a cubic and a quintic term. The idea now is to express the difference

$$f(\psi^{(2)}) - f(\psi^{(1)}) = F\left(\psi^{(1)}, \psi^{(2)}\right)\left(\psi^{(2)} - \psi^{(1)}\right)$$

via the mean value theorem and estimate $F$. We leave this as an exercise for the interested reader.

We reemphasize the point made previously about the relative simplicity of the Shatah-Struwe's argument presented above. It strictly relies on the use of real interpolation, Lorentz spaces, and endpoint Strichartz estimates, which are by now classical. We would like to conclude our discussion of Shatah-Struwe's result by remarking briefly on three items: higher regularity, the numerology involved in spatial dimensions $n \geq 5$, and open questions that stem from this work.

### 3.4.4.5 *Higher regularity*

Once the existence theorem has been established, we can obtain higher order regularity by differentiating the equation and applying the same estimates again. Here, the geometry helps if one observes that the natural geometric derivative of $\psi_\nu$ is $\nabla^A_\mu \psi_\nu$ and, moreover, the following commutation holds:

$$[\nabla^A_\mu, \Box^A] = F_\mu{}^\lambda \wedge \nabla^A_\lambda + \nabla^A_\lambda F_\mu{}^\lambda \wedge .$$

If we define $\omega_{\mu\nu} := \nabla^A_\mu \psi_\nu$, on the basis of $\Box^A \psi_\nu + K G_{\lambda\nu} \wedge \psi^\lambda = 0$ and $G_{\lambda\nu} = \psi_\nu \wedge \psi_\lambda$, we obtain the equation

$$\Box^A \omega_{\mu\nu} + K G_{\lambda\nu} \wedge \omega_\mu{}^\lambda + 2 K G_\mu{}^\lambda \wedge \omega_{\lambda\nu}$$
$$+ G(\omega_{\mu\nu} \wedge \psi_\lambda) \wedge \psi^\lambda + (\psi_\nu \wedge \omega_{\mu\lambda}) \wedge \psi^\lambda + K(\psi^\lambda \wedge \omega_{\lambda\mu}) \wedge \psi_\nu = 0 .$$

The salient point in the above equation is that we have

$$\Box^A \omega_{\mu\nu} = f_{\mu\nu} \qquad \text{with} \qquad f \sim \psi^2 \omega .$$

Thus, we can write $\Box \omega = \tilde{f}$, where the right-hand side is of the form

$$\tilde{f} \sim \psi^2 \omega + A\nabla\omega + (\nabla A)\omega + A^2 \omega.$$

Now, we can apply again endpoint Strichartz estimates in order to control $\|\omega\|_{L^2_t L^{(8,2)}_x}$. For example, one derives

$$\|\psi^2 \omega\|_{L^1_t L^2_x} \leq \|\psi\|^2_{L^2_t L^8_x} \|\omega\|_{L^\infty_t L^4_x},$$
$$\|A\nabla\omega\|_{L^1_t L^2_x} \leq \|A\|_{L^1_t L^\infty_x} \|\nabla\omega\|_{L^\infty_t L^2_x} \leq C\|\psi\|^2_{L^2_t L^{(8,2)}_x} \|\nabla\omega\|_{L^\infty_t L^2_x},$$

and other similar bounds. In order to balance the inequalities, we need to estimate $\|\nabla\omega\|_{L^\infty_t L^2_x}$, which should be part of an energy bound.

We can obtain an energy estimate in the following manner. The energy-momentum tensor

$$T_{\mu\nu} \stackrel{\text{def}}{=} \langle \nabla^A_\mu \omega, \nabla^A_\nu \omega \rangle - \frac{1}{2} g_{\mu\nu} \langle \nabla^A \omega, \nabla^A \omega \rangle,$$

satisfies $\nabla_\mu T^\mu_\nu \sim \psi^2 \omega \nabla^A \omega$. As a consequence, for the associated energy

$$\mathcal{E}_t(\omega) \stackrel{\text{def}}{=} \frac{1}{2} \int_{\{t\}\times\mathbb{R}^4} \{|\nabla^A_t \omega|^2 + |\nabla^A_x \omega|^2\} \, d\mathbf{x}$$

we deduce

$$\frac{d}{dt} \mathcal{E}_t(\omega) \sim \langle \psi^2 \omega, \nabla^A \omega \rangle,$$

which implies

$$\frac{d}{dt}\mathcal{E}_t(\omega) \le C\|\psi\|_{L_x^8}^2 \|\omega\|_{L_x^4} \|\nabla^A \omega\|_{L_x^2}. \tag{3.120}$$

In order to handle the right-hand side, we observe that $\nabla \omega = \nabla^A \omega - A \wedge \omega$ and, by squaring this identity, we obtain

$$|\nabla \omega|^2 = |\nabla^A \omega|^2 + |A \wedge \omega|^2 - 2\langle A \wedge \omega, \nabla^A \omega \rangle.$$

If we integrate, then it follows that

$$C_1\|\nabla^A \omega\|_{L_x^2}^2 - C_2\|A\|_{L_x^4}^2 \|\omega\|_{L_x^4}^2 \le \|\nabla \omega\|_{L_x^2}^2 \le C \left( \|\nabla^A \omega\|_{L_x^2}^2 + \|A\|_{L_x^4}^2 \|\omega\|_{L_x^4}^2 \right).$$

Due to

$$\|A\|_{L_t^\infty L_x^4} \le C\|\psi\|_{L_t^\infty \dot{H}_x^1}^2 \le C\epsilon$$

and the Sobolev inequality

$$\|\omega\|_{L_x^4} \le C\|\nabla \omega\|_{L_x^2},$$

we can infer, for a sufficiently small $\epsilon$, the equivalence

$$\|\nabla \omega\|_{L_x^2} \sim \|\nabla^A \omega\|_{L_x^2}$$

or

$$\mathcal{E}_t(\omega) \sim \|\partial_t \omega\|_{L_x^2}^2 + \|\omega\|_{\dot{H}_x^1}^2.$$

Hence, based on (3.120), we obtain

$$\frac{d}{dt}\mathcal{E}_t(\omega) \le C\|\psi\|_{L_x^8}^2 \mathcal{E}_t(\omega)$$

and an application of Grönwall's inequality gives the estimate

$$\|\partial_t \omega\|_{L_x^2}^2 + \|\omega\|_{\dot{H}_x^1}^2 \sim \mathcal{E}_t(\omega) \le \exp\left(C\|\psi\|_{L_t^2 L_x^8}^2\right) \mathcal{E}_0(\omega).$$

This bound can then be combined with the Strichartz estimate (3.111) applied for $\omega$, the end result being the control of the norm

$$\|\omega\|_{L_t^2 L_x^{(8,2)}} + \|\partial_t \omega\|_{L_x^2} + \|\omega\|_{\dot{H}_x^1},$$

provided $\|\psi\|_{L_t^2 L_x^{(8,2)}}$ is sufficiently small. We leave these details to the interested reader.

### 3.4.4.6   *The numerology of the high-dimensional case*

For the $n + 1$-dimensional case, an argument similar to the one which led to the derivation of (3.111) produces

$$\|\psi\|_{L_t^2 L_x^{(2n,2)}} + \|\psi\|_{L_t^\infty \dot{H}_x^{(n-2)/2}} + \|\partial_t \psi\|_{L_t^\infty \dot{H}_x^{(n-4)/2}}$$
$$\leq C \left( \|\psi(0)\|_{\dot{H}_x^{(n-2)/2}} + \|\partial_t \psi(0)\|_{\dot{H}_x^{(n-4)/2}} + \|\Box\psi\|_{L_t^1 \dot{H}_x^{(n-4)/2}} \right). \tag{3.121}$$

This estimate is the main ingredient for the following small data global regularity result, which is the counterpart of Theorem 3.8 (for small solutions) for dimensions $n \geq 4$.

**Theorem 3.9.** *Consider the Hodge system*

$$\nabla_\mu^A \psi^{a,\mu} = 0, \qquad\qquad \nabla_{[\mu}^A \psi_{\nu]}^a = 0, \qquad\qquad F_{\mu\nu}^a = KG_{\mu\nu}^a, \tag{3.122}$$

*with initial data prescribed at $t = 0$,*

$$\psi(0) := \psi_0, \qquad\qquad \partial_t\psi(0) := \psi_1, \tag{3.123}$$

*where $(\psi, A) = (\psi(t, \mathbf{x}), A(t, \mathbf{x}))$ are functions of $(t, \mathbf{x}) := (x^0, x^1, \ldots, x^n) \in \mathbb{R}^{n+1}$ and the indices $\mu$ and $\nu$ take values in the set $\{0, 1 \ldots, n\}$. Moreover, the gauge $A$ is constructed from $F_{\mu\nu} = KG_{\mu\nu}$ and the Coulomb gauge fixing $\nabla_j A^j = 0$, with $j = 1, 2, \ldots, n$ denoting the spatial indices. We assume that $\psi_0$ and $\psi_1$ satisfy the $n+1$-analog of the compatibility conditions $\tau(0) = \Lambda(0) = 0$ (see (3.83)).*

*Under these assumptions, there exists a sufficiently small $\epsilon > 0$ such that, if*

$$\|\psi_0\|_{\dot{H}^{(n-2/2)}} + \|\psi_1\|_{\dot{H}^{(n-4)/2}} \leq \epsilon \tag{3.124}$$

*then the system (3.122) admits a unique global solution satisfying*

$$\|\psi\|_{L_t^2 L_x^{(2n,2)}} + \|\psi\|_{L_t^\infty \dot{H}_x^{(n-2)/2}} + \|\partial_t\psi\|_{L_t^\infty \dot{H}_x^{(n-4)/2}} \leq C\epsilon. \tag{3.125}$$

**Proof.** We will only give a bare outline of the proof. In order to use (3.121), we need to estimate $\Box\psi$ in $L_t^1 \dot{H}_x^{(n-4)/2}$. This is slightly more complicated than in the $4 + 1$-dimensional case and, for this matter, we use the Leibnitz rule for fractional derivates. If we decompose $(n - 4)/2$ according to $(n - 4)/2 = k + s$, with $k$ an integer and $0 \leq s < 1$, and rely on the inequality

$$|\xi_1 + \xi_2|^s \leq C(|\xi_1|^s + |\xi_2|^s),$$

then, for purposes of estimating in $L^p$ spaces, we have

$$D^{(n-4)/2}(fg) \sim \sum_{k_1+k_2=k} C_{k_1}\left(\left(D^{k_1+s}f\right)\left(D^{k_2}g\right) + \left(D^{k_1}f\right)\left(D^{k_2+s}g\right)\right),$$

(3.126)

where $\widehat{Df}(\xi) := |\xi|\hat{f}(\xi)$.

Our nonlinearity has the schematic form

$$\Box\psi \sim \psi^3 + A(D\psi) + (DA)\psi + A^2\psi,$$

(3.127)

and we will only look at the most challenging term, which is $A(D\psi)$. It is for this term that we need to use the Lorentz spaces. Applying (3.126), we can infer that

$$D^{(n-4)/2}(AD\psi) = AD^{(n-2)/2}\psi + \text{lower order terms},$$

(3.128)

and for the first term on the right-hand side we have

$$\|AD^{(n-2)/2}\psi\|_{L_t^1 L_x^2} \leq \|A\|_{L_t^1 L_x^\infty}\|D^{(n-2)/2}\psi\|_{L_t^\infty L_x^2}.$$

Thus, we are confronted with estimating $A$ in the $L_t^1 L_x^\infty$ norm.

This goal can be achieved as follows. Recall our schematic equation for the gauge field,

$$A = R * \left(2A \wedge A - KG\right),$$

where

$$R \stackrel{\text{def}}{=} \frac{c}{|\mathbf{x}|^{n-1}} \in L_{\mathbf{x}}^{(\frac{n}{n-1},\infty)} \qquad \text{and} \qquad G \sim \psi^2.$$

Using the fact that $\left(L^{(n/(n-1),\infty)}\right)^* = L^{(n,1)}$, we can derive the estimate

$$\|A\|_{L_{\mathbf{x}}^\infty} \leq C\big(\|A\|^2_{L_{\mathbf{x}}^{(2n,2)}} + \|\psi\|^2_{L_{\mathbf{x}}^{(2n,2)}}\big) \leq C\left(\|A\|_{L_{\mathbf{x}}^n}\|A\|_{L_{\mathbf{x}}^\infty} + \|\psi\|^2_{L_{\mathbf{x}}^{(2n,2)}}\right).$$

We construct the gauge $A$ by iteration using the norm $\|A\|_{\dot{H}^{(n-2)/2}}$ and, assuming that $\|\psi\|_{\dot{H}^{(n-2)/2}} \leq \epsilon$, we deduce that $\|A\|_{\dot{H}^{(n-2)/2}} \leq C\epsilon$ if $\epsilon$ is small enough. Hence, on the account of the Sobolev inequality $\|A\|_{L^n} \leq C\|A\|_{\dot{H}^{(n-2)/2}}$, we finally obtain

$$\|A\|_{L_t^1 L_x^\infty} \leq C\|\psi\|^2_{L_t^2 L_x^{(2n,2)}},$$

which demonstrates that $\|AD^{(n-2)/2}\psi\|_{L_t^1 L_x^2}$ can be controlled by endpoint Strichartz estimates.

The rest of the terms in (3.128) and (3.127) can be estimated in a similar manner, using repeatedly Sobolev inequalities, the Leibnitz rule formula, and appropriate Hölder inequalities. We refer the reader for more details of the proof to the original paper of Shatah and Struwe [146]. Here, our goal was only to point out the salient features that make the method work. $\Box$

### 3.4.4.7 *Final comments*

Looking back again at the $4 + 1$-dimensional case, we notice that, except for the gauge invariance, we did not take into account any of the special structure for the nonlinear terms in the Hodge system. This being said, a careful review of the argument shows that the crucial quantity to be estimated is

$$\|G\|_{L_t^1 L_x^{(4,1)}} \, .$$

Moreover, the iteration norm (3.112) consists of two parts, one of which is the "energy", namely,

$$\|\psi\|_{L_t^\infty \dot{H}_x^1} + \|\partial_t \psi\|_{L_t^\infty L_x^2} \, .$$

As a matter of fact, the growth of the energy is controlled by the first quantity.

From the equation $\Box^A \psi_\nu + F_{\mu\nu} \wedge \psi^\mu = 0$, we can derive an energy-momentum tensor. If we freeze the index $\nu$ and define

$$T_{\lambda\mu(\nu)} \stackrel{\text{def}}{=} \langle \nabla_\lambda^A \psi_{(\nu)}, \nabla_\mu^A \psi_{(\nu)} \rangle - \frac{1}{2} g_{\lambda\mu} \langle \nabla_\alpha^A \psi_{(\nu)}, \nabla^{A,\alpha} \psi_{(\nu)} \rangle,$$

it follows that

$$\nabla_\lambda T^\lambda_{\ \mu(\nu)} = \langle F_{\lambda(\nu)} \wedge \psi^\lambda, \nabla_\mu^A \psi_{(\nu)} \rangle + \langle F_{\lambda\mu} \wedge \psi_{(\nu)}, \nabla^{A,\lambda} \psi_{(\nu)} \rangle \, .$$

The time component of the above tensor yields an estimate for the covariant energy

$$\mathcal{E}_t(\psi) \stackrel{\text{def}}{=} \int_{\{t\} \times \mathbb{R}^4} \{ |\nabla_0^A \psi|^2 + |\nabla_{\mathbf{x}}^A \psi|^2 \} \, d\mathbf{x} \, .$$

Using the fact that $F = KG$, we can derive the differential inequality,

$$\frac{d}{dt} \mathcal{E}_t(\psi) \leq \|G\|_{L_{\mathbf{x}}^4} \|\psi\|_{L_{\mathbf{x}}^4} \sqrt{\mathcal{E}_t(\psi)} \, .$$

If we would know the Sobolev-type inequality

$$\|\psi\|_{L_{\mathbf{x}}^4} \leq C \|\nabla_{\mathbf{x}}^A \psi\|_{L_{\mathbf{x}}^2},$$

then we can obtain the energy bound

$$\mathcal{E}_t(\psi) \leq \exp\left( C \|G\|_{L_t^1 L_{\mathbf{x}}^4} \right) \mathcal{E}_0(\psi) \, .$$

Thus, the issue is to establish an inequality of the form

$$\|\nabla_{\mathbf{x}} \psi\|_{L_{\mathbf{x}}^2} \leq C \|\nabla_{\mathbf{x}}^A \psi\|_{L_{\mathbf{x}}^2},$$

which, based on the identity

$$|\nabla\psi|^2 = |\nabla^A\psi|^2 + |A \wedge \psi|^2 - 2\langle A \wedge \psi, \nabla^A\psi\rangle,$$

holds true, provided that $\|A\|_{L_{\mathbf{x}}^4}$ can be made sufficiently small. Since we have the elliptic estimate $\|A\|_{L_{\mathbf{x}}^4} \le C\|\psi\|_{\dot{H}_{\mathbf{x}}^1}^2$, we can close our estimates if we make $\|G\|_{L_t^1 L_{\mathbf{x}}^{(4,1)}}$ sufficiently small and start with $\|\psi(0)\|_{\dot{H}_{\mathbf{x}}^1}$ small enough.

Alternatively, one can view Theorem 3.8 as a "weak" criterion for local regularity, in the sense that controlling $\|G\|_{L_t^1 L_{\mathbf{x}}^{(4,1)}([0,T]\times\mathbb{R}^4)}$ yields existence of solutions in the time interval $[0, T]$. Using the finite speed of propagation property, it follows that we can localize the problem. This means that we can consider the initial value problem with data supported in the ball $B_R(\mathbf{x}_0) := \{|\mathbf{x} - \mathbf{x}_0| \le R\}$ and satisfying $\|\psi(0)\|_{\dot{H}^1(B_R(\mathbf{x}_0))} \le \epsilon$, and construct a solution in the backward cone $C := \{(t, \mathbf{x}) \mid 0 \le t \le R\,,\ |\mathbf{x} - \mathbf{x}_0| \le R - t\}$. We would also need to localize the gauge field $A$ and construct the local version of the Coulomb gauge. After these constructions, we would have to patch-up the local solutions, with the end result being the establishment of local solutions defined on the time interval $[0, T]$, provided $\|G\|_{L_t^1 L_{\mathbf{x}}^{(4,1)}([0,T]\times\mathbb{R}^4)}$ is finite. This is just an outline, with details needing to be worked out. The issue now is how we can control or estimate such a quantity and this is an open question. Furthermore, our regularity criterion appears to be of the "weakest possible", in the sense that it was derived using only critical-type inequalities. This does not mean that there does not exist another, even weaker, criterion or that maybe another equivalent, but more amenable to estimation, criterion cannot be found.

Let us pause here to reflect on what we have achieved so far. The result seems to be a dead end. How can one proceed with the knowledge gained? The important point in such a construction is to isolate a quantity that is crucial for regularity. One can view this as a conditional regularity result in the sense that if a certain norm (or quantity) is finite, then the solution does not blow-up.

Here are some natural questions for the wave maps problem in $n + 1$ dimensions. Is there a difference in the behavior of solutions between the cases where the target manifold is either $\mathbb{S}^k$ or $\mathbb{H}^k$ (i.e., does the curvature play a role?)? Is it true that solutions of the wave maps problem when the target manifold is $\mathbb{H}^k$ are regular for all time? Can one give a "satisfactory" characterization of the possible set of singularities? As a matter of fact, for $n \ge 3$, the problem is energy-supercritical and, unfortunately, current techniques fail badly for such problems. The issue is that one cannot treat, in general, the nonlinear terms as a perturbation of the linear term $\Box \psi$. Understanding the solutions of supercritical problems is a major and completely open problem. Fortunately, for $n = 2$, the problem is energy-critical and in this case we have a more complete picture.

# Chapter 4

# Large data regularity and scattering for 2 + 1-dimensional wave maps

## 4.1 Introduction

After covering the small data problem, we now begin to discuss the large data regime, for which there is a noticeable a priori difference in the analyses of the $n = 2$ and $n \geq 3$ cases, respectively. This has to do with the fact that the Cauchy problem (3.2) is energy-critical for $n = 2$ and one can use the conservation of energy to control the critical Sobolev norm (i.e., $\dot{H}^1(\mathbb{R}^2)$). This is not the case for $n \geq 3$, when (3.2) is energy-supercritical and, according to Tataru [91], there are no current techniques to keep the $\dot{H}^{n/2}(\mathbb{R}^n)$ norm of wave maps bounded. In fact, it was also suggested in [91] as a possible research direction to study wave maps enjoying a uniform a priori critical Sobolev bound.

We focus in the sequel on the large data 2 + 1-dimensional problem, which received a lot of interest in recent years. The end result of this effort is that a comprehensive picture of the behavior of wave maps emerged. There are three major results, the first of which is Tao's impressive program completed in [172–176]. It addresses the case when the target is the unit hyperboloid in $\mathbb{R}^{m+1}$

$$N = \mathbb{H}^m = \{(y^0, \mathbf{y}) \in \mathbb{R}^{m+1}; \; y^0 = \sqrt{1 + |\mathbf{y}|^2}\},$$

with the Riemannian metric $h$ being the one induced by the Minkowski metric on $\mathbb{R}^{m+1}$ (the point is that the Minkowski metric restricted on the unit hyperboloid is negative definite). The hyperbolic space $\mathbb{H}^m$ is the prototype for an $m$-dimensional Riemannian manifold with constant negative curvature. Tao works with *classical data* and *classical wave maps*, which are slightly more general than the concepts of *smooth data* (described in Remark 3.8) and *classical wave maps* (previously defined in the discussion of his small data high-dimensional result), respectively. Here,

$(\phi(0), \partial_t\phi(0)) = (\phi_0, \phi_1)$ is a *classical data* if $\phi_0 : \mathbb{R}^2 \to \mathbb{H}^m$ differs from a constant by a Schwartz function and $\phi_1 : \mathbb{R}^2 \to T\mathbb{H}^m$ is a Schwartz map satisfying $\phi_1(\mathbf{x}) \in T_{\phi_0(\mathbf{x})}\mathbb{H}^m$ for all $\mathbf{x} \in \mathbb{R}^2$. Similarly, a *classical wave map* is defined to be a smooth map $\phi : I \times \mathbb{R}^2 \to \mathbb{H}^m$, where $I \subseteq \mathbb{R}$ is an interval, which solves (3.2) and differs from a constant by a function with Schwartz regularity in the spatial variables. In this setting, Tao proves that any *classical data* leads to a unique global *classical wave map*, which also enjoys certain scattering properties.

The second major result for this problem makes the subject of a recent book[1] by Krieger and Schlag [96] and it concerns the case when the target manifold is $\mathbb{H}^2$, the hyperbolic plane. Krieger and Schlag work with the original concepts of *smooth data* and *classical wave maps* defined in the previous section and prove global existence and regularity in this context. However, they are able to prove more than that as a byproduct of the method they rely on: the *Bahouri-Gérard concentration compactness method* [7], which was further developed by Kenig and Merle [80, 81]. Krieger and Schlag obtain relevant quantitative bounds concerning the asymptotic behavior of solutions, together with important concentration compactness properties for the same solutions.

Due to the length of Tao's work and because [96] contains a fairly detailed account of Krieger-Schlag's result, we concentrate our attention in this chapter on the third major contribution to this problem. This is due to Sterbenz and Tataru [160, 161] and it primarily addresses the case when the target $N$ is a compact manifold. Simply stated, Sterbenz-Tataru's main result says:

**Theorem 4.1.** *The Cauchy problem* (3.2) *exhibits global well-posedness and scattering for any initial data* $(\phi_0, \phi_1) \in \dot{H}^1 \times L^2(\mathbb{R}^2)$ *whose energy*

$$\mathcal{E}(\phi_0, \phi_1) = \frac{1}{2}\int_{\mathbb{R}^2} |\phi_1|_h^2 + |\nabla_\mathbf{x}\phi_0|_h^2 \, d\mathbf{x} \tag{4.1}$$

*is less than* $\mathcal{E}(N)$, *which is the minimal energy among the ones for all nontrivial harmonic maps (i.e., time-independent wave maps) from* $\mathbb{R}^2$ *to* $N$.

---

[1]This result was initially published in [95] on arXiv.org in 2009.

## 4.2   Informal discussion of Sterbenz-Tataru's results

We first describe the content of [160], in which Sterbenz and Tataru work with *large data finite energy wave maps*. Using Theorem 3.5, this concept can be defined as follows.

**Definition 4.1.** Let $I$ be an arbitrary time interval. We call $\phi$ a *large data finite energy wave map* on $I$ if

$$\phi \in C(I, \dot{H}^1(\mathbb{R}^2)) \cap \dot{C}^1(I, L^2(\mathbb{R}^2)),$$

and, for each $(t_0, \mathbf{x}_0) \in I \times \mathbb{R}^2$ and $R > 0$ such that $\phi[t_0]$ has small $\dot{H}^1 \times L^2(B_R(\mathbf{x}_0))$ norm, $\phi$ coincides with the solution given by ii) in Theorem 3.5 on $(I \times \mathbb{R}^2) \cap Q_R(t_0, \mathbf{x}_0)$.

For this type of wave maps, the relevant quantities are their *energy*, $\mathcal{E}[\phi]$, which is given by (3.3), and *energy dispersion*,

$$ED[\phi] \overset{\text{def}}{=} \sup_{\lambda \in 2^{\mathbb{Z}}} \|S_\lambda[\phi]\|_{L^\infty_{t,\mathbf{x}}(I \times \mathbb{R}^2)},$$

where $S_\lambda$ is the Littlewood-Paley projector localizing the spatial frequency to the region $\{\lambda/2 \leq |\xi| \leq 2\lambda\}$. With these prerequisites, we are able to formulate the main result in [160], which can be thought of as "*energy dispersion implies regularity*".

**Theorem 4.2.** *Let $N$ be a compact Riemannian manifold. For each $E > 0$, there exist two functions,*

$$1 \ll F = F(E) \qquad and \qquad 0 < \epsilon = \epsilon(E) \ll 1,$$

*such that any* large data finite energy wave map $\phi$ *on the open time interval $I$, with*

$$\mathcal{E}[\phi] = E \qquad and \qquad ED[\phi] \leq \epsilon(E),$$

*satisfies[2]*

$$\|\phi\|_{S[I]} \leq F(E).$$

*Moreover, this type of wave map extends regularly to a neighborhood for the closure of $I$.*

---

[2]Here, the function space $S$ is a slight modification of the one mentioned in Remark 3.26 (and used in proving Theorem 3.4). Precisely, the only modification is in the structure of the dyadic norm $\|\cdot\|_{S[\lambda]}$ (given by (3.27)) to which one adds $\|S_{2^k}\phi\|_{X_k(\mathbb{R}\times\mathbb{R}^2)}$, where $\lambda = 2^k$ and $\|\cdot\|_{X_k(\mathbb{R}\times\mathbb{R}^2)}$ is the Strichartz norm defined by (3.33).

**Remark 4.1.** This theorem is essential to the analysis of Sterbenz and Tataru in [161]. However, in [161], they use a variant in which the energy dispersion is measured by the slightly stronger norm

$$E_D[\phi] \stackrel{\text{def}}{=} \sup_{\lambda \in 2^{\mathbb{Z}}} \left\{ \left\| S_\lambda[\phi] \right\|_{L^\infty_{t,x}(I \times \mathbb{R}^2)} + \left\| \lambda^{-1} S_\lambda[\partial_t \phi] \right\|_{L^\infty_{t,x}(I \times \mathbb{R}^2)} \right\} . \quad (4.2)$$

Theorem 4.2 is certainly true if one replaces $ED[\phi]$ by $E_D[\phi]$ and we will use $E_D[\phi]$ instead of $ED[\phi]$ in our presentation.

One way to interpret this theorem is that large data energy dispersed wave maps persist globally. The argument has many similarities to the ones for small data results (e.g., [169], [180]): it is based on a bootstrap argument which uses frequency envelopes and involves a renormalization procedure. On the other hand, the functional setting and the multilinear estimates to be proved for this theorem are much more intricate than in the small data regime. Nevertheless, we believe the conscientious reader has by now the framework and the tools necessary to follow this argument by himself.

Another essential ingredient for the main argument in [161] is the $2 + 1$-dimensional case of Theorem 3.4, which can be thought of as a *"small energy implies regularity"* result and which is restated here for the convenience of the reader.

**Theorem 4.3.** *There exists a positive number denoted by $E_0$ and called perturbation energy such that, for any given smooth initial data $(\phi_0, \phi_1)$ satisfying*

$$\mathcal{E}(\phi_0, \phi_1) < E_0,$$

*the Cauchy problem (3.2) admits a unique global smooth solution $\phi \in S$. In addition, the solution operator $(\phi_0, \phi_1) \mapsto \phi$ extends to a continuous operator from $\dot{H}^1 \times L^2(\mathbb{R}^2) \mapsto S$, with the bound*

$$\|\phi\|_S \le C \|(\phi_0, \phi_1)\|_{\dot{H}^1 \times L^2(\mathbb{R}^2)}.$$

*Such solutions will be called "strong finite-energy wave maps".*

*Furthermore, for $-C(E_0) < s - 1 < 0$, the following weak Lipschitz stability estimate holds for two solutions, $\phi$ and $\psi$:*

$$\|\phi - \psi\|_{C(\mathbb{R}, \dot{H}^s(\mathbb{R}^2)) \cap \dot{C}^1(\mathbb{R}, \dot{H}^{s-1}(\mathbb{R}^2))} \le C \|\phi(0) - \psi(0)\|_{\dot{H}^s \times \dot{H}^{s-1}(\mathbb{R}^2)}. \quad (4.3)$$

As mentioned in Chapter 3, due to the finite speed of propagation property, one can obtain local versions of the above result, which are valid for initial data restricted to an arbitrary ball in $\mathbb{R}^2$ and with the corresponding domain of uniqueness being either a backward or a forward light cone. Using the translation invariance of the wave maps equation, one may assume the ball to be centered at the origin.

It is worth noticing that Theorem 4.2 is a subtle refinement of Theorem 4.3. They are both conditional regularity statements, but the condition in the former is weaker. We will use Theorem 4.2 later on to exclude what we will call "*the null concentration of energy scenario.*" This means the possibility that portions of the energy concentrate near the characteristic cone. In essence, Theorem 4.2 implies that non-concentration of energy inside a cone of aperture less than 1 implies regularity.

Now, we shift our focus to [161], in which Sterbenz and Tataru complete their large data global regularity theory. There, they prove unconditional versions of Theorem 4.2, thus sorting out when a large data finite energy wave map blows up and when it is global-in-time, with appropriate scattering properties. The following is an argument showing that if a wave map blows up, then the energy concentrates inside the backward light cone originating at the blow-up locus. Moreover, if it happens, failure of scattering also occurs inside a light cone. This result allows Sterbenz and Tataru to work only with finite energy wave maps inside light cones.

**Proposition 4.1 ([94]).** *Let the Cauchy problem* (3.2), *with $N$ being a uniformly isometrically embeddable manifold in a Euclidean space, be weakly globally well-posed for small data in $\dot{H}^1 \times L^2(\mathbb{R}^2)$ in the sense of Definition 3.2. Moreover, assume that there is no concentration of energy inside light cones, i.e.,*

$$\lim_{\substack{t \to t_0 \\ t < t_0}} \int_{|\mathbf{x}-\mathbf{x}_0|<t_0-t} |\nabla_{t,\mathbf{x}}\phi(t,\mathbf{x})|_h^2 \, d\mathbf{x} = 0, \qquad \forall \, (t_0, \mathbf{x}_0) \in \mathbb{R}^{2+1}. \qquad (4.4)$$

*Then the Cauchy problem* (3.2) *is weakly globally well-posed for arbitrary data in $\dot{H}^1 \times L^2(\mathbb{R}^2)$ in the sense of the same definition mentioned above.*

**Proof.** We reason by contradiction and assume that, for certain data $(\phi_0, \phi_1) \in \dot{H}^1 \times L^2(\mathbb{R}^2)$ which may be chosen to have compact support, (3.2) admits a solution $\phi \in C([0,T), \dot{H}^1(\mathbb{R}^2)) \cap C^1([0,T), L^2(\mathbb{R}^2))$ that blows up at time $0 < T < \infty$. We let $M > 0$ be the energy threshold appearing in Definition 3.2 and $n(2)$ be the minimal number of circles of radius 1 needed to cover a circle of radius 2 in $\mathbb{R}^2$.

Using (4.4), for a fixed $\mathbf{x}_0 \in \mathbb{R}^2$, we define

$$\delta(\mathbf{x}_0) \overset{\text{def}}{=} \sup \left\{ \delta > 0; \int_{|\mathbf{x}-\mathbf{x}_0|<\delta} |\nabla\phi(T-\delta, \mathbf{x})|_h^2 \, d\mathbf{x} < \frac{2M}{n(2)} \right\}.$$

We make the claim that

$$\inf_{\mathbf{x}_0 \in \mathbb{R}^2} \delta(\mathbf{x}_0) > 0. \tag{4.5}$$

The fact that the initial data have compact support, together with the finite speed of propagation property enjoyed by (3.1), implies that $\delta(\mathbf{x}_0) = T$ if $|\mathbf{x}_0|$ is sufficiently large. Therefore, if the previous claim does not hold, then there exist $R > 0$ and a sequence $(\mathbf{x}_0^n)_n \subset \mathbb{R}^2$ satisfying

$$|\mathbf{x}_0^n| < R \quad \text{and} \quad \delta(\mathbf{x}_0^n) \to 0.$$

Eventually restricting it to a subsequence, we can assume that $\mathbf{x}_0^n \to \bar{\mathbf{x}}_0$. Due to the regularity of $\phi$, we infer

$$\int_{|\mathbf{x}-\mathbf{x}_0^n|<\delta(\bar{\mathbf{x}}_0)} |\nabla\phi(T-\delta(\bar{\mathbf{x}}_0), \mathbf{x})|_h^2 \, d\mathbf{x} \to \int_{|\mathbf{x}-\bar{\mathbf{x}}_0|<\delta(\bar{\mathbf{x}}_0)} |\nabla\phi(T-\delta(\bar{\mathbf{x}}_0), \mathbf{x})|_h^2 \, d\mathbf{x}.$$

As $\delta(\mathbf{x}_0^n) \to 0$ and $\delta(\bar{\mathbf{x}}_0) > 0$, this yields a contradiction based on the local energy inequality.

Using (4.5), we deduce that there exists $0 < T_M < T$ such that

$$\int_{|\mathbf{x}-\mathbf{x}_0|<T-T_M} |\nabla\phi(T_M, \mathbf{x})|_h^2 \, d\mathbf{x} < \frac{2M}{n(2)}, \quad \forall \mathbf{x}_0 \in \mathbb{R}^2,$$

which, by our choice of $n(2)$, forces

$$\int_{|\mathbf{x}-\mathbf{x}_0|<2(T-T_M)} |\nabla\phi(T_M, \mathbf{x})|_h^2 \, d\mathbf{x} < 2M, \quad \forall \mathbf{x}_0 \in \mathbb{R}^2.$$

If we apply now Theorem 3.5 with $t_0 = T_M$ and initial data $(\phi(T_M), \partial_t \phi(T_M))$ having small energy in the ball $|\mathbf{x} - \mathbf{x}_0| < 2(T - T_M)$, it follows that we can extend the wave map $\phi$ to the cone

$$\{(t, \mathbf{x}); \; t - T_M + |\mathbf{x} - \mathbf{x}_0| < 2(T - T_M)\}, \quad \forall \mathbf{x}_0 \in \mathbb{R}^2.$$

Invoking the finite speed of propagation property, we derive that any two such extensions coincide on their intersection. This means that we are able to continue the solution past the blow-up time $T$, thus contradicting the original assumption. $\qquad\square$

The goal of the analysis in [161] is to obtain a result that claims a dichotomy; precisely one of the following two statements is true:

- *the solution is regular (i.e., there is no blow-up at the origin and, hence, no energy concentration occurs);*
- *there exists concentration of energy strictly inside the characteristic cone and, after an appropriate rescaling, a part of the wave map converges to a Lorentz transform of a nontrivial static solution (i.e., a nontrivial harmonic map).*

In order to make precise the above statements and state the main results of [161], we introduce next the necessary terminology. If $I$ is an arbitrary time interval and $t_0 \in \mathbb{R}$, then we recall our definitions of the characteristic cone, its subsets, and its time slices:

$$\mathcal{C} \stackrel{\text{def}}{=} \{(t, \mathbf{x}) \in \mathbb{R} \times \mathbb{R}^2;\ 0 \le |\mathbf{x}| \le t < \infty\},$$

$$\mathcal{C}_I \stackrel{\text{def}}{=} \mathcal{C} \cap (I \times \mathbb{R}^2), \qquad D_{t_0} \stackrel{\text{def}}{=} \mathcal{C} \cap (\{t_0\} \times \mathbb{R}^2).$$

For a wave map $\phi$ which is defined either on $\mathcal{C}$ or on one of its subsets, we denote its *local energy* or *energy of time slice* $D_t$ by

$$\mathcal{E}_{D_t}[\phi] \stackrel{\text{def}}{=} \frac{1}{2} \int_{D_t} |\nabla_{t,\mathbf{x}} \phi|^2 \, d\mathbf{x}.$$

As we shall see later on, by integrating the differential energy identity (4.15) over subsets of the characteristic cone described by (4.17), one obtains that $t \mapsto \mathcal{E}_{D_t}[\phi]$ is a nondecreasing function. Thus, given a wave map $\phi$ on $\mathcal{C}_{(0,1]}$, we deduce that $\lim_{t \to 0} \mathcal{E}_{D_t}[\phi]$ exists. We can now formulate Sterbenz-Tataru's result on the question of blow-up for large data finite energy wave maps, which, after a translation and rescaling, can be reduced to investigating the same question for smooth wave maps on $\mathcal{C}_{(0,1]}$.

**Theorem 4.4.** *Let $N$ be a compact Riemannian manifold and assume that $\phi : \mathcal{C}_{(0,1]} \to N$ is a $C^\infty$ wave map. Precisely one of the following two scenarios takes place:*

*(1) there exists a sequence of points $(t_n, \mathbf{x}_n)_n \subset \mathcal{C}_{(0,1]}$ and scales $(r_n)_n \subset \mathbb{R}$ satisfying*

$$\lim_{n \to \infty} (t_n, \mathbf{x}_n) = (0,0), \qquad \limsup_{n \to \infty} \frac{|\mathbf{x}_n|}{t_n} < 1, \qquad \lim_{n \to \infty} \frac{r_n}{t_n} = 0,$$

*such that the rescaled sequence of wave maps $(\phi^{(n)})_n$,*

$$\phi^{(n)}(t, \mathbf{x}) \stackrel{\text{def}}{=} \phi(t_n + r_n t, \mathbf{x}_n + r_n \mathbf{x}),$$

*converges strongly in $H^1_{loc}$ to a Lorentz transform of an entire harmonic map of nontrivial energy,*

$$\phi^{(\infty)} : \mathbb{R}^2 \to N, \qquad 0 < \|\phi^{(\infty)}\|_{\dot{H}^1(\mathbb{R}^2)} \le \lim_{t \to 0} \mathcal{E}_{D_t}[\phi]. \qquad (4.6)$$

(2) *for each $\epsilon > 0$, there exist $t_0 \in (0,1]$ and a wave map extension of $\phi$, denoted also $\phi$,*

$$\phi : (0, t_0] \times \mathbb{R}^2 \to N,$$

*satisfying*

$$\mathcal{E}[\phi] \leq (1 + \epsilon^8) \lim_{t \to 0} \mathcal{E}_{D_t}[\phi] \tag{4.7}$$

*and*

$$\sup_{t \in (0, t_0]} \sup_{\lambda \in 2^{\mathbb{Z}}} \left( \|S_\lambda[\phi](t)\|_{L^\infty(\mathbb{R}^2)} + \lambda^{-1} \|S_\lambda[\partial_t \phi](t)\|_{L^\infty(\mathbb{R}^2)} \right) \leq \epsilon. \tag{4.8}$$

An immediate consequence of this theorem is that, under the additional hypothesis that there are no nontrivial finite energy harmonic maps $\phi^{(\infty)}$ : $\mathbb{R}^2 \to N$, the only possible scenario is the second one, which implies the energy dispersion (4.8) for the wave map extension. Hence, we can apply Theorem 4.2 to derive:

**Theorem 4.5.** *If the target manifold $N$ is compact and there are no nontrivial finite energy harmonic maps $\phi^{(\infty)} : \mathbb{R}^2 \to N$, then, for any data $(\phi_0, \phi_1) \in \dot{H}^1 \times L^2(\mathbb{R}^2)$, the Cauchy problem (3.2) admits a global solution $\phi$ satisfying $\phi \in S((0, T))$ for all $T > 0$. Furthermore, any additional regularity of the initial data is inherited by $\phi$.*

Moreover, as a consequence of this theorem, Sterbenz and Tataru established the following useful result.

**Corollary 4.1.** *Let $\tilde{N}$ be a Riemannian manifold such that there exists a Riemannian covering map[3] $\pi : \tilde{N} \to N$, with $N$ being a manifold verifying the hypotheses of Theorem 4.5. Then, solutions to the Cauchy problem (3.2) in which $n = 2$, $\tilde{N}$ is the target manifold, and the initial data is smooth, are globally smooth in time.*

This can be seen by considering a wave map $\tilde{\phi} : \mathbb{R}^{2+1} \to \tilde{N}$ with smooth data and restricting $\phi = \pi \circ \tilde{\phi} : \mathbb{R}^{2+1} \to N$ to a sufficiently small time section of a cone, where a potential blow-up of $\tilde{\phi}$ occurs. As $\pi$ is a local isometry, the Levi-Civita connections of $\tilde{N}$ and $N$ correspond under $\pi$ (see Wilkins [188]) and, as a consequence, the restriction of $\phi$ is also a wave

---

[3]For two Riemannian manifolds $N$ and $\tilde{N}$, a map $\pi : \tilde{N} \to N$ is called a *Riemannian covering* if $\tilde{N}$ is path connected and locally path connected, $\pi$ is both surjective and a local isometry, and each point $p \in N$ has a connected neighborhood $U$ for which each component of $\pi^{-1}(U)$ is mapped diffeomorphically by $\pi$ onto $U$.

map. By Theorem 4.5, this restriction is smooth and its image lies in a simply connected set. Thus, one can invert the projection to obtain the smoothness of $\tilde{\phi}$ near the putative blow-up locus.

Sterbenz and Tataru noticed that this corollary and Theorem 4.5 provide a new proof for the global regularity of smooth wave maps into the hyperbolic spaces $\mathbb{H}^m$, which is also the result obtained independently by Tao and Krieger-Schlag. Indeed, an important class of manifolds verifying the hypotheses of the above theorem is the one formed by compact manifolds with negative curvature, which is a result due to Schoen and Yau [139] (see also Eells-Lemaire [40]). All that is left then is to invoke the classical Killing-Hopf theorem [69,83], which yields a Riemannian covering map from $\mathbb{H}^m$ to a corresponding compact manifold with constant negative curvature.

If the target manifold has nontrivial finite energy harmonic maps, then there exists such a map with minimal energy. This minimal energy is positive and is denoted by $\mathcal{E}(N)$ (see also Theorem 4.1). If the initial data has energy less than $\mathcal{E}(N)$, we deduce under the hypotheses of Theorem 4.4 that

$$\lim_{t \to 0} \mathcal{E}_{D_t}[\phi] < \mathcal{E}(N) \leq \|\phi^{(\infty)}\|_{\dot{H}^1(\mathbb{R}^2)}$$

holds for any entire harmonic map of nontrivial energy $\phi^{(\infty)} : \mathbb{R}^2 \to N$. Therefore, the first scenario in Theorem 4.4 is not possible, which leads to the following regularity result.

**Theorem 4.6.** *If the target manifold $N$ is compact and it has a lowest energy nontrivial finite energy harmonic map with energy $\mathcal{E}(N)$, then any finite energy initial data $(\phi_0, \phi_1) \in \dot{H}^1 \times L^2(\mathbb{R}^2)$ whose energy is below $\mathcal{E}(N)$ leads to a global solution $\phi$ for the Cauchy problem (3.2) satisfying $\phi \in S((0, T))$ for all $T > 0$.*

**Remark 4.2.** The previous theorem is sharp in the sense that, for $N = \mathbb{S}^2$, the stereographic projection

$$\phi^{(\infty)}(\mathbf{x}) \overset{\text{def}}{=} \left( \frac{2\mathbf{x}}{1 + |\mathbf{x}|^2}, \frac{1 - |\mathbf{x}|^2}{1 + |\mathbf{x}|^2} \right), \qquad \mathbf{x} \in \mathbb{R}^2,$$

is a nontrivial finite energy harmonic map with the lowest energy, and Krieger-Schlag-Tataru [97] showed that blow-up can occur for initial data whose energy is arbitrarily close to the one of $\phi^{(\infty)}$. As a matter of fact, the first scenario in Theorem 4.4 describes blow-up profiles constructed by Rodnianski-Sterbenz [138] and more recently by Raphaël-Rodnianski [136].

The combined results of Theorem 4.5, Corollary 4.1, and Theorem 4.6 establish a full regularity theory for large data wave maps, thus settling a long-standing conjecture due to Klainerman [85], also known as the *Threshold Conjecture*.

As previously mentioned, in addition to regularity results, Sterbenz and Tataru also show that global-in-time finite energy wave maps enjoy certain scattering properties. In their first paper (i.e., [160]), they prove that if such a map belongs to the function space $S$, then its behavior for large times resembles the one for a renormalization of a linear wave, which can be interpreted as scattering. Thus, their focus here is to investigate whether $\phi \in S$ when $\phi$ is the global solution in Theorems 4.5 and 4.6, respectively. Arguing as in the reduction of the regularity question to the proof of Theorem 4.4, we deduce that the scattering question is answered if one proves the following result.

**Theorem 4.7.** *Let $N$ be a compact Riemannian manifold and assume that $\phi : \mathcal{C}_{[1,\infty)} \to N$ is a $C^\infty$ wave map satisfying*

$$\lim_{t \to \infty} \mathcal{E}_{D_t}[\phi] < \infty. \tag{4.9}$$

*Precisely one of the following two scenarios takes place:*

*(1) there exists a sequence of points $(t_n, \mathbf{x}_n)_n \subset \mathcal{C}_{[1,\infty)}$ and scales $(r_n)_n \subset \mathbb{R}$ satisfying*

$$\lim_{n \to \infty} t_n = \infty, \qquad \limsup_{n \to \infty} \frac{|\mathbf{x}_n|}{t_n} < 1, \qquad \lim_{n \to \infty} \frac{r_n}{t_n} = 0,$$

*such that the rescaled sequence of wave maps $(\phi^{(n)})_n$,*

$$\phi^{(n)}(t, \mathbf{x}) \stackrel{\text{def}}{=} \phi(t_n + r_n t, \mathbf{x}_n + r_n \mathbf{x}),$$

*converges strongly in $H^1_{loc}$ to a Lorentz transform of an entire harmonic map of nontrivial energy,*

$$\phi^{(\infty)} : \mathbb{R}^2 \to N, \qquad 0 < \|\phi^{(\infty)}\|_{\dot{H}^1(\mathbb{R}^2)} \leq \lim_{t \to \infty} \mathcal{E}_{D_t}[\phi]. \tag{4.10}$$

*(2) for each $\epsilon > 0$, there exist $t_0 > 1$ and a wave map extension of $\phi$, denoted also $\phi$,*

$$\phi : [t_0, \infty) \times \mathbb{R}^2 \to N,$$

*verifying*

$$\mathcal{E}[\phi] \leq (1 + \epsilon^8) \lim_{t \to \infty} \mathcal{E}_{D_t}[\phi]$$

*and*

$$\sup_{t \in [t_0, \infty)} \sup_{\lambda \in 2^{\mathbb{Z}}} \left( \|S_\lambda[\phi](t)\|_{L^\infty(\mathbb{R}^2)} + \lambda^{-1} \|S_\lambda[\partial_t \phi](t)\|_{L^\infty(\mathbb{R}^2)} \right) \leq \epsilon. \tag{4.11}$$

If the second scenario takes place, an application of Theorem 4.2 yields $\phi \in S$. Therefore, if scattering fails, then the first scenario must be true. This implies that the only candidates for which scattering might not hold are finite energy wave maps $\phi : \mathbb{R}^{2+1} \to N$ satisfying

$$\mathcal{E}[\phi] \geq \mathcal{E}(N).$$

As a consequence, one can amend Theorems 4.5 and 4.6 by replacing *"satisfying $\phi \in S((0,T))$ for all $T > 0$"* with *"satisfying $\phi \in S$"*.

The remainder of this chapter is entirely dedicated to the detailed exposition of arguments for Theorems 4.4 and 4.7.

## 4.3   Basic setup of the problem

In this section, we recall the problem to be studied, together with some if its important features, such as standard energy estimates, the finite speed of propagation property, and geometrical invariances. We also introduce the framework and notations to be used from this point forth.

Our main object of investigation is the evolution of the wave maps problem in $2 + 1$ dimensions with data prescribed at time $t = 1$. Thus, we study maps $\phi : (\mathbb{R}^{2+1}, \mathrm{diag}(-1,1,1)) \to (N, h)$ satisfying the system

$$\Box \phi^a + \Gamma^a_{bc}(\phi)\langle \nabla \phi^b, \nabla \phi^c \rangle = 0 \tag{4.12}$$

and the initial data condition

$$\phi(1, \mathbf{x}) := \phi_0(\mathbf{x}), \qquad \partial_t \phi(1, \mathbf{x}) := \phi_1(\mathbf{x}),$$

which will be referred to as *wave maps*. Later, it will also be important to consider static solutions for (4.12), i.e.,

$$\Delta \phi^a + \Gamma^a_{bc}(\phi)\langle \nabla \phi^b, \nabla \phi^c \rangle = 0, \tag{4.13}$$

and we will call such solutions *harmonic maps*. Since constant maps are static solutions with zero energy, we will call them *trivial solutions*. Hence, a *nontrivial harmonic map* means a harmonic map with positive energy.

There is an alternative way to write the evolution of (4.12). If we employ Nash's embedding of the manifold $N$ in the Euclidean space $\mathbb{R}^d$ (for some sufficiently high dimension $d$), i.e., $\phi : \mathbb{R}^{2+1} \to N \subset \mathbb{R}^d$, then $\phi$ satisfies the system

$$\Box \phi^a + \mathcal{S}^a_{bc}(\phi)\langle \nabla \phi^b, \nabla \phi^c \rangle = 0, \tag{4.14}$$

where $S$ is called *the shape operator* or *the second fundamental form* of $N$. The shape operator can be understood as a map

$$S : TN \times TN \to \mathrm{Nor}(N),$$

$$\langle S(X,Y), Z \rangle = \langle \partial_X Z, Y \rangle, \qquad \forall\, (X,Y,Z) \in TN \times TN \times \mathrm{Nor}(N),$$

where $TN$ and $\mathrm{Nor}(N)$ denote the tangent and the normal bundles, respectively.

Let us make some remarks here concerning this alternative formulation. In our context, we are dealing with classical solutions and, as a consequence, it does not really matter which formulation we adopt. However, one of the main ingredients in our argument, Theorem 4.2, is stated in connection to the evolution described by (4.14), since the projection onto frequency $\lambda$ is for the vector $\phi \in \mathbb{R}^d$. As a matter of choice, one can work with the formulation given by (4.12), but the projection has to be stated then in terms of a distance function $x \to d(\phi_0, \phi(x))$ with respect to an arbitrary point $\phi_0 \in N$. For the energy estimates to be used in the argument, we decided to state them in the context of (4.12), keeping in mind that they hold for either formulation.

Next, we fix our notational conventions and the working framework. For the Minkowski metric $g = \mathrm{diag}(-1,1,1)$, we write $t := x^0$ for the time variable, $\mathbf{x} := (x^1, x^2)$ for the spatial ones, and simply $x$ or $x^\mu$ for general spacetime coordinates. We also rely on the short-hand notation $\vec{\nabla} := (\nabla_1, \nabla_2)$ for the spatial gradient. In what concerns the target manifold $N$, which is Riemannian, we use the convention $\left|\nabla_\mu \phi\right|^2 := h_{ab} \nabla_\mu \phi^a \nabla_\mu \phi^b = \langle \nabla_\mu \phi, \nabla_\mu \phi \rangle$, thus omitting for simplicity the dependence of the norm on the metric $h$. Since the evolution of wave maps is time-reversible, we choose to work with the backwards-in-time one, from initial data prescribed at time $t = 1$. Spacetime translation and scaling invariances allow us to assume that a possible singularity occurs at the origin $(0,0,0)$, which is simply labeled 0. Constants denoted by $C$, $C_1$, and $C_2$ are universal and may change from line to line.

Following these initial considerations, let us focus next on the conservation of energy exhibited by wave maps and the important information that can be deduced from it, both of which play a crucial role in our analysis. From Chapter 1, we recall the energy-momentum tensor

$$T_{\mu\nu} = \langle \nabla_\mu \phi, \nabla_\nu \phi \rangle - \frac{1}{2} g_{\mu\nu} \langle \nabla \phi, \nabla \phi \rangle$$

and its corresponding differential identity $\nabla_\mu \{ T^\mu_\nu \} = 0$. As the Lagrangian of this problem is invariant under time translations, it follows that the timelike Killing vector field $N^\nu = (1, 0, 0)$ generates the energy differential identity $\nabla_\mu P^\mu = 0$, where $P^\mu := T^\mu_\nu N^\nu$. This can be written in the more familiar form

$$\nabla_t e(\phi) - \nabla_j p^j(\phi) = 0, \tag{4.15}$$

where

$$P_0 = e(\phi) \stackrel{\text{def}}{=} \frac{1}{2}\left(|\nabla_t \phi|^2 + |\vec{\nabla}\phi|^2\right) \text{ and } P_j = p_j(\phi) \stackrel{\text{def}}{=} \langle \nabla_t \phi, \nabla_j \phi \rangle \tag{4.16}$$

denote the energy and momenta densities, respectively. In this context, we notice that covariant derivatives are ordinary partial derivatives. Henceforth, we will make frequent use of the energy density.

We would like to integrate (4.15) over a certain region in spacetime and for this purpose we define solid frustums of cones and their boundaries:

$$\mathcal{C}^s_{[t_0,t_1]} \stackrel{\text{def}}{=} \left\{ (t, \mathbf{x}) \mid |\mathbf{x}| \leq t - s \,,\, t_0 \leq t \leq t_1 \right\} \tag{4.17}$$

is a solid frustum with boundary,

$$\partial\mathcal{C}^s_{[t_0,t_1]} \stackrel{\text{def}}{=} \left\{ (t, \mathbf{x}) \mid |\mathbf{x}| = t - s \,,\, t_0 \leq t \leq t_1 \right\} \tag{4.18}$$

is a lateral frustum, and

$$D^s_{t_1} \stackrel{\text{def}}{=} \left\{ (t, \mathbf{x}) \mid t = t_1 \,,\, |\mathbf{x}| \leq t_1 - s \right\} \tag{4.19}$$

is a top disk. They are all subsets of the future-oriented cone $\mathcal{C}^s$, which has its tip at the point $(s, \mathbf{0})$. Whenever it is convenient, in order to simplify the notation, we will write $\mathcal{C}$, $\partial\mathcal{C}$, and $D_t$ instead of $\mathcal{C}^0$, $\partial\mathcal{C}^0$, and $D^0_t$, respectively. If we use the standard notational conventions of $r := |\mathbf{x}|$ for the radial variable,

$$\nabla_\theta \stackrel{\text{def}}{=} x^1 \nabla_2 - x^2 \nabla_1$$

for the angular derivative, and

$$\nabla_r \stackrel{\text{def}}{=} (x^1 \nabla_1 + x^2 \nabla_2)/|\mathbf{x}|$$

for the radial derivative, then we arrive, after integrating (4.15) over the frustum $\mathcal{C}_{[t_0,t_1]}$, at

$$\int_{D_{t_1}} \{e(\phi)\} d\mathbf{x} - \int_{D_{t_0}} \{e(\phi)\} d\mathbf{x}$$

$$= \frac{1}{2\sqrt{2}} \int_{\partial\mathcal{C}_{[t_0,t_1]}} \left\{ |(\nabla_t + \nabla_r)\phi|^2 + \frac{1}{r^2}|\nabla_\theta \phi|^2 \right\} d\sigma \tag{4.20}$$

$$= \frac{1}{2} \int_{t_0 \leq |\mathbf{x}| \leq t_1} \left\{ |(\nabla_t + \nabla_r)\phi(|\mathbf{x}|, \mathbf{x})|^2 + \frac{1}{r^2}|\nabla_\theta \phi(|\mathbf{x}|, \mathbf{x})|^2 \right\} d\mathbf{x},$$

which is usually called in the literature *the local energy identity*.

For a smooth wave map as in the hypothesis of Theorem 4.4, we introduce its *local energy* and *lateral flux* (or energy stored on the characteristic cone $\mathcal{C}$) to be defined by

$$\mathcal{E}_{D_{t_1}}(\phi) \overset{\text{def}}{=} \int_{D_{t_1}} \{e(\phi(t_1))\}d\mathbf{x}, \tag{4.21}$$

$$\delta_F(t_1, \phi) \overset{\text{def}}{=} \frac{1}{2} \int_{|\mathbf{x}| \leq t_1} \left\{ \left| (\nabla_t + \nabla_r)\phi(|\mathbf{x}|, \mathbf{x}) \right|^2 \right.$$
$$\left. + \frac{1}{r^2} |\nabla_\theta \phi(|\mathbf{x}|, \mathbf{x})|^2 \right\} d\mathbf{x}, \tag{4.22}$$

for all $0 < t_1 \leq 1$. The local energy identity implies that $t \to \mathcal{E}_{D_t}(\phi)$ is an increasing function and, hence, it has a definite limit as $t \to 0$. This yields that $t \to \delta_F(t, \phi)$ is well-defined and tends to zero as $t \to 0$. We summarize these two facts as

$$\lim_{t \to 0} \mathcal{E}_{D_t}(\phi) \overset{\text{def}}{=} E < \infty \quad \text{and} \quad \lim_{t \to 0} \delta_F(t, \phi) = 0. \tag{4.23}$$

Thus, possible energy concentration is described by $E > 0$, whereas the lateral flux can be made arbitrarily small by choosing $t$ sufficiently small. Moreover, using the previous definitions, the local energy identity can be written like

$$\mathcal{E}_{D_{t_1}}(\phi) = \mathcal{E}_{D_{t_0}}(\phi) + \delta_F(t_1, \phi) - \delta_F(t_0, \phi), \quad 0 < t_0 \leq t_1 \leq 1, \tag{4.24}$$

and, by further taking $t_0 \to 0$, we obtain

$$\mathcal{E}_{D_{t_1}}(\phi) = E + \delta_F(t_1, \phi), \quad 0 < t_1 \leq 1. \tag{4.25}$$

Similarly, for a wave map as in the hypothesis of Theorem 4.7, we keep the same definition (4.21) for its local energy and adjust its lateral flux to be given now by

$$\delta_F(t_1, \phi) \overset{\text{def}}{=} \frac{1}{2} \int_{|\mathbf{x}| \geq t_1} \left\{ \left| (\nabla_t + \nabla_r)\phi(|\mathbf{x}|, \mathbf{x}) \right|^2 \right.$$
$$\left. + \frac{1}{r^2} |\nabla_\theta \phi(|\mathbf{x}|, \mathbf{x})|^2 \right\} d\mathbf{x}, \tag{4.26}$$

for all $1 \leq t_1 < \infty$. Like before, it follows that the function $t \to \mathcal{E}_{D_t}(\phi)$ is increasing and, according to (4.9), has a definite limit as $t \to \infty$, which we denote also by $E$ (i.e., the one appearing in (4.23)). As a consequence, the above lateral flux is well-defined and $\delta_F(t, \phi) \to 0$ as $t \to \infty$. We write this as[4]

$$\lim_{t \to \infty} \mathcal{E}_{D_t}(\phi) \overset{\text{def}}{=} E < \infty \qquad \text{and} \qquad \lim_{t \to \infty} \delta_F(t, \phi) = 0. \qquad (4.27)$$

In this setting, the identity (4.24) takes the form

$$\mathcal{E}_{D_{t_1}}(\phi) = \mathcal{E}_{D_{t_0}}(\phi) + \delta_F(t_0, \phi) - \delta_F(t_1, \phi), \qquad 1 \le t_0 \le t_1 < \infty, \qquad (4.28)$$

whereas (4.25) changes to

$$E = \mathcal{E}_{D_{t_0}}(\phi) + \delta_F(t_0, \phi), \qquad 1 \le t_0 < \infty. \qquad (4.29)$$

The finite speed of propagation property implies that the solution inside the solid cone $\mathcal{C}_{(0,1]}$ is completely determined by the initial data $\phi_0 = \phi_0(\mathbf{x})$ and $\phi_1 = \phi_1(\mathbf{x})$ prescribed locally for $|\mathbf{x}| \le 1$. However, Theorem 4.2 considers the global problem and, because of this feature, we need to consider data defined for all $\mathbf{x} \in \mathbb{R}^2$.

The scaling invariance (i.e., if $\phi = \phi(x)$ is a solution to (4.12) or (4.14), then so is $\phi_\lambda = \phi(\lambda x)$) and the criticality of the problem entail that the energy of a scaled solution remains unchanged. We will later use this simple property to create from one wave map a sequence of rescaled wave maps $\{\phi^{(n)}\}$, with unchanged energy, whose limit, up to a subsequence, will be important in the analysis.

## 4.4  Outline of the argument proving Theorems 4.4 and 4.7

The main idea of the proof is to combine the results of Theorems 4.2 ("energy dispersion implies regularity") and 4.3 ("small energy implies regularity") with certain a priori estimates. In this section, we believe that it is helpful to provide the reader with a brief overview for the strategy used in the proof, together with a rough guide to the main steps of the argument. These are as follows:

### 4.4.1  *Extension and scaling (first step)*

We start by considering a wave map $\phi$ verifying the hypothesis of Theorem 4.4 (respectively Theorem 4.7), but which violates the second scenario with

---

[4]One reason for keeping the same notation here as in the previous case, both for the lateral flux and the definite limit of the local energy functional, is that the argument for Theorem 4.7 follows in the footsteps of the one for Theorem 4.4, with obvious modifications.

a threshold $\epsilon > 0$ that may be chosen sufficiently small. For this map, we construct an extension outside of the characteristic cone, whose energy in this region is small, of the order $\epsilon^8 E$. As a consequence, we deduce that the energy dispersion condition (4.8) (respectively (4.11)) does not hold, i.e., $E_D(\phi) > \epsilon$.

The extension construction sets out the two arguments (i.e., for Theorems 4.4 and 4.7, respectively) on quite similar paths and the careful reader can fill in the details for one of them, if he or she has a thorough understanding of the other. This is why, from here on, we will exclusively focus on the proof of the blow-up result.

The failure of (4.8) implies that, for a sequence of points $(t_n, \mathbf{x}_n)$ and corresponding frequencies $\lambda_n$, we have

$$\left|S_{\lambda_n}[\phi](t_n, \mathbf{x}_n)\right| + \lambda_n^{-1}\left|S_{\lambda_n}[\partial_t\phi](t_n, \mathbf{x}_n)\right| > \epsilon \tag{4.30}$$

with $t_n \to 0$. In a certain part of the argument, it will be easier to work with rescaled versions of $\phi$ in which $t_n$ shifts into 1. This is done by defining

$$\phi^{(n)}(x) \stackrel{\text{def}}{=} \phi(t_n x), \qquad x \in \mathcal{C}_{(0, t_n^{-1}]},$$

which produces a sequence of wave maps $\{\phi^{(n)}\}$ satisfying

$$\left|S_{\lambda_n t_n}[\phi^{(n)}](1, \mathbf{x}_n/t_n)\right| + (\lambda_n t_n)^{-1}\left|S_{\lambda_n t_n}[\partial_t\phi^{(n)}](1, \mathbf{x}_n/t_n)\right| > \epsilon.$$

Thus, relabeling $\lambda_n t_n$ by $\lambda_n$ and $\mathbf{x}_n/t_n$ by $\mathbf{x}_n$, we can write

$$\left|S_{\lambda_n}[\phi^{(n)}](1, \mathbf{x}_n)\right| + \lambda_n^{-1}\left|S_{\lambda_n}[\partial_t\phi^{(n)}](1, \mathbf{x}_n)\right| > \epsilon. \tag{4.31}$$

### 4.4.2    *Energy-type estimates (second step)*

In this step, we obtain energy-type bounds that are relevant for discussing concentration/non-concentration of energy scenarios. We consider a wave map $\phi$ in $\mathcal{C}_{(0,1]}$ and choose two positive parameters, $\epsilon_1 \ll 1$ and $\gamma_1$, such that $1 - \gamma_1^2 \approx \epsilon_1^6$. Hence, $\gamma_1 \approx 1$. Due to (4.23), one can find $\tilde{t} > 0$ satisfying

$$\sup_{t \in (0,\tilde{t}]} \delta_F(t, \phi) \ll \epsilon_1^{9/2}.$$

For each $t_1 \in (0, \tilde{t}]$, we will derive two a priori energy-type estimates, the first of which is

$$\int_{\gamma t_1 \leq |\mathbf{x}| \leq t_1} \left\{\left|(\nabla_t + \nabla_r)\phi(t_1)\right|^2 + \frac{1}{r^2}\left|\nabla_\theta \phi(t_1)\right|^2\right\} d\mathbf{x}$$

$$\leq C(1 - \gamma^2)^{1/2}\left(\epsilon_1^{3/2} + \frac{1}{T}\int_0^T \left\{\int_{|\mathbf{x}| \leq \gamma_1 t} e(\phi(t))d\mathbf{x}\right\} dt\right), \tag{4.32}$$

where $0 \leq \gamma \leq \gamma_1$ is arbitrary and

$$T = \frac{t_1 \delta_F(t_1, \phi)}{(1 - \gamma_1^2)^{3/4} E}.$$

The second bound takes the form

$$\int_{|\mathbf{x}| \leq t_1} \left\{ \sqrt{1 - \frac{r^2}{t_1^2}} \, e(\phi(t_1)) \right\} dx + \int_{\mathcal{R}_{\rho_1}^{t_1}} \left\{ \frac{1}{\rho} |X(\phi)|^2 \right\} dt \, d\mathbf{x}$$

$$\leq C \left( \epsilon_1^{3/2} + \frac{1}{T} \int_0^T \left\{ \int_{|\mathbf{x}| \leq \gamma_1 t} e(\phi(t)) dx \right\} dt \right), \tag{4.33}$$

where the domain of integration for the second integral is given by

$$\mathcal{R}_{\rho_1}^{t_1} \overset{\text{def}}{=} \{(t, \mathbf{x}) \mid \rho \geq \rho_1, \, t \leq t_1\}, \qquad \rho_1 \overset{\text{def}}{=} \frac{t_1 \delta_F(t_1, \phi)}{(1 - \gamma_1^2)^{1/4} E},$$

the weight $\rho$ and the vector field $X$ are defined by

$$\rho \overset{\text{def}}{=} \sqrt{-\langle x, x \rangle} = \sqrt{t^2 - |\mathbf{x}|^2}, \qquad X \overset{\text{def}}{=} \rho^{-1} x^\mu \nabla_\mu,$$

and $T$ is as before.

Moreover, due to (4.23), we also observe that

$$\frac{1}{T} \int_0^T \left\{ \int_{|\mathbf{x}| \leq \gamma_1 t} e(\phi(t)) dx \right\} dt \leq E + o(1), \tag{4.34}$$

where $o(1)$ designates a function of $T$ which has limit zero when $T \to 0$.

Subsequently, we distinguish two complementary cases:

(1) *non-concentration of energy inside a cone of aperture less than 1*, by which we mean that

$$\sup_{T \in (0,1]} \frac{1}{T} \int_0^T \left\{ \int_{|\mathbf{x}| \leq \gamma_1 t} e(\phi(t)) dx \right\} dt \leq \epsilon_1^{3/2}.$$

Choosing $\gamma < \gamma_1$ to satisfy $1 - \gamma^2 = \epsilon_1$, it follows, according to (4.32) and (4.33), that

$$\int_{\gamma t_1 \leq |\mathbf{x}| \leq t_1} \left\{ |(\nabla_t + \nabla_r)\phi(t_1)|^2 + \frac{1}{r^2} |\nabla_\theta \phi(t_1)|^2 \right\} dx$$

$$\leq C(1 - \gamma^2)^{1/2} \epsilon_1 \tag{4.35}$$

and

$$\int_{|\mathbf{x}| \leq \gamma t_1} \{ e(\phi(t_1)) \} dx \leq C \epsilon_1. \tag{4.36}$$

(2) *energy concentration inside a cone of aperture less than 1*, i.e., there exists a sequence of times $T_n \to 0$ with

$$\frac{1}{T_n} \int_0^{T_n} \left\{ \int_{|\mathbf{x}| \leq \gamma_1 t} e(\phi(t)) d\mathbf{x} \right\} dt > \epsilon_1^{3/2} \overset{\text{def}}{=} E_2 > 0. \qquad (4.37)$$

First, due to (4.34), we infer $\epsilon_1^{3/2} < CE$. Secondly, using the mean value theorem, we deduce the existence of a sequence $\tilde{t}_n \to 0$ satisfying

$$\int_{|\mathbf{x}| \leq \gamma_1 \tilde{t}_n} \left\{ e(\phi(\tilde{t}_n)) \right\} d\mathbf{x} > \epsilon_1^{3/2}. \qquad (4.38)$$

Finally, choosing $n$ sufficiently large, we also have

$$E \leq \mathcal{E}_{D_t}(\phi) \leq (1 + \epsilon^8) E, \qquad \forall 0 < t \leq \tilde{t}_n.$$

### 4.4.3 *Elimination of the null concentration scenario (third step)*

Here, the goal is to show that the first of the two complementary cases discussed in the second step is incompatible with $E_D[\phi] \geq \epsilon$, which represents the failure of energy dispersion assumed in the first step of the argument. For that, we calibrate the parameter $\epsilon_1$ in the second step to $\epsilon_1 \approx \epsilon^8$, with $\epsilon$ being the one in (4.30).

It is clear that if we are in the non-concentration case of the second step, then the energy of $\phi$ must concentrate near the characteristic cone, that is, in the region $\{\gamma t_1 \leq |\mathbf{x}| \leq t_1\}$. Moreover, according to (4.35), the angular and radial $\nabla_t + \nabla_r$ derivatives of $\phi$ are small in the same region. Thus, for $\phi^{(n)}$ (i.e., the rescaled version of $\phi$ constructed in the first step) we have, at time $t = 1$, the following estimates:

$$\int_{|\mathbf{x}| \leq 1} \left\{ e(\phi^{(n)}) \right\} d\mathbf{x} \approx E, \qquad \int_{|\mathbf{x}| \geq 1} \left\{ e(\phi^{(n)}) \right\} d\mathbf{x} \leq C\epsilon^8 E, \qquad (4.39)$$

$$\int_{|\mathbf{x}| \leq \gamma} \left\{ e(\phi^{(n)}) \right\} d\mathbf{x} \leq C\epsilon_1, \qquad (4.40)$$

$$\int_{\gamma \leq |\mathbf{x}| \leq 1} \left\{ |(\nabla_t + \nabla_r)\phi^{(n)}|^2 + \frac{1}{r^2} |\nabla_\theta \phi^{(n)}|^2 \right\} d\mathbf{x} \leq C\epsilon_1^2. \qquad (4.41)$$

Next, we focus on the energy dispersion condition (4.31) satisfied by $\phi^{(n)}$, which, for clarity purposes, is replaced by the simpler inequality
$$\left| S_{\lambda_n}[\phi^{(n)}](1, \mathbf{x}_n) \right| > \epsilon,$$
where
$$S_\lambda[\phi^{(n)}](1, \mathbf{x}) = \int \lambda^2 \, g\big(\lambda(\mathbf{x} - \mathbf{y})\big) \phi^{(n)}(1, \mathbf{y}) \, dy$$
and $g \in \mathcal{S}(\mathbb{R}^2)$ has its Fourier transform $\widehat{g} = \widehat{g}(\xi)$ supported on $\{1/2 \leq |\xi| \leq 2\}$. Subsequently, we construct a vector field $\mathbf{u}$ satisfying $\operatorname{div} \mathbf{u} = g$, that allows us to integrate by parts in the previous equality and derive
$$S_\lambda[\phi^{(n)}](1, \mathbf{x}) = \int \lambda u^j \big(\lambda(\mathbf{x} - \mathbf{y})\big) \nabla_j \phi^{(n)}(1, \mathbf{y}) \, dy.$$
This is done in order to relate, for any frequency $\lambda$, $S_\lambda[\phi^{(n)}](1, \mathbf{x})$ to the physical energy of $\phi^{(n)}$ through an application of the Cauchy-Schwarz inequality. A crucial observation is that one can take advantage of a gauge freedom in solving $\operatorname{div} \mathbf{u} = g$, in the sense that $\mathbf{u}$ *is unique modulo curl terms*. Hence, we can choose $\mathbf{u}$ to point in certain preferential directions for which we have good control through (4.39)-(4.41). The end result is an inequality of the form $\epsilon \leq C\epsilon^{9/8}$, that leads to a contradiction for $\epsilon$ sufficiently small.

### 4.4.4 Uniform time propagation of nontrivial energy (fourth step)

Based on the analysis done so far, we are left to address, under the failure of energy dispersion, the energy concentration case in the second step of the argument. Interestingly enough, in this part of the proof, we will not use at all the non-dispersion bound (4.30).

The estimate (4.38) describes the *trapping* for $\phi$ of a nontrivial amount of physical energy at time $t = \tilde{t}_n$, inside the disk $\{|\mathbf{x}| \leq \gamma_1 \tilde{t}_n\}$. Using this fact and a propagation result contained in Lemma 4.5, we show that a nontrivial amount of energy remains trapped inside a slightly larger cone $|\mathbf{x}| \leq \gamma_2 t$, for considerably smaller times. Namely, we prove that
$$\int_{|\mathbf{x}| \leq \gamma_2 t} \left\{ e\big(\phi(t)\big) \right\} dx \geq \tilde{E}_2 > 0, \qquad \forall t \in \left[ \delta_n^{1/2} \tilde{t}_n, \delta_n^{1/4} \tilde{t}_n \right],$$
where $\tilde{E}_2 > 0$ and $\delta_n \to 0$ as $n \to \infty$. Moreover, we can arrange for
$$\int_{\mathcal{C}^{\delta_n \tilde{t}_n}_{\left[\delta_n^{1/2} \tilde{t}_n, \delta_n^{1/4} \tilde{t}_n\right]}} \left\{ \frac{1}{\rho} |X(\phi)|^2 \right\} dt \, dx \leq CE$$

to hold too, which yields decay of the kinetic energy. This can be seen by subdividing $\left[\delta_n^{1/2}\tilde{t}_n, \delta_n^{1/4}\tilde{t}_n\right]$ into a large collection of intervals of the form $[t_{j,n}, T_n t_{j,n}]$, with $T_n \to \infty$, and then using the pigeonhole principle to find one such interval for which

$$\int_{\mathcal{C}^{\delta_n}_{[t_{j,n}, T_n t_{j,n}]}} \left\{\frac{1}{\rho}|X(\phi)|^2\right\} dt\, d\mathbf{x} \leq CE|\log(\delta_n)|^{-1/2}.$$

### 4.4.5　Final rescaling (fifth step)

In this step, we perform a rescaling procedure to shift $t_{j,n}$ to 1. Thus, we produce a wave map denoted by $\phi^{(n)}$, which lives inside the cone $\mathcal{C}_{[1,T_n]}$, where

$$T_n \approx e^{|\log(\delta_n)|^{1/2}} \to \infty,$$

and satisfies

$$\mathcal{E}_{D_t}(\phi^{(n)}) \approx E, \qquad \forall t \in [1, T_n],$$

$$\mathcal{E}_{D_t^{(1-\gamma_2)t}}(\phi^{(n)}) \geq E_2, \qquad \forall t \in [1, T_n],$$

$$\int_{\mathcal{C}^{\delta_n^{1/2}}_{[1,T_n]}} \left\{\frac{1}{\rho}|X(\phi^{(n)})|^2\right\} dt\, d\mathbf{x} \leq C|\log(\delta_n)|^{-1/2} \to 0.$$

In the above, by analogy with (4.21), we used the notation

$$\mathcal{E}_A(\phi) \stackrel{\text{def}}{=} \int_{(t,\mathbf{x})\in A} \{e(\phi(t))\}\, d\mathbf{x}, \tag{4.42}$$

where $A$ is an arbitrary subset of the time slice $\{t\} \times \mathbb{R}^2$.

### 4.4.6　Isolating concentration scales (sixth step)

In the final step of the argument we will rely on a compactness result, which applies only to wave maps with energy lower than $E_0$ (i.e., the perturbation energy introduced by Theorem 4.3). In preparation for this, we discuss here the possible scenarios that yield such maps, which follow as a result of further reasonings based on the pigeonhole principle.

　　The first such scenario is called *energy concentration at small scales*. It is described by the existence of a subsequence for the sequence of wave maps constructed in the previous step, which we still denote by $\{\phi^{(n)}\}$, and of a corresponding sequence of points $(t_n, \mathbf{x}_n) \to (t_0, \mathbf{x}_0) \in \mathcal{C}_{[1/2,\infty)}^{1/2}$ and shrinking scales $r_n \to 0$, such that

$$\mathcal{E}_{D_{r_n}^{t_n}(\mathbf{x}_n)}(\phi^{(n)}) = \frac{E_0}{10},$$

$$\mathcal{E}_{D_{r_n}^{t_n}(\mathbf{x})}(\phi^{(n)}) \le \frac{E_0}{10}, \qquad \forall \mathbf{x} \in D_r(\mathbf{x}_n),$$

and

$$\frac{1}{r_n} \int_{t_n - \frac{r_n}{2}}^{t_n + \frac{r_n}{2}} \int_{D_r(\mathbf{x}_n)} \left\{ \left| X(\phi^{(n)}) \right|^2 \right\} d\mathbf{x} \, dt \to 0 \quad \text{as} \quad n \to \infty,$$

where $D_r(\mathbf{x})$ is the closed disk of radius $r$ centered at $\mathbf{x}$ and, in the second and third estimates, $r > 0$ is independent of $n$. The first equation can be interpreted as *energy concentration at small scales*, the second one can be seen to describe *comparable energy nearby*, whereas the limit points to *decay of kinetic energy*.

The other possible scenario is called *non-concentration of uniform energy*. It is described, for each $j \in \mathbb{N}$, by the existence of $r_j > 0$ for which we have:

$$\mathcal{E}_{D_{r_j}^t(\mathbf{x})}(\phi^{(n)}) \le \frac{E_0}{10}, \qquad \forall (t, \mathbf{x}) \in \mathcal{C}_j^1,$$

$$\mathcal{E}_{D_t^{(1-\gamma_2)t}}(\phi^{(n)}) \ge E_2, \qquad \forall t \ge 1,$$

and

$$\int_{\mathcal{C}_j^1} \left\{ \left| X(\phi^{(n)}) \right|^2 \right\} dt \, d\mathbf{x} \to 0 \quad \text{as} \quad n \to \infty,$$

where $\mathcal{C}_j^1 := \mathcal{C}_{[1,\infty)}^1 \cap [2^j, 2^{j+1}] \times \mathbb{R}^2$ is a dyadic frustum. If the first bound expresses *non-concentration of energy at scales* $r_j$, the second one describes *nontrivial overall energy* with the limit indicating *decay of kinetic energy*, as before.

### 4.4.7 The compactness argument (final step)

In order to conclude the proof of Theorem 4.4, we derive a compactness result, contained in Lemma 4.8, which permits us to further analyze the two scenarios of the previous step.

For the first one, we perform a final rescaling,

$$\widetilde{\phi}^{(n)}(t, \mathbf{x}) \stackrel{\text{def}}{=} \phi^{(n)}(t_n + r_n t, \mathbf{x}_n + r_n \mathbf{x}),$$

and, based on the compactness result, we show that on a subsequence we have strong convergence in $\dot{H}^1$ to a wave map $\phi^\infty$, which is defined on $[-1/2, 1/2] \times \mathbb{R}^2$. We also obtain $X(t_0, \mathbf{x}_0)\phi^\infty = 0$, which implies that, after taking a Lorentz transformation, $\phi^\infty$ becomes a harmonic map. Using the fact that $r_n \to 0$, we infer that the original wave map blows up.

For the second scenario, we can get by without the rescaling procedure and, using again Lemma 4.8, we deduce that a subsequence of $\{\phi^{(n)}\}$ converges strongly in $\dot{H}^1(\mathbb{R}^2)$ to a wave map $\phi^\infty$. This satisfies now $X(\phi^\infty) = 0$, which implies that it is locally self-similar and, as a consequence, can be seen as a harmonic map in a portion of the hyperbolic plane, $\Omega \subset \mathbb{H}^2$. We can then fill in the entire hyperbolic plane with matching harmonic maps and thus obtain an instance which will be ruled out by the results of Lemma 4.1. Thus, the second scenario is not viable.

In conclusion, we obtain that, assuming the failure of the energy dispersion condition (4.8), the wave map $\phi$ in Theorem 4.4 behaves according to the first scenario in the same theorem, which describes a particular way that a wave map might blow up. Various instances of blow-up phenomena associated to wave maps will be addressed in the final chapter of the book.

## 4.5    The extension procedure

We begin in earnest the proof of Theorems 4.4 and 4.7 by discussing the extension construction mentioned in the previous section. Thus, for a $C^\infty$ map $\phi$ defined on $\mathcal{C}_{(0,1]}$ (respectively $\mathcal{C}_{[1,\infty)}$) and for a small parameter $\epsilon > 0$, we need to construct an extension of $\phi$ defined on $(0, t_0] \times \mathbb{R}^2$ (respectively $[t_0, \infty) \times \mathbb{R}^2$), with $t_0$ depending on both $\phi$ and $\epsilon$, which satisfies[5]

$$\mathcal{E}(\phi) \leq (1 + \epsilon^8) E,$$

where

$$\mathcal{E}(\phi) \stackrel{\text{def}}{=} \mathcal{E}(\phi)(t) = \frac{1}{2} \int\limits_{\mathbb{R}^2} \left\{ \left| \nabla_t \phi(t) \right|^2 + \left| \vec{\nabla} \phi(t) \right|^2 \right\} d\mathbf{x} \qquad (4.43)$$

is the total energy of $\phi$. This is needed in order for us to be able to use Theorem 4.2, which applies to nonlocalized wave maps.

---

[5]In the scattering case, the corresponding estimate is identical due to the notation in (4.27).

### 4.5.1    *The extension procedure for the blow-up result*

For a wave map as in the hypothesis of Theorem 4.4, its restriction to the characteristic cone $\partial \mathcal{C}_{(0,1]}$ can be written using polar coordinates like

$$\phi(x) = \phi(t, \mathbf{x}) = \phi(r, r, \theta), \qquad 0 < t = |\mathbf{x}| = r \leq 1.$$

Due to (4.22), the angular part of the flux satisfies

$$\frac{1}{2} \int_0^t \int_0^{2\pi} \left\{ \frac{1}{r} |\nabla_\theta \phi(r, r, \theta)|^2 \right\} d\theta \, dr \leq \delta_F(t, \phi), \qquad \forall \, 0 < t \leq 1.$$

Using the decay of flux in (4.23), it follows that we can pick $t_1 = t_1(\phi, \epsilon)$ sufficiently small such that

$$\int_0^{t_1} \int_0^{2\pi} \left\{ \frac{1}{r} |\nabla_\theta \phi(r, r, \theta)|^2 \right\} d\theta \, dr \leq 2\delta_F(t_1, \phi) \leq \epsilon^8 E.$$

Hence, relying on the mean value theorem, we derive the existence of $0 < t_0 < t_1$ such that

$$\int_0^{2\pi} \left\{ |\nabla_\theta \phi(t_0, t_0, \theta)|^2 \right\} d\theta \leq \frac{t_1}{t_0} \int_0^{2\pi} \left\{ |\nabla_\theta \phi(t_0, t_0, \theta)|^2 \right\} d\theta \leq \epsilon^8 E.$$

We claim that $\phi$ can then be extended to a wave map defined on $(0, t_0] \times \mathbb{R}^2$.

To achieve this, we first rescale $\phi$ to shift $t_0$ to 1 and thus work with

$$\int_0^{2\pi} \left\{ |\nabla_\theta \phi(1, 1, \theta)|^2 \right\} d\theta \leq \epsilon^8 E, \tag{4.44}$$

instead of the previous bound. An application of the Cauchy-Schwarz inequality shows that this estimate implies

$$\|\phi(1, 1, \cdot)\|_{\dot{C}_\theta^{1/2}([0, 2\pi])} \leq C\epsilon^4 \sqrt{E}.$$

As a consequence, we can find a fixed point $\Phi_0 \in N$ (as a reminder, $N$ is the target manifold) such that

$$\|\phi(1, 1, \cdot) - \Phi_0\|_{L_\theta^\infty([0, 2\pi])} \leq C\epsilon^4 \sqrt{E}.$$

For $\phi(1, \mathbf{x}) = \phi(1, r, \theta)$ originally defined for $r \leq 1$, let us consider its extension to the region $r > 1$ given by

$$\phi(1, r, \theta) \overset{\text{def}}{=} \begin{cases} (2 - r)\phi(1, 1, \theta) + (r - 1)\Phi_0, & \text{if } 1 < r \leq 2, \\ \Phi_0, & \text{if } r > 2. \end{cases}$$

For the time derivative $\nabla_t \phi(1, \mathbf{x})$, we simply set $\nabla_t \phi(1, \mathbf{x}) = 0$ if $r > 1$. We use now the original notation and denote $(\phi(1, \mathbf{x}), \nabla_t \phi(1, \mathbf{x}))$ by $(\phi_0, \phi_1)$. It is easy to see that

$$\frac{1}{2} \int_{\mathbb{R}^2} \left\{ |\phi_1|^2 + |\vec{\nabla} \phi_0|^2 \right\} dx \leq (1 + C\epsilon^8) E \tag{4.45}$$

and

$$\frac{1}{2} \int\limits_{|\mathbf{x}|>1} \left\{ |\phi_1|^2 + |\vec{\nabla}\phi_0|^2 \right\} d\mathbf{x} \leq C\epsilon^8 E, \qquad (4.46)$$

as both the radial and angular components of the energy for $\phi_0$ are small in the outside region, the former due to $\phi(1, 1, \theta)$ being close to $\Phi_0$.

With $(\phi_0, \phi_1)$ as initial data at time $t = 1$, we solve the wave maps equation (4.14) backwards in time. Inside the characteristic cone, the solution will coincide with the original wave map. The question is whether this solution survives outside the cone all the way to $t = 0$. To answer it, we integrate the energy differential identity (4.15) over the spacetime region $\{(t, \mathbf{x}) \mid |\mathbf{x}| \geq t , s_0 \leq t \leq s_1\}$, which is inside the solution's domain of existence, to derive

$$\mathcal{E}(\phi)(s_0) - \mathcal{E}_{D_{s_0}}(\phi) = \mathcal{E}(\phi)(s_1) - \mathcal{E}_{D_{s_1}}(\phi) + \delta_F(s_1, \phi) - \delta_F(s_0, \phi). \quad (4.47)$$

If we choose $s_1 = 1$, then, on the bases of (4.46) and the flux being of the order of $\epsilon^8 E$, it follows that, on each time slice, the energy of the solution outside of the cone is uniformly bounded above by $C\epsilon^8 E$. Hence, for $\epsilon$ sufficiently small depending on $E_0$, we can invoke the small data result (i.e., Theorem 4.3) to infer that the solution can be continued outside of the cone up to $t = 0$. For the unscaled version of the original map, this means that we have constructed an extension that lives in the region $(0, t_0] \times \mathbb{R}^2$, which will be our working object from here on.

### 4.5.2    *The extension procedure for the scattering result*

In this context, we are dealing with a wave map $\phi$ defined on $\mathcal{C}_{[1,\infty)}$, for which

$$\lim_{t \to \infty} \mathcal{E}_{D_t}(\phi) = E < \infty. \qquad (4.48)$$

We are following a similar approach to the blow-up case, in the sense that we would like first to extend $\phi$ to a full time slice $\{T\} \times \mathbb{R}^2$, with some $T > 1$, such that the energy of the extension outside the disk $D_T(0)$ is small. However, in this case, if we try and solve the wave maps equation forward in time, with the extension as initial data at time $t = T$, there is the possibility that the energy outside the disk will enter the region $\mathcal{C}_{[T,\infty)}$ at later times, thus contaminating the solution. Ideally, one would like to prescribe only outgoing data outside the disk; however, this is not easy to do and we need to go through a limiting procedure in order to achieve this goal.

We start by deriving from (4.26) that

$$\frac{1}{2} \int_t^\infty \int_0^{2\pi} \left\{ \frac{1}{r} \left| \nabla_\theta \phi(r,r,\theta) \right|^2 \right\} d\theta \, dr \le \delta_F(t,\phi), \qquad \forall \, 1 \le t < \infty.$$

Using now (4.27), we choose $t_0 = t_0(\phi, \epsilon)$ sufficiently large such that

$$\int_{t_0}^\infty \int_0^{2\pi} \left\{ \frac{1}{r} \left| \nabla_\theta \phi(r,r,\theta) \right|^2 \right\} d\theta \, dr \le 2\delta_F(t_0, \phi) \le \epsilon^8 E.$$

We claim that $\phi$ can be extended to a wave map defined on $[t_0, \infty) \times \mathbb{R}^2$. The convergence of the last integral implies the existence of a sequence $t_k \to \infty$, for which we have

$$\int_0^{2\pi} \left\{ \left| \nabla_\theta \phi(t_k, t_k, \theta) \right|^2 \right\} d\theta \to 0.$$

Next, this limit and the extension procedure from the blow-up case allow us to extend, for each $k$, the map $\mathbf{x} \mapsto (\phi(t_k, \mathbf{x}), \nabla_t \phi(t_k, \mathbf{x}))$ from the disk $D_{t_k}(0)$ to a map $\mathbf{x} \mapsto (\phi_0^{(k)}(\mathbf{x}), \phi_1^{(k)}(\mathbf{x}))$ defined on $\mathbb{R}^2$, such that

$$\frac{1}{2} \int_{\mathbb{R}^2} \left\{ |\phi_1^{(k)}|^2 + |\vec{\nabla} \phi_0^{(k)}|^2 \right\} d\mathbf{x} - \mathcal{E}_{D_{t_k}}(\phi) \to 0, \tag{4.49}$$

as $t_k \to \infty$. With data $(\phi_0^{(k)}, \phi_1^{(k)})$ prescribed at time $t = t_k$, we solve the wave maps equation backwards in time and call this solution $\phi^{(k)}$, which is obviously an extension of $\phi$. If $\phi^{(k)}$ is defined at time $t \in [t_0, t_k]$, then (4.28) implies that

$$\begin{aligned} \mathcal{E}(\phi^{(k)})(t) - \mathcal{E}_{D_t}(\phi^{(k)}) \\ = \mathcal{E}(\phi^{(k)})(t_k) - \mathcal{E}_{D_t}(\phi) + \delta_F(t, \phi) - \delta_F(t_k, \phi) \\ \le C\epsilon^8 E \end{aligned} \tag{4.50}$$

holds for $k$ sufficiently large. Therefore, for an $\epsilon$ sufficiently small depending on $E_0$, we can apply Theorem 4.3 and argue that $\phi^{(k)}$ is an extension for $\phi$ on $[t_0, t_k] \times \mathbb{R}^2$. Moreover, due to (4.181) and (4.49), we have that

$$\mathcal{E}(\phi^{(k)}) \to E, \tag{4.51}$$

as $t_k \to \infty$.

We are going to construct the extension of $\phi$ to $[t_0, \infty) \times \mathbb{R}^2$, denoted also by $\phi$, as the strong limit of a subsequence of $\{\phi^{(k)}\}$ in the energy norm

$$\Psi \mapsto \|\Psi\|_{L^\infty([t_0,\infty); \dot{H}^1(\mathbb{R}^2))} + \|\nabla_t \Psi\|_{L^\infty([t_0,\infty); L^2(\mathbb{R}^2))}.$$

We start by using (4.51) to infer that, for a fixed time $T > t_0$, the sequence[6] $\{(\phi^{(k)}(T), \nabla_t \phi^{(k)}(T))\}$ is uniformly bounded in $\dot{H}^1 \times L^2(\mathbb{R}^2)$. Due to the Banach-Alaoglu theorem, it follows that this sequence has a weakly convergent subsequence and its limit will be the extension of the original map $\phi$ to the time slice $\{T\} \times \mathbb{R}^2$.

As all $\phi^{(k)}$ coincide on the interior of $\mathcal{C}_{[t_0,\infty)}$ (being equal to $\phi$), it follows that the only relevant information from the previous convergence refers to the outside of the cone. There, all $\phi^{(k)}$ have small energy due to (4.50) and one can rely on weak stability bounds from the small energy theory (e.g., (4.3) in Theorem 4.3) to deduce that the extension obtained above is a regular wave map on $[t_0, \infty) \times \mathbb{R}^2$, with finite energy.

Using the weak convergence and (4.51), we derive that $\mathcal{E}(\phi) \leq E$. On the other hand, we have $\mathcal{E}(\phi) \geq E$ due to (4.181) and the obvious estimate $\mathcal{E}(\phi) \geq \mathcal{E}_{D_t}(\phi)$. Thus, we obtain

$$\mathcal{E}(\phi) = E = \lim_{k \to \infty} \mathcal{E}(\phi^{(k)}),$$

which, combined with the weak convergence, yields the pointwise convergence of the subsequence to $\phi$ in the $\dot{H}^1 \times L^2(\mathbb{R}^2)$ norm. We can further upgrade pointwise convergence to the strong convergence claimed at the beginning of the construction by applying the continuity of the energy in Theorem 4.3 outside the cone $\mathcal{C}_{[t_0,\infty)}$.

As mentioned in the outline of the argument, with this extension in hand, the interested reader can check that the proof of the scattering result follows exactly the same steps as the one for its blow-up counterpart, i.e., Theorem 4.4.

## 4.6   Generalized energy estimates and applications

In this section, we focus on deriving the energy-type estimates needed in our analysis. These are all extensions of the local energy identity (4.24), which is obtained by integrating (4.15) over frustums of the characteristic cone.

---

[6]As $\phi^{(k)}$ is defined on $[t_0, t_k]$ and $t_k \to \infty$, we have that $\phi^{(k)}(T)$ is well-defined only for $k$ sufficiently large, which is enough for our purposes.

### 4.6.1 *Energy-type integral identity on hyperboloids*

Using the same differential identity as before, we will first estimate the energy of a wave map which is stored on a hyperboloid. This will prove useful in several tasks, one of which being the exclusion of nontrivial, finite energy[7], self-similar wave maps in the context of Theorem 4.4.

For this purpose, let us recall from the outline section the hyperbolic radial coordinate

$$\rho = \rho(x) \overset{\text{def}}{=} \sqrt{-\langle x, x \rangle} = \sqrt{t^2 - |\mathbf{x}|^2}, \qquad (4.52)$$

which allows us to define the solid hyperboloid

$$\mathcal{R}_{\rho_0}^{t_1} \overset{\text{def}}{=} \{(t, \mathbf{x}) \mid \rho_0 \le \rho, \ t \le t_1\} \subseteq C_{[\rho_0, t_1]}^{\rho_0}, \qquad (4.53)$$

with fixed $0 < \rho_0 < t_1$. Its boundary consists of a round surface

$$\mathcal{H}_{\rho_0}^{t_1} \overset{\text{def}}{=} \{(t, \mathbf{x}) \mid \rho = \rho_0, \ t \le t_1\} \qquad (4.54)$$

and a flat top

$$\widetilde{D}_{t_1}^{\rho_0} \overset{\text{def}}{=} \{(t, \mathbf{x}) \mid t = t_1, \ |\mathbf{x}|^2 \le t_1^2 - \rho_0^2\}. \qquad (4.55)$$

Using the momenta densities defined in (4.16), we introduce the vector function $\mathbf{p}(\phi) := (p_1(\phi), p_2(\phi))$. An integration of (4.15) over $\mathcal{R}_{\rho_0}^{t_1}$ leads to the integral identity

$$
\int_{\widetilde{D}_{t_1}^{\rho_0}} \{e(\phi)\} d\mathbf{x} = \int_{\mathcal{H}_{\rho_0}^{t_1}} \left\{ \frac{te(\phi) + \mathbf{x} \cdot \mathbf{p}(\phi)}{t^2 + |\mathbf{x}|^2} \right\} d\sigma
$$
$$
= \int_{\mathcal{H}_{\rho_0}^{t_1}} \{te(\phi) + \mathbf{x} \cdot \mathbf{p}(\phi)\} \frac{d\mathbf{x}}{t}, \qquad (4.56)
$$

where $d\mathbf{x}/t$ is the Leray measure on the hyperboloid $\mathcal{H}_{\rho_0}^{t_1}$.

We would like to rewrite the above identity using hyperbolic coordinates and, for this purpose, we write

$$t = \rho \cosh(\chi), \qquad r = \rho \sinh(\chi), \qquad \theta = \theta. \qquad (4.57)$$

In the new coordinates, the Minkowski metric takes the form

$$g = -d\rho^2 + \rho^2 \big(d\chi^2 + \sinh^2(\chi) d\theta^2\big), \qquad (4.58)$$

while the measure of integration changes to

$$dt \, d\mathbf{x} = \rho^2 \sinh(\chi) d\rho d\chi d\theta. \qquad (4.59)$$

---

[7]A wave map $\phi$ defined on $C$ is said to have finite energy if $\sup_{t>0} \|\nabla \phi(t)\|_{L_{\mathbf{x}}^2(|\mathbf{x}|<t)} < \infty$.

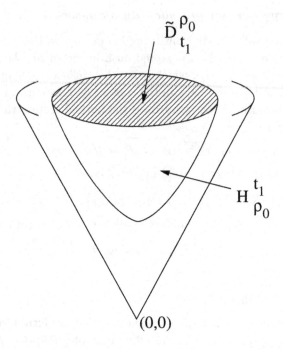

Fig. 4.1    The boundary of the solid hyperboloid $\mathcal{R}_{\rho_0}^{t_1}$.

Using the chain rule, we deduce

$$\rho \nabla_\rho \phi = t \nabla_t \phi + r \nabla_r \phi, \qquad \nabla_\chi \phi = r \nabla_t \phi + t \nabla_r \phi,$$

which imply

$$\nabla_t \phi = \cosh(\chi) \nabla_\rho \phi - \frac{\sinh(\chi)}{\rho} \nabla_\chi \phi, \quad \nabla_r \phi = -\sinh(\chi) \nabla_\rho \phi + \frac{\cosh(\chi)}{\rho} \nabla_\chi \phi.$$

Moreover, if we denote by $d\mu_h$ the measure on the unit hyperboloid, i.e.,

$$d\mu_h \overset{\text{def}}{=} \sinh(\chi) d\chi d\theta, \tag{4.60}$$

then for the Leray measure in (4.56) we have

$$\frac{d\mathbf{x}}{t} = \rho_0 d\mu_h.$$

With these prerequisites, the identity (4.56) can be written in the form

$$\int_{\tilde{D}_{t_1}^{\rho_0}} \{e(\phi)\}d\mathbf{x} = \int_{\mathcal{H}_{\rho_0}^{t_1}} \left\{ \frac{t}{2}(|\nabla_t\phi|^2 + |\vec{\nabla}\phi|^2) + r\langle\nabla_t\phi,\nabla_r\phi\rangle \right\} \frac{d\mathbf{x}}{t}$$

$$= \int_{0\le\chi\le\cosh^{-1}(t_1/\rho_0)} \left\{ \frac{1}{2}\cosh(\chi)\left(|\nabla_\chi\phi|^2 + \frac{1}{\sinh^2(\chi)}|\nabla_\theta\phi|^2 + \rho_0^2|\nabla_\rho\phi|^2\right) \right.$$

$$\left. - \rho_0\sinh(\chi)\langle\nabla_\chi\phi,\nabla_\rho\phi\rangle \right\}d\mu_h$$

$$= \int_{0\le\chi\le\cosh^{-1}(t_1/\rho_0)} \left\{ \frac{e^\chi}{4}|(\nabla_\chi - \rho_0\nabla_\rho)\phi|^2 + \frac{e^{-\chi}}{4}|(\nabla_\chi + \rho_0\nabla_\rho)\phi|^2 \right.$$

$$\left. + \frac{\cosh(\chi)}{2\sinh^2(\chi)}|\nabla_\theta\phi|^2 \right\}d\mu_h .$$

$$(4.61)$$

**Remark 4.3.** We consider it instructive to present another way of deriving the previous integral identity. We recall from Chapter 1 (more precisely, (1.25) and the discussion that follows it) that if

$$T_{\mu\nu} = \langle\nabla_\mu\phi,\nabla_\nu\phi\rangle - \frac{1}{2}g_{\mu\nu}\langle\nabla\phi,\nabla\phi\rangle \qquad (4.62)$$

is the energy-momentum tensor associated to (4.12) and $K$ is a Killing vector field, then one has

$$\nabla^\mu\{T_{\mu\nu}K^\nu\} = 0.$$

Choosing $K = \nabla_t$ and integrating on $\mathcal{R}_{\rho_0}^{t_1}$ this differential identity, we deduce

$$\int_{\tilde{D}_{t_1}^{\rho_0}} \{n^\mu T_{\mu\nu}K^\nu\}\,d\sigma + \int_{\mathcal{H}_{\rho_0}^{t_1}} \{n^\mu T_{\mu\nu}K^\nu\}\,d\sigma = 0, \qquad (4.63)$$

where $n$ is outward unit normal.

We already know that using the rectilinear coordinates one obtains

$$\int_{\tilde{D}_{t_1}^{\rho_0}} \{n^\mu T_{\mu\nu}K^\nu\}\,d\sigma = \int_{\tilde{D}_{t_1}^{\rho_0}} \{e(\phi)\}d\mathbf{x}. \qquad (4.64)$$

For the second integral, the computations are easier if they are based on hyperbolic coordinates. The outward unit normal is given by $n = -\nabla_\rho = (-1,0,0)$ and

$$\nabla_t = \cosh(\chi)\nabla_\rho - \frac{\sinh(\chi)}{\rho}\nabla_\chi = \left(\cosh(\chi), -\frac{\sinh(\chi)}{\rho}, 0\right),$$

which together imply

$$n^\mu T_{\mu\nu} K^\nu = -T_{\rho\rho} \cosh(\chi) + T_{\rho\chi} \frac{\sinh(\chi)}{\rho}.$$

Direct computations yield

$$T_{\rho\rho} = \frac{1}{2} \left\{ \frac{1}{\rho^2} \left( |\nabla_\chi \phi|^2 + \frac{1}{\sinh^2(\chi)} |\nabla_\theta \phi|^2 \right) + |\nabla_\rho \phi|^2 \right\}, \quad T_{\rho\chi} = \langle \nabla_\rho \phi, \nabla_\chi \phi \rangle,$$

and, also taking into account that the measure of integration on $\mathcal{H}_{\rho_0}^{t_1}$ is $d\sigma = \rho_0^2 d\mu_h$, it follows that

$$\int\limits_{\mathcal{H}_{\rho_0}^{t_1}} \{n^\mu T_{\mu\nu} K^\nu\} \, d\sigma$$

$$= - \int\limits_{0 \leq \chi \leq \cosh^{-1}(t_1/\rho_0)} \left\{ \frac{1}{2} \cosh(\chi) \left( |\nabla_\chi \phi|^2 + \frac{1}{\sinh^2(\chi)} |\nabla_\theta \phi|^2 + \rho_0^2 |\nabla_\rho \phi|^2 \right) \right.$$

$$\left. - \rho_0 \sinh(\chi) \langle \nabla_\chi \phi, \nabla_\rho \phi \rangle \right\} d\mu_h.$$

The identity (4.61) is then the result of combining the last equality with (4.63) and (4.64).

### 4.6.2 *The hyperbolic projection*

As mentioned earlier, we would like to use these newly-derived estimates to rule out finite energy, self-similar profiles from our analysis. To achieve this, we need some more prerequisites which are provided by the *hyperbolic projection*.

This is a map defined on the interior of the characteristic cone $\mathcal{C}$, with values in $\mathbb{R}^3$, and given by

$$x \mapsto y \overset{\text{def}}{=} n + \frac{2\rho}{\langle x - \rho n, x - \rho n \rangle} (x - \rho n), \qquad (4.65)$$

where $n := (-1, 0, 0)$ and $\rho$ is as in (4.52).

A routine computation reveals that

$$y = (y^0, y^1, y^2) = \left( 0, \frac{x^1}{t + \rho}, \frac{x^2}{t + \rho} \right)$$

and onward we are going to use the notation $\mathbf{y} := (y^1, y^2)$ and $p := |\mathbf{y}|$. It is also easy to see that the round surface $\mathcal{H}_{\rho_0}^{t_1}$ is mapped through the hyperbolic projection onto the closed disk $\left\{ |\mathbf{y}| \leq \sqrt{(t_1 - \rho_0)/(t_1 + \rho_0)} \right\}$.

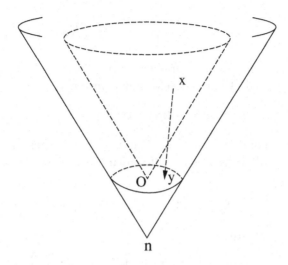

Fig. 4.2   The hyperbolic projection $x \mapsto y$.

In connection to this map, we can define a new set of coordinates for $\phi$,

$$(\rho, \mathbf{y}) = (\rho, y^1, y^2),$$

which are related to the hyperbolic ones $(\rho, \chi, \theta)$ by

$$y^1 = \tanh(\chi/2)\cos(\theta), \qquad y^2 = \tanh(\chi/2)\sin(\theta), \qquad p = \tanh(\chi/2).$$

Accordingly, in these coordinates, the Minkowski metric takes the form

$$g = -d\rho^2 + \rho^2 h(p)\big((dy^1)^2 + (dy^2)^2\big) = -d\rho^2 + \rho^2 h(p)\big(dp^2 + p^2 d\theta^2\big)$$

and the wave maps equation (4.12) can be written as

$$\left(\nabla_\rho^2 + \frac{2}{\rho}\nabla_\rho\right)\phi^a - \frac{1}{\rho^2 h(p)}\Delta_{\mathbf{y}}\phi^a + \Gamma^a_{bc}(\phi)\left\{-\langle\nabla_\rho\phi^b, \nabla_\rho\phi^c\rangle\right.$$
$$\left. + \frac{1}{\rho^2 h(p)}\delta^{jk}\langle\nabla_j\phi^b, \nabla_k\phi^c\rangle\right\} = 0, \tag{4.66}$$

where

$$h(p) = \frac{4}{(1 - p^2)^2}.$$

Other useful relations that can be deduced through direct calculations are as follows:

$$\cosh(\chi) = \frac{1 + p^2}{1 - p^2}, \qquad d\mu_h = \frac{4p}{(1 - p^2)^2}\, dp\, d\theta = h(p)d\mathbf{y},$$

$$|\nabla_\chi \phi|^2 + \frac{1}{\sinh^2(\chi)}|\nabla_\theta \phi|^2 = \frac{1}{h(p)}\left\{|\nabla_p \phi|^2 + \frac{1}{p^2}|\nabla_\theta \phi|^2\right\} = \frac{1}{h(p)}|\vec{\nabla}_{\mathbf{y}}\phi|^2,$$

where $\vec{\nabla}_{\mathbf{y}}$ denote the standard gradient of $\phi = \phi(\rho, \mathbf{y})$ with respect to $\mathbf{y}$.

We are now ready to discuss self-similar wave maps, i.e.,

$$\phi = \phi(t, \mathbf{x}) = \tilde{\phi}\left(\frac{\mathbf{x}}{t}\right),$$

for which it is immediate to verify that $\nabla_\rho \phi = 0$. Thus, due to (4.66), (4.61), and the above relations, self-similar wave maps on $\mathcal{C}$ satisfy the elliptic equation

$$-\Delta_{\mathbf{y}}\phi^a + \Gamma^a_{bc}(\phi)\langle \vec{\nabla}_{\mathbf{y}}\phi^b, \vec{\nabla}_{\mathbf{y}}\phi^c\rangle = 0 \qquad (4.67)$$

and the integral identity

$$\int_{\tilde{D}^{\rho_0}_{t_1}} \{e(\phi)\}d\mathbf{x} = \frac{1}{2}\int_{|\mathbf{y}|\leq\sqrt{\frac{t_1-\rho_0}{t_1+\rho_0}}} \left\{\frac{1+|\mathbf{y}|^2}{1-|\mathbf{y}|^2}|\vec{\nabla}_{\mathbf{y}}\phi|^2\right\}d\mathbf{y}.$$

If we take into account that $\tilde{D}^{\rho_0}_{t_1} \subseteq D_{t_1}$ and we allow for $\rho_0 \to 0$, we finally obtain

$$\frac{1}{2}\int_{|\mathbf{y}|\leq 1}\left\{\frac{1+|\mathbf{y}|^2}{1-|\mathbf{y}|^2}|\vec{\nabla}_{\mathbf{y}}\phi|^2\right\}d\mathbf{y} \leq \mathcal{E}_{D_{t_1}}(\phi). \qquad (4.68)$$

We can conclude that, in order to show that there are no nontrivial, finite energy, self-similar wave maps defined on $\mathcal{C}$, it will be enough to prove the absence of nontrivial solutions for (4.67), with a finite integral on the left-hand side of (4.68). We state this as a lemma and demonstrate it below.

**Lemma 4.1.** *The only solutions to (4.67), for which the integral on the left-hand side of (4.68) is finite, are the trivial ones (i.e., they are constant).*

**Proof.** We start by making the remark that solutions to (4.67) can be interpreted as harmonic maps from $D := \{|\mathbf{y}| < 1\}$ to $N$, thus being smooth up to the boundary. Working with the polar coordinates $(p, \theta)$ associated to $\mathbf{y}$, we deduce from (4.68) that

$$\int_0^1 \left\{\int_0^{2\pi} |\nabla_\theta \phi(p,\theta)|^2 d\theta\right\}\frac{1+p^2}{p(1-p^2)}dp < \infty,$$

which further implies the existence of a sequence of radii $p_n \to 1$ such that

$$\int_0^{2\pi} |\nabla_\theta \phi(p_n, \theta)|^2 d\theta \to 0.$$

As $\partial D = \mathbb{S}^1$, the trace theorem and the previous limit yield

$$\|\phi(1, \cdot)\|_{\dot{H}^{1/2}(\mathbb{S}^1)} = 0.$$

Hence, $\theta \mapsto \phi(1, \theta)$ is a constant function and, in particular, smooth. This allows us to apply a result due to Qing [135] and claim that $\phi \in C^\infty(\overline{D})$. Finally, due to $\phi$ being constant on $\mathbb{S}^1$, a classical uniqueness result for harmonic maps (e.g., Theorem 8.2.3 in [76]) forces $\phi$ to be constant throughout $D$. □

### 4.6.3 Partial non-concentration of energy and time decay estimates I

Next, we focus on deriving energy-type estimates which are useful in analyzing concentration/non-concentration of energy on various regions of the characteristic cone. We start by considering the vector field[8]

$$X = X^\mu \nabla_\mu \overset{\text{def}}{=} \frac{x^\mu}{\rho} \nabla_\mu, \tag{4.69}$$

with $\rho$ given by (4.52), and use the notation $X(\phi) := X^\mu \nabla_\mu \phi$ for its action on the wave map $\phi$. As derived in Chapter 1, the contraction of this vector field with the energy-momentum tensor (4.62) results in the identity

$$\nabla_\mu P^{X,\mu} = R^X, \tag{4.70}$$

where $P_\mu^X := T_{\mu\nu} X^\nu$ and $R^X := T_\nu^\mu \nabla_\mu X^\nu$. A direct computation shows that

$$\nabla_\mu X^\nu = \frac{\delta_\mu^\nu}{\rho} + \frac{x_\mu x^\nu}{\rho^3}$$

and, subsequently,

$$R^X = \frac{1}{\rho} \langle X^\mu \nabla_\mu \phi, X^\nu \nabla_\nu \phi \rangle \overset{\text{def}}{=} \frac{1}{\rho} |X(\phi)|^2. \tag{4.71}$$

The main benefit of the identity (4.70) is that $R^X$ is nonnegative, which one could employ to derive useful a priori estimates by integration. However, we would like to integrate (4.70) on regions of the characteristic cone and that poses a technical problem as $X$ is singular there (i.e., $\rho = 0$ on $\partial \mathcal{C}$).

---

[8]To our knowledge, this vector field was used for the first time in connection to energy-type bounds by Grillakis [62, 63].

To avoid this issue, we choose a timelike vector $x_0$, with $\langle x_0, x_0 \rangle = -\rho_0^2 < 0$, and replace $X$ by

$$X_y = X_y^\mu \nabla_\mu \overset{\text{def}}{=} \frac{y^\mu}{\rho(y)} \nabla_\mu = \frac{(x+x_0)^\mu}{\rho(x+x_0)} \nabla_\mu.$$

Calculations similar to the one above yield

$$\nabla_\mu P^{X_y,\mu} = R^{X_y} = \frac{1}{\rho(y)} \left| X_y(\phi) \right|^2$$

and, for $x \in \mathcal{C}$, we have

$$\rho(y) \geq \rho_0 > 0.$$

We can now integrate the last differential identity over the spacetime region $\mathcal{C}_{[t_0,t_1]}$ and deduce

$$\int_{D_{t_1}} \{ P_0^{X_y} \} d\mathbf{x} + \int_{\mathcal{C}_{[t_0,t_1]}} \left\{ \frac{1}{\rho(y)} \left| X_y(\phi) \right|^2 \right\} dt d\mathbf{x}$$

$$= - \int_{\partial \mathcal{C}_{[t_0,t_1]}} \{ n^\mu T_{\mu\nu} X_y^\nu \} d\sigma + \int_{D_{t_0}} \{ P_0^{X_y} \} d\mathbf{x}, \tag{4.72}$$

where

$$n = \frac{1}{\sqrt{2}} \left( -1, -\frac{\mathbf{x}}{|\mathbf{x}|} \right)$$

is a past-oriented null vector. As we shall see, the previous identity is written in such a way that all relevant integrands are nonnegative. Moreover, we work to find explicit forms for these integrands in order to appreciate the information that they provide.

In what concerns $P_0^{X_y}$, it is convenient to introduce the radial and angular derivatives with respect to $\mathbf{y} = \mathbf{x} + \mathbf{x}_0$:

$$\nabla_r^y \phi \overset{\text{def}}{=} \frac{1}{|\mathbf{y}|} \left( y^1 \nabla_1 + y^2 \nabla_2 \right) \phi,$$

$$\nabla_\theta^y \phi \overset{\text{def}}{=} \left( y^1 \nabla_2 - y^2 \nabla_1 \right) \phi.$$

With these conventions, we can write

$$P_0^{X_y} = \frac{y^0 + |\mathbf{y}|}{4\rho(y)} |(\nabla_0 + \nabla_r^y)\phi|^2 + \frac{\rho(y)}{4(y^0 + |\mathbf{y}|)} |(\nabla_0 - \nabla_r^y)\phi|^2$$

$$+ \frac{y^0}{2\rho(y)|\mathbf{y}|^2} |\nabla_\theta^y \phi|^2. \tag{4.73}$$

For the integrand on $\partial\mathcal{C}_{[t_0,t_1]}$, we complement the null vector $n$ with

$$n^\dagger = \frac{1}{\sqrt{2}}\left(-1, \frac{\mathbf{x}}{|\mathbf{x}|}\right) \quad \text{and} \quad \xi = \left(0, \frac{-x^2}{|\mathbf{x}|}, \frac{x^1}{|\mathbf{x}|}\right).$$

These three vectors form what is usually called a *null frame*, which satisfies

$$\langle n, n\rangle = \langle n^\dagger, n^\dagger\rangle = \langle n, \xi\rangle = \langle n^\dagger, \xi\rangle = 0,$$

$$\langle n, n^\dagger\rangle = -1, \qquad \langle \xi, \xi\rangle = 1.$$

Based on these relations, direct calculations yield

$$-n^\mu T_{\mu\nu} X_y^\nu = \langle X_y, n^\dagger\rangle\left|\nabla_n\phi - \frac{\langle X_y, \xi\rangle}{2\langle X_y, n^\dagger\rangle}\nabla_\xi\phi\right|^2 \\ + \frac{1}{4\langle X_y, n^\dagger\rangle}\left|\nabla_\xi\phi\right|^2, \tag{4.74}$$

where $\langle X_y, n^\dagger\rangle := X_y^\mu n_\mu^\dagger$, $\nabla_n\phi := n^\mu\nabla_\mu\phi$, and the other quantities have similar definitions. It is worth noticing that

$$\langle X_y, n^\dagger\rangle = \frac{y^0 + \mathbf{y} \cdot \frac{\mathbf{x}}{|\mathbf{x}|}}{\sqrt{2}\rho(y)}$$

and that on $\partial\mathcal{C}$ we have $\langle x, x\rangle = 0$, which implies $\rho(y) = \sqrt{\rho_0^2 - 2\langle x, x_0\rangle}$.

We will use the integral identity (4.72) with the simplest possible choice for the timelike vector $x_0$, i.e.,

$$x_0 = (T, 0, 0), \qquad T > 0,$$

so that $\rho_0 = T$. If we also define the optical functions

$$u \overset{\text{def}}{=} t - r \quad \text{and} \quad u^\dagger \overset{\text{def}}{=} t + r, \tag{4.75}$$

then, according to (4.73) and (4.74), we derive

$$P_0^{X_y} = \frac{\sqrt{u^\dagger + T}}{4\sqrt{u + T}}|(\nabla_t + \nabla_r)\phi|^2 + \frac{\sqrt{u + T}}{4\sqrt{u^\dagger + T}}|(\nabla_t - \nabla_r)\phi|^2 \\ + \frac{t + T}{2\sqrt{\rho^2 + T^2 + 2tT}}\frac{|\nabla_\theta\phi|^2}{r^2}$$

and

$$-n^\mu T_{\mu\nu}X_y^\nu = \frac{1}{2\sqrt{2}}\left(\frac{\sqrt{2r + T}}{\sqrt{T}}\left|(\nabla_t + \nabla_r)\phi\right|^2 + \frac{\sqrt{T}}{\sqrt{2r + T}}\frac{|\nabla_\theta\phi|^2}{r^2}\right).$$

As a consequence, we deduce

$$\lim_{t \to 0} \int_{D_t} P_0^{X_v} \, d\mathbf{x} = E.$$

Hence, allowing for $t_0 \to 0$ in (4.72), it follows that

$$\int_{D_{t_1}} \left\{ \frac{\sqrt{u^\dagger + T}}{4\sqrt{u + T}} \left| (\nabla_t + \nabla_r)\phi \right|^2 + \frac{\sqrt{u + T}}{4\sqrt{u^\dagger + T}} \left| (\nabla_t - \nabla_r)\phi \right|^2 \right.$$

$$\left. + \frac{t_1 + T}{2\sqrt{\rho^2 + T^2 + 2t_1 T}} \frac{|\nabla_\theta \phi|^2}{r^2} \right\} d\mathbf{x} + \int_{\mathcal{C}_{[0,t_1]}} \left\{ \frac{1}{\rho(y)} |X_y(\phi)|^2 \right\} dt \, d\mathbf{x}$$

$$= \frac{1}{2\sqrt{2}} \int_{\partial \mathcal{C}_{[0,t_1]}} \left\{ \frac{\sqrt{2r + T}}{\sqrt{T}} \left| (\nabla_t + \nabla_r)\,\phi \right|^2 + \frac{\sqrt{T}}{\sqrt{2r + T}} \frac{|\nabla_\theta \phi|^2}{r^2} \right\} d\sigma + E.$$

$$(4.76)$$

This integral identity allows us to obtain two useful energy-type estimates.

First, we work under the assumption that $0 < T < t_1$ and, directly from above, we easily infer that

$$\int_{\mathcal{C}_{[0,t_1]}} \left\{ \frac{1}{\rho(y)} |X_y(\phi)|^2 \right\} dt dx \le C \frac{\sqrt{t_1}}{\sqrt{T}} \delta_F(t_1, \phi) + E. \qquad (4.77)$$

Using the formula for $X_y$, we derive

$$X = \frac{\rho(y)}{\rho} X_y - \frac{T}{\rho} \nabla_t,$$

which implies

$$\frac{|X(\phi)|^2}{\rho} \le C \left( \left( \frac{\rho(y)}{\rho} \right)^3 \frac{|X_y(\phi)|^2}{\rho(y)} + \frac{T^2}{\rho^3} |\nabla_t \phi|^2 \right)$$

holds on a domain where $\rho > 0$. Such a domain is $\mathcal{C}_{[T,t_1]}^T \subset \mathcal{C}_{[0,t_1]}$, where a careful, yet simple, analysis reveals

$$1 \le \frac{\rho(y)}{\rho} \le 2 \quad \text{and} \quad \frac{T^2}{\rho^3} \le \frac{T^{1/2}}{t^{3/2}}.$$

Hence, it follows that

$$\int_{\mathcal{C}_{[T,t_1]}^T} \left\{ \frac{1}{\rho} |X(\phi)|^2 \right\} dt \, dx \le C \int_{\mathcal{C}_{[T,t_1]}^T} \left\{ \frac{1}{\rho(y)} |X_y(\phi)|^2 + \frac{T^{1/2}}{t^{3/2}} |\nabla_t \phi|^2 \right\} dt \, dx.$$

Combining this bound with (4.77) and taking into account (4.25), i.e.,

$$\mathcal{E}_{D_{t_1}}(\phi) = \delta_F(t_1, \phi) + E,$$

we deduce the subsequent result.

**Lemma 4.2.** *Assume that $\phi$ is a wave map in the cone $C_{(0,t_1]}$ and $0 < T < t_1$. Under the previous notational conventions, we have:*

$$\int\limits_{C^T_{[T,t_1]}} \left\{ \frac{1}{\rho} |X(\phi)|^2 \right\} dt\, d\mathbf{x} \leq C \left( \frac{\sqrt{t_1}}{\sqrt{T}} \delta_F(t_1, \phi) + E \right). \tag{4.78}$$

Another way in which we can use the identity (4.76) is by choosing $t_1$ and $T$ such that

$$\delta_F(t_1, \phi) \ll E \qquad \text{and} \qquad T = \frac{t_1 \delta_F^2(t_1, \phi)}{E^2}. \tag{4.79}$$

We can then control the integral on the right-hand side of (4.76) by

$$\frac{1}{2\sqrt{2}} \int\limits_{\partial C_{[0,t_1]}} \left\{ \frac{\sqrt{2r+T}}{\sqrt{T}} |(\nabla_t + \nabla_r)\phi|^2 + \frac{\sqrt{T}}{\sqrt{2r+T}} \frac{|\nabla_\theta \phi|^2}{r^2} \right\} d\sigma \leq CE.$$

Thus, we have obtained a new energy-type inequality, which is stated below as a lemma.

**Lemma 4.3.** *Assume that $\phi$ is a wave map in the cone $C_{(0,t_1]}$ and that $t_1$ and $T$ satisfy (4.79). With the previous notational conventions, the following estimate holds true:*

$$\int\limits_{D_{t_1}} \left\{ \frac{\sqrt{u^\dagger + T}}{4\sqrt{u+T}} |(\nabla_t + \nabla_r)\phi|^2 + \frac{\sqrt{u+T}}{4\sqrt{u^\dagger + T}} |(\nabla_t - \nabla_r)\phi|^2 \right.$$

$$\left. + \frac{t_1 + T}{2\sqrt{\rho^2 + T^2 + 2t_1 T}} \frac{|\nabla_\theta \phi|^2}{r^2} \right\} d\mathbf{x} \tag{4.80}$$

$$+ \int\limits_{C_{[0,t_1]}} \left\{ \frac{1}{\sqrt{\rho^2 + T^2 + 2tT}} |X_y(\phi)|^2 \right\} dt d\mathbf{x} \leq CE.$$

We make two remarks in connection to this result. The first one is that if we use the bound $\rho(y) \leq t + T$, then we deduce the rather crude estimate

$$\int_0^{t_1} \frac{1}{t+T} \left\{ \int\limits_{D_t} |X_y(\phi)|^2 d\mathbf{x} \right\} dt \leq CE.$$

As $t_1/T = E^2/\delta_F^2(t_1, \phi) \gg 1$, this implies that the function $t \mapsto \int_{D_t} |X_y(\phi)|^2 dx$ has to be small somewhere on the time interval $(0, t_1)$. We will further elaborate on this observation in the later stages of the argument.

The second remark is that the weights attached to the derivative terms $|(\nabla_t + \nabla_r)\phi|^2$ and $|\nabla_\theta \phi|^2/r^2$, in the integral on $D_{t_1}$, are large near $\partial D_{t_1}$, thus leading to non-concentration of these parts of the energy near the characteristic cone.

Our next goal is to refine the above estimate (i.e., (4.80)) and obtain energy-type bounds which involve, instead of the total energy, only the energy of the wave map inside a cone of aperture less than 1.

### 4.6.4 *Partial non-concentration of energy and time decay estimates II*

Here, we work with a modified version of the vector field $X$, which is given by

$$\tilde{X} = \tilde{X}^\mu \nabla_\mu \overset{\text{def}}{=} \begin{cases} \dfrac{x^\mu}{\rho} \nabla_\mu & \text{if } \rho_0 \leq \rho, \\ \dfrac{x^\mu}{\rho_0} \nabla_\mu & \text{if } 0 \leq \rho < \rho_0, \end{cases} \tag{4.81}$$

where $\rho_0$ is a small, positive fixed parameter to be specified later. Straightforward calculations yield

$$T_\nu^\mu \nabla_\mu \tilde{X}^\nu = R^{\tilde{X}} = \begin{cases} \dfrac{1}{\rho}|X(\phi)|^2 & \text{if } \rho_0 \leq \rho, \\ -\dfrac{1}{2\rho_0}\langle \nabla\phi, \nabla\phi \rangle & \text{if } 0 \leq \rho < \rho_0, \end{cases} \tag{4.82}$$

and we integrate the differential identity $\nabla_\mu \{P^{\tilde{X}, \mu}\} = R^{\tilde{X}}$ on the solid cone $\mathcal{C}_{[0, t_1]}$. The region of integration is split according to

$$\mathcal{C}_{[0, t_1]} = \mathcal{R}_{\rho_0}^{t_1} \cup \mathcal{Q}_{\rho_0}^{t_1},$$

where $\mathcal{R}_{\rho_0}^{t_1}$ is given by (4.53) and

$$\mathcal{Q}_{\rho_0}^{t_1} \overset{\text{def}}{=} \{(t, \mathbf{x}) \mid \rho \leq \rho_0 , t \leq t_1\}. \tag{4.83}$$

The integration results in the identity

$$\int_{D_{t_1}} \{P_0^{\tilde{X}}\} d\mathbf{x} + \int_{\mathcal{R}_{\rho_0}^{t_1}} \left\{ \frac{1}{\rho}|X(\phi)|^2 \right\} dt d\mathbf{x}$$

$$= \frac{1}{\sqrt{2}} \int_{\partial \mathcal{C}_{[0, t_1]}} \frac{t}{\rho_0} |(\nabla_t + \nabla_r)\phi|^2 d\sigma + \frac{1}{2\rho_0} \int_{\mathcal{Q}_{\rho_0}^{t_1}} \langle \nabla\phi, \nabla\phi \rangle dt d\mathbf{x}, \tag{4.84}$$

and, in order to obtain a useful energy-type bound, we need to estimate the last integral, which is indefinite.

For this purpose, we rely on the hyperbolic coordinates (4.57) used before, in particular (4.58), to deduce

$$
\begin{aligned}
\langle \nabla\phi, \nabla\phi \rangle &= \frac{1}{\rho^2}\left(\frac{1}{\sinh^2(\chi)}|\nabla_\theta\phi|^2 + |\nabla_\chi\phi|^2\right) - |\nabla_\rho\phi|^2 \\
&= \frac{1}{\rho^2\sinh^2(\chi)}|\nabla_\theta\phi|^2 + \frac{1}{\rho^2}\langle(\nabla_\chi + \rho\nabla_\rho)\phi, (\nabla_\chi - \rho\nabla_\rho)\phi\rangle.
\end{aligned}
\tag{4.85}
$$

This expression should be compared to the energy density on the unit hyperboloid, i.e.,

$$
\begin{aligned}
e_{\mathcal{H}}(\phi) \overset{\text{def}}{=} &\frac{e^{\chi}}{4}|(\nabla_\chi - \rho\nabla_\rho)\phi|^2 + \frac{e^{-\chi}}{4}|(\nabla_\chi + \rho\nabla_\rho)\phi|^2 \\
&+ \frac{\cosh(\chi)}{2\sinh^2(\chi)}|\nabla_\theta\phi|^2,
\end{aligned}
\tag{4.86}
$$

which appears in the identity (4.61). Based on (4.59), (4.60), the simple inequalities $\cosh(\chi) \geq 1$ and $u^2 + v^2 \geq 2uv$, and (4.61), we infer

$$
\begin{aligned}
&\frac{1}{2\rho_0}\int_{\mathcal{Q}_{\rho_0}^{t_1}} |\langle\nabla\phi, \nabla\phi\rangle|\, dtd\mathbf{x} \\
&\leq \frac{1}{\rho_0}\int_0^{\rho_0}\left(\int_{0\leq\chi\leq\cosh^{-1}(t_1/\rho)} e_{\mathcal{H}}(\phi)\, d\mu_h\right)d\rho \\
&\leq \frac{1}{\rho_0}\int_0^{\rho_0}\left(\int_{\tilde{D}_{t_1}^\rho} e(\phi)\, d\mathbf{x}\right)d\rho \\
&\leq \mathcal{E}_{D_{t_1}}(\phi) = \delta_F(t_1, \phi) + E.
\end{aligned}
\tag{4.87}
$$

In what concerns the first integral on the right-hand side of (4.84), we can easily estimate it by

$$
\frac{1}{\sqrt{2}}\int_{\partial\mathcal{C}_{[0,t_1]}} \frac{t}{\rho_0}|(\nabla_t + \nabla_r)\phi|^2 d\sigma \leq \frac{2t_1\delta_F(t_1, \phi)}{\rho_0}.
\tag{4.88}
$$

Therefore, if we choose $t_1$ to satisfy $\delta_F(t_1, \phi) \ll E$ and define

$$
\rho_0 \overset{\text{def}}{=} \frac{t_1\delta_F(t_1, \phi)}{E},
$$

then we finally derive

$$
\int_{D_{t_1}} \{P_0^{\tilde{X}}\}\, d\mathbf{x} + \int_{\mathcal{R}_{\rho_0}^{t_1}} \left\{\frac{1}{\rho}|X(\phi)|^2\right\}dtd\mathbf{x} \leq CE.
\tag{4.89}
$$

A closer inspection of this estimate reveals that it provides information similar to the one coming from (4.80) in Lemma 4.3.

Nevertheless, what we have done so far in this section serves as an important preliminary step for obtaining energy-type bounds involving cones of aperture less than 1. The next step in this direction is to introduce another fixed parameter, $\rho_1 := K\rho_0$, where $K \gg 1$ is a large constant to be specified later, and further modify the definition (4.81) of $\widetilde{X}$ by replacing $\rho_0$ with $\rho_1$. Thus, for the remainder of this section,

$$\widetilde{X} = \widetilde{X}^\mu \nabla_\mu \stackrel{\text{def}}{=} \begin{cases} \dfrac{x^\mu}{\rho} \nabla_\mu & \text{if } \rho_1 \leq \rho, \\[2mm] \dfrac{x^\mu}{\rho_1} \nabla_\mu & \text{if } 0 \leq \rho < \rho_1, \end{cases} \qquad (4.90)$$

and $R^{\widetilde{X}}$ changes accordingly. The integral identity (4.84) becomes now

$$\int_{D_{t_1}} \{P_0^{\widetilde{X}}\}d\mathbf{x} + \int_{\mathcal{R}_{\rho_1}^{t_1}} \left\{ \frac{1}{\rho}|X(\phi)|^2 \right\} dt d\mathbf{x}$$

$$= \frac{1}{\sqrt{2}} \int_{\partial \mathcal{C}_{[0,t_1]}} \frac{t}{\rho_1}|(\nabla_t + \nabla_r)\phi|^2 d\sigma + \frac{1}{2\rho_1} \int_{\mathcal{Q}_{\rho_1}^{t_1}} \langle \nabla\phi, \nabla\phi \rangle dt d\mathbf{x}, \qquad (4.91)$$

and the former main task of estimating the right-hand side remains the same.

However, in this context, we are going to separate the region of integration $\mathcal{Q}_{\rho_1}^{t_1}$ into further small regions and estimate each of the respective integrals. For that, we use another parameter $\gamma$ ($\gamma$ will be the aperture of a cone), with $0 < 1 - \gamma \ll 1$, and write

$$\mathcal{Q}_{\rho_1}^{t_1} = \mathcal{Q}_{\rho_0}^{t_1} \cup \mathcal{Q}_{[\rho_0,\rho_1]}^{t_1,\gamma} \cup \widetilde{\mathcal{Q}}_{[\rho_0,\rho_1]}^{t_1,\gamma},$$

where

$$\mathcal{Q}_{[\rho_0,\rho_1]}^{t_1,\gamma} \stackrel{\text{def}}{=} \{(t,\mathbf{x}) \mid \rho_0 \leq \rho \leq \rho_1 \,,\, t \leq t_1 \,,\, r/t \leq \gamma\}, \qquad (4.92)$$

$$\widetilde{\mathcal{Q}}_{[\rho_0,\rho_1]}^{t_1,\gamma} \stackrel{\text{def}}{=} \{(t,\mathbf{x}) \mid \rho_0 \leq \rho \leq \rho_1 \,,\, t \leq t_1 \,,\, \gamma < r/t \leq 1\}. \qquad (4.93)$$

In the above, $\rho$ and $r$ retain their previous meanings of hyperbolic radial and radial coordinates, respectively.

Arguing as in the derivation of (4.88) and (4.87) and taking into account that

$$\rho_1 = K\rho_0 = \frac{Kt_1\delta_F(t_1,\phi)}{E},$$

we easily deduce that

$$\frac{1}{\sqrt{2}} \int_{\partial \mathcal{C}_{[0,t_1]}} \frac{t}{\rho_1}|(\nabla_t + \nabla_r)\phi|^2 d\sigma \leq \frac{2E}{K} \qquad (4.94)$$

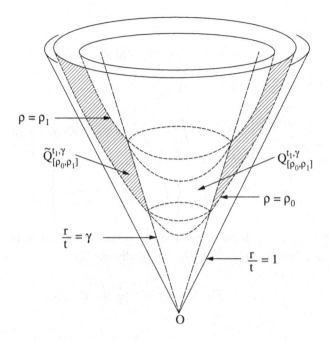

Fig. 4.3 The regions of integration $\mathcal{Q}_{[\rho_0,\rho_1]}^{t_1,\gamma}$ and $\widetilde{\mathcal{Q}}_{[\rho_0,\rho_1]}^{t_1,\gamma}$.

and

$$\frac{1}{2\rho_1}\int_{\mathcal{Q}_{\rho_0}^{t_1}}|\langle\nabla\phi,\nabla\phi\rangle|dtd\mathbf{x}\leq\frac{\mathcal{E}_{D_{t_1}}(\phi)}{K}\leq\frac{CE}{K}. \tag{4.95}$$

For the integral over the region $\mathcal{Q}_{[\rho_0,\rho_1]}^{t_1,\gamma}$, we notice first that

$$\mathcal{Q}_{[\rho_0,\rho_1]}^{t_1,\gamma}\subseteq\mathcal{Q}_{[0,\rho_1]}^{\gamma}\overset{\text{def}}{=}\{(t,\mathbf{x})\mid\rho\leq\rho_1\,,\ r/t\leq\gamma\}$$

and

$$r/t\leq\gamma\Longleftrightarrow0\leq\chi\leq\cosh^{-1}((1-\gamma^2)^{-1/2}).$$

This implies

$$\frac{1}{2\rho_1}\int_{\mathcal{Q}_{[\rho_0,\rho_1]}^{t_1,\gamma}}|\langle\nabla\phi,\nabla\phi\rangle|dtd\mathbf{x}$$

$$\leq\frac{1}{2\rho_1}\int_0^{\rho_1}\rho^2\left\{\int_{0\leq\chi\leq\cosh^{-1}((1-\gamma^2)^{-1/2})}|\langle\nabla\phi,\nabla\phi\rangle|d\mu_h\right\}d\rho.$$

Next, we rely on

$$\rho^2 |\langle \nabla\phi, \nabla\phi \rangle| \leq 2\, e_{\mathcal{H}}(\phi),$$

proven in connection to (4.87), and (4.61) to derive

$$\frac{1}{2\rho_1} \int_0^{\rho_1} \rho^2 \left\{ \int\limits_{0 \leq \chi \leq \cosh^{-1}((1-\gamma^2)^{-1/2})} |\langle \nabla\phi, \nabla\phi \rangle| d\mu_h \right\} d\rho$$

$$\leq \frac{1}{\rho_1} \int_0^{\rho_1} \left\{ \int_{\tilde{D}^\rho_{(1-\gamma^2)^{-1/2}\rho}} e(\phi)\, dx \right\} d\rho,$$

where $\tilde{D}^\rho_{\rho(1-\gamma^2)^{-1/2}}$ is defined according to (4.55). Thus, putting head-to-head the last two integral estimates, we finally obtain

$$\frac{1}{2\rho_1} \int\limits_{\mathcal{Q}^{t_1,\gamma}_{[\rho_0,\rho_1]}} |\langle \nabla\phi, \nabla\phi \rangle| dt dx \leq \frac{1}{T} \int_0^T \mathcal{E}_t^\gamma(\phi) dt, \qquad (4.96)$$

where $T = \rho_1 (1-\gamma^2)^{-1/2}$ and

$$\mathcal{E}_t^\gamma(\phi) \overset{\text{def}}{=} \int\limits_{|\mathbf{x}| \leq \gamma t} e(\phi(t))\, d\mathbf{x}$$

is the energy inside a cone of aperture $\gamma$.

We are left to analyze the integral over the domain $\tilde{\mathcal{Q}}^{t_1,\gamma}_{[\rho_0,\rho_1]}$, where $r/t > \gamma$ and, as a consequence,

$$e^\chi \geq \cosh(\chi) > (1-\gamma^2)^{-1/2}. \qquad (4.97)$$

Based on this bound, (4.85), and (4.86), we infer that

$$\rho^2 |\langle \nabla\phi, \nabla\phi \rangle| \leq 4(1-\gamma^2)^{1/2} e_{\mathcal{H}}(\phi) + 2 |\langle \rho\nabla_\rho\phi, (\nabla_\chi - \rho\nabla_\rho)\phi \rangle| \qquad (4.98)$$

holds in the region of integration. For the first term on the right-hand side, we invoke again the integral identity (4.61) to deduce

$$\frac{1}{2\rho_1} \int\limits_{\tilde{\mathcal{Q}}^{t_1,\gamma}_{[\rho_0,\rho_1]}} \frac{4(1-\gamma^2)^{1/2}}{\rho^2} e_{\mathcal{H}}(\phi) dt\, dx \leq 2(1-\gamma^2)^{1/2} \mathcal{E}_{D_{t_1}}(\phi) \leq C(1-\gamma^2)^{1/2} E.$$

For the second term, as $\nabla_\rho \phi = X(\phi)$, an application of the Cauchy-Schwarz inequality yields

$$\frac{1}{\rho_1} \int\limits_{\widetilde{\mathfrak{Q}}^{t_1,\gamma}_{[\rho_0,\rho_1]}} \frac{1}{\rho^2} \left| \langle \rho\nabla_\rho\phi, (\nabla_\chi - \rho\nabla_\rho)\phi \rangle \right| dt\, dx$$

$$\leq \left( \frac{1}{\rho_1} \int\limits_{\widetilde{\mathfrak{Q}}^{t_1,\gamma}_{[\rho_0,\rho_1]}} (\rho e^{-\chi}) \frac{1}{\rho} |X(\phi)|^2 dt\, dx \right.$$

$$\left. \cdot \frac{1}{\rho_1} \int\limits_{\widetilde{\mathfrak{Q}}^{t_1,\gamma}_{[\rho_0,\rho_1]}} \frac{e^\chi}{\rho^2} |(\nabla_\chi - \rho\nabla_\rho)\phi|^2 dt\, dx \right)^{1/2}.$$

Due to

$$e^\chi |(\nabla_\chi - \rho\nabla_\rho)\phi|^2 \leq 4\, e_{\mathcal{H}}(\phi),$$

we can argue as above and derive

$$\frac{1}{\rho_1} \int\limits_{\widetilde{\mathfrak{Q}}^{t_1,\gamma}_{[\rho_0,\rho_1]}} \frac{e^\chi}{\rho^2} |(\nabla_\chi - \rho\nabla_\rho)\phi|^2 dt\, dx \leq CE.$$

If we rely on the definition (4.93) and (4.97), it is easy to see that, on the domain of integration, we have

$$\widetilde{\mathfrak{Q}}^{t_1,\gamma}_{[\rho_0,\rho_1]} \subseteq R^{t_1}_{\rho_0}, \qquad \frac{\rho e^{-\chi}}{\rho_1} < (1-\gamma^2)^{1/2},$$

which, used in conjunction with the bound (4.89), yield

$$\frac{1}{\rho_1} \int\limits_{\widetilde{\mathfrak{Q}}^{t_1,\gamma}_{[\rho_0,\rho_1]}} (\rho e^{-\chi}) \frac{1}{\rho} |X(\phi)|^2 dt\, dx \leq C(1-\gamma^2)^{1/2} E.$$

Combining (4.98) with the integral estimates derived this analysis, we conclude that

$$\frac{1}{2\rho_1} \int\limits_{\widetilde{\mathfrak{Q}}^{t_1,\gamma}_{[\rho_0,\rho_1]}} \left| \langle \nabla\phi, \nabla\phi \rangle \right| dt dx \leq C(1-\gamma^2)^{1/4} E. \qquad (4.99)$$

Finally, if we choose $K := (1-\gamma^2)^{-1/4}$, then, together with (4.91), (4.94), (4.95), and (4.96), the previous estimate implies the following refinement of (4.80).

**Lemma 4.4.** *Assume that $\phi$ is a wave-map in the cone $C_{(0,t_1]}$, $\gamma$ is a fixed parameter with*

$$0 < \frac{\delta_F(t_1, \phi)}{E} \leq (1-\gamma^2)^{3/4} \ll 1,$$

*and*

$$\rho_1 \stackrel{\text{def}}{=} \frac{t_1 \delta_F(t_1, \phi)}{(1 - \gamma^2)^{1/4} E}, \qquad T \stackrel{\text{def}}{=} \frac{t_1 \delta_F(t_1, \phi)}{(1 - \gamma^2)^{3/4} E}.$$

*Under the notational conventions of this section, the next estimate holds true:*

$$\int_{D_{t_1}} \{P_0^{\tilde{X}}\} d\mathbf{x} + \int_{\mathcal{R}_{\rho_1}^{t_1}} \left\{ \frac{1}{\rho} |X(\phi)|^2 \right\} dt \, d\mathbf{x}$$

$$\le C(1 - \gamma^2)^{1/4} E + \frac{1}{T} \int_0^T \mathcal{E}_t^\gamma(\phi) dt. \tag{4.100}$$

The main improvement we achieved with this result over the one in Lemma 4.3 is that the right-hand side of (4.100) involves only the energy of the wave map inside a smaller cone (of aperture $\gamma < 1$) than the characteristic one and a quantity which can be made small by choosing $\gamma$ sufficiently close to 1. This suggests that the crucial information regarding regularity is the following statement:

*Non-concentration of energy inside a cone of aperture $\gamma < 1$ implies regularity.*

This will be proven by combining the above estimate with the energy dispersion result of Theorem 4.2.

In what concerns the left-hand side of (4.100), the first term provides information which is similar to the one given by the corresponding term in (4.80); i.e., the derivative terms $|(\nabla_t + \nabla_r)\phi|^2$ and $|\nabla_\theta \phi|^2 / r^2$ do not concentrate near the characteristic cone. For the second integral term, the obvious bound $\rho \le t$ implies the following time decay estimate:

$$\int_{\rho_1}^{t_1} \frac{1}{t} \left( \int_{\tilde{D}_t^{\rho_1}} |X(\phi)|^2 d\mathbf{x} \right) dt \le C(1 - \gamma^2)^{1/4} E + \frac{1}{T} \int_0^T \mathcal{E}_t^\gamma(\phi) dt.$$

Since $t_1 / \rho_1 = (1 - \gamma^2)^{1/4} E / \delta_F(t_1, \phi)$ can be made arbitrarily large by choosing $t_1$ sufficiently small, we deduce that the map $t \mapsto \int_{\tilde{D}_t^{\rho_1}} |X(\phi)|^2 d\mathbf{x}$ has to be small on a subinterval of $(\rho_1, t_1)$. We will give details on this heuristic later in the argument.

### 4.6.5    *Propagation of energy concentration*

In this subsection, our goal is to show that, if a wave map exhibits concentration of energy at a certain time strictly inside the characteristic cone, then, for the backward-in-time evolution, the same phenomenon will also happen for other, sufficiently small times.

In order to formalize this statement, let us consider a wave map $\phi$ defined on the cone $\mathcal{C}_{(0,t_1]}$ and a fixed, positive parameter $\gamma$ such that

$$\frac{\delta_F(t_1, \phi)}{E} \leq (1 - \gamma^2)^{3/4} \ll 1.$$

We choose

$$s \in \left( \frac{t_1 \delta_F(t_1, \phi)}{(1 - \gamma^2)^{3/4} E}, t_1 \right), \qquad p \geq \frac{t_1 \delta_F(t_1, \phi)}{(1 - \gamma^2)^{1/4} E},$$

and we define $\widetilde{X}_p$ to be the vector field given by (4.81), with $\rho_0$ replaced by $p$. As before, we have the differential identity $\nabla_\mu \{ P^{\widetilde{X}_p, \mu} \} = R^{\widetilde{X}_p}$, where $R^{\widetilde{X}_p}$ is prescribed by (4.82) with $\rho_0 = p$, and we integrate it over the frustum $\mathcal{C}_{[s,t_1]}$. We deduce that

$$\int_{D_{t_1}} \{ P_0^{\widetilde{X}_p} \} d\mathbf{x} + \int_{\mathcal{R}_p^{[s,t_1]}} \left\{ \frac{1}{\rho} |X(\phi)|^2 \right\} dt\, d\mathbf{x}$$

$$= \int_{D_s} \{ P_0^{\widetilde{X}_p} \} d\mathbf{x} + \frac{1}{2p} \int_{\mathcal{Q}_p^{[s,t_1]}} \langle \nabla \phi, \nabla \phi \rangle dt\, d\mathbf{x}$$

$$+ \frac{1}{\sqrt{2}} \int_{\partial \mathcal{C}_{[s,t_1]}} \frac{t}{p} |(\nabla_t + \nabla_r)\phi|^2 d\sigma,$$

where

$$\mathcal{R}_p^{[s,t_1]} \overset{\text{def}}{=} \left\{ (t, \mathbf{x}) \,\middle|\, p \leq \rho, \; s \leq t \leq t_1 \right\},$$

$$\mathcal{Q}_p^{[s,t_1]} \overset{\text{def}}{=} \left\{ (t, \mathbf{x}) \,\middle|\, 0 \leq \rho < p, \; s \leq t \leq t_1 \right\}.$$

In this context, we prove that, under the additional assumption

$$s \geq \frac{p}{(1 - \gamma^2)^{1/2}}, \tag{4.101}$$

the following *comparison estimate* holds:

$$\int_{D_s} \{ P_0^{\widetilde{X}_p} \} d\mathbf{x} \geq \int_{D_{t_1}} \{ P_0^{\widetilde{X}_p} \} d\mathbf{x} - C(1 - \gamma^2)^{1/4} E. \tag{4.102}$$

On the basis of the above integral identity, we achieve this if we show

$$\frac{1}{2p} \int_{\mathcal{Q}_p^{[s,t_1]}} \langle \nabla\phi, \nabla\phi \rangle dt\, dx + \frac{1}{\sqrt{2}} \int_{\partial\mathcal{C}_{[s,t_1]}} \frac{t}{p} |(\nabla_t + \nabla_r)\phi|^2 d\sigma \leq C(1-\gamma^2)^{1/4} E.$$

For the second integral, it is easy to see that

$$\frac{1}{\sqrt{2}} \int_{\partial\mathcal{C}_{[s,t_1]}} \frac{t}{p} |(\nabla_t + \nabla_r)\phi|^2 d\sigma \leq \frac{2t_1 \delta_F(t_1, \phi)}{p} \leq 2(1-\gamma^2)^{1/4} E.$$

In what concerns the integral over $\mathcal{Q}_p^{[s,t_1]}$, we notice first that (4.101) implies

$$\left(1 - r^2/t^2\right)^{1/2} < p/s \leq (1-\gamma^2)^{1/2},$$

which yields $r/t > \gamma$. This points us in the direction of an analysis similar to the one that produced (4.99) (i.e., the integral estimate corresponding to the domain $\widetilde{\mathcal{Q}}_{[\rho_0,\rho_1]}^{t_1,\gamma}$). Following the same steps, we derive

$$\frac{1}{2p} \int_{\mathcal{Q}_p^{[s,t_1]}} \langle \nabla\phi, \nabla\phi \rangle dt\, dx$$

$$\leq C_1(1-\gamma^2)^{1/2} E + C_2 E^{1/2} \left( \frac{1}{p} \cdot \int_{\mathcal{Q}_p^{[s,t_1]}} e^{-\chi} |X(\phi)|^2 dt\, dx \right)^{1/2}.$$

Unfortunately, the last term cannot be bounded like before, using $e^{-\chi} < (1-\gamma^2)^{1/2}$ and (4.89). This is due to the fact that the hyperbolic radial coordinate $\rho$ can be arbitrarily small on $\mathcal{Q}_p^{[s,t_1]}$ and, as a consequence, we cannot apply (4.89). Instead, we rely on (4.80) and infer that

$$\int_{\mathcal{Q}_p^{[s,t_1]}} \frac{1}{\sqrt{\rho^2 + T^2 + 2tT}} |X(\phi)|^2 dt\, dx \leq CE, \qquad \text{with} \quad T = \frac{t_1 \delta_F^2(t_1, \phi)}{E^2}.$$

However, it is a routine verification to check that, under our assumptions,

$$\frac{e^{-\chi}}{p} \leq C \frac{(1-\gamma^2)^{1/2}}{\sqrt{\rho^2 + T^2 + 2t_1 T}},$$

which implies

$$\frac{1}{p} \int_{\mathcal{Q}_p^{[s,t_1]}} e^{-\chi} |X(\phi)|^2 dt\, dx \leq C(1-\gamma^2)^{1/2} E.$$

This shows that

$$\frac{1}{2p} \int\limits_{\mathcal{Q}_p^{[s,t_1]}} \langle \nabla\phi, \nabla\phi \rangle dt\, dx \leq C(1-\gamma^2)^{1/4}E,$$

thus finishing the argument for (4.102).

The comparison estimate represents one of the key pieces in the argument for the next result, which makes precise our initial statement on the propagation of energy concentration.

**Lemma 4.5.** *Let $\phi$ be a wave map in $\mathcal{C}_{(0,t_1]}$, for which energy concentration is present inside a cone of aperture $\gamma_1 < 1$ at time $t = t_1$, i.e.,*

$$\int\limits_{|\mathbf{x}|\leq\gamma_1 t_1} \{e(\phi(t_1))\}d\mathbf{x} \geq E_1 > 0, \tag{4.103}$$

*and its lateral flux satisfies*

$$\delta_F(t_1, \phi) \ll (1-\gamma_1^2)^{3/2}.$$

*Then, there exist $\gamma_2 = \gamma_2(\gamma_1, E_1, E)$ and $E_2 = E_2(\gamma_1, E_1, E)$, which are two positive parameters satisfying*

$$\gamma_1 < \gamma_2 < 1$$

*and*

$$\int\limits_{|\mathbf{x}|\leq\gamma_2 t} \{e(\phi(t))\}d\mathbf{x} \geq E_2 > 0, \qquad \forall t \in \left[\frac{Ct_1\delta_F(t_1, \phi)}{(1-\gamma_1^2)^{3/2}E}, t_1\right]. \tag{4.104}$$

*Moreover, one can choose $\gamma_2$ to satisfy $(1-\gamma_2^2)^{1/2} \approx (1-\gamma_1^2)^{3/2}$.*

**Proof.** Apart from the comparison bound (4.102) mentioned above, another important ingredient of the argument is the ability to compare $P_0^{\tilde{X}}(\phi)$ to $e(\phi)$ in domains of interest to the estimates in the lemma.

We start by considering $\gamma_1 < \gamma < 1$, with $\gamma$ to be specified later, and by choosing

$$s \in \left[\frac{t_1\delta_F(t_1, \phi)}{(1-\gamma^2)^{3/4}E}, t_1\right] \qquad \text{and} \qquad p = s(1-\gamma^2)^{1/2}.$$

It is easy to check that, for $|\mathbf{x}| \leq \gamma_1 t_1$, we have

$$\rho(t_1, \mathbf{x}) = \sqrt{t_1^2 - |\mathbf{x}|^2} \geq p.$$

Hence, for the vector field $\widetilde{X}_p$ introduced in the context of the comparison estimate, we deduce that the following inequality holds in the domain $\{(t_1, \mathbf{x}) \mid |\mathbf{x}| \leq \gamma_1 t_1\}$:

$$
\begin{aligned}
P_0^{\widetilde{X}_p} &= \frac{t_1 + r}{4\rho} \left| (\nabla_t + \nabla_r)\phi \right|^2 + \frac{t_1 - r}{4\rho} \left| (\nabla_t - \nabla_r)\phi \right|^2 + \frac{t_1}{2\rho r^2} |\nabla_\theta \phi|^2 \\
&\geq \frac{t_1 - r}{\rho} \left[ \frac{1}{4} \left| (\nabla_t + \nabla_r)\phi \right|^2 + \frac{1}{4} \left| (\nabla_t - \nabla_r)\phi \right|^2 + \frac{1}{2r^2} |\nabla_\theta \phi|^2 \right] \\
&= \sqrt{\frac{t_1 - r}{t_1 + r}} \, e(\phi) \geq \frac{(1 - \gamma_1^2)^{1/2}}{2} e(\phi).
\end{aligned}
\tag{4.105}
$$

As a consequence, we infer based on (4.103) that

$$
\int_{D_{t_1}} \{P_0^{\widetilde{X}_p}\} d\mathbf{x} \geq \frac{(1 - \gamma_1^2)^{1/2}}{2} E_1.
$$

Due to the assumptions made so far on $\gamma$, $s$, and $p$, we can combine the previous bound with the comparison estimate (4.102) and derive

$$
\begin{aligned}
\int_{D_s} \{P_0^{\widetilde{X}_p}\} d\mathbf{x} &\geq \int_{D_{t_1}} \{P_0^{\widetilde{X}_p}\} d\mathbf{x} - C(1 - \gamma^2)^{1/4} E \\
&\geq \frac{(1 - \gamma_1^2)^{1/2}}{2} E_1 - C(1 - \gamma^2)^{1/4} E.
\end{aligned}
\tag{4.106}
$$

Next, we work with the vector field $\widetilde{X}$, which is given by (4.81) with $t_1$ replaced by $s$, i.e.,

$$
\widetilde{X} = \widetilde{X}^\mu \nabla_\mu \overset{\text{def}}{=} 
\begin{cases}
\dfrac{x^\mu}{\rho} \nabla_\mu & \text{if } \rho_0 \leq \rho, \\[2mm]
\dfrac{x^\mu}{\rho_0} \nabla_\mu & \text{if } 0 \leq \rho < \rho_0,
\end{cases}
$$

where

$$
\rho_0 = \frac{s\delta_F(s, \phi)}{E}.
$$

As $s \leq t_1$, we have that $\delta_F(s, \phi) \leq \delta_F(t_1, \phi) \ll E$ and, thus, we can rely on (4.89) to infer that

$$
\int_{D_s} \{P_0^{\widetilde{X}}\} d\mathbf{x} \leq CE.
\tag{4.107}
$$

Following this, we choose another parameter $\gamma_2$ such that $\gamma < \gamma_2 < 1$ and focus on the region

$$
\{(s, \mathbf{x}) \mid \gamma_2 s \leq |\mathbf{x}| \leq s\},
$$

where, based on our previous assumptions,
$$0 \leq \rho \leq s(1 - \gamma_2^2)^{1/2} < p.$$
If we denote by $\tilde{\rho}_v(x) := \max\{\rho(x), v\}$, then it is straightforward to argue that, in this domain, we have the estimate
$$P_0^{\tilde{X}_p} \leq \sup_{\{(s,\mathbf{x}) \mid \gamma_2 s \leq |\mathbf{x}| \leq s\}} \left\{ \frac{\tilde{\rho}_{\rho_0}(s, \mathbf{x})}{\tilde{\rho}_p(s, \mathbf{x})} \right\} P_0^{\tilde{X}} = \frac{(1 - \gamma_2^2)^{1/2}}{(1 - \gamma^2)^{1/2}} P_0^{\tilde{X}}.$$
Combining this bound with (4.107), we obtain
$$\int_{\gamma_2 s \leq |\mathbf{x}| \leq s} \{P_0^{\tilde{X}_p}\} d\mathbf{x} \leq C \frac{(1 - \gamma_2^2)^{1/2}}{(1 - \gamma^2)^{1/2}} E.$$
This new estimate and (4.106) then yield
$$\int_{|\mathbf{x}| \leq \gamma_2 s} \{P_0^{\tilde{X}_p}\} d\mathbf{x} \geq \frac{(1 - \gamma_1^2)^{1/2}}{2} E_1 - C_1 (1 - \gamma^2)^{1/4} E$$

$$\qquad (4.108)$$

$$- C_2 \frac{(1 - \gamma_2^2)^{1/2}}{(1 - \gamma^2)^{1/2}} E.$$
Finally, we show that, in the region $\{(s, \mathbf{x}) \mid |\mathbf{x}| \leq \gamma_2 s\}$, the energy density $e(\phi)$ dominates $P_0^{\tilde{X}_p}$. Indeed, by direct inspection, we can argue as follows:
$$e(\phi) \geq \frac{\rho}{s + r} P_0^{\tilde{X}_p} \geq \frac{(1 - \gamma_2^2)^{1/2}}{2} P_0^{\tilde{X}_p}.$$
As a consequence, on the basis of (4.108), we deduce that
$$\int_{|\mathbf{x}| \leq \gamma_2 s} \{e(\phi(s))\} d\mathbf{x} \geq \frac{(1 - \gamma_2^2)^{1/2}}{2} \left[ \frac{(1 - \gamma_1^2)^{1/2}}{2} E_1 - C_1 (1 - \gamma^2)^{1/4} E \right.$$

$$\left. - C_2 \frac{(1 - \gamma_2^2)^{1/2}}{(1 - \gamma^2)^{1/2}} E \right].$$
Now, all that is left to do in order to claim the estimate (4.104) is to choose appropriately the positive parameters $\gamma$ and $\gamma_2$ in accord with the assumptions made so far on them, and such that the right-hand side of the previous estimate is bounded from below by a positive constant for
$$\frac{C t_1 \delta_F(t_1, \phi)}{(1 - \gamma_1^2)^{3/2} E} \leq s \leq t_1.$$
This is done by first letting
$$(1 - \gamma^2)^{1/4} = \frac{(1 - \gamma_1^2)^{1/2} E_1}{C_1 M E},$$
with $M \gg 1$ such that $\gamma_1 < \gamma < 1$. We finish the proof by choosing
$$(1 - \gamma_2^2)^{1/2} = \frac{(\frac{1}{2} - \frac{1}{M})(1 - \gamma_1^2)^{3/2} E_1^3}{C_1^2 C_2 M^2 N E^3},$$
with $N \gg 1$ such that $\gamma < \gamma_2 < 1$. $\qquad \square$

### 4.6.6  *Energy concentration scenarios*

Here, we apply previously obtained energy-type estimates to split the argument in complementary cases related to the concentration of the energy. The key piece is Lemma 4.4, whose main result is the inequality (4.100). For notational convenience, we relabel the aperture $\gamma$ in the lemma by $\gamma_1$, and we recall that

$$\rho_1 = \frac{t_1 \delta_F(t_1, \phi)}{(1 - \gamma_1^2)^{1/4} E}, \qquad T = \frac{t_1 \delta_F(t_1, \phi)}{(1 - \gamma_1^2)^{3/4} E} \leq t_1. \qquad (4.109)$$

We focus on the first term on the left-hand side of (4.100), for which one has

$$P_0^{\widetilde{X}} = \frac{t_1 + r}{4 \widetilde{\rho}_{\rho_1}} \left| (\nabla_t + \nabla_r) \phi \right|^2 + \frac{t_1 - r}{4 \widetilde{\rho}_{\rho_1}} \left| (\nabla_t - \nabla_r) \phi \right|^2 + \frac{t_1}{2 \widetilde{\rho}_{\rho_1} \, r^2} \left| \nabla_\theta \phi \right|^2,$$

where $\widetilde{\rho}_{\rho_1}(x) := \max\{\rho(x), \rho_1\}$.

If we are in the region $\{|\mathbf{x}| \leq \gamma_1 t_1\}$ of $D_{t_1}$, then

$$\rho \geq (1 - \gamma_1^2)^{1/2} t_1 \geq \rho_1.$$

Hence, we deduce

$$P_0^{\widetilde{X}} \geq \sqrt{\frac{t_1 - r}{t_1 + r}} \, e(\phi),$$

which implies

$$\int_{|\mathbf{x}| \leq \gamma_1 t_1} \left\{ \sqrt{1 - \frac{r^2}{t_1^2}} \, e(\phi(t_1)) \right\} d\mathbf{x} \leq 2 \int_{|\mathbf{x}| \leq \gamma_1 t_1} P_0^{\widetilde{X}} \, d\mathbf{x}. \qquad (4.110)$$

On the other hand, if we are in the region $\{\gamma t_1 \leq |\mathbf{x}| \leq t_1\}$ of $D_{t_1}$, with $0 \leq \gamma \leq \gamma_1$ arbitrary, then

$$\rho \leq (1 - \gamma^2)^{1/2} t_1 \qquad \text{and} \qquad \rho_1 \leq (1 - \gamma_1^2)^{1/2} t_1 \leq (1 - \gamma^2)^{1/2} t_1.$$

It follows that

$$P_0^{\widetilde{X}} \geq \frac{(1 - \gamma^2)^{-1/2}}{4} \left[ \left| (\nabla_t + \nabla_r) \phi \right|^2 + \frac{1}{r^2} \left| \nabla_\theta \phi \right|^2 \right],$$

and, as a consequence,

$$\int_{\gamma t_1 \leq |\mathbf{x}| \leq t_1} \left\{ \left| (\nabla_t + \nabla_r) \phi(t_1) \right|^2 + \frac{1}{r^2} \left| \nabla_\theta \phi(t_1) \right|^2 \right\} d\mathbf{x}$$

$$\leq 4(1 - \gamma_1^2)^{1/2} \int_{\gamma t_1 \leq |\mathbf{x}| \leq t_1} P_0^{\widetilde{X}} \, d\mathbf{x}. \qquad (4.111)$$

Therefore, if we are under the hypotheses of Lemma 4.4 and we use the notation

$$(1 - \gamma_1^2)^{1/4} E = \epsilon_1^{3/2}, \qquad (4.112)$$

then we derive on the bases of (4.100), (4.110), and (4.111) that

$$\int_{|\mathbf{x}| \leq \gamma_1 t_1} \left\{ \sqrt{1 - \frac{r^2}{t_1^2}} \, e(\phi(t_1)) \right\} d\mathbf{x} + \int_{\mathcal{R}_{\rho_1}^{t_1}} \left\{ \frac{1}{\rho} |X(\phi)|^2 \right\} dt \, d\mathbf{x}$$

$$\leq C \left( \epsilon_1^{3/2} + \frac{1}{T} \int_0^T \mathcal{E}_t^{\gamma_1}(\phi) dt \right) \qquad (4.113)$$

and

$$\int_{\gamma t_1 \leq |\mathbf{x}| \leq t_1} \left\{ \left|(\nabla_t + \nabla_r)\phi(t_1)\right|^2 + \frac{1}{r^2} |\nabla_\theta \phi(t_1)|^2 \right\} d\mathbf{x}$$

$$\leq C(1 - \gamma_1^2)^{1/2} \left( \epsilon_1^{3/2} + \frac{1}{T} \int_0^T \mathcal{E}_t^{\gamma_1}(\phi) dt \right). \qquad (4.114)$$

These are precisely the two estimates, (4.33) and (4.32), claimed in the second step of the outline. Henceforth, the argument will be split according to the respective sizes for the two terms appearing on the right-hand side of the above bounds.

This is done by considering a wave map $\phi$ in $\mathcal{C}_{(0,1]}$, with $E > 0$ given by (4.23). We choose $\epsilon_1 > 0$ sufficiently small[9] such that we can define an aperture $0 < \gamma_1 < 1$ through (4.112). The decay of the lateral flux allows us to find $\tilde{t} \in (0, 1]$ satisfying

$$\sup_{t_1 \in (0, \tilde{t}]} \delta_F(t_1, \phi) \leq (1 - \gamma_1^2)^{3/4} E \approx \epsilon_1^{9/2}.$$

In this context, the first energy concentration scenario is:

**I. Uniform non-concentration of energy inside a cone of aperture less than 1.**

This is described by the existence of $\tilde{T} \in (0, 1]$ with

$$\sup_{T \in (0, \tilde{T}]} \frac{1}{T} \int_0^T \mathcal{E}_t^{\gamma_1}(\phi) \, dt \leq \epsilon_1^{3/2}. \qquad (4.115)$$

Therefore, for all $t_1 \leq \min\{\tilde{t}, \tilde{T}\}$, we can apply Lemma 4.4 and infer, due to (4.113) and (4.114), that both

$$\int_{|\mathbf{x}| \leq \gamma t_1} \{e(\phi(t_1))\} d\mathbf{x} \leq C(1 - \gamma^2)^{-1/2} \epsilon_1^{3/2}$$

---

[9]Later, we will calibrate $\epsilon_1 = \epsilon^8$, where $\epsilon$ is the energy dispersion threshold.

and
$$\int_{\gamma t_1 \leq |\mathbf{x}| \leq t_1} \left\{ \left|(\nabla_t + \nabla_r)\phi(t_1)\right|^2 + \frac{1}{r^2}\left|\nabla_\theta\phi(t_1)\right|^2 \right\} \, d\mathbf{x} \leq C(1 - \gamma_1^2)^{1/2}\epsilon_1^{3/2}$$

hold, with an arbitrary $0 \leq \gamma \leq \gamma_1$. Eventually readjusting the size of $\epsilon_1$, we can pick $\gamma$ such that $1 - \gamma^2 = \epsilon_1$ and thus obtain

$$\int_{|\mathbf{x}| \leq \gamma t_1} \{e(\phi(t_1))\} \, d\mathbf{x} \leq C\epsilon_1 \qquad (4.116)$$

and

$$\int_{\gamma t_1 \leq |\mathbf{x}| \leq t_1} \left\{ \left|(\nabla_t + \nabla_r)\phi(t_1)\right|^2 + \frac{1}{r^2}\left|\nabla_\theta\phi(t_1)\right|^2 \right\} d\mathbf{x} \qquad (4.117)$$
$$\leq C(1 - \gamma^2)^{1/2}\epsilon_1.$$

These two estimates will form the basis for the subsequent analysis of this case.

The complementary energy concentration scenario is:

*II. Energy concentration inside a cone of aperture less than 1.*

In this instance, we assume that there exists a sequence $(T_n)_n \in (0, 1]$ with $T_n \to 0$ and

$$\frac{1}{T_n} \int_0^{T_n} \mathcal{E}_t^{\gamma_1}(\phi) \, dt > \epsilon_1^{3/2}. \qquad (4.118)$$

Using the mean value theorem, we derive the existence of a sequence $(t_n)_n \in (0, 1]$ such that $0 < t_n < T_n$ and

$$\mathcal{E}_{t_n}^{\gamma_1}(\phi) = \int_{|\mathbf{x}| \leq \gamma_1 t_n} \{e(\phi(t_n))\} \, d\mathbf{x} \geq \epsilon_1^{3/2} = E_2 > 0.$$

Moreover, by choosing $n$ sufficiently large, we can guarantee that

$$E \leq \mathcal{E}_{D_t}(\phi) \leq (1 + \epsilon^8)E, \qquad \forall 0 < t \leq t_n.$$

Together with the energy-type estimate (4.78) and the energy propagation result (i.e., Lemma 4.5), these two bounds are the most important pieces in the investigation of this scenario.

## 4.7   Energy dispersion and non-concentration of energy

In this section, we show that the non-concentration of energy scenario is incompatible with the failure of energy dispersion, i.e., $E_D[\phi] > \epsilon$. We achieve this by assuming that both are true and showing that this leads to a contradiction for a small enough $\epsilon$. The main idea in the argument is to relate the energy dispersion to the physical energy.

### 4.7.1   *Preliminaries and formulation of the main goal*

Our initial setup is described by a wave map $\phi$ defined on $\mathcal{C}_{(0,1]}$ and satisfying

$$\lim_{t \to 0} \mathcal{E}_{D_t}(\phi) = E > 0,$$

for which an extension was constructed on $(0, t_0] \times \mathbb{R}^2$ such that

$$\mathcal{E}[\phi] \le (1 + \epsilon^8)E \qquad \text{and} \qquad E_D[\phi] > \epsilon.$$

The last bound, which indicates the lack of energy dispersion, implies the existence of a sequence of points $(t_n, \mathbf{x}_n) \in (0, t_0] \times \mathbb{R}^2$ and associated frequencies $\lambda_n \in 2^{\mathbb{Z}}$ that comply with $t_n \to 0$ and

$$\left| S_{\lambda_n}[\phi](t_n, \mathbf{x}_n) \right| + \lambda_n^{-1} \left| S_{\lambda_n}[\nabla_t \phi](t_n, \mathbf{x}_n) \right| > \epsilon.$$

If we are also in the non-concentration of energy scenario, then, for $n$ sufficiently large, we have access to the bounds (4.116) and (4.117), for the corresponding regions of $D_{t_n}$. Hence, the following four energy-type estimates are true:

$$\int_{|\mathbf{x}| \le t_n} \{e(\phi(t_n))\} d\mathbf{x} \approx E,$$

$$\int_{|\mathbf{x}| \ge t_n} \{e(\phi(t_n))\} d\mathbf{x} \le C\epsilon^8 E,$$

$$\int_{|\mathbf{x}| \le \gamma t_n} \{e(\phi(t_n))\} d\mathbf{x} \le C\epsilon_1,$$

$$\int_{\gamma t_n \le |\mathbf{x}| \le t_n} \left\{ \left| (\nabla_t + \nabla_r)\phi(t_n) \right|^2 + \frac{1}{r^2} \left| \nabla_\theta \phi(t_n) \right|^2 \right\} d\mathbf{x} \le C(1 - \gamma^2)^{1/2} \epsilon_1.$$

In order to simplify somewhat the subsequent analysis, we perform a rescaling procedure that shifts $t_n$ into 1. Thus, we obtain a sequence of wave maps $\phi^{(n)}$ given by

$$\phi^{(n)}(x) = \phi(t_n x), \qquad x \in \mathcal{C}_{(0, t_n^{-1}]} \cap \left( (0, t_n^{-1} t_0] \times \mathbb{R}^2 \right),$$

which, after relabeling $t_n \lambda_n$ by $\lambda_n$ and $t_n^{-1} \mathbf{x}_n$ by $\mathbf{x}_n$, satisfies

$$\left| S_{\lambda_n}[\phi^{(n)}](1, \mathbf{x}_n) \right| + \lambda_n^{-1} \left| S_{\lambda_n}[\nabla_t \phi^{(n)}](1, \mathbf{x}_n) \right| > \epsilon,$$

$$\int_{|\mathbf{x}|\leq 1} \{e(\phi^{(n)}(1))\}d\mathbf{x} \approx E,$$

$$\int_{|\mathbf{x}|\geq 1} \{e(\phi^{(n)}(1))\}d\mathbf{x} \leq C\epsilon^8 E,$$

$$\int_{|\mathbf{x}|\leq\gamma} \{e(\phi^{(n)}(1))\}d\mathbf{x} \leq C\epsilon_1,$$

$$\int_{\gamma\leq|\mathbf{x}|\leq 1} \left\{\left|(\nabla_t + \nabla_r)\phi^{(n)}(1)\right|^2 + \frac{1}{r^2}\left|\nabla_\theta\phi^{(n)}(1)\right|^2\right\} d\mathbf{x} \leq C(1-\gamma^2)^{1/2}\epsilon_1.$$

Before moving forward, we calibrate the parameters $\gamma$, $\epsilon$, and $\epsilon_1$, in the sense that we pick

$$1 - \gamma^2 = \epsilon_1 = \epsilon^8.$$

This calibration is not very sensitive, but it is allowed by the assumptions made on $\gamma$ and $\epsilon_1$ during the discussion of energy concentration scenarios. Under these conditions, the main goal is to prove that the four energy-type estimates imply

$$\|S_{\lambda_n}[\phi^{(n)}](1)\|_{L^\infty(\mathbb{R}^2)} + \lambda_n^{-1}\|S_{\lambda_n}[\nabla_t\phi^{(n)}](1)\|_{L^\infty(\mathbb{R}^2)} \ll \epsilon,$$

if $\epsilon$ is sufficiently small. This comes in contradiction to the pointwise bound satisfied by $\phi^{(n)}$, thus ruling out the non-concentration of energy scenario in the absence of energy dispersion.

As it turns out, the argument does not depend on the fact that $\phi^{(n)}$ are wave maps, nor on whether we are working with the whole sequence of wave maps or with just one of its terms. This is why, for what follows, we drop the index $n$ and use the original notation $\phi$ to designate any of these wave maps. We rewrite our main goal in terms of $\phi$ and state it as a proposition.

**Proposition 4.2.** *Let $\phi = \phi(t, \mathbf{x})$ be a wave map defined on $(0, T_0] \times \mathbb{R}^2$, with $T_0 > 1$, and satisfying*

$$\int_{|\mathbf{x}|\leq 1} \{e(\phi(1))\}d\mathbf{x} \approx E, \tag{4.119}$$

$$\int_{|\mathbf{x}|\geq 1} \{e(\phi(1))\}d\mathbf{x} \leq C\epsilon^8 E, \tag{4.120}$$

$$\int\limits_{|\mathbf{x}|\leq\gamma} \{e(\phi(1))\}d\mathbf{x} \leq C\epsilon^8, \tag{4.121}$$

$$\int\limits_{\gamma\leq|\mathbf{x}|\leq 1} \left\{ |(\nabla_t + \nabla_r)\phi(1)|^2 + \frac{1}{r^2}|\nabla_\theta\phi(1)|^2 \right\} d\mathbf{x} \leq C\epsilon^{12}, \tag{4.122}$$

*where* $1 - \gamma^2 = \epsilon^8$. *If* $\epsilon$ *is sufficiently small, then*

$$\|S_\lambda[\phi](1)\|_{L^\infty(\mathbb{R}^2)} + \lambda^{-1}\|S_\lambda[\nabla_t\phi](1)\|_{L^\infty(\mathbb{R}^2)} \ll \epsilon \tag{4.123}$$

*holds uniformly for* $\lambda \in 2^{\mathbb{Z}}$.

### 4.7.2   *A reduction step and initial gauge construction*

As a first step toward proving the previous proposition, we show that it is enough to argue that

$$\|S_\lambda[\phi](1)\|_{L^\infty(\mathbb{R}^2)} \ll \epsilon. \tag{4.124}$$

For this purpose, we recall the spatial Littlewood-Paley projector at frequency $\lambda \in 2^{\mathbb{Z}}$, which is defined using a smooth, radial function $g = g(\mathbf{x})$, whose Fourier transform $\widehat{g} = \widehat{g}(\xi)$ is also radial, nonnegative, and compactly supported in the region $\{1/2 \leq |\xi| \leq 2\}$. The projector is defined via the Fourier transform as

$$\widehat{S_\lambda[\psi]}(\xi) \stackrel{\text{def}}{=} \widehat{g}(\xi/\lambda)\widehat{\psi}(\xi), \tag{4.125}$$

while in physical space it takes the form of the convolution

$$S_\lambda[\psi](\mathbf{x}) = \int\limits_{\mathbb{R}^2} \lambda^2 g(\lambda(\mathbf{x} - \mathbf{y}))\psi(\mathbf{y})d\mathbf{y}. \tag{4.126}$$

It is important to notice that, due to the support of $\widehat{g}$, we have $\int_{\mathbb{R}^2} g = 0$, which implies

$$S_\lambda[\psi - c] = S_\lambda[\psi], \tag{4.127}$$

for any arbitrary constant $c$.

With these prerequisites, let us show that the $L^\infty$ estimate for $\lambda^{-1}S_\lambda[\nabla_t\phi](1)$ reduces to the one for $S_\lambda[\phi](1)$. We introduce a smooth cutoff function $b = b(r)$, supported in the interval $[1/2, 2]$, which helps us rewrite $\lambda^{-1}S_\lambda[\nabla_t\phi](1)$ in a convenient form. This is achieved by using the expression $b(|\mathbf{y}|)\nabla_r\phi(1,\mathbf{y})$ and integrating by parts as follows:

$$\lambda^{-1}S_\lambda[\nabla_t\phi](1,\mathbf{x})$$
$$= \int_{\mathbb{R}^2} \left\{ \lambda g(\lambda(\mathbf{x}-\mathbf{y}))\nabla_t\phi(1,\mathbf{y}) \right\} dy$$
$$= \int_{\mathbb{R}^2} \left\{ \lambda g(\lambda(\mathbf{x}-\mathbf{y})) \left(\nabla_t + b(|\mathbf{y}|)\nabla_r\right)\phi(1,\mathbf{y}) \right\} dy$$
$$- \int_{\mathbb{R}^2} \left\{ \lambda g(\lambda(\mathbf{x}-\mathbf{y}))b(|\mathbf{y}|)\nabla_r\phi(1,\mathbf{y}) \right\} dy$$
$$= \lambda^{-1}S_\lambda\left[(\nabla_t + b\,\nabla_r)\phi\right](1,\mathbf{x}) + \lambda^{-1}S_\lambda\left[(r^{-1}b + b_r)\phi\right](1,\mathbf{x})$$
$$- \int_{\mathbb{R}^2} \left\{ \lambda^2\nabla_j g(\lambda(\mathbf{x}-\mathbf{y}))\frac{b(|\mathbf{y}|)y^j}{|\mathbf{y}|}\phi(1,\mathbf{y}) \right\} dy.$$

For the term involving the directional derivative $\nabla_t + b\,\nabla_r$, we deduce based on (4.120)-(4.122) that

$$\|(\nabla_t + b\nabla_r)\phi(1)\|_{L^2}$$
$$\leq C\big(\|\nabla\phi(1)\|_{L^2(|\mathbf{x}|\leq\gamma)} + \|(\nabla_t + \nabla_r)\phi(1)\|_{L^2(\gamma\leq|\mathbf{x}|\leq 1)}$$
$$+ \|\nabla\phi(1)\|_{L^2(|\mathbf{x}|\geq 1)}\big)$$
$$\leq C\epsilon^4 E^{1/2}.$$

Hence, applying the Cauchy-Schwarz inequality, we derive

$$\lambda^{-1}\left|S_\lambda\left[(\nabla_t + b\,\nabla_r)\phi\right](1,\mathbf{x})\right| \leq \|g\|_{L^2}\|(\nabla_t + b\nabla_r)\phi(1)\|_{L^2}$$
$$\leq C\epsilon^4 E^{1/2}. \tag{4.128}$$

For the last two terms, the claim is that they can be treated with the general method we are going to use for $S_\lambda[\phi](1)$. The key facts are:

- the map $\phi$ appears in both of them without an explicit derivative attached to it;
- the term left in integral form, when compared to (4.126), can be seen as

$$- \left(S_{\lambda,1}\phi^1 + S_{\lambda,2}\phi^2\right),$$

where $S_{\lambda,j}$ is the Littlewood-Paley projector defined with $g$ replaced by $\nabla_j g$ and $\phi^j := by^j\phi/|\mathbf{y}|$;
- the functions $\nabla_j g$ have properties similar to the ones for $g$ (e.g., their Fourier transform is radial and compactly supported in $\{1/2 \leq |\xi| \leq 2\}$, and their integral on $\mathbb{R}^2$ is zero).

Thus, the analysis for $\lambda^{-1}S_\lambda[\nabla_t\phi](1)$ winds down to proving (4.125), the $L^\infty$ estimate for $S_\lambda[\phi](1)$.

With this reduction step out of the way, we focus next on constructing a vector field $\mathbf{u}$ whose divergence is $g$, i.e.,

$$\mathbf{u} = (u^1, u^2), \qquad \operatorname{div} \mathbf{u} = \nabla_1 u^1 + \nabla_2 u^2 = g, \qquad (4.129)$$

which is done in order to relate the projector to the energy. This can be seen through an integration by parts, that leads to

$$
\begin{aligned}
S_\lambda[\phi](1,\mathbf{x}) &= -\int_{\mathbb{R}^2} \lambda \nabla_j \left\{ u^j\left(\lambda(\mathbf{x} - \mathbf{y})\right) \right\} \phi(1,\mathbf{y}) dy \\
&= \int_{\mathbb{R}^2} \lambda\, u^j\left(\lambda(\mathbf{x} - \mathbf{y})\right) \nabla_j \phi(1,\mathbf{y}) dy,
\end{aligned}
\qquad (4.130)
$$

and by applying the Cauchy-Schwarz inequality in the last integral. Moreover, an important observation is that we have a gauge freedom in choosing $\mathbf{u}$, in the sense that we are free to prescribe its curl, $\vec{\nabla} \wedge \mathbf{u}$. This allows us to pick the vector field such that we have good control of the directional derivative $u^j \nabla_j \phi(1)$ through (4.119)-(4.122). The subsequent analysis is split into the complementary cases of low and high frequency, respectively. For the former, we use a simple choice of gauge, which is radial. For the high frequency case, we need to localize first and then consider a gauge pointing in certain preferred directions, together with a small correction.

A simple way to construct $\mathbf{u}$, which takes advantage of $g$ being radial, is to assume that $\mathbf{u}$ is radial too. Namely, making the ansatz

$$\mathbf{u}(\mathbf{x}) = a(|\mathbf{x}|)\,\mathbf{x},$$

we deduce

$$\operatorname{div} \mathbf{u} = |\mathbf{x}|a' + 2a = g,$$

and, thereafter,

$$\mathbf{u}(\mathbf{x}) = \left( \frac{1}{|\mathbf{x}|^2} \int_0^{|\mathbf{x}|} sg(s)\, ds \right) \mathbf{x}. \qquad (4.131)$$

In the above, we abused the notation for $g$ by writing $g(\mathbf{x}) = g(|\mathbf{x}|)$. The properties of function $g$ yield

$$\int_0^\infty rg(r) dr = 0 \qquad \text{and} \qquad |g(\mathbf{x})| \leq \frac{C_N}{(1 + |\mathbf{x}|)^N},$$

the latter being true for an arbitrary positive integer $N$. Based on these facts and the formula for $\mathbf{u}$, it easily follows that

$$|\mathbf{u}(\mathbf{x})| \leq \frac{C_N}{(1 + |\mathbf{x}|)^N}. \qquad (4.132)$$

Now, we have all the prerequisites to start the frequency-based argument.

### 4.7.3  The low frequency case

In this instance, we make the assumption $\lambda \leq C\epsilon^{-5}$ and we split the domain of integration in (4.130) according to

$$\mathbb{R}^2 = D_\gamma(\mathbf{0}) \cup \left(A_\gamma \cap D_{R/\lambda}(\mathbf{x})\right) \cup \left(A_\gamma \backslash D_{R/\lambda}(\mathbf{x})\right) \cup \{|\mathbf{y}| \geq 1\},$$

where

$$D_\gamma(\mathbf{0}) \stackrel{\text{def}}{=} \{|\mathbf{y}| \leq \gamma\}, \quad A_\gamma \stackrel{\text{def}}{=} \{\gamma \leq |\mathbf{y}| \leq 1\}, \quad D_{R/\lambda}(\mathbf{x}) \stackrel{\text{def}}{=} \{|\mathbf{y} - \mathbf{x}| \leq R/\lambda\},$$

and $R > 0$ is a fixed parameter to be chosen later.

If we apply the Cauchy-Schwarz inequality for each of the subdomains and use (4.119)-(4.121), then we can infer

$$\left| \int_{D_\gamma(\mathbf{0})} \lambda u^j \left(\lambda(\mathbf{x} - \mathbf{y})\right) \nabla_j \phi(1, \mathbf{y}) d\mathbf{y} \right| \leq C \|\mathbf{u}\|_{L^2} \epsilon^4,$$

$$\left| \int_{A_\gamma \cap D_{R/\lambda}(\mathbf{x})} \lambda u^j \left(\lambda(\mathbf{x} - \mathbf{y})\right) \nabla_j \phi(1, \mathbf{y}) d\mathbf{y} \right| \leq C \lambda \|\mathbf{u}\|_{L^\infty} \left| A_\gamma \cap D_{R/\lambda}(\mathbf{x}) \right|^{1/2} E^{1/2},$$

$$\left| \int_{A_\gamma \backslash D_{R/\lambda}(\mathbf{x})} \lambda u^j \left(\lambda(\mathbf{x} - \mathbf{y})\right) \nabla_j \phi(1, \mathbf{y}) d\mathbf{y} \right| \leq C \left( \int_{S_{\lambda,\gamma}} |\mathbf{u}|^2 d\mathbf{y} \right)^{1/2} E^{1/2},$$

and

$$\left| \int_{|\mathbf{y}| \geq 1} \lambda u^j \left(\lambda(\mathbf{x} - \mathbf{y})\right) \nabla_j \phi(1, \mathbf{y}) d\mathbf{y} \right| \leq C \|\mathbf{u}\|_{L^2} \epsilon^4 E^{1/2},$$

where $|\Omega|$ denotes the Lebesgue measure of the set $\Omega \subseteq \mathbb{R}^2$ and the region of integration $S_{\lambda,\gamma}$ is given by

$$S_{\lambda,\gamma} \stackrel{\text{def}}{=} \{\mathbf{y} \mid |\mathbf{y}| \geq R, \ \lambda\gamma \leq |\mathbf{y} - \lambda\mathbf{x}| \leq \lambda\}.$$

In connection to the second and third estimates, simple geometric arguments and (4.132) provide us with

$$\left| A_\gamma \cap D_{R/\lambda}(\mathbf{x}) \right| \leq C \frac{R(1 - \gamma)}{\lambda}$$

and

$$\int_{S_{\lambda,\gamma}} |\mathbf{u}|^2 d\mathbf{y} \leq \frac{C_N}{(1 + R)^{2N}} \lambda^2 (1 - \gamma^2).$$

If we combine these facts with the previous four bounds, the assumption $\lambda \leq C\epsilon^{-5}$, and the calibration $1 - \gamma^2 = \epsilon^8$, then we obtain

$$|S_\lambda[\phi](1, \mathbf{x})| \leq C(\epsilon^4 + R^{1/2}\epsilon^{3/2}) + \frac{C_N \epsilon^{-1}}{(1+R)^N}.$$

Choosing now $R = \epsilon^{-1/2}$ and $N = 5$, we finally deduce

$$|S_\lambda[\phi](1, \mathbf{x})| \leq C\epsilon^{5/4}, \tag{4.133}$$

which proves (4.124) in this case.

### 4.7.4   *The high frequency case*

Here, we work under the condition $\lambda > C\epsilon^{-5}$ and we first need to localize, in the sense that we perform the decomposition

$$g = g_{\text{int}} + g_{\text{ext}},$$

where $g_{\text{int}} = g_{\text{int}}(\mathbf{x})$ and $g_{\text{ext}} = g_{\text{ext}}(\mathbf{x})$ are functions supported in the regions $\{|\mathbf{y}| \leq R\}$ and $\{|\mathbf{y}| \geq R - 1\}$, respectively, and $R > 1$ is a fixed parameter to be specified later.

This is done by choosing

$$g_{\text{int}} = \chi_R g + m_R \qquad \text{and} \qquad g_{\text{ext}} = (1 - \chi_R)g - m_R,$$

with $\chi_R = \chi_R(\mathbf{x})$ being the characteristic function associated to the closed disk $\{|\mathbf{y}| \leq R\}$, and $m_R = m_R(\mathbf{x})$ acting as a small, radial correction, which is supported in the region $\{R - 1 \leq |\mathbf{y}| \leq R\}$. The correction is needed to ensure $\int_{\mathbb{R}^2} g_{\text{ext}} = 0$ and, as $\int_{\mathbb{R}^2} g = 0$, $\int_{\mathbb{R}^2} g_{\text{int}} = 0$ also. Moreover, due to the fast decay of $g$ at spatial infinity, we can pick the correction such that if $|g(\mathbf{x})| \leq C_N/(1 + |\mathbf{x}|)^N$, then

$$|m_R(\mathbf{x})| \leq \frac{C_N}{R^{N-1}} \chi_{[R-1,R]}(\mathbf{x}),$$

with $\chi_{[R-1,R]}$ denoting the characteristic function for the region $\{R - 1 \leq |\mathbf{y}| \leq R\}$.

Using the previous decomposition, we write

$$S_\lambda[\psi] = S_\lambda^{\text{int}}[\psi] + S_\lambda^{\text{ext}}[\psi],$$

where $S_\lambda^{\text{int}}[\psi]$ and $S_\lambda^{\text{ext}}[\psi]$ are obtained by replacing $g$ with $g_{\text{int}}$ and $g_{\text{ext}}$, respectively, in (4.126). In proving (4.124), we proceed by estimating $S_\lambda^{\text{int}}[\phi](1, \mathbf{x})$ and $S_\lambda^{\text{ext}}[\phi](1, \mathbf{x})$ separately.

In dealing with $S_\lambda^{\text{ext}}[\phi](1,\mathbf{x})$, we follow the approach from the low frequency case and construct

$$\mathbf{u}_{\text{ext}}(\mathbf{x}) \overset{\text{def}}{=} \left( \frac{1}{|\mathbf{x}|^2} \int_0^{|\mathbf{x}|} sg_{\text{ext}}(s)\, ds \right) \mathbf{x},$$

which can be easily seen to satisfy

$$|\mathbf{u}_{\text{ext}}(\mathbf{x})| \leq \frac{C_N}{|\mathbf{x}|(1+|\mathbf{x}|)^{N-3}} \chi_{[R-1,\infty)}(\mathbf{x}).$$

Together with (4.119) and (4.120), this implies

$$\begin{aligned}
\left| S_\lambda^{\text{ext}}[\phi](1,\mathbf{x}) \right| &= \left| \int_{\mathbb{R}^2} \lambda\, u_{\text{ext}}^j (\lambda(\mathbf{x}-\mathbf{y})) \nabla_j \phi(1,\mathbf{y}) d\mathbf{y} \right| \\
&\leq C \|\mathbf{u}_{\text{ext}}\|_{L^2} E^{1/2} \\
&\leq \frac{C_N}{(R-1)^{N-\frac{3}{2}}}.
\end{aligned} \tag{4.134}$$

It remains to estimate $S_\lambda^{\text{int}}[\phi](1,\mathbf{x})$, for which we split the analysis according to whether

$$D_{R/\lambda}(\mathbf{x}) \cap A_\gamma = \emptyset$$

or

$$D_{R/\lambda}(\mathbf{x}) \cap A_\gamma \neq \emptyset.$$

The former case is easy to deal with, by constructing yet again the radial gauge

$$\mathbf{u}_{\text{int}}(\mathbf{x}) \overset{\text{def}}{=} \left( \frac{1}{|\mathbf{x}|^2} \int_0^{|\mathbf{x}|} sg_{\text{int}}(s)\, ds \right) \mathbf{x}$$

and arguing that

$$|\mathbf{u}_{\text{int}}(\mathbf{x})| \leq CR\, \chi_R(\mathbf{x}).$$

Based on (4.120) and (4.121), it follows that

$$\begin{aligned}
\left| S_\lambda^{\text{int}}[\phi](1,\mathbf{x}) \right| &= \left| \int_{\mathbb{R}^2} \lambda\, u_{\text{int}}^j (\lambda(\mathbf{x}-\mathbf{y})) \nabla_j \phi(1,\mathbf{y}) d\mathbf{y} \right| \\
&\leq C \|\mathbf{u}_{\text{int}}\|_{L^2} \|\nabla\phi(1)\|_{L^2(\mathbb{R}^2 \backslash A_\gamma)} \\
&\leq CR^2 \epsilon^4.
\end{aligned} \tag{4.135}$$

Now, we need to analyze $S_\lambda^{\text{int}}[\phi](1,\mathbf{x})$ in the case when $D_{R/\lambda}(\mathbf{x}) \cap A_\gamma \neq \emptyset$. We notice first that

$$\big|\, |\mathbf{x}|-1 \,\big| \leq \big|\, |\mathbf{x}|-|\mathbf{y}| \,\big| + 1 - |\mathbf{y}| \leq \frac{R}{\lambda} + 1 - \gamma \leq CR\epsilon^5, \tag{4.136}$$

by choosing $\mathbf{y} \in D_{R/\lambda}(\mathbf{x}) \cap A_\gamma$ and taking into account that $\lambda > C\epsilon^{-5}$ and $1 - \gamma^2 = \epsilon^8$. Hence, if $R$ is chosen such that $R\epsilon^5 \ll 1$, then $|\mathbf{x}| \approx 1$. Moreover, given the built-in radial symmetry in our context, we can assume, without loss of generality, that $\mathbf{x} = (x^1, 0)$.

In order to obtain a useful estimate for $S_\lambda^{\text{int}}[\phi](1, \mathbf{x})$, we rely on the following auxiliary result.

**Lemma 4.6.** *Let $h = h(\mathbf{x})$ be a radially symmetric function, which is supported in the closed disk $\{|\mathbf{y}| \leq R\}$ and satisfies $\int_{\mathbb{R}^2} h = 0$. If we still denote by $S_\lambda$ the operator obtained by replacing $g$ with $h$ in (4.126), then the subsequent estimate holds:*

$$\left| S_\lambda[\psi](\mathbf{x}) \right|^2 \leq C \left( KR \int_{Q_\lambda^K} \left| \nabla_2 \psi(\mathbf{y}) \right|^2 d\mathbf{y} + \frac{R}{K} \int_{Q_\lambda^K} \left| \nabla_1 \psi(\mathbf{y}) \right|^2 d\mathbf{y} \right), \quad (4.137)$$

*where $K \geq R$ is arbitrary and the domain of integration is the rectangle*

$$Q_\lambda^K \overset{\text{def}}{=} \left\{ \mathbf{y} \;\middle|\; |y^1 - x^1| \leq \frac{R}{\lambda} , \; |y^2 - x^2| \leq 2\frac{K}{\lambda} \right\}.$$

**Proof.** The main idea in the argument is to construct a vector field $\mathbf{u}$ satisfying $\text{div } \mathbf{u} = h$ and the gauge condition $\mathbf{u}(\mathbf{y}) = \big(0, u(\mathbf{y})\big)$. Hence, we need to solve $\nabla_2 u = h$, which is achieved by choosing

$$u(\mathbf{y}) = u\big(y^1, y^2\big) \overset{\text{def}}{=} \int_0^{y^2} h(y^1, s) \, ds.$$

In the same context, we define

$$a(y^1) \overset{\text{def}}{=} \int_{\mathbb{R}} h(y^1, s) \, ds \quad \text{and} \quad v(y^1) \overset{\text{def}}{=} \int_{-\infty}^{y^1} a(s) \, ds.$$

The assumptions on $h$ easily implies the following facts:

i) if $|y^1| > R$, then $u(\mathbf{y}) = 0$, and if $|y^2| > R$, then $u(\mathbf{y}) = \text{sgn}(y^2)a(y^1)/2$;
ii) $a$ is supported in the compact interval $[-R, R]$ and $\int_{\mathbb{R}} a = 0$;
iii) $v$ is supported in the compact interval $[-R, R]$ and $v' = a$.

We also introduce a smooth, even cutoff function $b = b(s)$, which is identically equal to 1 on the interval $[-1, 1]$ and supported on the interval $[-2, 2]$.

With these prerequisites, we can now write

$$S_\lambda[\psi](\mathbf{x})$$

$$= \int_{\mathbb{R}^2} \lambda^2 h\big(\lambda(\mathbf{x} - \mathbf{y})\big) \psi(\mathbf{y})\, d\mathbf{y}$$

$$= \int_{\mathbb{R}^2} \lambda^2 h\big(\lambda(\mathbf{x} - \mathbf{y})\big) b\left(\frac{\lambda(x^2 - y^2)}{K}\right) \psi(\mathbf{y})\, d\mathbf{y}$$

$$= -\int_{\mathbb{R}^2} \lambda \nabla_2 u\big(\lambda(\mathbf{x} - \mathbf{y})\big) b\left(\frac{\lambda(x^2 - y^2)}{K}\right) \psi(\mathbf{y})\, d\mathbf{y}$$

$$= \int_{\mathbb{R}^2} \lambda u\big(\lambda(\mathbf{x} - \mathbf{y})\big) b\left(\frac{\lambda(x^2 - y^2)}{K}\right) \nabla_2 \psi(\mathbf{y})\, d\mathbf{y}$$

$$- \int_{\mathbb{R}^2} \frac{\lambda^2}{K} u\big(\lambda(\mathbf{x} - \mathbf{y})\big) b'\left(\frac{\lambda(x^2 - y^2)}{K}\right) \psi(\mathbf{y})\, d\mathbf{y}.$$

For the first integral, we use the properties we have on $u$ and $b$, coupled with the Cauchy-Schwarz inequality, to deduce

$$\left| \int_{\mathbb{R}^2} \lambda u\big(\lambda(\mathbf{x} - \mathbf{y})\big) b\left(\frac{\lambda(x^2 - y^2)}{K}\right) \nabla_2 \psi(\mathbf{y})\, d\mathbf{y} \right|^2$$

$$\leq \left( \int_{Q_K} \left| u(\mathbf{y}) b\left(\frac{y^2}{K}\right) \right|^2 d\mathbf{y} \right) \left( \int_{Q_\lambda^K} |\nabla_2 \psi(\mathbf{y})|^2 d\mathbf{y} \right) \qquad (4.138)$$

$$\leq CKR \int_{Q_\lambda^K} |\nabla_2 \psi(\mathbf{y})|^2 d\mathbf{y},$$

where

$$Q_K \overset{\text{def}}{=} \{ \mathbf{y} \mid |y^1| \leq R\,, \ |y^2| \leq 2K \}.$$

In what concerns the second integral, we rely first on the properties of $u$, $a$, $v$, and $b$, to rewrite it as

$$- \int_{\mathbb{R}^2} \frac{\lambda^2}{K} u\big(\lambda(\mathbf{x} - \mathbf{y})\big) b'\left(\frac{\lambda(x^2 - y^2)}{K}\right) \psi(\mathbf{y})\, d\mathbf{y}$$

$$= - \int_{\mathbb{R}^2} \frac{\lambda^2}{2K} \operatorname{sgn}(\lambda(x^2 - y^2))\, a\big(\lambda(x^1 - y^1)\big) b'\left(\frac{\lambda(x^2 - y^2)}{K}\right) \psi(\mathbf{y})\, d\mathbf{y}$$

$$= \int_{\mathbb{R}^2} \frac{\lambda}{2K} \nabla_1 \left\{ v\big(\lambda(x^1 - y^1)\big) \right\} \operatorname{sgn}(\lambda(x^2 - y^2)) b'\left(\frac{\lambda(x^2 - y^2)}{K}\right) \psi(\mathbf{y})\, d\mathbf{y}$$

$$= - \int_{\mathbb{R}^2} \frac{\lambda}{2K} v\big(\lambda(x^1 - y^1)\big) \operatorname{sgn}(\lambda(x^2 - y^2)) b'\left(\frac{\lambda(x^2 - y^2)}{K}\right) \nabla_1 \psi(\mathbf{y})\, d\mathbf{y}.$$

Then, we use the Cauchy-Schwarz inequality to derive the bound

$$\left| \int_{\mathbb{R}^2} \frac{\lambda^2}{K} u(\lambda(\mathbf{x} - \mathbf{y})) b' \left( \frac{\lambda(x^2 - y^2)}{K} \right) \psi(\mathbf{y}) \, d\mathbf{y} \right|^2$$

$$\leq \frac{C}{K^2} \left( \int_{Q_K} \left| v(y^1) b' \left( \frac{y^2}{K} \right) \right|^2 d\mathbf{y} \right) \left( \int_{Q_\lambda^K} \left| \nabla_1 \psi(\mathbf{y}) \right|^2 d\mathbf{y} \right) \quad (4.139)$$

$$\leq \frac{CR}{K} \int_{Q_\lambda^K} \left| \nabla_1 \psi(\mathbf{y}) \right|^2 d\mathbf{y}.$$

Finally, the estimate to be proved, (4.137), follows as a result of (4.138) and (4.139).

$\square$

**Remark 4.4.** This lemma can be modified to accommodate polar coordinates by solving $\nabla_\theta u = h$, instead of $\nabla_2 u = h$. Thus, one would obtain an estimate for $S_\lambda[\psi]$ in terms of $\nabla_r \psi$ and $\nabla_\theta \psi$, which is more readily applicable than (4.137) in our context. This is due to the control we have on the angular derivatives through (4.122). However, we do not pursue this idea here.

Using this result in our setting, which is described by

$$h = g_{\text{int}}, \qquad \psi = \phi(1), \qquad \mathbf{x} = (x^1, 0), \quad \text{and} \quad \lambda > C\epsilon^{-5},$$

and choosing

$$R = \epsilon^{-1/4}, \qquad K = \epsilon^{-5/2},$$

we infer that

$$\left| S_\lambda^{\text{int}}[\phi](1, \mathbf{x}) \right|^2$$

$$\leq C \left( \epsilon^{-11/4} \int_Q \left| \nabla_2 \phi(1, \mathbf{y}) \right|^2 d\mathbf{y} + \epsilon^{9/4} \int_Q \left| \nabla_1 \phi(1, \mathbf{y}) \right|^2 d\mathbf{y} \right), \quad (4.140)$$

where

$$Q = \left\{ \mathbf{y} \mid |y^1 - x^1| \leq C\epsilon^{19/4} , \ |y^2| \leq C\epsilon^{5/2} \right\}.$$

Next, we take advantage of (4.136) to claim $||x^1| - 1| \leq C\epsilon^{19/4}$, which implies, for all $\mathbf{y} \in Q$,

$$||\mathbf{y}| - 1| \leq C\epsilon^{5/2}.$$

As

$$\nabla_2 = \frac{y^2}{|\mathbf{y}|}\nabla_r + \frac{y^1}{|\mathbf{y}|^2}\nabla_\theta,$$

we obtain that

$$|\nabla_2\phi(1,\mathbf{y})|^2 \le C\left(\epsilon^5|\nabla_r\phi(1,\mathbf{y})|^2 + \frac{1}{r^2}|\nabla_\theta\phi(1,\mathbf{y})|^2\right)$$

holds on $Q$. Together with (4.140) and (4.119)-(4.122), this bound yields

$$\left|S_\lambda^{\mathrm{int}}[\phi](1,\mathbf{x})\right| \le C\epsilon^{9/8}. \tag{4.141}$$

Finally, if we pick $N = 6$ in (4.134) and we combine it with (4.135) and (4.141), then it follows that we have the estimate

$$|S_\lambda[\phi](1,\mathbf{x})| \le C\epsilon^{9/8}, \tag{4.142}$$

which proves (4.124) in the high frequency case. This finishes the proof of Proposition 4.2.

## 4.8   The concentration of energy scenario

After discussing the case of uniform non-concentration of energy, we focus on its counterpart, which is described by the existence of a sequence $(t_n)_n \in (0,1]$ such that $t_n \to 0$ and

$$\int_{|\mathbf{x}|\le\gamma_1 t_n} \{e(\phi(t_n))\}\,d\mathbf{x} \ge \epsilon_1^{3/2} > 0. \tag{4.143}$$

### 4.8.1   *Initial considerations and final rescaling*

As in the previous section, $\phi$ is a wave map defined on $\mathcal{C}_{(0,1]} \cap \left((0,t_0] \times \mathbb{R}^2\right)$, which satisfies

$$0 < \mathcal{E}[\phi] \le (1+\epsilon^8)E = (1+\epsilon^8)\lim_{t\to0}\mathcal{E}_{D_t}(\phi),$$

whereas $\gamma_1$ and $\epsilon_1$ are positive, fixed parameters, with

$$(1-\gamma_1^2)^{1/4}E = \epsilon_1^{3/2}.$$

It follows that, if we choose $n$ sufficiently large such that $t_n \le t_0$, then

$$\mathcal{E}_{D_t}(\phi) \approx E, \qquad \forall\, 0 < t \le t_n. \tag{4.144}$$

Due to $t_n \to 0$ and the flux decay (4.23), we can guarantee, eventually raising the value of $n$, that

$$\delta_F(t_n,\phi) \ll (1-\gamma_1^2)^{3/2}.$$

Together with (4.143), this allows us to apply the propagation of energy result, Lemma 4.5, for $\phi$ restricted to $\mathcal{C}_{(0,t_n]}$ and thus derive the existence of $\gamma_2 = \gamma_2(\gamma_1, \epsilon_1, E)$ and $E_2 = E_2(\gamma_1, \epsilon_1, E)$ for which

$$\int_{|\mathbf{x}| \leq \gamma_2 t} \{e(\phi(t))\} \, d\mathbf{x} \geq E_2 > 0, \qquad \forall t \in \left[ \frac{Ct_n \delta_F(t_n, \phi)}{(1 - \gamma_1^2)^{3/2} E}, t_n \right]. \qquad (4.145)$$

Next, we define

$$\delta_n \stackrel{\text{def}}{=} \left( \frac{C\delta_F(t_n, \phi)}{(1 - \gamma_1^2)^{3/2} E} \right)^2$$

and notice that $\delta_n \to 0$. We use the energy-type estimate (4.78) with $T := \delta_n t_n$ and $t_1 := t_n$ to deduce

$$\int_{\mathcal{C}^{\delta_n t_n}_{[\delta_n t_n, t_n]}} \left\{ \frac{1}{\rho} |X(\phi)|^2 \right\} dt \, d\mathbf{x} \leq C \left( \delta_n^{-1/2} \delta_F(t_n, \phi) + E \right) \leq CE.$$

In particular, this bound implies

$$\int_{\mathcal{C}^{\delta_n t_n}_{[\delta_n^{1/2} t_n, \delta_n^{1/4} t_n]}} \left\{ \frac{1}{\rho} |X(\phi)|^2 \right\} dt \, d\mathbf{x} \leq CE, \qquad (4.146)$$

for which the claim is that it indicates decay of the kinetic energy.

To see this, for fixed $n$, we decompose the time interval $\left[ \delta_n^{1/2} t_n, \delta_n^{1/4} t_n \right]$ according to

$$\left[ \delta_n^{1/2} t_n, \delta_n^{1/4} t_n \right] = \bigcup_{k=1}^{N_n} [s_{k,n}, s_{k+1,n}],$$

where $(s_{k,n})$ is a geometric progression with $s_{1,n} = \delta_n^{1/2} t_n$ and $s_{N_n+1,n} = \delta_n^{1/4} t_n$. If we denote the rate of the progression by $T_n$, then we have

$$T_n^{N_n} = \delta_n^{-1/4}.$$

Using the additivity of the integral with respect to the domain of integration and the pigeonhole principle, it follows that there exists $1 \leq k_0 = k_0(n) \leq N_n$ such that

$$\int_{\mathcal{C}^{\delta_n t_n}_{[s_{k_0,n}, s_{k_0+1,n}]}} \left\{ \frac{1}{\rho} |X(\phi)|^2 \right\} dt \, d\mathbf{x} \leq \frac{CE}{N_n}. \qquad (4.147)$$

We choose

$$N_n \approx \left( \log (1/\delta_n) \right)^{1/2}$$

and, as a consequence,

$$T_n \approx e^{(\log(1/\delta_n))^{1/2}/4} \to \infty.$$

Summarizing (4.144), (4.145), and (4.147), we obtain that the wave map $\phi$ satisfies, for $n$ sufficiently large:

$$\int_{|\mathbf{x}| \leq t} \{e(\phi(t))\}\, d\mathbf{x} \approx E, \qquad 0 < t \leq t_n, \tag{4.148}$$

$$\int_{|\mathbf{x}| \leq \gamma_2 t} \{e(\phi(t))\}\, d\mathbf{x} \geq E_2, \qquad \delta_n^{1/2} t_n < t \leq t_n, \tag{4.149}$$

$$\int_{\mathcal{C}^{\delta_n t_n}_{[s_{k_0,n}, s_{k_0+1,n}]}} \left\{ \frac{1}{\rho} |X(\phi)|^2 \right\} dt\, d\mathbf{x} \leq \frac{CE}{(\log(1/\delta_n))^{1/2}}. \tag{4.150}$$

This is the moment in the argument when we perform a final rescaling by defining, for a fixed $n$,

$$\phi^{(n)}(x) \overset{\text{def}}{=} \phi(s_{k_0,n} x), \qquad x \in \mathcal{C}_{(0, t_n/s_{k_0,n}]}.$$

Due to

$$\delta_n^{1/2} t_n \leq s_{k_0,n} < T_n s_{k_0,n} = s_{k_0,n+1} \leq \delta_n^{1/4} t_n,$$

we infer

$$\mathcal{C}_{[1,T_n]} \subset \mathcal{C}_{(0, t_n/s_{k_0,n}]} \qquad \text{and} \qquad \mathcal{C}^{\delta_n^{1/2}}_{[1,T_n]} \subset \mathcal{C}^{\delta_n t_n/s_{k_0,n}}_{[1,T_n]}.$$

Hence, we derive a sequence of wave maps $\{\phi^{(n)}\}$, defined on corresponding conical regions $\mathcal{C}_{[1,T_n]}$, which, on the basis of (4.148)-(4.150), satisfies

$$\int_{\{|\mathbf{x}| \leq t\}} \{e(\phi^{(n)}(t))\}\, d\mathbf{x} \approx E, \qquad 1 \leq t \leq T_n, \tag{4.151}$$

$$\int_{\{|\mathbf{x}| \leq \gamma_2 t\}} \{e(\phi^{(n)}(t))\}\, d\mathbf{x} \geq E_2, \qquad 1 \leq t \leq T_n, \tag{4.152}$$

$$\int_{\mathcal{C}^{\delta_n^{1/2}}_{[1,T_n]}} \left\{ \frac{1}{\rho} |X(\phi^{(n)})|^2 \right\} dt\, d\mathbf{x} \leq \frac{CE}{(\log(1/\delta_n))^{1/2}}, \tag{4.153}$$

with $T_n \to \infty$ and $\delta_n \to 0$ as $n \to \infty$. Moreover, $\mathcal{E}[\phi^{(n)}] = \mathcal{E}[\phi]$. In formal terms, the previous three estimates tell us that these maps exhibit:

- almost constant total energy inside the characteristic cone;
- energy concentration inside a smaller cone;
- decay of the average of the kinetic energy.

For the rest of the argument, we will be strictly working with the sequence $\{\phi^{(n)}\}$ and (4.151)-(4.153).

### 4.8.2   *Refined concentration scenarios*

Let us first introduce the notation used in this subsection. Due to $T_n \to \infty$, we work with the cones $C^1_{[1,\infty)}$ and $C^{1/2}_{[1/2,\infty)}$, which are partitioned into the dyadic pieces

$$C_j^1 \stackrel{\text{def}}{=} C^1_{[1,\infty)} \cap \left( I_j \times \mathbb{R}^2 \right), \qquad C_j^{1/2} \stackrel{\text{def}}{=} C^{1/2}_{[1/2,\infty)} \cap \left( I_j \times \mathbb{R}^2 \right),$$

where $j$ is an arbitrary nonnegative integer and $I_j := \left[ 2^j, 2^{j+1} \right]$. If $D_r(\mathbf{x})$ denotes the closed disk of radius $r$ centered at $\mathbf{x}$, then

$$D_r^t(\mathbf{x}) \stackrel{\text{def}}{=} \{t\} \times D_r(\mathbf{x}).$$

For $A \subset \{t\} \times \mathbb{R}^2$, we recall the notation (4.42), i.e.,

$$\mathcal{E}_A(\psi) \stackrel{\text{def}}{=} \int_{(t,\mathbf{x}) \in A} \{e(\psi(t))\} \, d\mathbf{x}.$$

Related to this, if $B \subset I \times \mathbb{R}^2$, where $I$ is an arbitrary time interval, then $B(t) := B \cap \left( \{t\} \times \mathbb{R}^2 \right)$ and

$$\mathcal{E}_B(\psi) \stackrel{\text{def}}{=} \int_I \mathcal{E}_{B(t)}(\psi) \, dt.$$

We also remind the reader that we denoted by $E_0$ the perturbation energy threshold in Theorem 4.3.

With these notations, the following result identifies, on a subsequence of $\{\phi^{(n)}\}$, two refined energy concentration scenarios.

**Lemma 4.7.** *Let $\phi^{(n)}$ be the previously constructed sequence of wave maps satisfying (4.151)-(4.153). If $j$ is a fixed nonnegative integer, then, on a subsequence of $\{\phi^{(n)}\}$ which is still denoted $\{\phi^{(n)}\}$, we are in one of the following two contexts:*

   **(i)** *there exist a sequence of points $(t_n, \mathbf{x}_n) \in C_j^{1/2}$, a corresponding sequence of scales $r_n \to 0$, and a fixed radius $r = r(j) \in (0, 1/4)$ depending solely on $j$ such that*

$$\mathcal{E}_{D_{r_n}^{t_n}(\mathbf{x}_n)}(\phi^{(n)}) = \frac{E_0}{10}, \tag{4.154}$$

$$\mathcal{E}_{D_{r_n}^{t_n}(\mathbf{x})}(\phi^{(n)}) \leq \frac{E_0}{10}, \qquad \forall \mathbf{x} \in D_r(\mathbf{x}_n), \tag{4.155}$$

$$\lim_{n \to \infty} \frac{1}{r_n} \int_{t_n - \frac{r_n}{2}}^{t_n + \frac{r_n}{2}} \int_{D_r(\mathbf{x}_n)} \left\{ \left| X(\phi^{(n)}) \right|^2 \right\} \, d\mathbf{x} \, dt = 0; \tag{4.156}$$

**(ii)** *there exists a fixed radius* $r = r(j) \in (0, 1/4)$ *depending solely on* $j$
*such that*

$$\mathcal{E}_{D_r^t(\mathbf{x})}(\phi^{(n)}) \leq \frac{E_0}{10}, \qquad \forall (t, \mathbf{x}) \in \mathcal{C}_j^1, \tag{4.157}$$

$$\mathcal{E}_{D_t^{(1-\gamma_2)t}}(\phi^{(n)}) \geq E_2, \qquad \forall t \geq 1, \tag{4.158}$$

$$\lim_{n \to \infty} \int_{\mathcal{C}_j^1} \left\{ \left| X(\phi^{(n)}) \right|^2 \right\} dt \, d\mathbf{x} = 0. \tag{4.159}$$

**Remark 4.5.** The two scenarios described in the above lemma can be referred to as *concentration of nontrivial energy at small scales* and *non-concentration of energy at uniform scales*, respectively. For the first one, we have substantial, but not too large, concentration of energy at some point on smaller and smaller scales and the average of the kinetic energy decays on the same scale. In the second scenario, we have nontrivial energy inside a cone of aperture strictly less than one, but the energy does not concentrate on small scales. At the same time, the kinetic energy decays.

**Remark 4.6.** As it turns out, a compactness argument will rule out the second scenario, leaving only the first one as a possible behavior for the subsequence of $\phi^{(n)}$. This agrees with what is known about the blow-up behavior of wave-maps.

**Proof.** The argument relies on standard energy estimates and several applications of the pigeonhole principle. Given these tools, it is quite generic in nature, presumably having a wide relevancy. As the approaches for different values of $j$ are very similar, its value is fixed for the remainder of the proof. Moreover, it is clear from the statement of the two scenarios that it is enough to prove the lemma by switching from $\mathcal{C}_j^{1/2}$ in **(i)** and $\mathcal{C}_j^1$ in **(ii)** to

$$\mathcal{C}_{j,k}^{1/2} = \mathcal{C}_j^{1/2} \cap (I_{j,k} \times \mathbb{R}^2) \qquad \text{and} \qquad \mathcal{C}_{j,k}^1 = \mathcal{C}_j^1 \cap (I_{j,k} \times \mathbb{R}^2),$$

respectively, where $1 \leq k \leq 10N \cdot 2^j$,

$$I_{j,k} \overset{\text{def}}{=} \left[ 2^j + \frac{k-1}{10N} \, , \, 2^j + \frac{k}{10N} \right],$$

and $N$ is a large integer parameter to be specified later. One notices that

$$I_j = \bigcup_{k=1}^{10N \cdot 2^j} I_{j,k}.$$

The first step in the argument consists of building an *energy barrier*, which is a region inside $\mathcal{C}_{j,k}^{1/2}$ where the energy of $\phi^{(n)}$ is considerably smaller than $E_0$. For this, we start by decomposing the annular region $\mathcal{C}_{j,k}^{1/2} \setminus \mathcal{C}_{j,k}^1$ according to

$$\mathcal{C}_{j,k}^{1/2} \setminus \mathcal{C}_{j,k}^1 = \bigcup_{l=1}^{N} \mathcal{C}_{j,k,l} = \bigcup_{l=1}^{N} \left( \mathcal{C}_{j,k}^{1/2} \cap \left\{ \frac{l-1}{2N} \leq t - |\mathbf{x}| - \frac{1}{2} \leq \frac{l}{2N} \right\} \right).$$

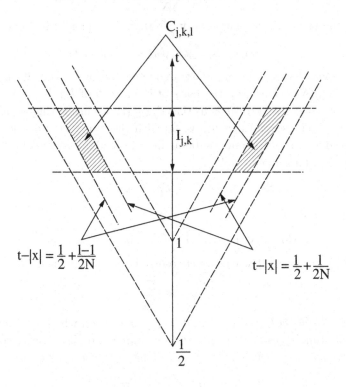

Fig. 4.4   The annular region $\mathcal{C}_{j,k,l}$.

As the energy density is nonnegative and the length of the time interval $I_{j,k}$ is $1/(10N)$, it follows that

$$\sum_{l=1}^{N} \mathcal{E}_{\mathcal{C}_{j,k,l}}(\phi^{(n)}) \leq \mathcal{E}_{\mathcal{C}_{j,k}^{1/2}}(\phi^{(n)}) \leq \frac{\mathcal{E}[\phi^{(n)}]}{10N} = \frac{\mathcal{E}[\phi]}{10N}.$$

Choosing $N$ to be divisible by 3 and applying the pigeonhole principle, we deduce the existence of $2 \leq l_n \leq N - 1$ such that

$$\mathcal{E}_{\mathcal{C}_{j,k,l_n-1}}(\phi^{(n)}) + \mathcal{E}_{\mathcal{C}_{j,k,l_n}}(\phi^{(n)}) + \mathcal{E}_{\mathcal{C}_{j,k,l_n+1}}(\phi^{(n)}) \leq \frac{3\mathcal{E}[\phi]}{10N^2}.$$

Next, we use the mean value theorem to infer that there exists $t_n \in I_{j,k}$ satisfying

$$\mathcal{E}_{\mathcal{C}_{j,k,l_n-1}(t_n)}(\phi^{(n)}) + \mathcal{E}_{\mathcal{C}_{j,k,l_n}(t_n)}(\phi^{(n)}) + \mathcal{E}_{\mathcal{C}_{j,k,l_n+1}(t_n)}(\phi^{(n)}) \le \frac{3\mathcal{E}[\phi]}{N}.$$

We make the claim that this estimate implies

$$\mathcal{E}_{\mathcal{C}_{j,k,l_n}(t)}(\phi^{(n)}) \le \frac{3\mathcal{E}[\phi]}{N}, \qquad \forall\, t \in I_{j,k}. \tag{4.160}$$

This is due to two facts. The first one is that the region of influence for

$$\mathcal{C}_{j,k,l_n-1}(t_n) \cup \mathcal{C}_{j,k,l_n}(t_n) \cup \mathcal{C}_{j,k,l_n+1}(t_n)$$

is a diamond-shaped domain whose diameter and overall height are both equal to $3/(2N)$. Hence, on the basis of the length of $I_{j,k}$ being $1/(10N)$, elementary solid geometry tells us that $\mathcal{C}_{j,k,l_n}(t)$ lies inside this domain, for all $t \in I_{j,k}$. The second fact has to do with the integration of the energy differential identity (4.15) over subdomains of the above region of influence, which yields, for a fixed $t \in I_{j,k}$,

$$\mathcal{E}_{\mathcal{C}_{j,k,l_n}(t)}(\phi^{(n)}) \le \mathcal{E}_{\tilde{\mathcal{C}}_{j,k,l_n}(t_n)}(\phi^{(n)}),$$

with

$$\tilde{\mathcal{C}}_{j,k,l_n}(t_n) \subset \mathcal{C}_{j,k,l_n-1}(t_n) \cup \mathcal{C}_{j,k,l_n}(t_n) \cup \mathcal{C}_{j,k,l_n+1}(t_n).$$

This is the moment where we choose $N$ large enough such that

$$\frac{3\mathcal{E}(\phi)}{N} \le \frac{E_0}{20}.$$

Thus, due to (4.160), we derive that $\mathcal{C}_{j,k,l_n}$ is a region where the energy of $\phi^{(n)}$ on its time slices is very small, when compared to $E_0$. This region acts as a barrier in $\mathcal{C}_{j,k}^{1/2}$ between an interior domain, which is given by

$$\mathcal{C}_{j,k,>l_n} \stackrel{\text{def}}{=} \left( \bigcup_{l=l_n+1}^{N} \mathcal{C}_{j,k,l} \right) \cup \mathcal{C}_{j,k}^1,$$

and an exterior one, defined as

$$\mathcal{C}_{j,k,<l_n} \stackrel{\text{def}}{=} \bigcup_{l=1}^{l_n-1} \mathcal{C}_{j,k,l}.$$

We will also use the notation $\mathcal{C}_{j,k,\ge l_n} := \mathcal{C}_{j,k,>l_n} \cup \mathcal{C}_{j,k,l_n}$. If we pick $0 < r_0 < 1/(8N)$ arbitrary, then it is easy to see that

$$\{t\} \times D_{4r_0}(\mathbf{x}) \subset \mathcal{C}_{j,k,\ge l_n}, \qquad \forall\, (t,\mathbf{x}) \in \mathcal{C}_{j,k,>l_n}. \tag{4.161}$$

We recognize that $r_0$ is independent of $n$, as the choice for $N$ depends only on the ratio $\mathcal{E}(\phi)/E_0$. This concludes the construction of the energy barrier.

The second step focuses on defining two sequences of functions which are relevant to analyzing energy concentration on disks for $\{\phi^{(n)}\}$. The first sequence measures the actual energy concentration on disks, whereas the second one measures the spatial scale on which this concentration occurs. To start, we introduce

$$f_n : I_{j,k} \times [0, r_0] \to \mathbb{R}^+, \quad f_n(t, r) \stackrel{\text{def}}{=} \sup_{D_r^t(\mathbf{x}) \subset \mathcal{C}_{j,k, \geq l_n}(t)} \mathcal{E}_{D_r^t(\mathbf{x})}(\phi^{(n)}), \quad (4.162)$$

and notice, due to $\phi^{(n)}$ being smooth, that these are continuous functions. Moreover, $r \mapsto f_n(t, r)$ is monotone increasing. Next, we define

$$r_n : I_{j,k} \to (0, r_0],$$

$$r_n(t) \stackrel{\text{def}}{=} \begin{cases} \inf\{r \in (0, r_0] \, , \, f_n(t, r) \geq E_0/10\} & \text{if } f_n(t, r_0) \geq E_0/10, \\ r_0 & \text{if } f_n(t, r_0) < E_0/10. \end{cases} \quad (4.163)$$

An important property for these functions is that

$$|r_n(t) - r_n(s)| \leq |t - s|, \qquad \forall 0 < t, s \leq r_0. \quad (4.164)$$

Given the symmetry of this bound, in order to prove it, it would be enough to show

$$r_n(s) \geq r_n(t) - |t - s|, \qquad \forall 0 < t, s \leq r_0. \quad (4.165)$$

This is obvious if $r_n(t) \leq |t - s|$, as $r_n(s) \geq 0$. The case when $r_n(t) > |t - s|$ can be justified by first integrating (4.15) over the domain of influence for $D_r^t(\mathbf{x})$ and deducing

$$\mathcal{E}_{D_{r-|t-s|}^s(\mathbf{x})}(\phi^{(n)}) \leq \mathcal{E}_{D_r^t(\mathbf{x})}(\phi^{(n)}), \qquad \text{if } |t - s| \leq r, \quad (4.166)$$

which, in our setting, implies

$$f_n(t, r_n(t)) \geq f_n(s, r_n(t) - |t - s|).$$

Based on the definitions (4.162) and (4.163), we infer

$$f_n(u, r_n(u)) = \min\left\{\frac{E_0}{10}, f_n(u, r_0)\right\}, \qquad \forall u \in I_{j,k}. \quad (4.167)$$

Hence, if $f_n(s, r_n(s)) = E_0/10$, then, due to the last estimate, we deduce

$$f_n(s, r_n(s)) > f_n(s, r_n(t) - |t - s|),$$

unless

$$f_n(s, r_n(s)) = f_n(t, r_n(t)) = f_n(s, r_n(t) - |t - s|) = \frac{E_0}{10},$$

which is a case left to be investigated by the interested reader. Using now the monotonicity of $r \mapsto f_n(s, r)$, the desired estimate follows. If $f_n(s, r_n(s)) < E_0/10$, then $r_n(s) = r_0$, and (4.165) holds as $r_n(t) \leq r_0$.

In the final step, we perform an analysis for the values of the concentration scales functions $\{r_n\}$, which yields the two scenarios in the lemma. We start by considering the case when all these functions are bounded from below by a positive value, i.e.,

$$\inf_{I_{j,k}} r_n \geq r_1 > 0, \qquad \forall n \geq 1,$$

and show that this corresponds to scenario **(ii)**. By choosing $r := \min\{r_0, r_1\}$, we obtain

$$\{t\} \times D_r(\mathbf{x}) \subset C_{j,k,\geq l_n}, \qquad \forall (t, x) \in \mathcal{C}^1_{j,k},$$

and

$$f_n(t, r) \leq \frac{E_0}{10}, \qquad \forall t \in I_{j,k},$$

for any $n \geq 1$. Hence, (4.157) holds. It is clear that (4.158) is nothing but a restatement of (4.152). In what concerns (4.159), we notice that by choosing $n$ sufficiently large such that $2^{j+1} \leq T_n$, we have

$$\mathcal{C}^1_{j,k} \subset \mathcal{C}^{\delta_n^{1/2}}_{[1.T_n]}.$$

As $\rho \leq 2^{j+1}$ on $\mathcal{C}^1_{j,k}$, we derive

$$\int_{\mathcal{C}^1_{j,k}} \left\{ |X(\phi^{(n)})|^2 \right\} dt\, d\mathbf{x} \leq 2^{j+1} \int_{\mathcal{C}^{\delta_n^{1/2}}_{[1,T_n]}} \left\{ \frac{1}{\rho} |X(\phi^{(n)})|^2 \right\} dt\, d\mathbf{x} \qquad (4.168)$$

and (4.159) follows due to (4.153).

We are left to investigate the case when

$$\liminf_n \inf_{I_{j,k}} r_n = 0, \qquad (4.169)$$

for which we prove that it leads to scenario **(i)**. Eventually considering a subsequence of $\{r_n\}$, we can both replace lim inf in the above expression with lim and assure that $\inf_{I_{j,k}} r_n < r_0$ for all $n$. In what follows, we rely on the notation

$$\alpha_n^2 \stackrel{\text{def}}{=} \int_{\mathcal{C}^{1/2}_{j,k}} \left\{ |X(\phi^{(n)})|^2 \right\} dt\, d\mathbf{x}.$$

An almost identical argument to the one leading to (4.168) yields, for $n$ sufficiently large,

$$\int_{\mathcal{C}_{j,k}^{1/2}} \left\{ |X(\phi^{(n)})|^2 \right\} dt\, d\mathbf{x} \leq 2^{j+1} \int_{\mathcal{C}_{[1,T_n]}^{\delta_n^{1/2}}} \left\{ \frac{1}{\rho} |X(\phi^{(n)})|^2 \right\} dt\, d\mathbf{x}. \qquad (4.170)$$

Due to (4.153), this implies that the sequence $\{\alpha_n\}$ converges to 0. After these preliminary remarks, we split the analysis for this case into three subcases, which are determined by the relative position of $\{\alpha_n\}$ with respect to $\{\inf_{I_{j,k}} r_n\}$ and $\{\sup_{I_{j,k}} r_n\}$. Therefore, up to a subsequence, we can assume one of the following possibilities.

**I.** $\alpha_n < \inf_{I_{j,k}} r_n$, **for all** $n$. Using the properties of $r_n$, we choose $t_n \in I_{j,k}$ such that

$$r_n(t_n) = \inf_{I_{j,k}} r_n \to 0.$$

Hence, if we rely on (4.167), then we deduce

$$f_n(t_n, r_n(t_n)) = \frac{E_0}{10}.$$

Based on (4.162), we pick $\mathbf{x}_n$ satisfying

$$D_{r_n(t_n)}^{t_n}(\mathbf{x}_n) \subset \mathcal{C}_{j,k,\geq l_n}(t_n), \quad f_n(t_n, r_n(t_n)) = \mathcal{E}_{D_{r_n(t_n)}^{t_n}(\mathbf{x}_n)}(\phi^{(n)}). \qquad (4.171)$$

We claim that the first scenario in the lemma holds with the sequence of points $\{(t_n, \mathbf{x}_n)\}$ thus obtained, the corresponding sequence of scales $\{r_n(t_n)\}$, and by choosing $r(j, k) = r_0$.

The first condition, (4.154), is immediate from the previous considerations. For the second one, we argue that $\mathbf{x}_n$ must be, on the time slice $t = t_n$, at least $2r_0$ away from the outer boundary of $\mathcal{C}_{j,k,l_n}(t_n)$. Otherwise, on the basis of (4.161), we have

$$D_{r_n(t_n)}^{t_n}(\mathbf{x}_n) \subset \mathcal{C}_{j,k,l_n}(t_n)$$

and, using (4.160), it would follow that

$$\frac{E_0}{10} = \mathcal{E}_{D_{r_n(t_n)}^{t_n}(\mathbf{x}_n)}(\phi^{(n)}) \leq \frac{E_0}{20},$$

which is a contradiction. Hence, if $\mathbf{x} \in D_{r_0}(\mathbf{x}_n)$, then we infer

$$D_{r_n(t_n)}^{t_n}(\mathbf{x}) \subset \mathcal{C}_{j,k,\geq l_n}(t_n),$$

and, as a consequence of (4.162), we further obtain

$$\mathcal{E}_{D_{r_n(t_n)}^{t_n}(\mathbf{x})}(\phi^{(n)}) \leq f_n(t_n, r_n(t_n)) = \frac{E_0}{10},$$

which proves (4.155).

In order to show (4.156), it is enough to prove that

$$\left[ t_n - \frac{r_n(t_n)}{2}, t_n + \frac{r_n(t_n)}{2} \right] \times D_{r_0}(\mathbf{x}_n) \subset C_{j,k}^{1/2} \tag{4.172}$$

holds for $n$ sufficiently large. Indeed, if that is true, then we have

$$\frac{1}{r_n(t_n)} \int_{t_n - \frac{r_n(t_n)}{2}}^{t_n + \frac{r_n(t_n)}{2}} \int_{D_{r_0}(\mathbf{x}_n)} \left\{ \left| X(\phi^{(n)}) \right|^2 \right\} d\mathbf{x} \, dt \leq \frac{\alpha_n^2}{r_n(t_n)} \leq \alpha_n,$$

with the last inequality being motivated by the assumption made for this subcase. Due to

$$\left[ t_n - \frac{r_n(t_n)}{2}, t_n + \frac{r_n(t_n)}{2} \right] \subset I_{j,k},$$

for showing (4.172), it is sufficient to argue that

$$|t - t_n| \leq \frac{r_n(t_n)}{2} \qquad \text{and} \qquad |\mathbf{x} - \mathbf{x}_n| \leq r_0$$

implies

$$t - |\mathbf{x}| \geq \frac{1}{2}.$$

Using the relative position of $(t_n, \mathbf{x}_n)$ with respect to the outer boundary of $C_{j,k,l_n}(t_n)$, we derive

$$|\mathbf{x}_n| + 2r_0 \leq t_n - \frac{1}{2} - \frac{l_n - 1}{2^N}.$$

It follows that, for $(t, \mathbf{x})$ as above, we have

$$t - |\mathbf{x}| \geq t_n - \frac{r_n(t_n)}{2} - |\mathbf{x}_n| - r_0 \geq \frac{1}{2} + \frac{l_n - 1}{2^N} + r_0 - \frac{r_n(t_n)}{2}, \tag{4.173}$$

which implies the desired lower bound on the basis of $r_n(t_n) \to 0$. This finishes the discussion of this subcase.

**II. For all $n$, there exists $t_n \in I_{j,k}$ such that $\alpha_n = r_n(t_n)$.** The analysis for this subcase is very similar to the previous one, in the sense that the only modification is that we work with the sequence $\{t_n\}$ given by the assumption and not the one minimizing $\{r_n\}$ on $I_{j,k}$. Here, $r_n(t_n) \to 0$ follows as a consequence of $\alpha_n \to 0$, and not because of (4.169). The rest of the argument remains the same.

**III. $\sup_{I_{j,k}} r_n < \alpha_n$, for all $n$.** For this subcase, we start by introducing

$$g : I_{j,k} \to \mathbb{R}^+, \qquad g(t) \stackrel{\text{def}}{=} \int_{C_{j,k,\geq l_n}(t)} \left\{ \left| X(\phi^{(n)}) \right|^2 \right\} d\mathbf{x},$$

which, from the definition of $\alpha_n$, satisfies

$$\int_{I_{j,k}} g(t)\,dt \leq \alpha_n^2.$$

Furthermore, if we denote by $I_{j,k,1/3}$ the middle interval obtained by splitting $I_{j,k}$ into thirds, then, using the assumption made in this subcase and $\alpha_n \to 0$, the following average is well-defined at least for $n$ sufficiently large:

$$\mathcal{I} \overset{\text{def}}{=} \int_{I_{j,k,1/3}} \left\{ \frac{1}{r_n(t)} \int_{t-\frac{r_n(t)}{2}}^{t+\frac{r_n(t)}{2}} g(s)\,ds \right\} dt. \qquad (4.174)$$

For $|s - t| \leq r_n(t)/2$, the Lipschitz property (4.164) implies

$$\frac{r_n(t)}{2} \leq r_n(s) \leq \frac{3r_n(t)}{2}$$

and, subsequently,

$$s - r_n(s) \leq t \leq s + r_n(s).$$

Relying on these estimates, Tonelli's theorem, and $g \geq 0$, we infer

$$\mathcal{I} \leq 2 \int_{I_{j,k,1/3}} \left\{ \int_{t-\frac{r_n(t)}{2}}^{t+\frac{r_n(t)}{2}} \frac{g(s)}{r_n(s)}\,ds \right\} dt$$

$$\leq 2 \int_{I_{j,k}} \left\{ \int_{s-r_n(s)}^{s+r_n(s)} dt \right\} \frac{g(s)}{r_n(s)}\,ds$$

$$= 4 \int_{I_{j,k}} g(s)\,ds$$

$$\leq 4\alpha_n^2.$$

If we apply now the mean value theorem in (4.174), we obtain the existence of $t_n \in I_{j,k,1/3}$ such that

$$\frac{1}{r_n(t_n)} \int_{t_n-\frac{r_n(t_n)}{2}}^{t_n+\frac{r_n(t_n)}{2}} \left\{ \int_{\mathcal{C}_{j,k,\geq l_n}(t)} \left| X(\phi^{(n)}) \right|^2 d\mathbf{x} \right\} dt$$

$$= \frac{1}{r_n(t_n)} \int_{t_n-\frac{r_n(t_n)}{2}}^{t_n+\frac{r_n(t_n)}{2}} g(t)\,dt$$

$$\leq 120N\alpha_n^2. \qquad (4.175)$$

As in the first two subcases, $\{r_n(t_n)\}$ represents the sequence of concentrating scales, $r(j, k) = r_0$, and we choose $\{\mathbf{x}_n\}$ according to (4.171).

Due to $\sup_{I_{j,k}} r_n < \alpha_n \to 0$, it follows that $r_n(t_n) \to 0$ and, consequently, $f_n(t_n, r_n(t_n)) = E_0/10$. Thus, we can run the argument from the first subcase and derive (4.154)-(4.155). To conclude (4.156) on the basis of (4.175), we just need to verify that

$$D_{r_0}(\mathbf{x}_n) \subset \mathcal{C}_{j,k,\geq l_n}(t), \qquad \forall\, |t - t_n| \leq \frac{r_n(t_n)}{2}.$$

We claim that this has already been proven in (4.173), on the account that (4.173) implies in fact

$$t - |\mathbf{x}| \geq \frac{1}{2} + \frac{l_n - 1}{2^N},$$

if $|t - t_n| \leq r_n(t_n)/2$, $|\mathbf{x} - \mathbf{x}_n| \leq r_0$, and $n$ is sufficiently large. This ends the analysis of this subcase and the argument for the lemma.

$\square$

## 4.9    The final compactness argument

In this section, we conclude the proof of Theorems 4.4 and 4.7 by further analyzing the two scenarios obtained in Lemma 4.7, using a simple, yet quite general, compactness result for sequences of wave maps. This will show that, in fact, the *concentration of nontrivial energy at small scales* scenario is the only possible one, thus leading to a rescaled sequence of wave maps connected to $\phi$, which converges strongly to the Lorentz transform of a nontrivial harmonic map. This is precisely what one should obtain as a result of Theorems 4.4 and 4.7 if the energy dispersion condition (i.e., (4.8) and (4.11)) fails to be true.

We begin by stating and proving the compactness result.

**Lemma 4.8.** *Let $Q$ denote a cube in $\mathbb{R}^{2+1}$ whose side has length comparable to 1 and consider $\{\psi_n\}$ to be a sequence of wave maps defined on $3Q$ satisfying*

$$\mathcal{E}(\psi_n) \leq \frac{E_0}{10}, \qquad \forall\, n \geq 1, \tag{4.176}$$

*and*

$$\lim_{n \to \infty} \|K\psi_n\|_{L^2(3Q)} = 0, \tag{4.177}$$

*where $K$ is a smooth, timelike vector field. There exists then a subsequence of $\{\psi_n\}$ convergent in $\dot{H}^1_{t,\mathbf{x}}(Q)$ to a wave map $\psi$, with $K\psi = 0$ and $\mathcal{E}(\psi) \leq E_0/10$.*

**Proof.** The approach we use here follows the blueprint of the one taken by Struwe [164] in analyzing the heat flow associated to harmonic maps, for which compactness is obtained as a result of the improved regularity given by integration in time.

Due to the small energy condition satisfied by $\{\psi_n\}$, we can apply the local result contained in Theorem 3.5 to deduce the uniform bound

$$\|\chi\psi_n\|_{\dot{X}_\infty^{1,1/2}} \le C, \tag{4.178}$$

where $\chi$ is an arbitrary smooth function compactly supported in $3Q$, and

$$\|\Psi\|_{\dot{X}_\infty^{1,1/2}} \stackrel{\text{def}}{=} \left(\sum_{\lambda\in 2^{\mathbb{Z}}} \sup_{\mu\in 2^{\mathbb{Z}}} \mu\|\nabla B_\mu S_\lambda\Psi\|_{L^2(\mathbb{R}^{2+1})}^2\right)^{1/2}$$

in the notation used for (3.27). According to (3.26) and (3.27), one has the embedding $S \subset \dot{X}_\infty^{1,1/2}$, with $S$ being the solution space for Theorems 3.4 and 3.5.

The next step consists of constructing a microlocal spacetime decomposition adapted to the cutoff function $\chi$, which is taken to be equal to 1 in $2Q$, and the vector field $K$. For that, we introduce the Fourier variables $\tau$ and $\xi$ corresponding to $t$ and $\mathbf{x}$, respectively, and, using the classical formalism of pseudo-differential operators (e.g., Section 8.4 in Hörmander [73]), if $K = K^\mu\nabla_\mu$, then $K \in \operatorname{Op} S^1$ and its symbol is given by

$$K(t,\mathbf{x},\tau,\xi) = i\left(K^0(t,\mathbf{x})\tau + K^1(t,\mathbf{x})\xi_1 + K^2(t,\mathbf{x})\xi_2\right).$$

As $K$ is timelike, we have

$$\inf_{3Q}\left\{|K^0|^2 - |K^1|^2 - |K^2|^2\right\} > 0.$$

Hence, for a region of the type $\{\tau > (1-2\delta)|\xi|\}$, for some $\delta > 0$, we deduce

$$\begin{aligned}
|K(t,\mathbf{x},\tau,\xi)| &\ge |K^0||\tau| - \left(|K^1|^2 + |K^2|^2\right)^{1/2}|\xi| \\
&\ge |K^0|(|\tau| - |\xi|) \\
&\ge C(|\tau| + |\xi|),
\end{aligned}$$

which implies that the symbol for $K$ is elliptic in this region. We can then apply classical results on such operators (e.g., Theorem 8.4.6 in [73]) and "invert" $K$. Also taking into account the properties of $\chi$, this leads to the following decomposition:

$$\chi = Q_{-1}(x,\nabla)K + Q_0(x,\nabla) + R(x,\nabla),$$

where

- $Q_{-1} \in \operatorname{Op} S^{-1}$ with its symbol supported in $3Q \times \{\tau > (1 - 2\delta)|\xi|\}$,
- $Q_0 \in \operatorname{Op} S^0$ with its symbol supported in $3Q \times \{\tau < (1 - \delta)|\xi|\}$,
- $R \in \operatorname{Op} S^{-\infty}$ with its symbol spatially supported in $3Q$.

Based on this information and the mapping properties related to Sobolev spaces for these operators (e.g., Theorem 8.4.5 in [73]), we infer that

$$\left\| \left( Q_0(x, \nabla) + R(x, \nabla) \right) \psi_n \right\|_{\dot{H}^{3/2}(3Q)} \leq C \|\chi \psi_n\|_{\dot{X}^{1,1/2}_\infty}$$

and

$$\left\| Q_{-1}(x, \nabla) K \psi_n \right\|_{\dot{H}^1(3Q)} \leq C \| K \psi_n \|_{L^2(3Q)}.$$

In the context of (4.177) and (4.178), these two estimates imply the existence of a subsequence for $\{\chi \psi_n\}$, for which we keep the same notation, that is strongly convergent in $\dot{H}^1(3Q)$ to a map $\psi$. Moreover, we have

$$\| K \psi_n - K \psi \|_{L^2(2Q)} \leq C \| \psi_n - \psi \|_{\dot{H}^1(2Q)},$$

with $C$ being a constant depending on the size for the coefficients of $K$ on the cube $2Q$. Hence, on the basis of (4.177), we derive that $K\psi = 0$ on $2Q$. All that is left to argue is that $\psi$ is a *strong finite energy wave map* in $Q$, according to the definition in Theorem 4.3.

For that purpose, we show first that we can find a time section of $2Q$ (i.e., $\{t_0\} \times 2Q_\mathbf{x}$, where $2Q_\mathbf{x}$ is the projection of $2Q$ on the **x**-plane) such that it is close to the center of $2Q$ and for which, up to a subsequence,

$$\lim_{n \to \infty} \left\| \psi_n(t_0) - \psi(t_0) \right\|_{\dot{H}^1 \times L^2(2Q_\mathbf{x})} = 0. \tag{4.179}$$

The location of the time section is chosen such that its domain of influence contains $Q$. For all time intervals $I$ satisfying $I \times 2Q_\mathbf{x} \subset 2Q$, it is easy to infer that

$$\int_I \left\| \psi_n(t) - \psi(t) \right\|^2_{\dot{H}^1 \times L^2(2Q_\mathbf{x})} \, dt \leq C \left\| \psi_n - \psi \right\|^2_{\dot{H}^1(2Q)}.$$

Thus, if we choose $I$ sufficiently close to the center of $2Q$, use the convergence of $\{\chi \psi_n\}$ to $\psi$ in $\dot{H}^1(3Q)$, and invoke Fatou's lemma, then the previous claim follows.

Combining (4.176) and (4.179), we also obtain

$$\left\| \psi(t_0) \right\|_{\dot{H}^1 \times L^2(2Q_\mathbf{x})} \leq \frac{E_0}{10}. \tag{4.180}$$

This allows us to apply Theorem 3.5 and solve the wave maps equation locally with $\psi[t_0]$ as initial data on $\{t_0\} \times 2Q_\mathbf{x}$. As mentioned above, the resulting wave map, denoted $\widetilde{\psi}$, is defined on $Q$ and, using the same theorem jointly with (4.179), we deduce

$$\lim_{n \to \infty} \left\| \psi_n - \widetilde{\psi} \right\|_{L^\infty \dot{H}^1(Q)} + \left\| \nabla_t \psi_n - \nabla_t \widetilde{\psi} \right\|_{L^\infty L^2(Q)} = 0.$$

Given that the side of $Q$ has length comparable to 1, this easily implies that $\{\psi_n\}$ is strongly convergent to $\widetilde{\psi}$ in $\dot{H}^1(Q)$. Hence, $\psi = \widetilde{\psi}$ on $Q$ and, as a consequence, $\psi$ is a wave map there. Moreover, (4.180) and the local energy estimate (4.166) yield that the energy of $\widetilde{\psi}$ on any time section of $Q$ is at most $E_0/10$, thus finishing the argument. □

With this result in hand, we can now start investigating the two energy concentrating profiles.

### 4.9.1 *Analysis of scenario (ii) in Lemma 4.7*

We show that this scenario is not possible. According to (4.157) and (4.159), we can apply Lemma 4.8 for the sequence $\{\phi^{(n)}\}$, $K = X$, and cubes that can be fit into balls of radius $r = r(j)$ which lie inside $\mathcal{C}_j^1$. Given that the hypothesis in the compactness lemma refers to the cube $3Q$ and the conclusion is for the cube $Q$, it follows that successive applications of this result yield, up to a subsequence,

$$\lim_{n \to \infty} \phi^{(n)} = \phi^\infty \quad \text{in } \dot{H}^1_{\text{loc}}(\mathcal{C}^2_{[2,\infty)}),$$

where $\phi^\infty$ is a wave map in $\mathcal{C}^2_{[2,\infty)}$ satisfying $X\phi^\infty = 0$.

We also claim that

$$0 < E_2 \le \sup_{t \ge 2} \mathcal{E}_{D_t^2}(\phi^\infty) \le \mathcal{E}(\phi), \tag{4.181}$$

which says that $\phi^\infty$ is a nontrivial wave map of finite energy. This can be justified using the $\dot{H}^1$ convergence, the local energy estimate (4.166), and the energy concentration bound (4.158). First, for an arbitrary $T \ge 2$ and with $n$ sufficiently large such that $T + 1 \le T_n$ (as a reminder, $\phi^{(n)}$ is defined on $\mathcal{C}_{[1,T_n]}$), one has

$$\int_T^{T+1} \mathcal{E}_{D_t^2}\left(\phi^{(n)}\right) dt \le \mathcal{E}(\phi^{(n)}) = \mathcal{E}(\phi).$$

On the other hand, the $\dot{H}^1$ convergence implies

$$\int_T^{T+1} \mathcal{E}_{D_t^2}(\phi^\infty) dt = \lim_{n \to \infty} \int_T^{T+1} \mathcal{E}_{D_t^2}\left(\phi^{(n)}\right) dt.$$

Finally, (4.166) tells us that $t \mapsto \mathcal{E}_{D_t^2}(\phi^\infty)$ is monotone increasing. Hence, due to these last three facts, we derive

$$\mathcal{E}_{D_T^2}(\phi^\infty) \leq \mathcal{E}(\phi),$$

which proves half of (4.181). For the other half, we first choose $T$ large enough such that $(1 - \gamma_2)T \geq 2$. Then, on the basis of (4.158), the $\dot{H}^1$ convergence, and the monotonicity of $t \mapsto \mathcal{E}_{D_t^2}(\phi^\infty)$, we infer that

$$
\begin{aligned}
E_2 &\leq \lim_{n \to \infty} \int_T^{T+1} \mathcal{E}_{D_t^{(1-\gamma_2)t}}\left(\phi^{(n)}\right) dt \\
&= \int_T^{T+1} \mathcal{E}_{D_t^{(1-\gamma_2)t}}(\phi^\infty) \, dt \\
&\leq \int_T^{T+1} \mathcal{E}_{D_t^2}(\phi^\infty) \, dt \\
&\leq \mathcal{E}_{D_{T+1}^2}(\phi^\infty),
\end{aligned}
$$

which finishes the proof of (4.181).

Following this, we place the wave map $\phi^\infty$ in the context of Subsection 4.6.2, which addressed the hyperbolic projection. Instead of the full cone $\mathcal{C}$, $\phi^\infty$ is defined on $\mathcal{C}_{[2,\infty)}^2$. However, choosing in that context $t_1 = T \geq 2$ and

$$\rho_0 = \sqrt{T^2 - (T-2)^2} = 2\sqrt{T-1},$$

which is the largest value of the hyperbolic coordinate $\rho$ on $D_t^2$, it follows that $\mathcal{H}_{\rho_0}^{t_1} \subset \mathcal{C}_{[2,\infty)}^2$. Therefore, as $X\phi^\infty = 0$, one can rework through that argument and show that, in the set of coordinates $(\rho, \mathbf{y})$, $\phi^\infty$ satisfies $\nabla_\rho \phi^\infty = 0$, the elliptic equation (4.67), and the identity

$$\mathcal{E}_{D_T^2}(\phi^\infty) = \frac{1}{2} \int_{|\mathbf{y}| \leq \frac{\sqrt{T-1}-1}{\sqrt{T-1}+1}} \left\{ \frac{1+|\mathbf{y}|^2}{1-|\mathbf{y}|^2} |\vec{\nabla}_\mathbf{y} \phi^\infty|^2 \right\} d\mathbf{y}.$$

As a consequence, by taking $T \to \infty$ and relying on (4.181), we deduce that

$$\frac{1}{2} \int_{|\mathbf{y}| \leq 1} \left\{ \frac{1+|\mathbf{y}|^2}{1-|\mathbf{y}|^2} |\vec{\nabla}_\mathbf{y} \phi^\infty|^2 \right\} d\mathbf{y} < \infty.$$

According to Lemma 4.1, it follows that $\phi^\infty$ is a trivial map, which leads to a contradiction. This finishes the discussion of this energy scenario.

### 4.9.2 Analysis of scenario (i) in Lemma 4.7

We start by fixing a nonnegative integer $j$ for which this alternative is possible. As $(t_n, \mathbf{x}_n) \in \mathcal{C}_j^{1/2}$ for all $n$, this sequence of points has a convergent subsequence, for which we keep the same notation and whose limit is denoted by $(t_0, \mathbf{x}_0)$. We rescale the associated sequence of wave maps according to

$$\psi^{(n)}(t, \mathbf{x}) \overset{\text{def}}{=} \phi^{(n)}(t_n + r_n t, \mathbf{x}_n + r_n \mathbf{x}).$$

For a fixed $n$, $\psi^{(n)}$ is a wave map defined on the domain $[-1/2, 1/2] \times D_{r/r_n}(\mathbf{0})$, which, due to $\mathcal{E}(\phi^{(n)}) = \mathcal{E}(\phi)$ and (4.154)-(4.155), satisfies the following conditions:

$$\mathcal{E}(\psi^{(n)})(t) \leq \mathcal{E}(\phi), \qquad \forall t \in \left[-\frac{1}{2}, \frac{1}{2}\right], \tag{4.182}$$

$$\mathcal{E}_{D_1^0(\mathbf{0})}(\psi^{(n)}) = \frac{E_0}{10}, \tag{4.183}$$

$$\mathcal{E}_{D_1^0(\mathbf{x})}(\psi^{(n)}) \leq \frac{E_0}{10}, \qquad \forall \mathbf{x} \in D_{r/r_n - 1}(\mathbf{0}). \tag{4.184}$$

Using the local energy estimate (4.166) and (4.184), we infer that

$$\mathcal{E}_{D_{1/2}^t(\mathbf{x})}(\psi^{(n)}) \leq \frac{E_0}{10}, \qquad \forall (t, \mathbf{x}) \in [-1/2, 1/2] \times D_{r/r_n - 1}(\mathbf{0}). \tag{4.185}$$

Moreover, (4.156) implies

$$\lim_{n \to \infty} \int_{[-1/2, 1/2] \times D_{r/r_n}(\mathbf{0})} \left\{ \left| X^{(n)}(\psi^{(n)}) \right|^2 \right\} dt \, d\mathbf{x} = 0,$$

where

$$X^{(n)} = \frac{(t_n + r_n t)\nabla_t + (\mathbf{x}_n + r_n \mathbf{x})\nabla_x}{\rho(t_n + r_n t, \mathbf{x}_n + r_n \mathbf{x})}$$

is a smooth, timelike vector field. A direct, local argument that combines this limit with $(t_n, \mathbf{x}_n) \to (t_0, \mathbf{x}_0)$ and (4.182) yields

$$\lim_{n \to \infty} \int_{[-1/2, 1/2] \times D_{1/2}(\mathbf{y})} \left\{ \left| X^{(0)}(\psi^{(n)}) \right|^2 \right\} dt \, d\mathbf{x} = 0, \qquad \forall \mathbf{y} \in \mathbb{R}^2, \tag{4.186}$$

with

$$X^{(0)} = \frac{t_0 \nabla_t + \mathbf{x}_0 \nabla_x}{\rho(t_0, \mathbf{x}_0)}$$

being a timelike vector field with constant coefficients.

The previous limit and the estimate (4.185) permit us to apply the compactness lemma locally in $[-1/2, 1/2] \times \mathbb{R}^2$ and derive that, on a subsequence,

$$\lim_{n \to \infty} \psi^{(n)} = \psi \quad \text{in } \dot{H}^1_{\text{loc}}([-1/2, 1/2] \times \mathbb{R}^2),$$

where $\psi$ is a wave map satisfying $X^{(0)}\psi = 0$. Furthermore, based on (4.182) and (4.183), we also have

$$\frac{E_0}{10} \leq \mathcal{E}(\psi) \leq \mathcal{E}(\phi).$$

Such a map can be extended uniquely to $\mathbb{R}^{2+1}$ by transporting its values along the flow of the vector field $X^{(0)}$. The end result is that, after performing a Lorentz transformation that maps $X^{(0)}$ into $\nabla_t$, we obtain a nontrivial, static wave map whose total energy is bounded by $\mathcal{E}(\phi)$. This is exactly the content of the first scenario in Theorem 4.4 and its proof is thus complete.

# Chapter 5

# General well-posedness issues for Skyrme and Skyrme-like models

## 5.1 Relevant results

Static properties for the Skyrme and Skyrme-like models introduced in Chapter 1 have been heavily investigated for many years. A nonexhaustive list of such results include:

- numerical computations of various physical quantities (e.g., mass, magnetic moment) for topological solitons by Adkins-Nappi [1] and Adkins-Nappi-Witten [2];

- scattering of topological solitons by Peyrard-Piette-Zakrzewski [130, 131];

- computer simulations for topological solitons of various indices by Faddeev-Niemi [45] and Battye-Sutcliffe [11,12];

- theoretical arguments for the topologically-constrained energy minimization problem by Vakulenko-Kapitanskiĭ [182], Dolbeault [39], Esteban [42], and Lin-Yang [109,110];

- geometric features of topological solitons by Manton [112].

On the other hand, there are considerably fewer studies concerning the time-dependent case of these theories, possibly due to the complex structure of the associated Euler-Lagrange systems (e.g., (1.8) or (1.10)). In this chapter, we focus only on those studies addressing the general time evolution; i.e., no symmetry assumptions (like radial or equivariant) on the involved maps are made. To our knowledge, the most notable results are:

- the Skyrme model exhibits instabilities generated by the existence of a non-hyperbolic region in the phase space of the model, which is due to Crutchfield-Bell [33];

- the *Einstein-Hilbert stress-energy tensor* (or *energy-momentum tensor*) for the Skyrme model satisfies the *dominant energy condition*, a fact proven by Gibbons [56];

- Wong [191] showed that, in certain regimes, the Skyrme model has an *ultrahyperbolic-type breakdown of hyperbolicity*, while in others it exhibits *regular hyperbolicity*, thus being locally well-posed for almost stationary initial data;

- for the $2+1$-dimensional Faddeev model, compactly supported, smooth initial data evolve into global classical solutions, which was proved by Lei-Lin-Zhou [103].

In what follows, we describe in fairly formal details the last two results, our motivation being twofold. First, in what concerns the loss of hyperbolicity for the Skyrme model, [191] can be seen as an improvement over [33]. Moreover, Wong's paper reproves Gibbons' result in the more general context of Lagrangian field theories. The second motivation stems from the fact that Lei-Lin-Zhou's argument relies heavily on the use of a specific family of vector fields, which are generators for the Poincaré group[1]. These yield both generalized energy bounds, as in Klainerman's vector field method, and time decay estimates that are essential in a perturbative analysis of the Faddeev system (1.10). We believe that techniques based on invariant vector fields are quite powerful for nonlinear equations, especially quasilinear ones, where one does not have direct access to useful expressions for solutions and is forced to work strictly with the equation.

## 5.2   Wong's result

The main themes of this article are the concepts of *dominant energy condition* and *regular hyperbolicity* for a family of maps associated to various Lagrangians and the often perceived strong connection between these concepts. The usual heuristic supporting this connection is that the dominant energy condition is a key ingredient for a large number of matter fields to have a domain of dependence property (see Hawking-Ellis [67]). This property is usually related to the well-posedness for the time evolution of the matter field, which in turn is most of the time taken as a definition of hyperbolicity.

However, this is not always the norm. On one hand, we have the case

---

[1]The Poincaré group is the group of Minkowski spacetime isometries.

of the semilinear wave equation

$$\Box u = -u^3,$$

which is locally well-posed for smooth, compactly supported data in $\mathbb{R}^3$ and has its energy-momentum tensor violating the dominant energy condition in certain regimes depending on the size of $u$. On the other hand, for the Skyrme model, the energy-momentum tensor satisfies the dominant energy condition, while the time evolution fails to be regularly hyperbolic when the strain tensor admits a timelike eigenvector corresponding to an eigenvalue with large enough norm. These two facts will be the focus of our exposition here.

### 5.2.1 Preliminaries

If $\phi : (M, g) \to (N, h)$ is a continuously differentiable map from an $m + 1$-dimensional Lorentzian manifold with signature $(m, 1)$ to an $n$-dimensional Riemannian manifold, then its *strain tensor* $D^\phi$ is a $(1, 1)$-type tensor defined as the composition of the inverse metric $g^{-1}$ with the pullback $\phi^*h$ of metric $h$ by $\phi$:

$$(D^\phi)^a_b \stackrel{\text{def}}{=} g^{ac} (\phi^*h)_{cb} = g^{ac} h_{AB}(\phi) \, \partial_c\phi^A \, \partial_b\phi^B = \langle \partial^a\phi, \partial_b\phi \rangle_h,$$

where $0 \le a, b, c \le m$ and $1 \le A, B \le n$. An immediate consequence is

$$\begin{aligned}
g(D^\phi X, Y) &= g(X, D^\phi Y) \\
&= \langle d\phi \cdot X, d\phi \cdot Y \rangle_h \qquad\qquad (5.1) \\
&= \phi^*h(X, Y), \qquad \forall X, Y \in TM.
\end{aligned}$$

For $p \in M$ fixed, the strain tensor determines a self-adjoint linear operator on the tangent space $T_pM$, whose nonzero eigenvalues, counted with multiplicity, are labeled as $\{\lambda_1, \ldots, \lambda_k\}$. Hence, with 0 having implicitly multiplicity $m + 1 - k$, an application of the rank-nullity theorem implies

$$k \le \min\{m + 1, n\}.$$

One can then introduce the elementary symmetric polynomials for these eigenvalues, which are given by

$$\sigma_0(D^\phi) \stackrel{\text{def}}{=} 1,$$

$$\sigma_j(D^\phi) \stackrel{\text{def}}{=} \sum_{1 \le \alpha_1 < \ldots < \alpha_j \le k} \lambda_{\alpha_1} \cdot \ldots \cdot \lambda_{\alpha_j}, \qquad \forall 1 \le j \le k,$$

$$\sigma_j(D^\phi) \stackrel{\text{def}}{=} 0, \qquad \forall j > k.$$

We make the remark that, even though some of these eigenvalues may be complex, all these polynomials are real-valued.

As his main object of study, Wong investigates formal critical points for actions described by

$$S = \int_M L(s(\phi), \sigma_1(D^\phi), \ldots, \sigma_{m+1}(D^\phi)) \, dg, \tag{5.2}$$

where the Lagrangian is understood to depend on $\phi$ strictly through the symmetric polynomials and the nonnegative scalar function $s : N \to \mathbb{R}$. One recognizes immediately that the actions for the nonlinear sigma and Skyrme models (i.e., (1.1) and (1.7), respectively) both fit the above profile. Indeed, routine calculations yield

$$\sigma_1(D^\phi) = \mathrm{tr}(D^\phi) = \langle \partial^a \phi, \partial_a \phi \rangle_h \tag{5.3}$$

and

$$\sigma_2(D^\phi) = \frac{1}{2} \left[ (\mathrm{tr}(D^\phi))^2 - \mathrm{tr}((D^\phi)^2) \right]$$
$$= \frac{1}{2} \left[ \langle \partial^a \phi, \partial_a \phi \rangle_h^2 - \langle \partial^a \phi, \partial^b \phi \rangle_h \langle \partial_a \phi, \partial_b \phi \rangle_h \right], \tag{5.4}$$

which imply that

$$L = \frac{1}{2} \sigma_1(D^\phi) \tag{5.5}$$

for (1.1) and[2]

$$L = \frac{1}{2} \sigma_1(D^\phi) + \frac{\Lambda^2}{2} \sigma_2(D^\phi) \tag{5.6}$$

for (1.7). Born-Infeld-type models [19] are other field theories with actions of the type given by (5.2).

### 5.2.2  *Dominant energy condition*

In the context described above, *the Einstein-Hilbert stress-energy tensor*, also called *the energy-momentum tensor*, is a $(0, 2)$-type tensor defined as

$$T \stackrel{\text{def}}{=} \frac{1}{\sqrt{-\det g}} \frac{\delta(L\sqrt{-\det g})}{\delta g^{-1}} = \frac{\delta L}{\delta g^{-1}} - \frac{1}{2} Lg, \tag{5.7}$$

---

[2]Here, we stick to our choice in Chapter 1 of considering only the massless version of the Skyrme model. However, even if one includes the term giving the $\pi$ mesons a mass, which has the form $s(\phi) = m |\phi - \phi_0|_h^2$, the Skyrme Lagrangian would still match the profile in (5.2).

where the derivative on the right-hand side is a variational one. For this tensor, the dominant energy condition is the following pointwise constraint.

**Definition 5.1.** The energy-momentum tensor obeys the dominant energy condition at $p \in M$ if for every future-directed timelike vector $X \in T_p M$, i.e.,

$$X^0 > 0 \qquad \text{and} \qquad g(X,X) = g_{ab} X^a X^b < 0,$$

then

$$Y^a \stackrel{\text{def}}{=} -T_b^a X^b = -g^{ac} T_{cb} X^b$$

is a future-directed timelike or null vector, i.e.,

$$Y^0 > 0 \qquad \text{and} \qquad g(Y,Y) \leq 0.$$

**Remark 5.1.** Equivalent formulations for the previous definition include:

- if $X$ and $Y$ are future-directed timelike vectors, then $T(X,Y) \geq 0$;
- in all local Lorentz frames, $T_{00} \geq |T_{ab}|$ for all pairs of indices $(a,b)$;
- if one can diagonalize $T$ with respect to $g$ such that

$$g = \text{diag}\,(-1, 1, \ldots, 1) \qquad \text{and} \qquad T = \text{diag}\,(T_{00}, T_{11}, \ldots, T_{mm}),$$

then $T_{00} \geq |T_{ii}|$, for all $1 \leq i \leq m$.

The main result proved by Wong in this direction is:

**Theorem 5.1.** *For Lagrangian theories given by* (5.2), *the energy-momentum tensor satisfies the dominant energy condition if the function*

$$F = L(x_0, x_1, \ldots, x_{m+1}) : \mathbb{R}^{m+2} \to \mathbb{R}$$

*is continuously differentiable with nonnegative first order partial derivatives, it is concave, and* $L(0) \geq 0$.

It is clear that this theorem implies the same statement for the nonlinear sigma and Skyrme models as, according to (5.5) and (5.6), one has $F = x_1/2$ for (1.1) and $F = (x_1 + \Lambda^2 x_2)/2$, respectively, which are allowable functions. In proving the theorem, Wong demonstrates two independent facts, which together yield the desired result:

- if the real-valued function $F : \mathbb{R}^k \to \mathbb{R}$ satisfies the constraints in the theorem and $(L_i)_{1 \leq i \leq k}$ is a collection of Lagrangians whose energy-momentum tensors each obey the dominant energy condition, then the energy-momentum tensor for $L = F(L_1, \ldots, L_k)$ also satisfies the dominant energy condition;

- for either $L = s(\phi)$ or $L = \sigma_j(D^\phi)$ $(1 \leq j \leq m+1)$, the energy-momentum tensor obeys the dominant energy condition.

If the former claim has a direct computational proof relying in a simple way on the properties of $F$, the argument for the latter is more intricate, using advanced linear algebra results combined with tensorial analysis. For this reason and for pedagogical purposes, we choose to present Gibbons's result [56], whose approach is specific to the Skyrme model and, in our opinion, is more transparent.

Thus, for the Skyrme model, we work with

$$(M, g) = (\mathbb{R}^{3+1}, \mathrm{diag}(-1, 1, 1, 1)), \qquad N = \mathbb{S}^3 \subset \mathbb{R}^4,$$

and $h$ is the round metric induced on $\mathbb{S}^3$ by the Euclidean one on $\mathbb{R}^4$. In order to facilitate computations, for a fixed $p \in M$, we first construct orthonormal bases (relative to $g$) in $T_p M$, i.e.,

$$(e_a)_{0 \leq a \leq 3} \subset T_p M, \qquad (g(e_a, e_b))_{0 \leq a, b \leq 3} = \mathrm{diag}\,(-1, 1, 1, 1),$$

which diagonalizes or almost diagonalizes the positive semidefinite bilinear form associated to $\phi^* h$.

**Lemma 5.1.** *In the above setting, if the kernel of the linear transformation*

$$d\phi : T_p M \to T_{\phi(p)} N$$

*is a non-degenerate subspace[3] of $T_p M$, then there exists an orthonormal basis which diagonalizes $\phi^* h$. Otherwise, there exists an orthonormal basis for which $(\phi^* h(e_a, e_b))_{0 \leq a, b \leq 3}$ is either diagonal or of the type*

$$\begin{pmatrix} 1/2 & -1/2 & 0 & 0 \\ -1/2 & 1/2 & 0 & 0 \\ 0 & 0 & \lambda_2 & 0 \\ 0 & 0 & 0 & \lambda_3 \end{pmatrix}.$$

**Proof.** We use a result in O'Neill [126] (exercise 19 in Chapter 9) which, applied to the self-adjoint linear operator $D^\phi$ on the Lorentz vector space $T_p M$, yields

- *either the existence of an orthonormal basis relative to which the matrix of $D^\phi$ is diagonal or of the type*

$$\begin{pmatrix} a & b & 0 & 0 \\ -b & a & 0 & 0 \\ 0 & 0 & \lambda_2 & 0 \\ 0 & 0 & 0 & \lambda_3 \end{pmatrix}$$

---

[3] $V \subseteq T_p M$ is a non-degenerate subspace if $X \in V$ and $g(X, Y) = 0$ for all $Y \in V$ implies $X = 0$.

with $b \neq 0$,

- or the existence of a basis $\{u, v, e_2, e_3\}$ with all scalar products zero except for

$$g(u,v) = g(e_2, e_2) = g(e_3, e_3) = 1,$$

relative to which the matrix of $D^\phi$ is of the type

$$\begin{pmatrix} \lambda & 0 & 0 & 0 \\ \pm 1 & \lambda & 0 & 0 \\ 0 & 0 & \lambda_2 & 0 \\ 0 & 0 & 0 & \lambda_3 \end{pmatrix}$$

or

$$\begin{pmatrix} \lambda & 0 & 1 & 0 \\ 0 & \lambda & 0 & 0 \\ 0 & 1 & \lambda & 0 \\ 0 & 0 & 0 & \lambda_3 \end{pmatrix}.$$

Using (5.1), we can eliminate two of the above matrix profiles. First, if we have an orthonormal basis relative to which the matrix of $D^\phi$ is of the type

$$\begin{pmatrix} a & b & 0 & 0 \\ -b & a & 0 & 0 \\ 0 & 0 & \lambda_2 & 0 \\ 0 & 0 & 0 & \lambda_3 \end{pmatrix},$$

then

$$-a = g(D^\phi e_0, e_0) = |d\phi \cdot e_0|_h^2 \geq 0$$

and

$$a = g(D^\phi e_1, e_1) = |d\phi \cdot e_1|_h^2 \geq 0.$$

This implies $a = 0$, which yields $d\phi \cdot e_0 = 0$. Hence, $D^\phi e_0 = 0$, thus forcing $b = 0$, which is a contradiction.

Second, if we have a basis $\{u, v, e_2, e_3\}$ as described above relative to which the matrix of $D^\phi$ is of the type

$$\begin{pmatrix} \lambda & 0 & 1 & 0 \\ 0 & \lambda & 0 & 0 \\ 0 & 1 & \lambda & 0 \\ 0 & 0 & 0 & \lambda_3 \end{pmatrix},$$

then

$$|d\phi \cdot u|_h^2 = g(D^\phi u, u) = 0.$$

This implies $d\phi \cdot u = 0$, which yields $D^\phi u = 0$. Hence, $\lambda u + e_2 = 0$, thus contradicting $\{u, v, e_2, e_3\}$ being a basis.

Let us assume now that the matrix of $D^\phi$ is of the type

$$\begin{pmatrix} \lambda & 0 & 0 & 0 \\ \pm 1 & \lambda & 0 & 0 \\ 0 & 0 & \lambda_2 & 0 \\ 0 & 0 & 0 & \lambda_3 \end{pmatrix}.$$

Like before, we derive $u \in \ker(d\phi)$ and $D^\phi u = 0$, which implies $\lambda = 0$. As

$$\pm 1 = g(D^\phi v, v) = |d\phi \cdot v|_h^2 \geq 0,$$

it follows that $D^\phi v = u$. Therefore, the new updated profile of the matrix is

$$\begin{pmatrix} 0 & 0 & 0 & 0 \\ 1 & 0 & 0 & 0 \\ 0 & 0 & \lambda_2 & 0 \\ 0 & 0 & 0 & \lambda_3 \end{pmatrix}. \tag{5.8}$$

If we also suppose that $\ker(d\phi)$ is a non-degenerate subspace of $T_pM$, a classical result yields the existence of a null vector $w \in \ker(d\phi)$, for which $\{u, w\}$ is linearly independent. Thus, for $w = au + bv + ce_2 + de_3$, we have

$$0 = D^\phi w = aD^\phi u + bD^\phi v + cD^\phi e_2 + dD^\phi e_3 = bu + c\lambda_2 e_2 + d\lambda_3 e_3.$$

We deduce $b = 0$, which further implies

$$0 = g(w, w) = c^2 + d^2.$$

Hence, $c = d = 0$ and, as a result, $w = au$. This contradicts $\{u, w\}$ being linearly independent and tells us that, for the non-degenerate case, the only allowable matrix profile is the diagonal one.

In the degenerate case, we have either the diagonal profile or the one given by (5.8). For the latter, we construct the orthonormal basis

$$\left\{ e_0 = \frac{1}{\sqrt{2}}(u - v), e_1 = \frac{1}{\sqrt{2}}(u + v), e_2, e_3 \right\}$$

relative to which the matrix of $D^\phi$ becomes

$$\begin{pmatrix} -1/2 & -1/2 & 0 & 0 \\ 1/2 & 1/2 & 0 & 0 \\ 0 & 0 & \lambda_2 & 0 \\ 0 & 0 & 0 & \lambda_3 \end{pmatrix}$$

and, as a consequence, $(\phi^* h(e_a, e_b))_{0 \leq a, b \leq 3}$ is of the type described in the lemma. $\qquad\square$

In what follows, we show Gibbons's argument when one can diagonalize the bilinear form associated to $\phi^*h$ relative to $g$ and leave the proof for the other matrix profile in the lemma to the interested reader. Thus, for a fixed $p \in M$, there exists an orthonormal basis $(e_a)_{0 \le a \le 3} \subset T_pM$ such that

$$(\phi^*h(e_a, e_b))_{0 \le a,b \le 3} = \operatorname{diag}(\lambda_0, \lambda_1, \lambda_2, \lambda_3).$$

All of the $\lambda$'s are nonnegative due to $\phi^*h$ being positive semidefinite. With respect to this frame, straightforward computations based on (5.3) and (5.4) yield

$$\sigma_1(D^\phi) = -\lambda_0 + \lambda_1 + \lambda_2 + \lambda_3$$

and

$$\sigma_2(D^\phi) = -\lambda_0(\lambda_1 + \lambda_2 + \lambda_3) + \lambda_1\lambda_2 + \lambda_2\lambda_3 + \lambda_3\lambda_1.$$

Next, we use (5.7) to find the energy-momentum tensor for the Skyrme model. First, we rely again on (5.3) and (5.4) to infer that, in our special frame,

$$\frac{\delta(\sigma_1(D^\phi))}{\delta g^{ab}} = \langle \partial_a\phi, \partial_b\phi \rangle_h = \operatorname{diag}(\lambda_0, \lambda_1, \lambda_2, \lambda_3)$$

and

$$\frac{\delta(\sigma_2(D^\phi))}{\delta g^{ab}} = \sigma_1(D^\phi)\langle \partial_a\phi, \partial_b\phi \rangle_h - \langle \partial_a\phi, \partial^c\phi \rangle_h \langle \partial_b\phi, \partial_c\phi \rangle_h$$

$$= \operatorname{diag}(\lambda_0(\lambda_1 + \lambda_2 + \lambda_3), \lambda_1(-\lambda_0 + \lambda_2 + \lambda_3),$$
$$\lambda_2(-\lambda_0 + \lambda_3 + \lambda_1), \lambda_3(-\lambda_0 + \lambda_1 + \lambda_2)).$$

On account of (5.6) and (5.7), we derive

$$T = \frac{1}{4}\operatorname{diag}\big(\lambda_0 + \lambda_1 + \lambda_2 + \lambda_3,$$
$$\lambda_0 + \lambda_1 - \lambda_2 - \lambda_3,$$
$$\lambda_0 + \lambda_2 - \lambda_3 - \lambda_1,$$
$$\lambda_0 + \lambda_3 - \lambda_1 - \lambda_2\big)$$

$$+ \frac{\Lambda^2}{4}\operatorname{diag}\big(\lambda_0(\lambda_1 + \lambda_2 + \lambda_3) + \lambda_1\lambda_2 + \lambda_2\lambda_3 + \lambda_3\lambda_1,$$
$$(\lambda_0 + \lambda_1)(\lambda_2 + \lambda_3) - \lambda_0\lambda_1 - \lambda_2\lambda_3,$$
$$(\lambda_0 + \lambda_2)(\lambda_3 + \lambda_1) - \lambda_0\lambda_2 - \lambda_3\lambda_1,$$
$$(\lambda_0 + \lambda_3)(\lambda_1 + \lambda_2) - \lambda_0\lambda_3 - \lambda_1\lambda_2\big).$$

Therefore, we can diagonalize the energy-momentum tensor relative to the Minkowski metric and, as a consequence, we can check the dominant energy condition by using the third characterization from Remark 5.1. But this is obviously satisfied due to all $\lambda$'s being nonnegative, which finishes the argument.

### 5.2.3    *Regular hyperbolicity*

Another focal point of Wong's work is the question of hyperbolicity for the
Euler-Lagrange system corresponding to (5.2), which takes the quasilinear
form

$$m^{ab}_{AB}(p, \phi, \partial\phi) \, \nabla^2_{ab}\phi^B \; = \; F_A(p, \phi, \partial\phi). \tag{5.9}$$

In this context, $\nabla$ denotes the covariant derivative with respect to the
metric $g$ and

$$m^{ab}_{AB} \stackrel{\text{def}}{=} \frac{\delta^2 L}{\delta(\partial_a \phi^A)\, \delta(\partial_b \phi^B)} \tag{5.10}$$

represents the second variational derivative of the Lagrangian with respect
to the field velocity. As previously mentioned, hyperbolicity for a system
of partial differential equations usually means the well-posedness of the associated Cauchy problem with smooth initial data. More precise versions
of hyperbolicity exist, like the one applicable to linear systems with constant coefficients (e.g., sections 5.4-5.6 in Hörmander [71]), Leray's strict
hyperbolicity [107], or the symmetric hyperbolicity due to Friedrichs [48].
One point made by Wong is that there are significant technical difficulties
implementing these concepts to (5.9).

As an alternative, he argues for the use of Christodoulou's regular hyperbolicity framework [28], which has already been successful in dealing
with a similar problem; i.e., Speck's global well-posedness result [157] for
the Maxwell-Born-Infeld system. The mathematical language in [28] exhibits many challenges and this is why we will be quite informal in our
exposition here. Generally speaking, *regular hyperbolicity* means the ability
of proving energy estimates which can be used in an iterative argument
to show well-posedness. Moreover, the concept of regular hyperbolicity is
especially suited for Euler-Lagrange systems and works directly with the
Lagrangians that generated them. Perhaps as a minus for this concept, like
the strict and symmetric hyperbolicities, regular hyperbolicity is only a sufficient condition for the well-posedness of the Cauchy problem in question.

In the basic setting when $(M, g) = (\mathbb{R}^{m+1}, \text{diag}(-1, 1, \ldots, 1))$, the main
goal of this framework is to deduce energy estimates of the type

$$\|\phi(t)\|_{H^k(\mathbb{R}^m)} \lesssim \|\phi(0)\|_{H^k(\mathbb{R}^m)}\, e^{Ct}, \qquad \forall t \in (0, T), \tag{5.11}$$

where $k \geq 0$ is an integer, $0 < T < \infty$, and $C$ is a constant depending
solely on $T$ and $\|\phi(0)\|_{H^k(\mathbb{R}^m)}$. Typically, they are derived by constructing

vector fields $J = J(t, x, \phi, \nabla\phi, \dots, \nabla^k\phi)$ involving derivatives of $\phi$ of order up to $k$, which satisfy

$$\int_{\mathbb{R}^m} \operatorname{div} J(t, x)\, dx \lesssim \|\phi(t)\|_{H^k(\mathbb{R}^m)} \tag{5.12}$$

and

$$\int_{\mathbb{R}^m} dt \circ J(t, x)\, dx \simeq \|\phi(t)\|_{H^k(\mathbb{R}^m)}, \tag{5.13}$$

followed by an application of Gauss's theorem combined with Grönwall's lemma.

First, we discuss (5.12), for which an important issue is that div $J$ depends a priori on derivatives of $\phi$ of order up to $k + 1$ and thus would be difficult, if not impossible, to control by a quantity involving derivatives of lesser order, such as $\|\phi(t)\|_{H^k}$. Hence, one expects here to use the equations for $\phi$ in order to trade higher-order derivatives for lower-order ones. In constructing vector fields satisfying (5.12), two key facts come into play:

- the coefficients of the higher-order derivatives in (5.9) enjoy the symmetry property

$$m_{AB}^{ab} = m_{BA}^{ba}, \qquad \forall\, 0 \le a, b \le m,\; 1 \le A, B \le n, \tag{5.14}$$

  which is an immediate consequence of their definition (5.10);
- arbitrary partial derivatives of $\phi$ satisfy quasilinear equations similar to (5.9), i.e.,

$$m_{AB}^{ab}(p, \phi, \partial\phi)\, \partial_{ab}^2 \nabla^\alpha \phi^B = F_A^\alpha(p, \phi, \partial\phi, \dots, \partial^{|\alpha|+1}\phi), \tag{5.15}$$

with $\partial^\beta$ denoting the collection of all partial derivatives of order $\beta$ and $\nabla^\alpha$ denoting the partial derivative corresponding to the multi-index $\alpha$.

The vector fields in question are defined using *the canonical stress tensor* associated to solutions of (5.9), which is given by

$$Q[\psi]_d^c \stackrel{\text{def}}{=} (m_{AB}^{ac}\, \delta_d^b + m_{AB}^{cb}\, \delta_d^a - m_{AB}^{ab}\, \delta_d^c)\, \partial_a \psi^A\, \partial_b \psi^B. \tag{5.16}$$

In the above, the $m$'s depend on $\phi$ (i.e., the solution of (5.9)), while $\psi$ is a function taking values in $\mathbb{R}^n$. A routine computation based on (5.14) and (5.15) yields

$$\begin{aligned}
\partial_c\, &Q[\nabla^\alpha\phi]_d^c \\
&= \partial_c \left\{ m_{AB}^{ac}\, \delta_d^b + m_{AB}^{cb}\, \delta_d^a - m_{AB}^{ab}\, \delta_d^c \right\} \partial_a \nabla^\alpha \phi^A\, \partial_b \nabla^\alpha \phi^B \\
&\quad + 2\partial_d \nabla^\alpha \phi^A\, F_A^\alpha(p, \phi, \partial\phi, \dots, \partial^{|\alpha|+1}\phi),
\end{aligned} \tag{5.17}$$

with the right-hand side containing derivatives of order no higher than $|\alpha| + 1$. This implies that, for any fixed vector $X = (X^a)_a$, the vector field defined by

$$J = J^c \partial_c \stackrel{\text{def}}{=} \left( |\phi|^2 X^c + \sum_{|\alpha| \leq k-1} Q[\nabla^\alpha \phi]_d^c X^d \right) \partial_c \qquad (5.18)$$

and its divergence depend both on derivatives of $\phi$ of order less than or equal to $k$. It is easy to see then by applying Hölder estimates and Sobolev embeddings that (5.12) holds for this vector field. We notice that the hyperbolic nature of (5.9) does not play any role in the previous construction; the argument is made possible by these equations being the Euler-Lagrange equations corresponding to a Lagrangian theory.

Next, we focus on (5.13), for which we need to introduce the notions of *time function* and *observer field*. In the general context, a function $t = t(p)$ ($p \in M$) is called a time function for a Lagrangian if, given a foliation of $M$ by its level sets, the matrix

$$(m_{AB}^{ab}(dt)_a(dt)_b)_{A,B} \qquad (5.19)$$

is negative definite. For the same foliation, the vector $Y = (Y^a)_a$ is called an observer field if the matrix

$$(m_{AB}^{ab} \omega_a \omega_b)_{A,B} \qquad (5.20)$$

is positive definite for all nonzero covectors $\omega = (\omega_a)_a$ such that

$$Y(\omega) = Y^a \omega_a = 0.$$

In [28], Christodoulou proved that (5.13) holds for vector fields $J$ given by (5.18), if there exists a time function for which $X$ is an observer field. Furthermore, current work in progress by Speck and Wong will address the claim made in [28] that this is all that is needed for a local well-posedness result concerning the initial value problem for (5.9), with data on a level set of the time function. Thus, one is entitled to say that regular hyperbolicity fails if there is no existence for a time function and an observer field. Related to this, Wong distinguishes:

- an *elliptic-type breakdown of hyperbolicity* if a time function does not exist even locally;
- an *ultrahyperbolic-type breakdown of hyperbolicity* if one has existence of a time function, but cannot construct an observer field.

This terminology has to do with the elliptic equation

$$\partial_{tt}^2 u + \Delta u = 0$$

and the ultrahyperbolic equation

$$-\partial_{tt}^2 u - 2\partial_{11}^2 u + \Delta u = 0$$

fitting the above two profiles, respectively.

Wong's article contains two main results concerning regular hyperbolicity and breakdowns of hyperbolicity for Lagrangian theories described by (5.2). The first one states that if the Lagrangian function in (5.2) is a linear combination with nonnegative coefficients of its variables, then the resulting Euler-Lagrange equations cannot have a genuine elliptic-type breakdown of hyperbolicity. In particular, Wong shows that if the coefficient of $\sigma_1(D^\phi)$ is positive (which is the case of the Skyrme model Lagrangian (5.6)), then any time function relative to $g$ is also a time function for (5.2).

However, it is the second result which is more relevant for the content of this book and which is presented in the remainder of this section. It addresses the Skyrme model with $\Lambda^2 = 1$ in (1.7) (and (5.6)), i.e., the Lagrangian is given by

$$
\begin{aligned}
L &= \frac{1}{2}\sigma_1(D^\phi) + \frac{1}{2}\sigma_2(D^\phi) \\
&= \frac{1}{2}g^{ab}\,h_{AB}\,\partial_a\phi^A\,\partial_b\phi^B \\
&\quad + \frac{1}{4}g^{ab}\,g^{cd}\,(h_{AB}\,h_{CD} - h_{AD}\,h_{BC})\partial_a\phi^A\,\partial_b\phi^B\,\partial_c\phi^C\,\partial_d\phi^D,
\end{aligned}
\tag{5.21}
$$

and it states that:

**Theorem 5.2.** *The Skyrme model is regularly hyperbolic if either one of the following three conditions hold.*

*(1) The kernel of $d\phi$ contains a timelike vector.*
*(2) The kernel of $d\phi$ is a degenerate subspace.*
*(3) The timelike eigenvector for $D^\phi$ has a corresponding eigenvalue with absolute value less than 1.*

*If the timelike eigenvector for $D^\phi$ has a corresponding eigenvalue with absolute value greater than 1, then the Skyrme model experiences an ultrahyperbolic-type breakdown of hyperbolicity.*

According to Christodoulou's algorithm, we first compute

$$
\begin{aligned}
m_{AB}^{ab} &= g^{ab}\,h_{AB} + \big[g^{ab}\,g^{cd}(h_{AB}\,h_{CD} - h_{AC}\,h_{BD}) \\
&\quad + g^{ac}\,g^{bd}(2h_{AC}\,h_{BD} - h_{AB}\,h_{CD} - h_{AD}\,h_{BC})\big]\partial_c\phi^C\,\partial_d\phi^D.
\end{aligned}
\tag{5.22}
$$

As a warmup for further computations, let us verify directly that if $t$ is the canonical time function of the $\mathbb{R}^{3+1}$ Minkowski spacetime, then it is also a time function for the Skyrme Lagrangian. This corroborates the fact proven by Wong concerning time functions for Lagrangians in which $\sigma_1(D^\phi)$ is relevant. It is easy to see that

$$m_{AB}^{ab}(dt)_a(dt)_b = -h_{AB} + (h_{AC}h_{BD} - h_{AB}h_{CD})\nabla_x\phi^C \cdot \nabla_x\phi^D,$$

which, due to the Cauchy-Schwarz inequality, implies

$$m_{AB}^{ab}(dt)_a(dt)_b\xi^A\xi^B = -|\xi|_h^2 + \sum_{1\leq i\leq 3}\left(\langle\xi,\partial_i\phi\rangle_h^2 - |\xi|_h^2|\partial_i\phi|_h^2\right) \leq -|\xi|_h^2,$$

thus proving the claim.

Hence, in order to decide upon regular hyperbolicity or ultrahyperbolic-type breakdown of hyperbolicity, one needs to investigate the existence of an observer field. As was the case for Gibbons' result on the dominant energy condition, we rely on Lemma 5.1 to work with orthonormal frames that have "diagonalization" properties relative to $\phi^*h$. To streamline the presentation, we only show the analysis for the case when we can diagonalize $\phi^*h$ relative to $g$ and leave the one for the non-diagonal case of Lemma 5.1 to the avid reader.

Thus, we work with an orthonormal basis $(e_a)_{0\leq a\leq 3} \subset T_pM$ such that

$$(\phi^*h(e_a,e_b))_{0\leq a,b\leq 3} = \text{diag}(\mu_0^2,\mu_1^2,\mu_2^2,\mu_3^2).$$

The notation for the diagonal entries is motivated by the bilinear form associated to $\phi^*h$ being positive semidefinite. Moreover, at least one of these entries is zero, which is justified as follows. Since $(d\phi \cdot e_a)_{0\leq a\leq 3} \in T_{\phi(p)}N$ and $\dim T_{\phi(p)}N = 3$, there exist $(\alpha_a)_{0\leq a\leq 3} \subset \mathbb{R}$, not all zero, such that

$$\sum_{0\leq a\leq 3} \alpha_a \, d\phi \cdot e_a = 0.$$

Taking the norm of the left-hand side with respect to $h$, we derive

$$\sum_{0\leq a\leq 3} \alpha_a^2\mu_a^2 = 0,$$

which proves the claim.

Next, based on these properties for $(e_a)_{0\leq a\leq 3}$, we can choose $(d_a)_{0\leq a\leq 3} \subset T_{\phi(p)}N$ to be made up of the zero vector and an orthonormal basis for $T_{\phi(p)}N$ relative to $h$, and to satisfy

$$d\phi \cdot e_a = \mu_a d_a, \qquad \forall 0 \leq a \leq 3. \qquad (5.23)$$

This can be done by prescribing $d_a = \mu_a^{-1} d\phi \cdot e_a$ when $\mu_a \neq 0$, while for $\mu_a = 0$ the choice for $d_a$ is quite flexible, as the previous condition is automatically satisfied. For example, if precisely two of the $\mu_a$'s are zero, then we pick one of the corresponding $d_a$'s to be the zero vector and the other one to be a unit vector, which, together with the other two $d_a$'s, forms an orthonormal basis for $T_{\phi(p)}N$.

Finally, we consider the dual families in the cotangent spaces for $(e_a)_{0 \leq a \leq 3}$ and $(d_a)_{0 \leq a \leq 3}$, with the understanding that the covector corresponding to $d_a = 0$ is set to vanish itself. These families of covectors are denoted by $(E_a)_{0 \leq a \leq 3} \subset T_p^* M$ and $(D_a)_{0 \leq a \leq 3} \subset T_{\phi(p)}^* N$, respectively. One deduces easily that

$$X = \sum_{0 \leq a \leq 3} (-1)^{\delta_{0a}} \langle X, e_a \rangle_g \, e_a = \sum_{0 \leq a \leq 3} X(E_a) \, e_a, \quad \forall X \in T_p M, \quad (5.24)$$

$$Y = \sum_{0 \leq a \leq 3} \langle Y, d_a \rangle_h \, d_a = \sum_{0 \leq a \leq 3} Y(D_a) \, d_a, \quad \forall Y \in T_{\phi(p)} N, \quad (5.25)$$

$$\omega = \sum_{0 \leq a \leq 3} (-1)^{\delta_{0a}} \langle \omega, E_a \rangle_{g^{-1}} E_a = \sum_{0 \leq a \leq 3} e_a(\omega) \, E_a, \quad \forall \omega \in T_p^* M. \quad (5.26)$$

Using this framework, we start computing the individual terms appearing in (5.22). Due to (5.26), for two covectors in $T_p^* M$, we obtain

$$g^{ab} \omega_a^1 \omega_b^2 = \langle \omega^1, \omega^2 \rangle_{g^{-1}}$$

$$= \sum_{0 \leq a,b \leq 3} e_a(\omega^1) \, e_b(\omega^2) \, \langle E_a, E_b \rangle_{g^{-1}}$$

$$= \sum_{0 \leq a,b \leq 3} e_a(\omega^1) \, e_b(\omega^2) \, \langle e_a, e_b \rangle_g$$

$$= -e_0(\omega_1) \, e_0(\omega_2) + e_1(\omega_1) \, e_1(\omega_2) + e_2(\omega_1) \, e_2(\omega_2) + e_3(\omega_1) \, e_3(\omega_2),$$

which implies, using the tensorial notation,

$$g^{ab} = (-e_0 \otimes e_0 + e_1 \otimes e_1 + e_2 \otimes e_2 + e_3 \otimes e_3)^{ab}. \quad (5.27)$$

Next, based on (5.25), we derive for two vectors in $T_{\phi(p)} N$ that

$$h_{AB} X^A Y^B = \langle X, Y \rangle_h$$

$$= \sum_{0 \leq a,b \leq 3} X(D_a) Y(D_b) \, \langle d_a, d_b \rangle_h$$

$$= \sum_{0 \leq a \leq 3} X(D_a) Y(D_a),$$

which yields

$$h_{AB} = (D_0 \otimes D_0 + D_1 \otimes D_1 + D_2 \otimes D_2 + D_3 \otimes D_3)_{AB}. \quad (5.28)$$

Finally, with the help of (5.24) and (5.23), we deduce for a vector in $T_p M$ and a covector in $T^*_{\phi(p)} N$ that

$$\partial_a \phi^A X^a \omega_A = (d\phi \cdot X)(\omega)$$

$$= \sum_{0 \le a \le 3} X(E_a)(d\phi \cdot e_a)(\omega)$$

$$= \sum_{0 \le a \le 3} \mu_a X(E_a) d_a(\omega),$$

which produces

$$\partial_a \phi^A = (\mu_0 E_0 \otimes d_0 + \mu_1 E_1 \otimes d_1 + \mu_2 E_2 \otimes d_2 + \mu_3 E_3 \otimes d_3)^A_a. \quad (5.29)$$

In order to find out whether an observer field exists or not for the Skyrme Lagrangian, Wong focuses his computations almost exclusively on

$$m^{00}_{AB} \stackrel{\text{def}}{=} m^{ab}_{AB}(E_0)_a(E_0)_b, \qquad m^{30}_{AB} \stackrel{\text{def}}{=} m^{ab}_{AB}(E_3)_a(E_0)_b,$$

$$m^{31}_{AB} \stackrel{\text{def}}{=} m^{ab}_{AB}(E_3)_a(E_1)_b, \qquad m^{33}_{AB} \stackrel{\text{def}}{=} m^{ab}_{AB}(E_3)_a(E_3)_b.$$

This has to do with the existing symmetries in (5.27)-(5.29) relative to the index $1 \le a \le 3$ and with the ability[4] of choosing $\omega \in \text{span}\{E_0, E_3\}$ for (5.20). For such a covector, $\omega = \alpha E_0 + \beta E_3$, one obtains

$$m^{ab}_{AB} \omega_a \omega_b = \alpha^2 m^{00}_{AB} + \alpha\beta \left(m^{30}_{AB} + m^{30}_{BA}\right) + \beta^2 m^{33}_{AB}. \quad (5.30)$$

Thus, direct computations relying on (5.22) and (5.27)-(5.29) yield:

$$m^{00}_{AB}$$

$$= -h_{AB} - (h_{AB} h_{CD} - h_{AC} h_{BD}) g^{cd} \partial_c \phi^C \partial_d \phi^D$$

$$\quad + (2h_{AC} h_{BD} - h_{AB} h_{CD} - h_{AD} h_{BC}) g^{ac}(E_0)_a \partial_c \phi^C g^{bd}(E_0)_b \partial_d \phi^D$$

$$= -h_{AB} - (h_{AB} h_{CD} - h_{AC} h_{BD}) \sum_{0 \le a \le 3} (-1)^{\delta_{0a}} (d\phi \cdot e_a)^C (d\phi \cdot e_a)^D$$

$$\quad + (2h_{AC} h_{BD} - h_{AB} h_{CD} - h_{AD} h_{BC})(d\phi \cdot e_0)^C (d\phi \cdot e_0)^D$$

$$= -h_{AB} - (h_{AB} h_{CD} - h_{AC} h_{BD}) \sum_{1 \le a \le 3} \mu_a^2 (d_a)^C (d_a)^D$$

$$= -(1 + \mu_1^2 + \mu_2^2 + \mu_3^2) h_{AB} + \sum_{1 \le a \le 3} \mu_a^2 (D_a \otimes D_a)_{AB}$$

$$= -(1 + \mu_1^2 + \mu_2^2 + \mu_3^2)(D_0 \otimes D_0)_{AB} - (1 + \mu_2^2 + \mu_3^2)(D_1 \otimes D_1)_{AB}$$

$$\quad - (1 + \mu_1^2 + \mu_3^2)(D_2 \otimes D_2)_{AB} - (1 + \mu_1^2 + \mu_2^2)(D_3 \otimes D_3)_{AB},$$

$$(5.31)$$

---

[4]For an observer field $X$, $\dim(\ker X) = 3$, while $\dim(\text{span}\{E_0, E_3\}) = 2$. As both are subspaces of the 4-dimensional vector space $T^*_p M$, their intersection is a 1-dimensional subspace.

$m_{AB}^{30}$

$$= (2h_{AC}\,h_{BD} - h_{AB}\,h_{CD} - h_{AD}\,h_{BC})g^{ac}(E_3)_a\,\partial_c\phi^C\,g^{bd}(E_0)_b\,\partial_d\phi^D$$
$$= -(2h_{AC}\,h_{BD} - h_{AB}\,h_{CD} - h_{AD}\,h_{BC})(d\phi\cdot e_3)^C(d\phi\cdot e_0)^D$$
$$= -\mu_0\mu_3(2h_{AC}\,h_{BD} - h_{AB}\,h_{CD} - h_{AD}\,h_{BC})(d_3)^C(d_0)^D$$
$$= -\mu_0\mu_3\left[2(D_3\otimes D_0)_{AB} - (D_0\otimes D_3)_{AB}\right],$$

$$(5.32)$$

$m_{AB}^{31}$

$$= (2h_{AC}\,h_{BD} - h_{AB}\,h_{CD} - h_{AD}\,h_{BC})g^{ac}(E_3)_a\,\partial_c\phi^C\,g^{bd}(E_1)_b\,\partial_d\phi^D$$
$$= (2h_{AC}\,h_{BD} - h_{AB}\,h_{CD} - h_{AD}\,h_{BC})(d\phi\cdot e_3)^C(d\phi\cdot e_1)^D$$
$$= \mu_1\mu_3(2h_{AC}\,h_{BD} - h_{AB}\,h_{CD} - h_{AD}\,h_{BC})(d_3)^C(d_1)^D$$
$$= \mu_1\mu_3\left[2(D_3\otimes D_1)_{AB} - (D_1\otimes D_3)_{AB}\right],$$

$$(5.33)$$

$m_{AB}^{33}$

$$= h_{AB} + (h_{AB}\,h_{CD} - h_{AC}\,h_{BD})g^{cd}\,\partial_c\phi^C\,\partial_d\phi^D$$
$$\quad + (2h_{AC}\,h_{BD} - h_{AB}\,h_{CD} - h_{AD}\,h_{BC})g^{ac}(E_3)_a\,\partial_c\phi^C\,g^{bd}(E_3)_b\,\partial_d\phi^D$$
$$= h_{AB} + (h_{AB}\,h_{CD} - h_{AC}\,h_{BD})\sum_{0\leq a\leq 3}(-1)^{\delta_{0a}}(d\phi\cdot e_a)^C(d\phi\cdot e_a)^D$$
$$\quad + (2h_{AC}\,h_{BD} - h_{AB}\,h_{CD} - h_{AD}\,h_{BC})(d\phi\cdot e_3)^C(d\phi\cdot e_3)^D$$
$$= h_{AB} + (h_{AB}\,h_{CD} - h_{AC}\,h_{BD})\sum_{0\leq a\leq 2}(-1)^{\delta_{0a}}\mu_a^2(d_a)^C(d_a)^D$$
$$= (1 - \mu_0^2 + \mu_1^2 + \mu_2^2)h_{AB} + \sum_{0\leq a\leq 2}(-1)^{\delta_{0a}+1}\mu_a^2(D_a\otimes D_a)_{AB}$$
$$= (1 + \mu_1^2 + \mu_2^2)(D_0\otimes D_0)_{AB} + (1 - \mu_0^2 + \mu_2^2)(D_1\otimes D_1)_{AB}$$
$$\quad + (1 - \mu_0^2 + \mu_1^2)(D_2\otimes D_2)_{AB} + (1 - \mu_0^2 + \mu_1^2 + \mu_2^2)(D_3\otimes D_3)_{AB}.$$

$$(5.34)$$

We now have all that is needed for a formal discussion of Theorem 5.2. Given that at least one of the $\mu_a$'s vanishes, we split the argument into the cases when $\mu_0 = 0$, $\mu_3 = 0 \neq \mu_0$, and $\mu_1 = 0 \neq \mu_0\mu_3$, respectively, which are exhaustive based on the symmetries explained before.

For $\mu_0 = 0$, we show that $e_0$ is an observer field and, as a result, the Skyrme model is regularly hyperbolic. This corresponds to the first condition in Theorem 5.2 ensuring regular hyperbolicity. If $\omega$ is a covector verifying $e_0(\omega) = 0$, then it follows that

$$\omega = \alpha_1 E_1 + \alpha_2 E_2 + \alpha_3 E_3,$$

which implies

$$m_{AB}^{ab}\, \omega_a\, \omega_b = \sum_{1 \le a \le 3} \alpha_a^2\, m_{AB}^{aa} + \sum_{1 \le a < b \le 3} \alpha_a \alpha_b \left( m_{AB}^{ab} + m_{BA}^{ba} \right).$$

Due to (5.33), (5.34), and other similar formulas, we deduce for an arbitrary real vector $\xi = (\xi^A)_{1 \le A \le 3}$ that

$$
\begin{aligned}
m_{AB}^{ab}\, &\omega_a\, \omega_b\, \xi^A \xi^B \\
&= \left[ \alpha_1^2(1 + \mu_2^2 + \mu_3^2) + \alpha_2^2(1 + \mu_1^2 + \mu_3^2) + \alpha_3^2(1 + \mu_1^2 + \mu_2^2) \right] |\xi(D_0)|^2 \\
&+ \left[ \alpha_1^2(1 + \mu_2^2 + \mu_3^2) + \alpha_2^2(1 + \mu_3^2) + \alpha_3^2(1 + \mu_2^2) \right] |\xi(D_1)|^2 \\
&+ \left[ \alpha_1^2(1 + \mu_3^2) + \alpha_2^2(1 + \mu_1^2 + \mu_3^2) + \alpha_3^2(1 + \mu_1^2) \right] |\xi(D_2)|^2 \\
&+ \left[ \alpha_1^2(1 + \mu_2^2) + \alpha_2^2(1 + \mu_1^2) + \alpha_3^2(1 + \mu_1^2 + \mu_2^2) \right] |\xi(D_3)|^2 \\
&+ 2 \sum_{1 \le a < b \le 3} \alpha_a\, \alpha_b\, \mu_a\, \mu_b\, \xi(D_a)\, \xi(D_b),
\end{aligned}
$$

where we denoted $\xi(D_a) = \xi^A (D_a)_A$. Using the obvious inequality $u^2 + v^2 + 2uv = (u + v)^2 \ge 0$, we infer

$$
\begin{aligned}
m_{AB}^{ab}\, &\omega_a\, \omega_b\, \xi^A \xi^B \\
&\ge (\alpha_1^2 + \alpha_2^2 + \alpha_3^2)(|\xi(D_0)|^2 + |\xi(D_1)|^2 + |\xi(D_2)|^2 + |\xi(D_3)|^2) \\
&= (\alpha_1^2 + \alpha_2^2 + \alpha_3^2)|\xi|_h^2,
\end{aligned}
$$

which proves that $m_{AB}^{ab}\, \omega_a\, \omega_b$ is positive definite, thus ending the argument for this case.

If $\mu_3 = 0 \ne \mu_0$, let us choose a covector $\omega = \alpha_0 E_0 + \alpha_3 E_3$, which is in the kernel of a potential observer field $Y$. On the basis of (5.30), (5.31), (5.32), and (5.34), we derive for a 3-dimensional real vector

$$
\begin{aligned}
m_{AB}^{ab}\, \omega_a\, \omega_b\, \xi^A \xi^B &= (\alpha_3^2 - \alpha_0^2)(1 + \mu_1^2 + \mu_2^2)|\xi(D_0)|^2 \\
&+ \left[ -\alpha_0^2(1 + \mu_2^2) + \alpha_3^2(1 - \mu_0^2 + \mu_2^2) \right] |\xi(D_1)|^2 \\
&+ \left[ -\alpha_0^2(1 + \mu_1^2) + \alpha_3^2(1 - \mu_0^2 + \mu_1^2) \right] |\xi(D_2)|^2 \\
&+ \left[ -\alpha_0^2(1 + \mu_1^2 + \mu_2^2) + \alpha_3^2(1 - \mu_0^2 + \mu_1^2 + \mu_2^2) \right] |\xi(D_3)|^2.
\end{aligned}
$$

In this case, the construction of the family $(D_a)_{0 \le a \le 3}$ implies that at least one of $D_1$ and $D_2$ doesn't vanish and, without loss of generality, let us assume that $D_1 \ne 0$. By choosing $\xi$ such that $\xi(D_1) \ne 0 = \xi(D_a)$, for all $a \ne 1$, we obtain

$$m_{AB}^{ab}\, \omega_a\, \omega_b\, \xi^A \xi^B = \left[ -\alpha_0^2(1 + \mu_2^2) + \alpha_3^2(1 - \mu_0^2 + \mu_2^2) \right] |\xi|_h^2.$$

Then it is clear that, for $\mu_0^2 > 1 + \mu_2^2$, $m_{AB}^{ab} \omega_a \omega_b$ cannot be positive definite. It follows that no observer field exists and the Skyrme model exhibits an ultrahyperbolic-type breakdown of hyperbolicity. For a detailed explanation on how $\mu_0^2 > 1 + \mu_2^2$ relates to the condition in Theorem 5.2 on the eigenvalue corresponding to the timelike eigenvector of $D^\phi$, we refer the reader to Remarks 18 and 19 in Wong's paper.

Finally, for $\mu_1 = 0 \neq \mu_0 \mu_3$, we proceed as in the previous case to deduce

$$
\begin{aligned}
m_{AB}^{ab} \omega_a \omega_b \xi^A \xi^B &= \left[ -\alpha_0^2 (1 + \mu_2^2 + \mu_3^2) + \alpha_3^2 (1 + \mu_2^2) \right] |\xi(D_0)|^2 \\
&\quad + \left[ -\alpha_0^2 (1 + \mu_2^2 + \mu_3^2) + \alpha_3^2 (1 - \mu_0^2 + \mu_2^2) \right] |\xi(D_1)|^2 \\
&\quad + \left[ -\alpha_0^2 (1 + \mu_3^2) + \alpha_3^2 (1 - \mu_0^2) \right] |\xi(D_2)|^2 \\
&\quad + \left[ -\alpha_0^2 (1 + \mu_2^2) + \alpha_3^2 (1 - \mu_0^2 + \mu_2^2) \right] |\xi(D_3)|^2 \\
&\quad - 2\alpha_0 \alpha_3 \mu_0 \mu_3 \xi(D_0) \xi(D_3).
\end{aligned}
$$

If $\mu_2 \neq 0$, then $D_2 \neq 0$ and choosing $\xi$ to satisfy $\xi(D_2) \neq 0 = \xi(D_a)$, for all $a \neq 2$, we derive

$$
m_{AB}^{ab} \omega_a \omega_b \xi^A \xi^B = \left[ -\alpha_0^2 (1 + \mu_3^2) + \alpha_3^2 (1 - \mu_0^2) \right] |\xi|_h^2.
$$

Hence, an ultrahyperbolic-type breakdown of hyperbolicity occurs for $\mu_0^2 > 1$. If $\mu_2 = 0$, then, as $D_3 \neq 0$ (due to $\mu_3 \neq 0$), one selects $\xi$ with $\xi(D_3) \neq 0 = \xi(D_a)$, for all $a \neq 3$, to infer

$$
m_{AB}^{ab} \omega_a \omega_b \xi^A \xi^B = \left[ -\alpha_0^2 + \alpha_3^2 (1 - \mu_0^2) \right] |\xi|_h^2.
$$

Thus, we obtain yet again a breakdown of hyperbolicity for $\mu_0^2 > 1$. This is in accord with the condition in Theorem 5.2 assuring loss of hyperbolicity, because $D^\phi e_0 = -\mu^2 e_0$ and $e_0$ is timelike, and it concludes the discussion of this theorem.

## 5.3   Lei-Lin-Zhou's result

This work is concerned with the $2 + 1$-dimensional Faddeev model, which is the study of formal critical points $\mathbf{n} : (\mathbb{R}^{2+1}, g = \mathrm{diag}(1, -1, -1)) \longrightarrow \mathbb{S}^2$ for the action[5]

$$
S = \int_{\mathbb{R}^{2+1}} \frac{1}{2} \partial_\mu \mathbf{n} \cdot \partial^\mu \mathbf{n} - \frac{1}{4} (\partial_\mu \mathbf{n} \wedge \partial_\nu \mathbf{n}) \cdot (\partial^\mu \mathbf{n} \wedge \partial^\nu \mathbf{n}) \, dg. \tag{5.35}
$$

The associated variational equations are given by

$$
\mathbf{n} \wedge \partial_\mu \partial^\mu \mathbf{n} + (\partial_\mu [\mathbf{n} \cdot (\partial^\mu \mathbf{n} \wedge \partial^\nu \mathbf{n})]) \partial_\nu \mathbf{n} = 0, \tag{5.36}
$$

---

[5] We use here the original notation in [103] and not the one in Chapter 1, in order to facilitate a smooth transition for the reader between the article and the book. This is why this action differs by a $-$ sign from the one in (1.9).

which, by taking the cross product to the left with $\mathbf{n}$, become

$$(\mathbf{n} \cdot \partial_\mu \partial^\mu \mathbf{n})\,\mathbf{n} - (\mathbf{n} \cdot \mathbf{n})\, \partial_\mu \partial^\mu \mathbf{n} + (\partial_\mu [\mathbf{n} \cdot (\partial^\mu \mathbf{n} \wedge \partial^\nu \mathbf{n})])\,\mathbf{n} \wedge \partial_\nu \mathbf{n} = 0.$$

Using the geometry of the target manifold in the form of $\mathbf{n} \cdot \mathbf{n} = 1$, it follows that $\mathbf{n} \cdot \partial_\mu \mathbf{n} = 0$ and, subsequently,

$$\mathbf{n} \cdot \partial_\mu \partial^\mu \mathbf{n} + \partial^\mu \mathbf{n} \cdot \partial_\mu \mathbf{n} = 0.$$

These relations allow us to rewrite (5.36) as

$$\partial_\mu \partial^\mu \mathbf{n} + (\partial^\mu \mathbf{n} \cdot \partial_\mu \mathbf{n})\,\mathbf{n} + (\partial_\mu [\mathbf{n} \cdot (\partial^\mu \mathbf{n} \wedge \partial^\nu \mathbf{n})])\, \partial_\nu \mathbf{n} \wedge \mathbf{n} = 0, \tag{5.37}$$

that can be recognized to have a quasilinear wave structure.

We are interested in the Cauchy problem associated to (5.36), for which the constraint $\mathbf{n} = (n_1, n_2, n_3) \in \mathbb{S}^2$ permits us to consider initial data prescribed by

$$\mathbf{n}(0, x) = (n_1(0, x), n_2(0, x), n_3(0, x))$$
$$= \left( n_{10}(x), n_{20}(x), \sqrt{1 - n_{10}^2(x) - n_{20}^2(x)} \right), \tag{5.38}$$

$$\partial_t \mathbf{n}(0, x) = (\partial_t n_1(0, x), \partial_t n_2(0, x), \partial_t n_3(0, x))$$
$$= \left( n_{11}(x), n_{21}(x), -\frac{n_{11}(x) n_{10}(x) + n_{21}(x) n_{20}(x)}{\sqrt{1 - n_{10}^2(x) - n_{20}^2(x)}} \right), \tag{5.39}$$

with $n_{10}$, $n_{20}$, $n_{11}$, and $n_{21} \in C_0^\infty(\mathbb{R}^2)$. The following is the main result in [103].

**Theorem 5.3.** *There exists $\epsilon > 0$ such that if $s \geq 9$ is integer-valued and*

$$\max \left\{ \|n_{10}\|_{H^{s+2}(\mathbb{R}^2)}, \|n_{20}\|_{H^{s+2}(\mathbb{R}^2)}, \right.$$
$$\left. \|n_{11}\|_{H^{s+1}(\mathbb{R}^2)}, \|n_{21}\|_{H^{s+1}(\mathbb{R}^2)} \right\} \leq \epsilon, \tag{5.40}$$

*then the initial value problem formed by (5.36), (5.38), and (5.39) admits a unique global classical solution. Moreover, if $\epsilon > 0$ is sufficiently small and $R > 0$ is such that*

$$\bigcup_{1 \leq i \leq 2,\, 0 \leq j \leq 1} supp\, n_{ij} \subseteq \{x \in \mathbb{R}^2;\ |x| \leq R\}, \tag{5.41}$$

*then there exist two positive constants $M = M(\epsilon, R)$ and $\delta = \delta(\epsilon, R)$ satisfying*

$$\|\nabla \mathbf{n}(t)\|_{\Gamma, s, L^2(\mathbb{R}^2)} + \|\nabla^2 \mathbf{n}(t)\|_{\Gamma, s, L^2(\mathbb{R}^2)} \leq M\epsilon(1 + t)^\delta, \tag{5.42}$$

$$\|n_1(t), n_2(t)\|_{\Gamma, s, L^2(\mathbb{R}^2)} \leq M\epsilon(1 + t)^{1/2 + 2\delta}, \tag{5.43}$$

$$\|n_1(t), n_2(t)\|_{\Gamma, s-2, L^\infty(\mathbb{R}^2)} \leq M\epsilon(1 + t)^{-1/2}, \tag{5.44}$$

*for all $0 \leq t < \infty$.*

**Notation 5.1.** In the above,

$$\nabla^2 \overset{\text{def}}{=} \{\partial_{tt}, \partial_{t1}, \partial_{t2}, \partial_{11}, \partial_{12}, \partial_{22}\},$$

$\Gamma$ denotes the family of vector fields

$$\Gamma \overset{\text{def}}{=} \{\nabla, S, L_1, L_2, \Omega_{12}\}$$
$$= \{\partial_t, \partial_1, \partial_2, t\partial_t + x_1\partial_1 + x_2\partial_2, t\partial_1 + x_1\partial_t, t\partial_2 + x_2\partial_t, x_1\partial_2 - x_2\partial_1\},$$

and, for any function $u = u(t, x)$,

$$\|u(t)\|_{\Gamma, s, L^p(\mathbb{R}^2)} \overset{\text{def}}{=} \sum_{|\alpha| \leq s} \|\Gamma^\alpha u(t)\|_{L^p(\mathbb{R}^2)}.$$

The proof of this theorem is based on the classical local well-posedness result for general quasilinear wave equations due to Hughes-Kato-Marsden [74], and on a continuity argument in the time variable. Precisely, [74] guarantees the existence of a local classical solution for the Cauchy problem formed by (5.36), (5.38), and (5.39) if $s > 1$. Moreover, this implies that (5.42)-(5.44) hold for sufficiently small values of $t$ and for large enough values of $M$, which depend only on the initial data. If we denote by $[0, T]$ the largest time interval on which (5.42)-(5.44) are true, then our goal is to prove that $T = \infty$. In fact, we will show that $T < \infty$ implies

$$\|\nabla\mathbf{n}(t)\|_{\Gamma, s, L^2(\mathbb{R}^2)} + \|\nabla^2\mathbf{n}(t)\|_{\Gamma, s, L^2(\mathbb{R}^2)} < M\epsilon(1 + t)^\delta, \tag{5.45}$$

$$\|n_1(t), n_2(t)\|_{\Gamma, s, L^2(\mathbb{R}^2)} < M\epsilon(1 + t)^{1/2+2\delta}, \tag{5.46}$$

and

$$\|n_1(t), n_2(t)\|_{\Gamma, s-2, L^\infty(\mathbb{R}^2)} < M\epsilon(1 + t)^{-1/2} \tag{5.47}$$

are also true for $0 \leq t \leq T$. By the continuous dependence of the solution map (also featured in [74]), it will follow that (5.42)-(5.44) hold in fact for a larger time interval than $[0, T]$, thus contradicting the maximality of $T$. Therefore, we deduce that $T = \infty$ and, thus, the solution is global in time. To summarize, we have reduced the proof of Theorem 5.3 to arguing that if $0 < T < \infty$ and $\mathbf{n}$ is a classical solution on $[0, T]$ to the previously described initial value problem satisfying (5.42)-(5.44), then $\mathbf{n}$ also satisfies (5.45)-(5.47).

As a first step toward proving (5.45)-(5.47), we use (5.37) and identify the individual wave equations satisfied by $n_1$ and $n_2$, respectively. For $n_1$, we have

$$\partial_\mu\partial^\mu n_1 + (\partial^\mu\mathbf{n}\cdot\partial_\mu\mathbf{n}) \, n_1 + (\partial_\mu[\mathbf{n}\cdot(\partial^\mu\mathbf{n}\wedge\partial^\nu\mathbf{n})]) \, (n_3\partial_\nu n_2 - n_2\partial_\nu n_3) = 0. \tag{5.48}$$

However, for purposes which will become apparent later, we want to transform the previous equation into one that has, as much as possible, no terms involving derivatives of $n_3$. Relying on $\mathbf{n} \cdot \partial_\mu \mathbf{n} = 0$, we derive

$$\partial_\mu n_3 = -\frac{n_1 \partial_\mu n_1 + n_2 \partial_\mu n_2}{n_3}$$

and a similar formula for $\partial^\mu n_3$. By direct computations, these yield

$$\partial^\mu \mathbf{n} \cdot \partial_\mu \mathbf{n} = \left(1 + \frac{n_1^2}{n_3^2}\right) \partial^\mu n_1 \, \partial_\mu n_1 + \left(1 + \frac{n_2^2}{n_3^2}\right) \partial^\mu n_2 \, \partial_\mu n_2$$
$$+ \frac{2 n_1 n_2}{n_3^2} \partial^\mu n_1 \, \partial_\mu n_2,$$

$$\partial_\mu [\mathbf{n} \cdot (\partial^\mu \mathbf{n} \wedge \partial^\nu \mathbf{n})] = \partial_\mu \left(\frac{\partial^\mu n_1 \, \partial^\nu n_2 - \partial^\nu n_1 \, \partial^\mu n_2}{n_3}\right),$$

and

$$n_3 \partial_\nu n_2 - n_2 \partial_\nu n_3 = \frac{(1 - n_1^2)\partial_\nu n_2 + n_1 n_2 \partial_\nu n_1}{n_3}.$$

Plugging these expressions into (5.48), we obtain

$$\partial_\mu \partial^\mu n_1$$
$$+ \left[\left(1 + \frac{n_1^2}{n_3^2}\right) \partial^\mu n_1 \, \partial_\mu n_1 + \left(1 + \frac{n_2^2}{n_3^2}\right) \partial^\mu n_2 \, \partial_\mu n_2 + \frac{2 n_1 n_2}{n_3^2} \partial^\mu n_1 \, \partial_\mu n_2\right] n_1$$
$$+ \partial_\mu \left(\frac{\partial^\mu n_1 \, \partial^\nu n_2 - \partial^\nu n_1 \, \partial^\mu n_2}{n_3}\right) \frac{(1 - n_1^2)\partial_\nu n_2 + n_1 n_2 \partial_\nu n_1}{n_3} = 0,$$

$$(5.49)$$

which has the desired structure as a nonlinear wave equation for $n_1$. A similar calculation for $n_2$ leads to

$$\partial_\mu \partial^\mu n_2$$
$$+ \left[\left(1 + \frac{n_1^2}{n_3^2}\right) \partial^\mu n_1 \, \partial_\mu n_1 + \left(1 + \frac{n_2^2}{n_3^2}\right) \partial^\mu n_2 \, \partial_\mu n_2 + \frac{2 n_1 n_2}{n_3^2} \partial^\mu n_1 \, \partial_\mu n_2\right] n_2$$
$$- \partial_\mu \left(\frac{\partial^\mu n_1 \, \partial^\nu n_2 - \partial^\nu n_1 \, \partial^\mu n_2}{n_3}\right) \frac{(1 - n_2^2)\partial_\nu n_1 + n_1 n_2 \partial_\nu n_2}{n_3} = 0.$$

$$(5.50)$$

In accord with the choice of initial data in (5.38) and taking into account (5.44), we will be working with

$$n_3 \stackrel{\text{def}}{=} \sqrt{1 - n_1^2 - n_2^2},$$

which allows us to recast (5.49) and (5.50) together in the more compact form

$$\Box\tilde{\mathbf{n}} = F(\tilde{\mathbf{n}}, \nabla\tilde{\mathbf{n}}, \nabla^2\tilde{\mathbf{n}}), \qquad \tilde{\mathbf{n}} \overset{\text{def}}{=} (n_1, n_2).$$

One can check immediately that $F(0, 0, \nabla^2\tilde{\mathbf{n}}) = 0$ and, as a consequence (e.g., invoking Corollary 2.3 in Sogge [156]), a form of Huygens' principle is available for this type of system. Given that $n_1$ and $n_2$ are classical solutions for (5.49) and (5.50), respectively, with compactly supported smooth data, it follows that

$$\text{supp } \tilde{\mathbf{n}}(t) \cup \text{supp } \partial_t\tilde{\mathbf{n}}(t) \subseteq \{x \in \mathbb{R}^2; |x| \leq t + R\}, \qquad (5.51)$$

where $R > 0$ verifies (5.41). If we use (5.44) and ask for $M\epsilon$ to be sufficiently small, we deduce the uniform asymptotic

$$n_3(t, x) \simeq 1, \qquad \forall \, (t, x) \in [0, T] \times \mathbb{R}^2. \qquad (5.52)$$

Coupled with $\mathbf{n} \cdot \partial_\mu\mathbf{n} = 0$ and (5.51), it implies

$$\text{supp } \nabla n_3(t) \subseteq \{x \in \mathbb{R}^2; |x| \leq t + R\}. \qquad (5.53)$$

We discuss next the specific tools employed by Lei-Lin-Zhou in demonstrating (5.45)-(5.47), which are available precisely because of (5.51) and (5.53). For proving (5.45), they rely on the basic energy identity,

$$\frac{d}{dt}\left[\frac{1}{2}\|\nabla u(t)\|^2_{L^2(\mathbb{R}^2)}\right] = -\int_{\mathbb{R}^2} \partial_t u(t, x) \Box u(t, x) \, dx, \qquad (5.54)$$

while for showing (5.46), they use their own $L^2$-based estimate,

$$\|u(t)\|_{L^2(\mathbb{R}^2)} - \|u(0)\|_{L^2(\mathbb{R}^2)}$$

$$\lesssim (1+t)^{1/2}\Big(\|\partial_t u(0)\|_{L^{4/3}(\mathbb{R}^2)} + \int_0^t \big\{\|\Box u(\tau)\|_{L^{4/3}(|x|\leq 1+\tau/2)} \qquad (5.55)$$

$$+ (1+\tau)^{-1/2}\|\Box u(\tau)\|_{L^{1,2}(|x|>1+\tau/2)}\big\} \, d\tau\Big).$$

In the above, for $1 \leq p < \infty$, the last norm is defined by

$$\|v\|_{L^{p,q}(\mathbb{R}^2)} \overset{\text{def}}{=} \left(\int_0^\infty \|v(r\xi)\|^p_{L^q(\mathbb{S}^1)} \, r \, dr\right)^{1/p}, \qquad \text{with } v = v(x) = v(r\xi).$$

Finally, for concluding (5.47), Lei-Lin-Zhou work with an $L^\infty - L^1$ bound, which is a combination of estimates previously obtained by Klainerman [84] and Hörmander [72]:

$$\|u(t)\|_{L^\infty(\mathbb{R}^2)}$$

$$\lesssim (1+t)^{-1/2}\Big(\|u(0)\|_{W^{2,1}(\mathbb{R}^2)} + \|\partial_t u(0)\|_{W^{1,1}(\mathbb{R}^2)} \qquad (5.56)$$

$$+ \int_0^t (1+\tau)^{-1/2}\|\Box u(\tau)\|_{\Gamma, 1, L^1(\mathbb{R}^2)} \, d\tau\Big),$$

where the classical Sobolev spaces $W^{k,p}$ with integer index $k$ are defined by

$$W^{k,p}(\mathbb{R}^2) \overset{\text{def}}{=} \left\{ u \in L^p(\mathbb{R}^2); \ \nabla^\alpha u \in L^p(\mathbb{R}^2), \ \forall |\alpha| \leq k \right\}.$$

After a quick glance at the tools presented above, someone may wonder whether they are strong enough to handle the quasilinear system (5.36), even if one chooses to work with its perturbative form (5.37), i.e.,

$$\Box \mathbf{n} = F(\mathbf{n}, \nabla \mathbf{n}, \nabla^2 \mathbf{n}).$$

*We emphasize here that the analysis goes through precisely because of the null structure present in the perturbative part of the system.* Apart from the null form[6]

$$\partial^\mu \mathbf{n} \cdot \partial_\mu \mathbf{n} = Q(\mathbf{n}, \mathbf{n}) \tag{5.57}$$

inherited from the wave map system, the quartic part of the nonlinearity also contains factors having null structure. This can be seen more clearly in (5.49) or (5.50), where

$$\partial^\mu n_1 \, \partial^\nu n_2 - \partial^\nu n_1 \, \partial^\mu n_2 = \pm Q_{\mu\nu}(n_1, n_2),$$

with

$$Q_{ij}(\phi, \psi) = \partial_i \phi \, \partial_j \psi - \partial_j \phi \, \partial_i \psi, \qquad 0 \leq i < j \leq 2, \qquad \partial_0 = \partial_t, \tag{5.58}$$

being the other classical null forms associated to the linear wave equation.

In line with our goals for this book, we worked to streamline the argument of Lei-Lin-Zhou and to make the reasoning behind certain steps more transparent. In what follows, after a section of preliminary facts, we present proofs for (5.45) and (5.47) and argue that (5.46) can be demonstrated in a similar fashion to (5.47).

### 5.3.1 *Preliminaries*

First, we want to address how we can control the size of $\mathbf{n}(0)$ in function spaces relevant to the analysis, solely based on the Sobolev norms of the initial data in (5.40). For that, we prove:

**Proposition 5.1.** *Under the assumptions of Theorem 5.3, for any multi-index $\alpha$,*

$$\Gamma^\alpha \tilde{\mathbf{n}}(0) \overset{\text{def}}{=} (\Gamma^\alpha n_1(0), \Gamma^\alpha n_2(0))$$

*can be expressed entirely in terms of*

---

[6]As we choose to keep Lei-Lin-Zhou's original setting, with the Minkowski metric being $\mathrm{diag}(1, -1, -1)$ instead of $\mathrm{diag}(-1, 1, 1)$, the null form $Q$ here is in fact the opposite of the similar null form defined in Chapter 3.

- *the family of monomials $\{x^\beta;\ \beta \le \alpha\}$*

*and*

- *the spatial derivatives of order at most $|\alpha|$ for the functions $n_{10}$, $n_{20}$, $n_{11}$, and $n_{21}$.*

**Proof.** Given the structure of $\Gamma$ as a family of vector fields, it is enough to prove the statement for $\partial_{tt}$, because then an induction scheme on the number of time derivates present in $\Gamma^\alpha$ settles the argument. Easy algebraic manipulations turn (5.49) and (5.50) into

$$\partial_{tt}n_1 \left(1 - \frac{|\nabla_x n_2|^2(1 - n_1^2) + (\nabla_x n_1 \cdot \nabla_x n_2)n_1 n_2}{1 - n_1^2 - n_2^2}\right)$$

$$+ \partial_{tt}n_2 \left(\frac{(\nabla_x n_1 \cdot \nabla_x n_2)(1 - n_1^2) + |\nabla_x n_1|^2 n_1 n_2}{1 - n_1^2 - n_2^2}\right)$$

$$= \frac{F_1(\tilde{n}, \nabla\tilde{n}, \nabla_x\nabla\tilde{n})}{(1 - n_1^2 - n_2^2)^2}$$

and

$$\partial_{tt}n_1 \left(\frac{(\nabla_x n_1 \cdot \nabla_x n_2)(1 - n_2^2) + |\nabla_x n_2|^2 n_1 n_2}{1 - n_1^2 - n_2^2}\right)$$

$$+ \partial_{tt}n_2 \left(1 - \frac{|\nabla_x n_1|^2(1 - n_2^2) + (\nabla_x n_1 \cdot \nabla_x n_2)n_1 n_2}{1 - n_1^2 - n_2^2}\right)$$

$$= \frac{F_2(\tilde{n}, \nabla\tilde{n}, \nabla_x\nabla\tilde{n})}{(1 - n_1^2 - n_2^2)^2},$$

respectively, where $F_1$ and $F_2$ are polynomials of degree at most 7 in their variables.

For $t = 0$, it is easy to see that the coefficients of $\partial_{tt}n_1(0)$ and $\partial_{tt}n_2(0)$ and the right-hand side terms in both equations involve only $n_{10}$, $n_{20}$, $n_{11}$, $n_{21}$, and their spatial derivatives of order up to 2. Due to the assumptions on the initial data in (5.40) and the Sobolev embedding $H^2(\mathbb{R}^2) \subset L^\infty(\mathbb{R}^2)$, it follows that

$$|n_{10}\, n_{20}| \le n_{10}^2 + n_{20}^2 = O(\epsilon^2)$$

and

$$|\nabla_x n_{10} \cdot \nabla_x n_{20}| \le |\nabla_x n_{10}|^2 + |\nabla_x n_{20}|^2 = O(\epsilon^2).$$

This implies that we can solve through Cramer's rule the system formed by these two equations, with $\partial_{tt}n_1(0)$ and $\partial_{tt}n_2(0)$ as unknowns, thus finishing the proof. $\qquad\square$

A consequence of this proposition is the following corollary, which contains the desired bounds for $\mathbf{n}(0)$.

**Corollary 5.1.** *Under the assumptions of Theorem 5.3, we have*[7]

$$\|\nabla \mathbf{n}(0)\|_{\Gamma,s,L^2(\mathbb{R}^2)} + \|\nabla^2 \mathbf{n}(0)\|_{\Gamma,s,L^2(\mathbb{R}^2)} \leq C(R)\,\epsilon, \qquad (5.59)$$

*and*

$$\|\Gamma^\alpha \tilde{\mathbf{n}}(0)\|_{W^{2,1}(\mathbb{R}^2)} + \|\partial_t \Gamma^\alpha \tilde{\mathbf{n}}(0)\|_{W^{1,1}(\mathbb{R}^2)} \leq C(R)\,\epsilon, \quad \forall |\alpha| \leq s. \quad (5.60)$$

**Proof.** The computations done for $\partial_{tt}\tilde{\mathbf{n}}(0)$ in Proposition 5.1 show that one has in fact a more qualitative result than the one stated; i.e., the expressions for $\Gamma^\alpha \tilde{\mathbf{n}}(0)$ are rational functions of the monomials and spatial derivatives in question, which can be easily estimated using the hypotheses of the main theorem. Thus, with a little bit more effort and also taking into account (5.41), we deduce

$$\|\tilde{\mathbf{n}}(0)\|_{\Gamma,s+2,L^2(\mathbb{R}^2)} \lesssim C(R)\big(\|n_{10}\|_{H^{s+2}(\mathbb{R}^2)} + \|n_{20}\|_{H^{s+2}(\mathbb{R}^2)}$$
$$+ \|n_{11}\|_{H^{s+1}(\mathbb{R}^2)} + \|n_{21}\|_{H^{s+1}(\mathbb{R}^2)}\big). \qquad (5.61)$$

If we factor in now the constraint $\mathbf{n}(0) \in \mathbb{S}^2$ and (5.52), the estimate (5.59) is immediate. We notice that this can be considered a better bound than (5.42) at $t = 0$. Finally, (5.60) follows if we combine (5.61) with the Cauchy-Schwarz inequality. □

Next, we prove bounds for $n_3$, which are similar in spirit to (5.43) and (5.44).

**Proposition 5.2.** *Under the bootstrap assumptions (5.42)-(5.44), the following inequalities hold for $M\epsilon$ sufficiently small:*

$$\|\Gamma\, n_3(t)\|_{\Gamma,s-1,L^2(\mathbb{R}^2)} + \|\Gamma\,(1/n_3)(t)\|_{\Gamma,s-1,L^2(\mathbb{R}^2)}$$
$$\lesssim (M\epsilon)^2(1+t)^{2\delta}, \qquad (5.62)$$

$$\|\Gamma\, n_3(t)\|_{\Gamma,s-3,L^\infty(\mathbb{R}^2)} + \|\Gamma\,(1/n_3)(t)\|_{\Gamma,s-3,L^\infty(\mathbb{R}^2)}$$
$$\lesssim (M\epsilon)^2(1+t)^{-1}, \qquad (5.63)$$

*for all $0 \leq t \leq T$.*

---

[7]In the sequel, $C(R)$ will denote a constant that depends only on $R$, which may change its value from line to line.

**Proof.** For the first norms in both estimates, one uses (5.43), (5.44), (5.52), and an argument by induction on $|\alpha|$ in the context of the identity

$$\sum_{\alpha_1+\alpha_2=\alpha} \Gamma^{\alpha_1} n_3 \Gamma^{\alpha_2} n_3 = - \sum_{\beta_1+\beta_2=\alpha} \Gamma^{\beta_1} n_1 \Gamma^{\beta_2} n_1 - \sum_{\gamma_1+\gamma_2=\alpha} \Gamma^{\gamma_1} n_2 \Gamma^{\gamma_2} n_2.$$

The estimates for $1/n_3$ are then easy to deduce, knowing that $n_3$ is uniformly bounded from below. $\qquad\square$

Following this, we record commutator formulas, which are used extensively in the argument and can be checked by direct computations.

**Proposition 5.3.** *i)* If $\alpha$ *is an arbitrary multi-index, then*

$$[\nabla, \Gamma^\alpha] = \sum_{\beta<\alpha} A_{\alpha\beta}\, \Gamma^\beta \nabla \qquad (5.64)$$

*and*

$$[\square, \Gamma^\alpha] = \sum_{\beta<\alpha} B_{\alpha\beta}\, \Gamma^\beta \square, \qquad (5.65)$$

*where $A_{\alpha\beta}$ and $B_{\alpha\beta}$ are absolute constants.*

*ii) For the null forms described by (5.57) and (5.58), we have*

$$[\Gamma, Q] = C\, Q + \sum_{0\leq k<l\leq 2} D_{kl}\, Q_{kl} \qquad (5.66)$$

*and*

$$[\Gamma, Q_{ij}] = E\, Q + \sum_{0\leq k<l\leq 2} F_{ijkl}\, Q_{kl}, \qquad (5.67)$$

*with $C$, $D_{kl}$, $E$, and $F_{ijkl}$ being absolute constants. In the above,*

$$[\Gamma, Q](\phi, \psi) \stackrel{\text{def}}{=} \Gamma(Q(\phi,\psi)) - Q(\Gamma\phi, \psi) - Q(\phi, \Gamma\psi)$$

*and a similar formula holds for $[\Gamma, Q_{ij}]$.*

Next, we present time decay estimates satisfied by the null forms, which are an integral part of the argument.

**Proposition 5.4.** *The following inequality is true:*

$$|Q(\phi, \psi)| + \sum_{0\leq i<j\leq 2} |Q_{ij}(\phi, \psi)| \lesssim (1+|t|)^{-1} (|\Gamma\phi||\nabla\psi| + |\nabla\phi||\Gamma\psi|). \qquad (5.68)$$

**Proof.** We can verify that
$$Q_{01}(\phi, \psi) = \frac{\partial_t \phi L_1 \psi - L_1 \phi \partial_t \psi}{t},$$
which yields
$$|Q_{01}(\phi, \psi)| \lesssim |t|^{-1} (|\Gamma \phi||\nabla \psi| + |\nabla \phi||\Gamma \psi|).$$
Similar estimates hold for the other null forms. It follows that (5.68) is true when $|t| \gtrsim 1$. The case $|t| \ll 1$ is immediate. $\qquad\square$

As a consequence of the last three results, we derive null form estimates in function spaces relevant to the analysis.

**Proposition 5.5.** *Under the bootstrap assumptions* (5.42)-(5.44), *if $T$ is any of the null forms featured in* (5.57) *and* (5.58) *and* $1 \le i, j \le 3$, *then*

$$\|T(n_i, n_j)(t)\|_{\Gamma, s-3, L^2(\mathbb{R}^2)} + \|T(\nabla n_i, n_j)(t)\|_{\Gamma, s-3, L^2(\mathbb{R}^2)}$$
$$\lesssim (M\epsilon)^2 (1 + t)^{-3/2+\delta}, \tag{5.69}$$

$$\|T(n_i, n_j)(t)\|_{\Gamma, s-1, L^1(\mathbb{R}^2)} + \|T(\nabla n_i, n_j)(t)\|_{\Gamma, s-1, L^1(\mathbb{R}^2)}$$
$$\lesssim (M\epsilon)^2 (1 + t)^{-1/2+3\delta}, \tag{5.70}$$

*hold true uniformly for $0 \le t \le T$.*

**Proof.** Applying successively (5.66), (5.67), (5.68) and (5.64), we deduce

$$|\Gamma^\alpha T(\phi, \psi)| \lesssim \left| \sum_{\beta+\gamma \le \alpha} T(\Gamma^\beta \phi, \Gamma^\gamma \psi) \right|$$

$$\lesssim (1+t)^{-1} \sum_{\beta+\gamma \le \alpha} (|\Gamma\Gamma^\beta \phi||\nabla\Gamma^\gamma \psi| + |\nabla\Gamma^\beta \phi||\Gamma\Gamma^\gamma \psi|)$$

$$\lesssim (1+t)^{-1} \sum_{\beta+\gamma \le \alpha} (|\Gamma^\beta\Gamma \phi||\Gamma^\gamma \nabla \psi| + |\Gamma^\beta \nabla \phi||\Gamma^\gamma\Gamma \psi|).$$

The desired conclusion follows if we rely now on Hölder's inequality and take into account (5.42)-(5.44) and (5.62)-(5.63). $\qquad\square$

Finally, we prove a bilinear bound which is needed in order to infer certain optimal energy estimates.

**Proposition 5.6.** *If $\rho > 0$ and $u = u(t, x)$ and $v = v(t, x)$ are real-valued functions on $\mathbb{R}_+ \times \mathbb{R}^2$ satisfying*

$$supp\, u \cup supp\, v \subseteq \{(t, x) \in \mathbb{R}_+ \times \mathbb{R}^n; |x| \le t + \rho\},$$

*then*

$$\|u(t)\, \nabla v(t)\|_{L^2(\mathbb{R}^2)} \le C(\rho)\, \|\nabla_x u(t)\|_{L^2(\mathbb{R}^2)}\, \|\Gamma\, v(t)\|_{L^\infty(\mathbb{R}^2)}, \tag{5.71}$$

*holds for all $t \ge 0$.*

**Proof.** Using the identities

$$\partial_t = \frac{tS - x_1 L_1 - x_2 L_2}{t^2 - |x^2|},$$

$$\partial_{x_1} = \frac{tL_1 - x_1 S - x_2 \Omega_{12}}{t^2 - |x^2|},$$

$$\partial_{x_2} = \frac{tL_2 - x_2 S - x_1 \Omega_{12}}{t^2 - |x^2|},$$

and the information on the support of $v$, we deduce

$$|\nabla v(t,x)| \lesssim \frac{1}{1 + |t - |x||} |\Gamma v(t,x)| \lesssim \frac{\rho + 1}{\rho + |t - |x||} |\Gamma v(t,x)|.$$

On the other hand, if we factor in the information on the support of $u$ and integrate by parts using polar coordinates[8], we obtain

$$\left\| \frac{1}{\rho + |t - |x||} u(t) \right\|_{L^2} \lesssim \|\nabla_x u(t)\|_{L^2}.$$

Hence, it follows that

$$\|u(t) \nabla v(t)\|_{L^2} \lesssim (\rho + 1) \left\| \frac{1}{\rho + |t - |x||} |u(t)| \, |\Gamma v(t)| \right\|_{L^2}$$

$$\lesssim (\rho + 1) \|\nabla_x u(t)\|_{L^2} \|\Gamma v(t)\|_{L^\infty}.$$

$\square$

### 5.3.2 *Proof of* (5.45)

In proving (5.45), Lei-Lin-Zhou claim that it is enough to show that

**Proposition 5.7.** *Under the bootstrap assumptions* (5.42)-(5.44), *if $a = 0$ or $1$, $0 < \delta < 1/2$, and $M\epsilon$ is sufficiently small, then the map $\mathbf{n} : \mathbb{R}^{2+1} \to \mathbb{S}^2$ satisfies the following fixed time estimates, uniformly for $0 \le t \le T$:*

$$\frac{d}{dt} \left\{ \|\nabla \Gamma^\alpha \nabla^a \mathbf{n}(t)\|_{L^2}^2 \right\} \le C(R)(M\epsilon)^4 (1+t)^{-1+2\delta}, \quad \forall \, 0 \le |\alpha| \le s - 1,$$

$$(5.72)$$

$$\frac{d}{dt} \left\{ \|\nabla \Gamma^\alpha \nabla^a \mathbf{n}(t)\|_{L^2}^2 + \|\mathbf{n} \cdot (\Gamma^\alpha \nabla^a \partial_t \mathbf{n} \wedge \nabla_x \mathbf{n})(t)\|_{L^2}^2 \right.$$

$$\left. - \|\mathbf{n} \cdot (\partial_t \mathbf{n} \wedge \Gamma^\alpha \nabla^a \nabla_x \mathbf{n})(t)\|_{L^2}^2 + \|\mathbf{n} \cdot \Gamma^\alpha \nabla^a (\partial_1 \mathbf{n} \wedge \partial_2 \mathbf{n})(t)\|_{L^2}^2 \right\}$$

$$\le C(R)(M\epsilon)^4 (1+t)^{-1+2\delta}, \quad \forall \, |\alpha| = s.$$

$$(5.73)$$

---

[8] For more precise details, we refer the reader to Lemma 4.1 in [103].

Indeed, if we integrate (5.72) first and use (5.64) and (5.59) in an induction scheme on $|\alpha|$, then we derive

$$\|\nabla \mathbf{n}(t)\|_{\Gamma,s-1,L^2} + \|\nabla^2 \mathbf{n}(t)\|_{\Gamma,s-1,L^2}$$
$$\leq C(R)(\epsilon + (M\epsilon)^2 \delta^{-1/2}(1+t)^\delta), \quad \forall 0 \leq t \leq T.$$

Integrating now (5.73) and relying on the previous bound combined with (5.64), (5.59), (5.42), (5.44), and (5.63), we deduce

$$\|\nabla \mathbf{n}(t)\|_{\Gamma,s,L^2} + \|\nabla^2 \mathbf{n}(t)\|_{\Gamma,s,L^2}$$
$$\leq C(R)(\epsilon + (M\epsilon)^2(1+t)^{-1/2+\delta} + (M\epsilon)^2 \delta^{-1/2}(1+t)^\delta), \quad \forall 0 \leq t \leq T.$$

It is clear that with extra adjustments in the values of $M$ and $\delta$, depending solely on $\epsilon$ and $R$, one derives (5.45).

The starting point in demonstrating the previous proposition is inferring from (5.37), with the help of (5.65), that

$$\Box \Gamma^\alpha \nabla^a \mathbf{n} = \sum_{\beta \leq \alpha} C_{\alpha\beta} \left( \Gamma^\beta \nabla^a N_1 + \Gamma^\beta \nabla^a N_2 \right),$$

where $\alpha$ is an arbitrary multi-index with $0 \leq |\alpha| \leq s$, $a = 0$ or $1$, $C_{\alpha\beta}$ are absolute constants, and

$$N_1 \overset{\text{def}}{=} (\partial^\mu \mathbf{n} \cdot \partial_\mu \mathbf{n}) \, \mathbf{n} = Q(\mathbf{n}, \mathbf{n}) \, \mathbf{n},$$

$$N_2 \overset{\text{def}}{=} (\partial_\mu [\mathbf{n} \cdot (\partial^\mu \mathbf{n} \wedge \partial^\nu \mathbf{n})]) \, \partial_\nu \mathbf{n} \wedge \mathbf{n}.$$

If we multiply the previous equation by $\partial_t \Gamma^\alpha \nabla^a \mathbf{n}$ and integrate by parts in the spatial variables (which is allowed by (5.51) and (5.53)), we deduce

$$\frac{d}{dt} \left\{ \frac{1}{2} \|\nabla \Gamma^\alpha \nabla^a \mathbf{n}(t)\|_{L^2}^2 \right\}$$
$$= - \sum_{\beta \leq \alpha} C_{\alpha\beta} \int_{\mathbb{R}^2} \left( \partial_t \Gamma^\alpha \nabla^a \mathbf{n} \cdot \Gamma^\beta \nabla^a N_1 + \partial_t \Gamma^\alpha \nabla^a \mathbf{n} \cdot \Gamma^\beta \nabla^a N_2 \right)(t,x) \, dx.$$

### 5.3.2.1   *Analysis of the $N_1$-integrand term*

We will prove that

$$\left| \int_{\mathbb{R}^2} \partial_t \Gamma^\alpha \nabla^a \mathbf{n} \cdot \Gamma^\beta \nabla^a N_1 \, dx \right| \leq C(R)(M\epsilon)^4 (1+t)^{-1+2\delta}, \quad \forall \beta \leq \alpha. \quad (5.74)$$

One rewrites the integrand using Leibniz's rule, i.e.,

$$\partial_t \Gamma^\alpha \nabla^a \mathbf{n} \cdot \Gamma^\beta \nabla^a N_1$$
$$= \Gamma^\beta \nabla^a Q(\mathbf{n}, \mathbf{n}) \, (\partial_t \Gamma^\alpha \nabla^a \mathbf{n} \cdot \mathbf{n})$$
$$+ \sum_{\gamma \leq \beta, \, b \leq a, \, |\gamma|+b>0} \Gamma^{\beta-\gamma} \nabla^{a-b} Q(\mathbf{n}, \mathbf{n}) \, (\partial_t \Gamma^\alpha \nabla^a \mathbf{n} \cdot \Gamma^\gamma \nabla^b \mathbf{n}),$$

and estimates its integral by

$$\left| \int_{\mathbb{R}^2} \partial_t \Gamma^\alpha \nabla^a \mathbf{n} \cdot \Gamma^\beta \nabla^a N_1 \, dx \right|$$

$$\lesssim \|\Gamma^\beta \nabla^a Q(\mathbf{n}, \mathbf{n})\|_{L^2} \|\partial_t \Gamma^\alpha \nabla^a \mathbf{n} \cdot \mathbf{n}\|_{L^2}$$

$$+ \|\partial_t \Gamma^\alpha \nabla^a \mathbf{n}\|_{L^2} \sum_{\gamma \leq \beta, \, b \leq a, \, 0 < |\gamma| + b \leq \frac{|\beta| + a}{2}} \|\Gamma^{\beta - \gamma} \nabla^{a-b} Q(\mathbf{n}, \mathbf{n})\|_{L^2} \|\Gamma^\gamma \nabla^b \mathbf{n}\|_{L^\infty}$$

$$+ \|\partial_t \Gamma^\alpha \nabla^a \mathbf{n}\|_{L^2} \sum_{\gamma \leq \beta, \, b \leq a, \, |\gamma| + b > \frac{|\beta| + a}{2}} \|\Gamma^{\beta - \gamma} \nabla^{a-b} Q(\mathbf{n}, \mathbf{n})\|_{L^\infty} \|\Gamma^\gamma \nabla^b \mathbf{n}\|_{L^2}.$$

$$(5.75)$$

We notice that $0 \leq a \leq 1$ combined with successive applications of (5.66) and (5.67) allows one to derive

$$\Gamma^\beta \nabla^a Q(\mathbf{n}, \mathbf{n}) \simeq \Gamma^\beta Q(\nabla^a \mathbf{n}, \mathbf{n})$$

$$= \sum_{\gamma_1 + \gamma_2 \leq \beta} \left[ C_{\gamma_1 \gamma_2} Q(\Gamma^{\gamma_1} \nabla^a \mathbf{n}, \Gamma^{\gamma_2} \mathbf{n}) + \sum_{0 \leq i < j \leq 2} C_{\gamma_1 \gamma_2 ij} Q_{ij}(\Gamma^{\gamma_1} \nabla^a \mathbf{n}, \Gamma^{\gamma_2} \mathbf{n}) \right].$$

If we factor in now the commutator relation (5.64) and the pointwise bound (5.68), it follows that

$$|\Gamma^\beta \nabla^a Q(\mathbf{n}, \mathbf{n})(t)|$$

$$\lesssim (1 + t)^{-1} \sum_{\gamma_1 + \gamma_2 \leq \beta} \Big( |\Gamma^{\gamma_1} \nabla^a \Gamma \mathbf{n}(t)| \, |\Gamma^{\gamma_2} \nabla \mathbf{n}(t)| \qquad (5.76)$$

$$+ |\Gamma^{\gamma_1} \nabla^{1+a} \mathbf{n}(t)| \, |\Gamma^{\gamma_2} \Gamma \mathbf{n}(t)| \Big).$$

A similar estimate is available for $\Gamma^{\beta - \gamma} \nabla^{a-b} Q(\mathbf{n}, \mathbf{n})$. These bounds form the basis for estimating the norms involving the null form.

In the case when $|\gamma| + b > \frac{|\beta| + a}{2}$ and $M\epsilon$ is sufficiently small, we obtain

$$\|\Gamma^{\beta - \gamma} \nabla^{a-b} Q(\mathbf{n}, \mathbf{n})(t)\|_{L^\infty} \lesssim (1 + t)^{-1} \|\Gamma \, \mathbf{n}(t)\|_{\Gamma, s-3, L^\infty}^2 \qquad (5.77)$$

$$\lesssim (M\epsilon)^2 (1 + t)^{-2},$$

due to (5.44) and (5.63). To deduce useful bounds for the $L^2$ norms involving the null form, the strategy is to use (5.76) and estimate:

- the factor with the lower order derivative in $L^\infty$;
- the factor with the higher order derivative in $L^2$.

For the former, the bound is provided by

$$\|\Gamma \, \mathbf{n}(t)\|_{\Gamma, s-3, L^\infty} \lesssim M\epsilon (1 + t)^{-1/2}.$$

For the latter, in the majority of cases, we can use as an upper bound

$$\max\left\{\|\nabla\mathbf{n}(t)\|_{\Gamma,s,L^2(\mathbb{R}^2)} + \|\nabla^2\mathbf{n}(t)\|_{\Gamma,s,L^2(\mathbb{R}^2)}, \|\Gamma\,\mathbf{n}(t)\|_{\Gamma,s-1,L^2}\right\}$$
$$\lesssim M\epsilon(1+t)^{1/2+2\delta}, \tag{5.78}$$

based on (5.42), (5.43), and (5.62). However, there are two profiles of factors that appear in extremal cases, i.e.,

$$\Gamma^{\tilde{\beta}}\mathbf{n} \quad\text{and}\quad \Gamma^{\tilde{\beta}}\nabla\mathbf{n}, \quad\text{with}\quad |\tilde{\beta}| = s+1,$$

for which we do not have direct good bounds. Instead, we rely on (5.51) and (5.53) to derive

$$|\Gamma\,\mathbf{n}(t,x)| \le (1+t+|x|)\,|\nabla\mathbf{n}(t,x)| \le C(R)(1+t)\,|\nabla\mathbf{n}(t,x)|,$$

which implies

$$\|\Gamma^{\tilde{\beta}}\mathbf{n}(t)\|_{L^2} + \|\Gamma^{\tilde{\beta}}\nabla\mathbf{n}(t)\|_{L^2}$$
$$\le C(R)(1+t)(\|\nabla\mathbf{n}(t)\|_{\Gamma,s,L^2} + \|\nabla\mathbf{n}(t)\|_{\Gamma,s,L^2})$$
$$\le C(R)M\epsilon(1+t)^{1+\delta},$$

for any multi-index $\tilde{\beta}$ with $|\tilde{\beta}| = s+1$. This is obviously a weaker upper bound than (5.78). In conclusion, we obtain

$$\|\Gamma^{\beta}\nabla^a Q(\mathbf{n},\mathbf{n})(t)\|_{L^2} \le C(R)(M\epsilon)^2(1+t)^{-1/2+\delta} \tag{5.79}$$

and the same estimate holds for $\Gamma^{\beta-\gamma}\nabla^{a-b}Q(\mathbf{n},\mathbf{n})$ when

$$0 < |\gamma|+b \le \frac{|\beta|+a}{2}.$$

We can now finish the analysis for the two summation terms appearing on the right-hand side of (5.75). Their common factor can be easily estimated using (5.64) and (5.42) by

$$\|\partial_t\Gamma^{\alpha}\nabla^a\mathbf{n}(t)\|_{L^2} \lesssim \|\nabla^{1+a}\mathbf{n}(t)\|_{\Gamma,s,L^2} \lesssim M\epsilon(1+t)^{\delta}. \tag{5.80}$$

For the first summation term, we argue that

$$\|\Gamma^{\gamma}\nabla^b\mathbf{n}(t)\|_{L^\infty} \lesssim \|\Gamma\,\mathbf{n}(t)\|_{\Gamma,s-3,L^\infty} \lesssim M\epsilon(1+t)^{-1/2}.$$

For the second one, we deduce

$$\|\Gamma^{\gamma}\nabla^b\mathbf{n}(t)\|_{L^2} \lesssim \max\{\|\Gamma\,\mathbf{n}(t)\|_{\Gamma,s-1,L^2}, \|\nabla\,\mathbf{n}(t)\|_{\Gamma,s,L^2}\} \lesssim M\epsilon(1+t)^{1/2+2\delta}.$$

If we combine the last three bounds with (5.77) and (5.79), we derive

$$\|\partial_t\Gamma^{\alpha}\nabla^a\mathbf{n}\|_{L^2} \sum_{\gamma\le\beta,\,b\le a,\,0<|\gamma|+b\le\frac{|\beta|+a}{2}} \|\Gamma^{\beta-\gamma}\nabla^{a-b}Q(\mathbf{n},\mathbf{n})\|_{L^2}\,\|\Gamma^{\gamma}\nabla^b\mathbf{n}\|_{L^\infty}$$

$$+ \|\partial_t\Gamma^{\alpha}\nabla^a\mathbf{n}\|_{L^2} \sum_{\gamma\le\beta,\,b\le a,\,|\gamma|+b>\frac{|\beta|+a}{2}} \|\Gamma^{\beta-\gamma}\nabla^{a-b}Q(\mathbf{n},\mathbf{n})\|_{L^\infty}\,\|\Gamma^{\gamma}\nabla^b\mathbf{n}\|_{L^2}$$

$$\le C(R)(M\epsilon)^4(1+t)^{-1+2\delta}. \tag{5.81}$$

Based on the previous estimate and (5.79), all which is left to do in order to finish the proof of (5.74) is to show that

$$\|\partial_t \Gamma^\alpha \nabla^a \mathbf{n} \cdot \mathbf{n}\|_{L^2} \leq C(R)(M\epsilon)^2 (1+t)^{-1/2+\delta}. \tag{5.82}$$

It is clear that a direct approach like

$$\|\partial_t \Gamma^\alpha \nabla^a \mathbf{n} \cdot \mathbf{n}\|_{L^2} \leq \|\partial_t \Gamma^\alpha \nabla^a \mathbf{n}\|_{L^2} \|\mathbf{n}\|_{L^\infty} \lesssim \|\nabla^{1+a}\mathbf{n}(t)\|_{\Gamma,s,L^2} \lesssim M\epsilon(1+t)^\delta$$

does not yield the desired result. Instead, relying on the geometry of the target which implies $\nabla \mathbf{n} \cdot \mathbf{n} = 0$ and on (5.64), we infer

$$\|\partial_t \Gamma^\alpha \nabla^a \mathbf{n} \cdot \mathbf{n}\|_{L^2} \lesssim \sum_{\tilde{\beta} \leq \alpha} \|\Gamma^{\tilde{\beta}} \nabla^a \partial_t \mathbf{n} \cdot \mathbf{n}\|_{L^2}$$

$$\lesssim \sum_{\gamma \leq \tilde{\beta} \leq \alpha, \, b \leq a, \, 0 < |\gamma|+b} \|\Gamma^{\tilde{\beta}-\gamma} \nabla^{a-b} \partial_t \mathbf{n} \cdot \Gamma^\gamma \nabla^b \mathbf{n}\|_{L^2}.$$

For most terms in the above summation, an upper bound equal to $(M\epsilon)^2(1+t)^{-1/2+\delta}$ is easily obtained by arguing as in previous $L^2$ analyses. However, for the extremal terms

$$\nabla^a \partial_t \mathbf{n} \cdot \Gamma^\gamma \mathbf{n} \quad \text{and} \quad \Gamma \nabla^a \partial_t \mathbf{n} \cdot \Gamma^\gamma \mathbf{n}, \quad \text{with} \quad s-1 \leq |\gamma| \leq s,$$

such a technique does not work. This is the moment in the argument when we need the bilinear estimate (5.71), which, together with (5.51), (5.53), (5.64), (5.42), (5.44), and (5.63), yields

$$\|\nabla^a \partial_t \mathbf{n} \cdot \Gamma^\gamma \mathbf{n}\|_{L^2} \leq C(R) \|\Gamma \nabla^a \mathbf{n}\|_{L^\infty} \|\nabla_x \Gamma^\gamma \mathbf{n}\|_{L^2}$$

$$\leq C(R) \|\Gamma \mathbf{n}(t)\|_{\Gamma,s-3,L^\infty} \|\nabla \mathbf{n}(t)\|_{\Gamma,s,L^2}$$

$$\leq C(R)(M\epsilon)^2 (1+t)^{-1/2+\delta}$$

and the same bound for the other extremal term. This finishes the proof of (5.82) and consequently of (5.74).

### 5.3.2.2 *Analysis of the $N_2$-integrand term*

We start the discussion of this term by claiming that

$$\left| \int_{\mathbb{R}^2} \partial_t \Gamma^\alpha \nabla^a \mathbf{n} \cdot \Gamma^\beta \nabla^a N_2 \, dx \right| \lesssim (M\epsilon)^4 (1+t)^{-1+2\delta}, \quad \forall |\beta| \leq s-1. \tag{5.83}$$

We can explain this bound informally as follows. Due to the definition of $N_2$, $\Gamma^\beta \nabla^a N_2$ contains two types of terms that can be written generically as[9]

$$\Gamma^\beta \nabla^a N_{21} = \Gamma^{\beta_1} \nabla^{1+a_1} \mathbf{n} \, \Gamma^{\beta_2} \nabla^{1+a_2} \mathbf{n} \, \Gamma^{\beta_3} \nabla^{1+a_3} \mathbf{n} \, \Gamma^{\beta_4} \nabla^{1+a_4} \mathbf{n} \, \Gamma^{\beta_5} \nabla^{a_5} \mathbf{n}$$

---

[9]We choose not to write either the cross product or the dot product between the factors in these terms, because, for the purposes of estimating them in Lebesgue spaces, it does not matter which product appears in an application of Hölder's inequality.

and

$$\Gamma^\beta \nabla^a N_{22} = \Gamma^{\beta_1} \nabla^{2+a_1} \mathbf{n} \; \Gamma^{\beta_2} \nabla^{1+a_2} \mathbf{n} \; \Gamma^{\beta_3} \nabla^{1+a_3} \mathbf{n} \; \Gamma^{\beta_4} \nabla^{a_4} \mathbf{n} \; \Gamma^{\beta_5} \nabla^{a_5} \mathbf{n}$$

with

$$\sum_{i=1}^5 \beta_i = \beta \qquad \text{and} \qquad \sum_{i=1}^5 a_i = a.$$

For the first type of terms, a careful analysis shows that we can estimate three of the first four factors in $L^\infty$ with the help of (5.44) and (5.63). The remaining two factors are bounded as before: i.e., the one with the higher order derivative in $L^2$ using either (5.42) or (5.43) and (5.62), and the one with the low order derivative in $L^\infty$ relying on either $\|\mathbf{n}\|_{L^\infty} = 1$ or (5.44) and (5.63). Also taking into account (5.80) and choosing $M\epsilon(1+T) \lesssim 1$, we deduce

$$\left| \int_{\mathbb{R}^2} \partial_t \Gamma^\alpha \nabla^a \mathbf{n} \cdot \Gamma^\beta \nabla^a N_{21} \, dx \right| \lesssim (M\epsilon)^5 (1+t)^{-3/2+2\delta}, \tag{5.84}$$

which is a better bound than the one on the right-hand side of (5.83) if $\delta < 1/2$.

For the second type of term, one conducts a similar analysis by placing the factor with the higher order derivative in $L^2$ and the remaining four factors in $L^\infty$. If $M\epsilon(1+T) \lesssim 1$, a case-by-case investigation of all possible profiles yields

$$\left| \int_{\mathbb{R}^2} \partial_t \Gamma^\alpha \nabla^a \mathbf{n} \cdot \Gamma^\beta \nabla^a N_{22} \, dx \right| \lesssim (M\epsilon)^4 (1+t)^{-1+2\delta}, \tag{5.85}$$

which combined with (5.84) derives (5.83). An example of a profile that can be estimated precisely by the previous bound corresponds to

$$\left| \int_{\mathbb{R}^2} \left[ \partial_t \Gamma^\alpha \nabla^a \mathbf{n} \cdot (\nabla \mathbf{n} \wedge \mathbf{n}) \right] \left[ \mathbf{n} \cdot (\Gamma^\beta \nabla^2 \mathbf{n} \wedge \nabla \mathbf{n}) \right] dx \right|$$

$$\lesssim \| \partial_t \Gamma^\alpha \nabla^a \mathbf{n} \|_{L^2} \| \Gamma^\beta \nabla^2 \mathbf{n} \|_{L^2} \| \nabla \mathbf{n} \|_{L^\infty}^2 \| \mathbf{n} \|_{L^\infty}^2$$

$$\lesssim \| \nabla^{1+a} \mathbf{n}(t) \|_{\Gamma,s,L^2} \| \nabla \mathbf{n}(t) \|_{\Gamma,s,L^2} \| \Gamma \mathbf{n}(t) \|_{\Gamma,s-3,L^\infty}^2 \| \mathbf{n}(t) \|_{L^\infty}^2$$

$$\lesssim (M\epsilon)^4 (1+t)^{-1+2\delta}.$$

This finishes the argument for (5.72).

Therefore, the only thing left to do for proving (5.73) is to investigate the $N_2$-integrand term when $\alpha = \beta$ and $|\alpha| = s$. Moreover, in this case, we claim that if we apply Leibniz's rule for the differential operator $\Gamma^\alpha \nabla^a$ with respect to $N_2$ and denote

$$R(N_2) \stackrel{\text{def}}{=} \Gamma^\alpha \nabla^a N_2 - \partial_\mu [\mathbf{n} \cdot \Gamma^\alpha \nabla^a (\partial^\mu \mathbf{n} \wedge \partial^\nu \mathbf{n})] \partial_\nu \mathbf{n} \wedge \mathbf{n},$$

then

$$\left| \int_{\mathbb{R}^2} \partial_t \Gamma^\alpha \nabla^a \mathbf{n} \cdot R(N_2) \, dx \right| \lesssim (M\epsilon)^4 (1+t)^{-1+2\delta}. \tag{5.86}$$

This can be argued by checking that all of the terms in $R(N_2)$ have profiles similar to ones that have been previously discussed. We do not pursue this further and leave the verification of this fact to the reader.

Thus, we only need to analyze

$$\int_{\mathbb{R}^2} \partial_t \Gamma^\alpha \nabla^a \mathbf{n} \cdot (\partial_\nu \mathbf{n} \wedge \mathbf{n}) \, \partial_\mu [\mathbf{n} \cdot \Gamma^\alpha \nabla^a (\partial^\mu \mathbf{n} \wedge \partial^\nu \mathbf{n})] \, dx,$$

which, after applying the properties of the scalar triple product and integrating by parts (allowed by (5.51) and (5.53)), can be written as the sum of the following four integrals

$$\mathrm{I} = - \int_{\mathbb{R}^2} \mathbf{n} \cdot (\partial_t \Gamma^\alpha \nabla^a \mathbf{n} \wedge \nabla_x \mathbf{n}) \, \partial_t [\mathbf{n} \cdot \Gamma^\alpha \nabla^a (\partial_t \mathbf{n} \wedge \nabla_x \mathbf{n})] \, dx,$$

$$\mathrm{II} = \int_{\mathbb{R}^2} \nabla_x [\mathbf{n} \cdot (\partial_t \mathbf{n} \wedge \partial_t \Gamma^\alpha \nabla^a \mathbf{n})] \, \mathbf{n} \cdot \Gamma^\alpha \nabla^a (\partial_t \mathbf{n} \wedge \nabla_x \mathbf{n}) \, dx,$$

$$\mathrm{III} = - \int_{\mathbb{R}^2} \partial_1 [\mathbf{n} \cdot (\partial_t \Gamma^\alpha \nabla^a \mathbf{n} \wedge \partial_2 \mathbf{n})] \, \mathbf{n} \cdot \Gamma^\alpha \nabla^a (\partial_1 \mathbf{n} \wedge \partial_2 \mathbf{n}) \, dx,$$

$$\mathrm{IV} = - \int_{\mathbb{R}^2} \partial_2 [\mathbf{n} \cdot (\partial_1 \mathbf{n} \wedge \partial_t \Gamma^\alpha \nabla^a \mathbf{n})] \, \mathbf{n} \cdot \Gamma^\alpha \nabla^a (\partial_1 \mathbf{n} \wedge \partial_2 \mathbf{n}) \, dx.$$

Due to (5.86), it is clear that (5.73) holds if we show that

$$\left| \mathrm{I} + \mathrm{II} + \frac{1}{2} \frac{d}{dt} \Big\{ \| \mathbf{n} \cdot (\Gamma^\alpha \nabla^a \partial_t \mathbf{n} \wedge \nabla_x \mathbf{n})(t) \|_{L^2}^2 \right.$$
$$\left. - \| \mathbf{n} \cdot (\partial_t \mathbf{n} \wedge \Gamma^\alpha \nabla^a \nabla_x \mathbf{n})(t) \|_{L^2}^2 \Big\} \right| \lesssim (M\epsilon)^4 (1+t)^{-1+2\delta} \tag{5.87}$$

and

$$\left| \mathrm{III} + \mathrm{IV} + \frac{1}{2} \frac{d}{dt} \Big\{ \| \mathbf{n} \cdot \Gamma^\alpha \nabla^a (\partial_1 \mathbf{n} \wedge \partial_2 \mathbf{n})(t) \|_{L^2}^2 \Big\} \right|$$
$$\lesssim (M\epsilon)^4 (1+t)^{-1+2\delta}. \tag{5.88}$$

The arguments for the two estimates are very similar and this is why we choose to present only the one for the latter bound, which is also slightly more transparent.

Using the symmetries of the scalar triple product, we infer

$$\text{III} + \text{IV} + \frac{1}{2}\frac{d}{dt}\left\{\|\mathbf{n}\cdot\Gamma^\alpha\nabla^a(\partial_1\mathbf{n}\wedge\partial_2\mathbf{n})\|_{L^2}^2\right\}$$

$$= \int_{\mathbb{R}^2} \mathbf{n}\cdot\Gamma^\alpha\nabla^a(\partial_1\mathbf{n}\wedge\partial_2\mathbf{n})\Big(\partial_t\mathbf{n}\cdot\Gamma^\alpha\nabla^a(\partial_1\mathbf{n}\wedge\partial_2\mathbf{n})$$

$$+ 2\partial_t\Gamma^\alpha\nabla^a\mathbf{n}\cdot(\partial_1\mathbf{n}\wedge\partial_2\mathbf{n}) + \mathbf{n}\cdot\Big[\partial_t\Gamma^\alpha\nabla^a(\partial_1\mathbf{n}\wedge\partial_2\mathbf{n})$$

$$- (\partial_1\mathbf{n}\wedge\partial_t\partial_2\Gamma^\alpha\nabla^a\mathbf{n}) - (\partial_t\partial_1\Gamma^\alpha\nabla^a\mathbf{n}\wedge\partial_2\mathbf{n})\Big]\Big)\,dx.$$

Next, we rely on Leibniz's rule, (5.42), (5.44), and (5.63) to derive

$$\|\Gamma^\alpha\nabla^a(\partial_1\mathbf{n}\wedge\partial_2\mathbf{n})(t)\|_{L^2} \lesssim \|\nabla^{1+a}\mathbf{n}(t)\|_{\Gamma,s,L^2}\|\Gamma\mathbf{n}(t)\|_{\Gamma,s-3,L^\infty}$$

$$\lesssim (M\epsilon)^2(1+t)^{-1/2+\delta},$$

which implies

$$\|\mathbf{n}\cdot\Gamma^\alpha\nabla^a(\partial_1\mathbf{n}\wedge\partial_2\mathbf{n})(t)\|_{L^2} \lesssim (M\epsilon)^2(1+t)^{-1/2+\delta} \tag{5.89}$$

and

$$\|\partial_t\mathbf{n}\cdot\Gamma^\alpha\nabla^a(\partial_1\mathbf{n}\wedge\partial_2\mathbf{n})(t)\|_{L^2} \lesssim (M\epsilon)^3(1+t)^{-1+\delta}. \tag{5.90}$$

Moreover, we obtain based on (5.64), (5.42), (5.44), and (5.63) that

$$\|\partial_t\Gamma^\alpha\nabla^a\mathbf{n}\cdot(\partial_1\mathbf{n}\wedge\partial_2\mathbf{n})\|_{L^2} \lesssim \|\nabla^{1+a}\mathbf{n}(t)\|_{\Gamma,s,L^2}\|\Gamma\mathbf{n}(t)\|_{\Gamma,s-3,L^\infty}^2$$
$$\lesssim (M\epsilon)^3(1+t)^{-1+\delta}. \tag{5.91}$$

Finally, we can write

$$\partial_t\Gamma^\alpha\nabla^a(\partial_1\mathbf{n}\wedge\partial_2\mathbf{n}) - (\partial_1\mathbf{n}\wedge\partial_t\partial_2\Gamma^\alpha\nabla^a\mathbf{n}) - (\partial_t\partial_1\Gamma^\alpha\nabla^a\mathbf{n}\wedge\partial_2\mathbf{n})$$

$$= \partial_t\Gamma^\alpha\nabla^a(\partial_1\mathbf{n}\wedge\partial_2\mathbf{n}) - (\partial_1\mathbf{n}\wedge\partial_t\Gamma^\alpha\nabla^a\partial_2\mathbf{n}) - (\partial_t\Gamma^\alpha\nabla^a\partial_1\mathbf{n}\wedge\partial_2\mathbf{n})$$

$$+ (\partial_1\mathbf{n}\wedge\partial_t[\Gamma^\alpha,\partial_2]\nabla^a\mathbf{n}) + (\partial_t[\Gamma^\alpha,\partial_1]\nabla^a\mathbf{n}\wedge\partial_2\mathbf{n}).$$

For the first three terms, Leibniz's rule implies

$$\partial_t\Gamma^\alpha\nabla^a(\partial_1\mathbf{n}\wedge\partial_2\mathbf{n}) - (\partial_1\mathbf{n}\wedge\partial_t\Gamma^\alpha\nabla^a\partial_2\mathbf{n}) - (\partial_t\Gamma^\alpha\nabla^a\partial_1\mathbf{n}\wedge\partial_2\mathbf{n})$$

$$= \sum_{\substack{d\leq 1,\gamma\leq\alpha,b\leq a \\ (0,0,0)\neq(d,\gamma,b)\neq(1,\alpha,a)}} C_{d\gamma b}\,(\partial_t^d\Gamma^\gamma\nabla^b\partial_1\mathbf{n}\wedge\partial_t^{1-d}\Gamma^{\alpha-\gamma}\nabla^{a-b}\partial_2\mathbf{n})$$

and a careful analysis of the various possible values for $(d,\gamma,b)$ together with (5.64), (5.42), (5.44), and (5.63) reveal that

$$\|\partial_t^d\Gamma^\gamma\nabla^b\partial_1\mathbf{n}\wedge\partial_t^{1-d}\Gamma^{\alpha-\gamma}\nabla^{a-b}\partial_2\mathbf{n}\|_{L^2}$$

$$\lesssim (\|\nabla\mathbf{n}(t)\|_{\Gamma,s,L^2} + \|\nabla^2\mathbf{n}(t)\|_{\Gamma,s,L^2})\|\Gamma\mathbf{n}(t)\|_{\Gamma,s-3,L^\infty} \tag{5.92}$$

$$\lesssim (M\epsilon)^2(1+t)^{-1/2+\delta}.$$

Invoking the same four statements as before, we also deduce

$$\|\partial_1 \mathbf{n} \wedge \partial_t [\Gamma^\alpha, \partial_2] \nabla^a \mathbf{n}\|_{L^2} + \|\partial_t [\Gamma^\alpha, \partial_1] \nabla^a \mathbf{n} \wedge \partial_2 \mathbf{n}\|_{L^2}$$

$$\lesssim \|\nabla^{1+a} \mathbf{n}(t)\|_{\Gamma, s, L^2} \|\Gamma \mathbf{n}(t)\|_{\Gamma, s-3, L^\infty} \quad (5.93)$$

$$\lesssim (M\epsilon)^2 (1+t)^{-1/2+\delta}.$$

We can then conclude that (5.88) is the result of combining (5.89)-(5.93).

### 5.3.3 *Proof of* (5.47)

Throughout this section, we will be working with the notation introduced in the first section, i.e.,

$$\tilde{\mathbf{n}} = (n_1, n_2),$$

which permits us to write (5.49) and (5.50) in the generic compact form

$$\Box \tilde{\mathbf{n}} = \left( C_1 \tilde{\mathbf{n}} + C_2 \frac{\tilde{\mathbf{n}}^3}{n_3^2} \right) Q(\tilde{\mathbf{n}}, \tilde{\mathbf{n}}) + \partial_\mu \left( \frac{Q_{\mu\nu}(\tilde{\mathbf{n}}, \tilde{\mathbf{n}})}{n_3} \right) \frac{(C_3 + C_4 \tilde{\mathbf{n}}^2) \partial_\nu \tilde{\mathbf{n}}}{n_3},$$

with $(C_i)_{1 \leq i \leq 4}$ being absolute constants. For a multi-index $\alpha$ satisfying $|\alpha| \leq s - 2$, the combined application of (5.56) and (5.65) to the previous equation yields

$$\|\Gamma^\alpha \tilde{\mathbf{n}}(t)\|_{L^\infty}$$

$$\lesssim (1+t)^{-1/2} \bigg( \|\Gamma^\alpha \tilde{\mathbf{n}}(0)\|_{W^{2,1}} + \|\partial_t \Gamma^\alpha \tilde{\mathbf{n}}(0)\|_{W^{1,1}} \quad (5.94)$$

$$+ \int_0^t (1+\tau)^{-1/2} \|\Box \tilde{\mathbf{n}}(\tau)\|_{\Gamma, s-1, L^1} \, d\tau \bigg).$$

Our goal is to show that:

**Proposition 5.8.** *Under the bootstrap assumptions* (5.42)-(5.44), *if* $0 < \delta < 1/6$ *and* $M\epsilon$ *is sufficiently small, then the following fixed time estimate holds uniformly for* $0 \leq t \leq T$,

$$\|\Gamma^\alpha \tilde{\mathbf{n}}(t)\|_{L^\infty} \leq C(R) \left( \epsilon + \frac{(M\epsilon)^3}{1 - 6\delta} \right) (1+t)^{-1/2}. \quad (5.95)$$

If this is the case, then with further adjustments in the values of $M$ and $\delta$, depending entirely on $\epsilon$ and $R$, and choosing $\epsilon$ small enough, we conclude that (5.47) holds.

By (5.60), we already know that

$$\|\Gamma^\alpha \tilde{\mathbf{n}}(0)\|_{W^{2,1}} + \|\partial_t \Gamma^\alpha \tilde{\mathbf{n}}(0)\|_{W^{1,1}} \leq C(R)\epsilon. \quad (5.96)$$

Thus, we need to bound the integrand on the right-hand side of (5.94), for which we treat

$$\tilde{N}_1 \overset{\text{def}}{=} \left( C_1 \tilde{n} + C_2 \frac{\tilde{n}^3}{n_3^2} \right) Q(\tilde{n}, \tilde{n})$$

and

$$\tilde{N}_2 \overset{\text{def}}{=} \partial_\mu \left( \frac{Q_{\mu\nu}(\tilde{n}, \tilde{n})}{n_3} \right) \frac{(C_3 + C_4 \tilde{n}^2) \partial_\nu \tilde{n}}{n_3}$$

separately. If we prove that

$$\|\tilde{N}_1(t)\|_{\Gamma, s-1, L^1} + \|\tilde{N}_2(t)\|_{\Gamma, s-1, L^1} \lesssim (M\epsilon)^3 (1+t)^{-1+3\delta} \tag{5.97}$$

is true uniformly for $0 \le t \le T$, then a simple integration yields (5.95).

### 5.3.3.1   *Analysis of the $\tilde{N}_1$ term*

If we apply Leibniz's rule, we can infer that

$$\|\tilde{N}_1\|_{\Gamma, s-1, L^1} \lesssim \left\| C_1 \tilde{n} + C_2 \frac{\tilde{n}^3}{n_3^2} \right\|_{\Gamma, s-1, L^2} \|Q(\tilde{n}, \tilde{n})\|_{\Gamma, \lfloor \frac{s-1}{2} \rfloor, L^2}$$
$$+ \left\| C_1 \tilde{n} + C_2 \frac{\tilde{n}^3}{n_3^2} \right\|_{\Gamma, \lfloor \frac{s-1}{2} \rfloor, L^\infty} \|Q(\tilde{n}, \tilde{n})\|_{\Gamma, s-1, L^1}. \tag{5.98}$$

According to (5.43), (5.44), (5.69), and (5.70), we have

$$\|\tilde{n}(t)\|_{\Gamma, s-1, L^2} \lesssim M\epsilon (1+t)^{1/2+2\delta}, \tag{5.99}$$

$$\|\tilde{n}(t)\|_{\Gamma, \lfloor \frac{s-1}{2} \rfloor, L^\infty} \lesssim M\epsilon (1+t)^{-1/2}, \tag{5.100}$$

$$\|Q(\tilde{n}, \tilde{n})(t)\|_{\Gamma, \lfloor \frac{s-1}{2} \rfloor, L^2} \lesssim (M\epsilon)^2 (1+t)^{-3/2+\delta}, \tag{5.101}$$

$$\|Q(\tilde{n}, \tilde{n})(t)\|_{\Gamma, s-1, L^1} \lesssim (M\epsilon)^2 (1+t)^{-1/2+3\delta}. \tag{5.102}$$

The only expression for which we do not have a direct previous estimate is $\frac{\tilde{n}^3}{n_3^2}$.

However, arguing again on the basis of Leibniz's rule, we derive

$$\left\| \frac{\tilde{n}^3}{n_3^2} \right\|_{\Gamma, s-1, L^2} \lesssim \|\tilde{n}\|_{\Gamma, s-1, L^2} \|\tilde{n}\|_{\Gamma, \lfloor \frac{s-1}{2} \rfloor, L^\infty}^2 \|1/n_3\|_{\Gamma, \lfloor \frac{s-1}{2} \rfloor, L^\infty}^2$$
$$+ \|\tilde{n}\|_{\Gamma, \lfloor \frac{s-1}{2} \rfloor, L^\infty}^3 \|1/n_3\|_{\Gamma, \lfloor \frac{s-1}{2} \rfloor, L^\infty} \|\Gamma(1/n_3)\|_{\Gamma, s-2, L^2}$$

and

$$\left\| \frac{\tilde{n}^3}{n_3^2} \right\|_{\Gamma, \lfloor \frac{s-1}{2} \rfloor, L^\infty} \lesssim \|\tilde{n}\|_{\Gamma, \lfloor \frac{s-1}{2} \rfloor, L^\infty}^3 \|1/n_3\|_{\Gamma, \lfloor \frac{s-1}{2} \rfloor, L^\infty}^2.$$

Due to (5.52) and (5.63), as $M\epsilon$ can be made sufficiently small,

$$\|1/n_3(t)\|_{\Gamma,\lfloor\frac{s-1}{2}\rfloor,L^\infty} \simeq 1. \tag{5.103}$$

On the other hand, (5.62) implies

$$\|\Gamma(1/n_3)(t)\|_{\Gamma,s-2,L^2} \lesssim (M\epsilon)^2(1+t)^{2\delta}. \tag{5.104}$$

Hence, relying also on (5.99) and (5.100), we deduce

$$\left\|\frac{\tilde{\mathbf{n}}^3}{n_3^2}(t)\right\|_{\Gamma,s-1,L^2} \lesssim (M\epsilon)^3(1+t)^{-1/2+2\delta} \tag{5.105}$$

and

$$\left\|\frac{\tilde{\mathbf{n}}^3}{n_3^2}(t)\right\|_{\Gamma,\lfloor\frac{s-1}{2}\rfloor,L^\infty} \lesssim (M\epsilon)^3(1+t)^{-3/2}. \tag{5.106}$$

Combining the last two estimates with (5.99)-(5.102), we conclude on the basis of (5.98) that

$$\|\tilde{N}_1(t)\|_{\Gamma,s-1,L^1} \lesssim (M\epsilon)^3(1+t)^{-1+3\delta}. \tag{5.107}$$

### 5.3.3.2  *Analysis of the $\tilde{N}_2$ term*

For this part of the integrand, we write it as

$$\tilde{N}_2 \simeq \frac{Q_{\mu\nu}(\partial_\mu\tilde{\mathbf{n}},\tilde{\mathbf{n}})(C_3+C_4\tilde{\mathbf{n}}^2)\partial_\nu\tilde{\mathbf{n}}}{n_3^2}$$
$$+ \frac{Q_{\mu\nu}(\tilde{\mathbf{n}},\tilde{\mathbf{n}})(C_3+C_4\tilde{\mathbf{n}}^2)\partial_\nu\tilde{\mathbf{n}}}{n_3}\partial_\mu\left(\frac{1}{n_3}\right), \tag{5.108}$$

and we treat the two new terms independently.

In the case of the former, if we apply as before Leibniz's rule, then we can infer that

$$\left\|\frac{Q_{\mu\nu}(\partial_\mu\tilde{\mathbf{n}},\tilde{\mathbf{n}})(C_3+C_4\tilde{\mathbf{n}}^2)\partial_\nu\tilde{\mathbf{n}}}{n_3^2}\right\|_{\Gamma,s-1,L^1}$$
$$\lesssim \|Q_{\mu\nu}(\partial_\mu\tilde{\mathbf{n}},\tilde{\mathbf{n}})\|_{\Gamma,s-1,L^1}\left\|\frac{(C_3+C_4\tilde{\mathbf{n}}^2)\partial_\nu\tilde{\mathbf{n}}}{n_3^2}\right\|_{\Gamma,\lfloor\frac{s-1}{2}\rfloor,L^\infty}$$
$$+ \|Q_{\mu\nu}(\partial_\mu\tilde{\mathbf{n}},\tilde{\mathbf{n}})\|_{\Gamma,\lfloor\frac{s-1}{2}\rfloor,L^2}\left\|\frac{(C_3+C_4\tilde{\mathbf{n}}^2)\partial_\nu\tilde{\mathbf{n}}}{n_3^2}\right\|_{\Gamma,s-1,L^2}.$$

Using (5.69) and (5.70), it follows that

$$\|Q(\partial_\mu\tilde{\mathbf{n}},\tilde{\mathbf{n}})(t)\|_{\Gamma,\lfloor\frac{s-1}{2}\rfloor,L^2} \lesssim (M\epsilon)^2(1+t)^{-3/2+\delta}, \tag{5.109}$$

$$\|Q(\partial_\mu\tilde{\mathbf{n}},\tilde{\mathbf{n}})(t)\|_{\Gamma,s-1,L^1} \lesssim (M\epsilon)^2(1+t)^{-1/2+3\delta}. \tag{5.110}$$

The above $L^\infty$ norm can be easily estimated by

$$\left\|\frac{(C_3 + C_4\tilde{\mathbf{n}}^2)\partial_\nu\tilde{\mathbf{n}}}{n_3^2}\right\|_{\Gamma,\lfloor\frac{s-1}{2}\rfloor,L^\infty}$$

$$\lesssim (1 + \|\tilde{\mathbf{n}}\|^2_{\Gamma,\lfloor\frac{s-1}{2}\rfloor,L^\infty})\|\nabla\tilde{\mathbf{n}}\|_{\Gamma,\lfloor\frac{s-1}{2}\rfloor,L^\infty}\|1/n_3\|^2_{\Gamma,\lfloor\frac{s-1}{2}\rfloor,L^\infty},$$

which, on the account of (5.44) and (5.103), implies

$$\left\|\frac{(C_3 + C_4\tilde{\mathbf{n}}^2)\partial_\nu\tilde{\mathbf{n}}}{n_3^2}(t)\right\|_{\Gamma,\lfloor\frac{s-1}{2}\rfloor,L^\infty} \lesssim M\epsilon(1+t)^{-1/2}. \qquad (5.111)$$

We are left to find a suitable bound for

$$\left\|\frac{(C_3 + C_4\tilde{\mathbf{n}}^2)\partial_\nu\tilde{\mathbf{n}}}{n_3^2}\right\|_{\Gamma,s-1,L^2} \lesssim \left\|\frac{\partial_\nu\tilde{\mathbf{n}}}{n_3^2}\right\|_{\Gamma,s-1,L^2} + \left\|\frac{\tilde{\mathbf{n}}^2\partial_\nu\tilde{\mathbf{n}}}{n_3^2}\right\|_{\Gamma,s-1,L^2}.$$

For the second norm on the right-hand side, which is slightly more intricate, we argue that

$$\left\|\frac{\tilde{\mathbf{n}}^2\partial_\nu\tilde{\mathbf{n}}}{n_3^2}\right\|_{\Gamma,s-1,L^2}$$

$$\lesssim \|\nabla\tilde{\mathbf{n}}\|_{\Gamma,s-1,L^2}\|\tilde{\mathbf{n}}\|^2_{\Gamma,\lfloor\frac{s-1}{2}\rfloor,L^\infty}\|1/n_3\|^2_{\Gamma,\lfloor\frac{s-1}{2}\rfloor,L^\infty}$$

$$+ \|\nabla\tilde{\mathbf{n}}\|_{\Gamma,\lfloor\frac{s-1}{2}\rfloor,L^\infty}\|\tilde{\mathbf{n}}\|_{\Gamma,s-1,L^2}\|\tilde{\mathbf{n}}\|_{\Gamma,\lfloor\frac{s-1}{2}\rfloor,L^\infty}\|1/n_3\|^2_{\Gamma,\lfloor\frac{s-1}{2}\rfloor,L^\infty}$$

$$+ \|\nabla\tilde{\mathbf{n}}\|_{\Gamma,\lfloor\frac{s-1}{2}\rfloor,L^\infty}\|\tilde{\mathbf{n}}\|^2_{\Gamma,\lfloor\frac{s-1}{2}\rfloor,L^\infty}\|1/n_3\|_{\Gamma,\lfloor\frac{s-1}{2}\rfloor,L^\infty}\|\Gamma(1/n_3)\|_{\Gamma,s-2,L^2}.$$

Thus, if we rely on (5.42), (5.43), (5.44), (5.103), and (5.104), we derive

$$\left\|\frac{\tilde{\mathbf{n}}^2\partial_\mu\tilde{\mathbf{n}}}{n_3^2}(t)\right\|_{\Gamma,s-1,L^2} \lesssim (M\epsilon)^3(1+t)^{-1/2+2\delta}.$$

An easier and similar analysis produces

$$\left\|\frac{\partial_\nu\tilde{\mathbf{n}}}{n_3^2}(t)\right\|_{\Gamma,s-1,L^2} \lesssim M\epsilon(1+t)^\delta.$$

Due to $\delta < 1/2$ and $M\epsilon$ being sufficiently small, we obtain

$$\left\|\frac{(C_3 + C_4\tilde{\mathbf{n}}^2)\partial_\nu\tilde{\mathbf{n}}}{n_3^2}(t)\right\|_{\Gamma,s-1,L^2} \lesssim M\epsilon(1+t)^\delta. \qquad (5.112)$$

Combining this estimate with (5.109), (5.110), and (5.111), we derive

$$\left\|\frac{Q_{\mu\nu}(\partial_\mu\tilde{\mathbf{n}},\tilde{\mathbf{n}})(C_3 + C_4\tilde{\mathbf{n}}^2)\partial_\nu\tilde{\mathbf{n}}}{n_3^2}(t)\right\|_{\Gamma,s-1,L^1} \lesssim (M\epsilon)^3(1+t)^{-1+3\delta}. \qquad (5.113)$$

For the second term in (5.108), an almost identical approach (which is left for verification to the interested reader) with the one we have just finished yields

$$\left\| \frac{Q_{\mu\nu}(\tilde{\mathbf{n}}, \tilde{\mathbf{n}})(C_3 + C_4\tilde{\mathbf{n}}^2)\partial_\nu\tilde{\mathbf{n}}}{n_3} \partial_\mu\left(\frac{1}{n_3}\right)(t)\right\|_{\Gamma, s-1, L^1} \tag{5.114}$$
$$\lesssim (M\epsilon)^5(1+t)^{-2+3\delta}.$$

Hence, we can conclude based on $M\epsilon$ being sufficiently small that

$$\|\tilde{N}_2(t)\|_{\Gamma, s-1, L^1} \lesssim (M\epsilon)^3(1+t)^{-1+3\delta}, \tag{5.115}$$

which, together with (5.107), finishes the proof for (5.97).

# Chapter 6

# Equivariant results

## 6.1 Prologue

After discussing well-posedness results for the general case of the wave map system and of the Skyrme and Skyrme-like models, we focus now on the *equivariant case* of these problems, which corresponds to adding a natural rotational symmetry assumption. The motivation to present this case is twofold. On one hand, for wave maps, the first known global regularity results for the $2+1$-dimensional problem were proven in this context, while the most recent blow-up analyses (to be presented in the last chapter) have strong connections with equivariant maps. On the other hand, for Skyrme and Skyrme-like models, the equivariant problem recently experienced a strong record of global regularity results, both for smooth and rough data.

According to the definition given in Section 8.1 of Shatah-Struwe [145], one needs to assume for the equivariant case:

- the target manifold $(N, h)$ to be a rotationally symmetric Riemannian manifold, i.e.,

$$(u, \Omega) \in N = [0, u^*) \times \mathbb{S}^{n-1}, \quad u^* \in \mathbb{R}^+ \cup \{\infty\}, \quad h = du^2 + f^2(u)d\Omega^2,$$

  where $f : \mathbb{R} \to \mathbb{R}$ is a smooth, odd function, with $f(0) = 0$ and $f'(0) = 1$;

- the map $\phi : (\mathbb{R}^{n+1}, g) \to (N, h)$ to be described by

$$\phi(t, r, \omega) = (u, \Omega) = (u(t, r), \Omega(\omega)),$$

  where

$$g = -dt^2 + dr^2 + r^2 d\omega^2, \quad (t, r, \omega) \in \mathbb{R} \times \mathbb{R}^+ \times \mathbb{S}^{n-1},$$

  and $\omega \in \mathbb{S}^{n-1} \mapsto \Omega(\omega) \in \mathbb{S}^{n-1}$ is a harmonic polynomial map (i.e., it is the restriction of a map from $\mathbb{R}^n$ to $\mathbb{R}^n$, whose components are all harmonic homogeneous polynomials of positive degree).

For equivariant wave maps, Section 8.1 of [145] also contains a detailed account of a 2 + 1-dimensional global regularity result for smooth data when $N$ is *geodesically convex* (i.e., $f(u)f'(u) > 0$ if $u > 0$). Moreover, Section 8.3.2 of the same book provides a thorough exposition of the relevant bibliography on this topic, up to that point in time. What we choose to present here is unpublished work by Grillakis [60], building on results from [59,61], which extends the above mentioned geodesically convex result and for which a substantial part of the argument survives even without the equivariant assumption.

On the side of equivariant Skyrme and Skyrme-like models, the first important result is due to Bizoń-Chmaj-Rostworowski [17], which addresses numerically the stability[1] of the skyrmion with unit charge. This paper also conjectures that all skyrmions behave like global attractors for the evolution of smooth, finite energy data in the equivariant Skyrme model. This was followed by an impressive number of works in a relatively short period of time, which focused on the regularity of the equivariant Skyrme, Adkins-Nappi, and 2 + 1-dimensional Faddeev problems[2]. Chronologically, they are as follows:

- Geba-Rajeev [53, 54] proved non-concentration of the Adkins-Nappi energy;
- Geba-da Silva [50] showed non-concentration of the energy for 2 + 1-dimensional Faddeev model;
- Li [108] demonstrated global regularity for large smooth data in the case of the Skyrme model;
- Geba-Nakanishi-Rajeev [51] proved global well-posedness for small data in critical Sobolev-Besov spaces for the Adkins-Nappi and Skyrme models;
- Geba-Nakanishi-Zhang [52] showed the counterpart of Geba-Nakanishi-Rajeev's result for the 2 + 1-dimensional Faddeev model;
- Creek [31] demonstrated a result similar to Li's but for the 2 + 1-dimensional Faddeev model;
- Lawrie [99] proved conditional global existence for large data in the case of the Adkins-Nappi model.

---

[1] See also recent work of Creek-Donninger-Schlag-Snelson [32].

[2] We make the remark that, based on the equivariant assumption introduced before, it is natural to study the 2 + 1-dimensional Faddeev model and not the original one. For maps from $\mathbb{R}^{3+1}$ to $\mathbb{S}^2$, an "equivariant" assumption would have to take into account their Hopf index (1.11) and use toroidal coordinates instead of polar ones, as in [4].

In this chapter, the selection of topics we decided to present is as follows. First, we discuss the non-concentration of energy component in Grillakis's argument [60] for large energy equivariant wave maps, but choose to skip its small energy global regularity component. This is because, next, we focus on a synthesis of small data global well-posedness results [51, 52] for the equivariant Skyrme, Adkins-Nappi, and 2 + 1-dimensional Faddeev problems. These arguments have a natural common theme and generalize the ones for wave maps. Finally, we discuss Creek's large data equivariant result for the Faddeev problem, which, from a pedagogical point of view, is more direct in certain aspects than its counterpart for the Skyrme model proved by Li.

## 6.2 Non-concentration of energy for classical equivariant wave maps

Our focus here is on the study of 2 + 1-dimensional equivariant wave maps, for which, under the assumptions made in the prologue, the wave map system (1.21) takes the form:

$$\Box u + f(u)f'(u)\nabla_\mu \Omega \nabla^\mu \Omega = 0,$$
$$\nabla_\mu \big(f^2(u)\nabla^\mu \Omega\big) = 0,$$

with $\Box = \partial_{tt} - \partial_{rr} - \partial_r/r$ being the radial d'Alembertian. If we further assume that $\Omega = k\omega$, where $k$ is an integer, then the first equation becomes

$$\Box u + \frac{k^2}{r^2} f(u)f'(u) = 0, \tag{6.1}$$

whereas the second one is automatically satisfied. Thus, we move forward with the investigation of the semilinear wave equation for $u$, which has as special cases $f(u) = \sin(u)$ when $N = \mathbb{S}^2$ and $f(u) = \sinh(u)$ corresponding to $N = \mathbb{H}^2$.

In what follows, we intend to cover Grillakis's regularity result [60] for (6.1), that was preceded by work due to Shatah-Zadeh [147] and was later refined by Struwe [165]. The motivation for this presentation has to do with the simpler nature of the equivariant problem, when compared to the original one, and also with the derivation of certain a priori estimates which are of wider interest. These estimates are called in the literature Morawetz-type bounds and are obtained using Friedrichs's *ABC* multiplier method [49].

Next, we formulate the setting in which we analyze the equation (6.1). This is a Goursat problem in the diamond-shaped region

$$D_T \overset{\text{def}}{=} \{(t,r) \mid 0 \le r \le t, \ t+r \le 2T, \ 0 \le t \le 2T\}, \qquad (6.2)$$

with data prescribed on the conical surface

$$K_T \overset{\text{def}}{=} \{(t,r) \mid t+r = 2T, \ T \le t \le 2T\}, \qquad (6.3)$$

where $T > 0$ is fixed.

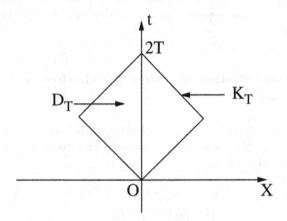

Fig. 6.1   The diamond-shaped region $D_T$.

Accordingly, for this data, we use the notation $u(2T - r, r) := u_0(r)$ with $0 \le r \le T$. We take for granted the local existence of smooth solutions for (6.1) and assume that such a solution exists in the domain $D_T \setminus \{(0,0)\}$, with the origin being the point where it may lose regularity. Our goal is to examine under what conditions on the target manifold this cannot happen.

### 6.2.1   *Differential identities*

In proving regularity, one of the main ingredients in the argument is estimates deduced from integrating differential identities, which are, in turn, derived from working with the energy-momentum tensor. If we denote $(t, \mathbf{x}) = (x^0, x^1, x^2)$ and $g = m = \text{diag}(-1, 1, 1)$, then the energy-momentum tensor in its general form is given by

$$T_{\mu\nu} = \nabla_\mu u \nabla_\nu u + f^2(u) \nabla_\mu \Omega \nabla_\nu \Omega - \frac{1}{2} m_{\mu\nu} \left( \nabla_\alpha u \nabla^\alpha u + f^2(u) \nabla_\alpha \Omega \nabla^\alpha \Omega \right).$$

Using the equivariant assumption $(u, \Omega) = (u(t, r), k\omega)$, we compute

$$\nabla_0 u = \partial_t u, \qquad \nabla_j u = \frac{x_j}{r} \partial_r u, \qquad \nabla_0 \Omega = 0, \qquad \nabla_j \Omega = k\epsilon_{jl} \frac{x^l}{r^2},$$

where $\epsilon_{jl}$ is the 2-dimensional Levi-Civita tensor. This implies that the components of the energy-momentum tensor have the formulas

$$T_{00} = \frac{1}{2} \left\{ (\partial_t u)^2 + (\partial_r u)^2 + \frac{k^2}{r^2} f^2(u) \right\}, \tag{6.4}$$

$$T_{0j} = T_{j0} = \frac{x_j}{r} (\partial_t u)(\partial_r u), \tag{6.5}$$

$$\begin{aligned} T_{jk} = &\frac{x_j x_k}{r^2} (\partial_r u)^2 + \frac{\epsilon_{jl} x^l \epsilon_{km} x^m}{r^2} k^2 \frac{f^2(u)}{r^2} \\ &- \frac{1}{2} \delta_{jk} \left\{ (\partial_r u)^2 - (\partial_t u)^2 + \frac{k^2}{r^2} f^2(u) \right\}, \end{aligned} \tag{6.6}$$

with $1 \leq j, k \leq 2$. Furthermore, we notice that

$$\mathrm{tr}(T_{jk}) = (\partial_t u)^2, \qquad T_{jk} \frac{x^j x^k}{r^2} = \frac{1}{2} \left\{ (\partial_t u)^2 + (\partial_r u)^2 - \frac{k^2}{r^2} f^2(u) \right\}.$$

We recall from Chapter 1 that the energy-momentum tensor satisfies the structure equations

$$\nabla_\mu T^\mu_\nu = 0,$$

which lead, after contracting them with an arbitrary vector field $X_\nu$, to

$$\nabla_\mu \left\{ T^{\mu\nu} X_\nu \right\} + R = 0, \tag{6.7}$$

$$R \overset{\mathrm{def}}{=} -T^{\mu\nu} \nabla_\mu X_\nu. \tag{6.8}$$

If we choose the vector field to have $X_0 = 1$ and $X_j = 0$, then $R = 0$ and we obtain the well-known differential identity

$$\partial_t e(u) - \partial_j p^j(u) = 0 \tag{6.9}$$

where

$$e(u) \overset{\mathrm{def}}{=} \frac{1}{2} \left\{ (\partial_t u)^2 + (\partial_r u)^2 + \frac{k^2}{r^2} f^2(u) \right\}, \tag{6.10}$$

$$p^j(u) \overset{\mathrm{def}}{=} \frac{x^j}{r} \partial_t u \, \partial_r u. \tag{6.11}$$

This will be useful in deriving certain a priori local energy estimates, but it will not be enough to establish regularity.

For this purpose, we need one more identity deduced with the help of a vector field given by

$$X_0 = 1, \qquad X_j = -\lambda^{-\alpha} \frac{x_j}{t}, \qquad (6.12)$$

where $\lambda := r/t$ and $\alpha \in [0,1]$. Direct calculations provide us with

$$\nabla_\alpha X_0 = 0, \quad \nabla_0 X_j = -\frac{(1-\alpha)}{t} \lambda^{1-\alpha} \frac{x_j}{r}, \quad \nabla_j X_k = -\frac{\lambda^{-\alpha}}{t} \left\{ \delta_{jk} - \alpha \frac{x_j x_k}{r^2} \right\},$$

which imply, according to (6.8),

$$R = \frac{\lambda^{-\alpha}}{t} \left\{ \frac{2-\alpha}{2} (\partial_t u)^2 - \frac{\alpha}{2} (\partial_r u)^2 + \frac{\alpha}{2} \frac{k^2}{r^2} f^2(u) \right\}$$
$$+ \frac{1-\alpha}{t} \lambda^{1-\alpha} \partial_t u \, \partial_r u. \qquad (6.13)$$

However, by itself, this expression is not useful as it does not have a definite sign and the idea is to combine it with dilations on the target manifold. This is the $C$ part in Friedrichs's $ABC$ method, for which we define the auxiliary function

$$b = b(t,r) \overset{\text{def}}{=} \frac{\alpha \lambda^{-\alpha} + (1-\alpha)}{t}, \qquad (6.14)$$

and multiply our main equation (6.1) by $bu/2$. The outcome is a differential identity written in the abstract form

$$-\nabla_\mu L^\mu + Q = 0, \qquad (6.15)$$

where

$$L^\mu \overset{\text{def}}{=} \frac{1}{4} \nabla^\mu (bu^2) - \nabla^\mu (b) \frac{u^2}{2}, \qquad (6.16)$$

$$Q \overset{\text{def}}{=} \frac{b}{2} \left\{ \nabla_\mu u \nabla^\mu u + \frac{k^2}{r^2} f(u) f'(u) u \right\} + \Box b \frac{u^2}{4}. \qquad (6.17)$$

Straightforward computations yield

$$\partial_t b = -\frac{1-\alpha}{r^2} \left( \lambda^2 + \alpha \lambda^{2-\alpha} \right), \qquad \partial_j b = -\frac{\alpha^2}{r^2} \lambda^{1-\alpha} \frac{x_j}{r}, \qquad (6.18)$$

$$\Box b = \frac{1-\alpha}{t^3} \left\{ 2 + \alpha \left( 2 - \alpha \right) \lambda^{-\alpha} \right\} - \frac{\alpha^3}{tr^2} \lambda^{-\alpha}. \qquad (6.19)$$

With their help and the fact that

$$\nabla_\mu u \nabla^\mu u = (\partial_r u)^2 - (\partial_t u)^2,$$

we deduce, by adding (6.7) (corresponding to (6.12)) to (6.15):

$$\nabla_\mu P^\mu + S = 0, \qquad (6.20)$$

with

$$P^\mu \overset{\text{def}}{=} T^{\mu\nu} X_\nu - L^\mu, \qquad S \overset{\text{def}}{=} R + Q. \tag{6.21}$$

We can explicitly calculate the components of $P^\mu$ as

$$P^0 = \frac{1}{2}\left\{ (\partial_t u)^2 + (\partial_r u)^2 + \frac{k^2}{r^2} f^2(u) \right\} + \lambda^{1-\alpha} \partial_t u \, \partial_r u$$
$$+ \partial_t \left\{ \frac{bu^2}{4} \right\} - \frac{u^2}{2} \partial_t b, \tag{6.22}$$

$$P^j = -\frac{x^j}{r}\left\{ \partial_t u \, \partial_r u + \frac{\lambda^{1-\alpha}}{2}\left( (\partial_t u)^2 + (\partial_r u)^2 - \frac{k^2}{r^2} f^2(u) \right) \right\}$$
$$- \partial_j \left\{ \frac{bu^2}{4} \right\} + \frac{u^2}{2} \partial_j b. \tag{6.23}$$

Furthermore, it is convenient to write the error term like

$$S = S^{(0)} + S^{(1)} + S^{(2)}, \tag{6.24}$$

where

$$S^{(0)} \overset{\text{def}}{=} \frac{1-\alpha}{2t}\left\{ (2\lambda^{-\alpha} - 1)(\partial_t u)^2 + (\partial_r u)^2 + 2\lambda^{1-\alpha}(\partial_t u)(\partial_r u) \right\}, \tag{6.25}$$

$$S^{(1)} \overset{\text{def}}{=} \frac{k^2 \lambda^{-\alpha}}{2tr^2}\left\{ (1-\alpha)\lambda^\alpha f(u) f'(u) u \right.$$
$$\left. + \alpha \left( f(u) f'(u) u + f^2(u) - \alpha^2 \frac{u^2}{2k^2} \right) \right\}, \tag{6.26}$$

$$S^{(2)} \overset{\text{def}}{=} \frac{1-\alpha}{t^3}\left\{ 2 + \alpha\,(2-\alpha)\,\lambda^{-\alpha} \right\} \frac{u^2}{4}. \tag{6.27}$$

Based on the assumption $\alpha \in [0,1]$ and that $\lambda \le 1$ on $D_T$ (i.e., the region on which we want to integrate (6.20)), we observe immediately that $S^{(2)} \ge 0$ and the same is true about $S^{(0)}$ due to

$$S^{(0)} = \frac{1-\alpha}{2t}\left\{ \frac{1+\lambda^{1-\alpha}}{2}(\partial_t u + \partial_r u)^2 + \frac{1-\lambda^{1-\alpha}}{2}(\partial_t u - \partial_r u)^2 \right\}$$
$$+ \frac{1-\alpha}{t}(\lambda^{-\alpha} - 1)(\partial_t u)^2. \tag{6.28}$$

Thus, the remaining issue is under what conditions on $f = f(u)$ is $S^{(1)}$ also nonnegative.

### 6.2.2 *Standard energy estimates and applications*

We start this subsection by introducing, for $0 < \sigma_0 < \sigma_1 \leq T$, the following domain and surfaces in $\mathbb{R}^{2+1}$,

$$D_{[\sigma_0,\sigma_1]} \overset{\text{def}}{=} \left\{(t,r) \mid 0 \leq r \leq t,\ 2\sigma_0 \leq r+t \leq 2\sigma_1\right\}, \tag{6.29}$$

$$K_{\sigma_0} \overset{\text{def}}{=} \left\{(t,r) \mid r+t = 2\sigma_0,\ \sigma_0 \leq t \leq 2\sigma_0\right\}, \tag{6.30}$$

$$\mathcal{C}_{[\sigma_0,\sigma_1]} \overset{\text{def}}{=} \left\{(t,r) \mid r = t,\ \sigma_0 \leq t \leq \sigma_1\right\}, \tag{6.31}$$

for which one recognizes that

$$\partial D_{[\sigma_0,\sigma_1]} = \mathcal{C}_{[\sigma_0,\sigma_1]} \cup K_{\sigma_0} \cup K_{\sigma_1}.$$

Given that we are working with radially symmetric functions, the previous domain and surfaces can be thought of as a domain and line segments in the $(r,t)$ plane, respectively. As such, we can parametrize $K_{\sigma_0}$ and $\mathcal{C}_{[\sigma_0,\sigma_1]}$ using the radial variable according to

$$K_{\sigma_0} = \left\{(2\sigma_0 - r, r) \mid 0 \leq r \leq \sigma_0\right\}, \tag{6.32}$$

$$\mathcal{C}_{[\sigma_0,\sigma_1]} = \left\{(r,r) \mid \sigma_0 \leq r \leq \sigma_1\right\}. \tag{6.33}$$

Hence, for functions $v = v(t,r)$, we can slightly abuse the notation and write:

$$\int_{D_{[\sigma_0,\sigma_1]}} v\, dt d\mathbf{x} = 2\pi \int_{D_{[\sigma_0,\sigma_1]}} v\, r dr dt, \tag{6.34}$$

$$\int_{K_{\sigma_0}} v\, dS = 2\sqrt{2}\pi \int_{K_{\sigma_0}} v\, r dr, \tag{6.35}$$

$$\int_{\mathcal{C}_{[\sigma_0,\sigma_1]}} v\, dS = 2\sqrt{2}\pi \int_{\mathcal{C}_{[\sigma_0,\sigma_1]}} v\, r dr. \tag{6.36}$$

If we rely on these notational conventions and integrate the energy differential identity (6.9) over the domain $D_{[\sigma_0,\sigma_1]}$, then we derive

$$\int_{K_{\sigma_1}} \left\{ e(u) - \frac{x^j}{r} p_j(u) \right\} r dr = \int_{\mathcal{C}_{[\sigma_0,\sigma_1]}} \left\{ e(u) + \frac{x^j}{r} p_j(u) \right\} r dr$$

$$+ \int_{K_{\sigma_0}} \left\{ e(u) - \frac{x^j}{r} p_j(u) \right\} r dr,$$

which leads, due to (6.10) and (6.11), to

$$\frac{1}{2} \int_{K_{\sigma_1}} \left\{ (\partial_t u - \partial_r u)^2 + \frac{k^2}{r^2} f^2(u) \right\} r dr$$

$$= \frac{1}{2} \int_{C_{[\sigma_0, \sigma_1]}} \left\{ (\partial_t u + \partial_r u)^2 + \frac{k^2}{r^2} f^2(u) \right\} r dr \qquad (6.37)$$

$$+ \frac{1}{2} \int_{K_{\sigma_0}} \left\{ (\partial_t u - \partial_r u)^2 + \frac{k^2}{r^2} f^2(u) \right\} r dr.$$

Next, we define the associated energy and flux, and their respective radial and angular components, as

$$\mathcal{E}(\sigma) \stackrel{\text{def}}{=} \mathcal{E}_{\text{rad}}(\sigma) + \mathcal{E}_{\text{ang}}(\sigma), \qquad (6.38)$$

$$\mathcal{E}_{\text{rad}}(\sigma) \stackrel{\text{def}}{=} \frac{1}{2} \int_{K_\sigma} \left\{ (\partial_t u - \partial_r u)^2 \right\} r dr, \qquad (6.39)$$

$$\mathcal{E}_{\text{ang}}(\sigma) \stackrel{\text{def}}{=} \frac{1}{2} \int_{K_\sigma} \left\{ \frac{k^2}{r^2} f^2(u) \right\} r dr, \qquad (6.40)$$

$$\mathcal{F}(\sigma_0, \sigma_1) \stackrel{\text{def}}{=} \mathcal{F}_{\text{rad}}(\sigma_0, \sigma_1) + \mathcal{F}_{\text{ang}}(\sigma_0, \sigma_1), \qquad (6.41)$$

$$\mathcal{F}_{\text{rad}}(\sigma_0, \sigma_1) \stackrel{\text{def}}{=} \frac{1}{2} \int_{C_{[\sigma_0, \sigma_1]}} \left\{ (\partial_t u + \partial_r u)^2 \right\} r dr, \qquad (6.42)$$

$$\mathcal{F}_{\text{ang}}(\sigma_0, \sigma_1) \stackrel{\text{def}}{=} \frac{1}{2} \int_{C_{[\sigma_0, \sigma_1]}} \left\{ \frac{k^2}{r^2} f^2(u) \right\} r dr, \qquad (6.43)$$

and notice that both $\mathcal{E}(\sigma)$ and $\mathcal{F}(\sigma_0, \sigma_1)$ are nonnegative. Using these definitions, (6.37) takes the abstract form

$$\mathcal{E}(\sigma_1) = \mathcal{F}(\sigma_0, \sigma_1) + \mathcal{E}(\sigma_0), \qquad (6.44)$$

which implies the next lemma.

**Lemma 6.1.** *For a smooth solution of* (6.1) *in* $D_T \setminus \{(0,0)\}$, *the map* $\sigma \mapsto \mathcal{E}(\sigma)$ *is increasing over* $(0, T]$ *and the limit*

$$\lim_{\tau \to 0} \mathcal{F}(\tau, \sigma) \stackrel{\text{def}}{=} \mathcal{F}(\sigma)$$

*exists and is finite. Moreover, we have* $\mathcal{F}(\sigma) \leq \mathcal{E}(\sigma)$ *and*

$$\lim_{\sigma \to 0} \mathcal{F}(\sigma) = 0. \qquad (6.45)$$

**Proof.** The argument is straightforward, with the monotonicity of $\sigma \mapsto \mathcal{E}(\sigma)$ following immediately from (6.44). Therefore, $\lim_{\sigma \to 0} \mathcal{E}(\sigma)$ exists and is finite and, as a consequence, the same can be inferred about $\lim_{\tau \to 0} \mathcal{F}(\tau, \sigma)$. One can pass then to the limit in (6.44) with $\sigma_0 \to 0$ and derive

$$\mathcal{E}(\sigma_1) = \mathcal{F}(\sigma_1) + \lim_{\sigma \to 0} \mathcal{E}(\sigma),$$

which easily yields the last two claims in the lemma. $\qquad\qquad\square$

**Remark 6.1.** A similar line of reasoning also proves that

$$\lim_{\tau \to 0} \mathcal{F}_{\mathrm{rad}}(\tau, \sigma) \overset{\text{def}}{=} \mathcal{F}_{\mathrm{rad}}(\sigma) \quad \text{and} \quad \lim_{\tau \to 0} \mathcal{F}_{\mathrm{ang}}(\tau, \sigma) \overset{\text{def}}{=} \mathcal{F}_{\mathrm{ang}}(\sigma)$$

exist, are finite, and

$$\lim_{\sigma \to 0} \mathcal{F}_{\mathrm{rad}}(\sigma) = \lim_{\sigma \to 0} \mathcal{F}_{\mathrm{ang}}(\sigma) = 0. \tag{6.46}$$

**Remark 6.2.** At this point, we recognize that regularity at the origin follows from non-concentration of energy, i.e.,

$$\lim_{\sigma \to 0} \mathcal{E}(\sigma) = 0.$$

Howeve, regularity can also be obtained through a weaker condition, namely the non-concentration of the angular part of the energy, $\mathcal{E}_{\mathrm{ang}}(\sigma)$.

This is the moment in the argument where, in addition to the original assumptions on $f$ made in the prologue, we ask for

$$f(u) > 0, \qquad \forall\, u \in (0, U_r), \tag{6.47}$$

where $U_r$ is the first nonzero root of $f$. This extra piece of information will allow us to control the size of the solution through its energy. Due to $f$ being odd, if we define the potential function

$$F(u) \overset{\text{def}}{=} \int_0^u f(s)\, ds, \qquad \forall\, u \in [-U_r, U_r], \tag{6.48}$$

then $F$ is even, increasing on $[0, U_r]$, and we can uniquely identify $U(\sigma) \in [0, U_r]$ satisfying

$$F(U(\sigma)) \overset{\text{def}}{=} \frac{1}{k} \mathcal{E}(\sigma)$$

provided $\mathcal{E}(\sigma) \leq kF(U_r)$. Based on this terminology, we can now state the result which makes precise the control on the size of the solution mentioned above.

**Lemma 6.2.** *Let $u$ be a smooth solution of* (6.1) *in $D_T \setminus \{(0,0)\}$, whose initial data satisfies $u(2T, 0) = 0$ and the energy bound*

$$\frac{1}{k} \mathcal{E}(T) < F(U_r). \tag{6.49}$$

*Then, for all $0 < \sigma \leq T$, the following estimates hold true:*

$$\sup_{K_\sigma} |F(u)| \leq \frac{2}{k} \mathcal{E}_{\mathrm{rad}}^{1/2}(\sigma) \mathcal{E}_{\mathrm{ang}}^{1/2}(\sigma) \leq \frac{1}{k} \mathcal{E}(\sigma) = F(U(\sigma)) < F(U_r), \qquad (6.50)$$

$$\sup_{\mathcal{C}_{[0,\sigma]}} |F(u)| \leq \frac{2}{k} \mathcal{F}_{\mathrm{rad}}^{1/2}(\sigma) \mathcal{F}_{\mathrm{ang}}^{1/2}(\sigma) \leq \frac{1}{k} \mathcal{F}(\sigma). \qquad (6.51)$$

*As a consequence, the size of $u$ is controlled through*

$$\sup_{K_\sigma} |u| \leq U(\sigma) < U_r, \qquad \lim_{\sigma \to 0} \sup_{\mathcal{C}_{[0,\sigma]}} |u| = 0. \qquad (6.52)$$

**Proof.** We start by observing that, due to (6.32) and (6.33), one has

$$\partial \left( u\big|_{K_\sigma} \right) = \left( -\partial_t u + \partial_r u \right)\big|_{K_\sigma}, \qquad \partial \left( u\big|_{\mathcal{C}_{[0,\sigma]}} \right) = \left( \partial_t u + \partial_r u \right)\big|_{\mathcal{C}_{[0,\sigma]}},$$

and, subsequently,

$$\partial \left( F(u)\big|_{K_\sigma} \right) = f(u)\big|_{K_\sigma} \left( -\partial_t u + \partial_r u \right)\big|_{K_\sigma},$$

$$\partial \left( F(u)\big|_{\mathcal{C}_{[0,\sigma]}} \right) = f(u)\big|_{\mathcal{C}_{[0,\sigma]}} \left( \partial_t u + \partial_r u \right)\big|_{\mathcal{C}_{[0,\sigma]}}.$$

Moreover, as $\mathcal{E}(\sigma) < \infty$, it follows that $f(u(2\sigma, 0)) = 0$ and, based on $u(2T, 0) = 0$ and the smoothness of $u$ in $D_T \setminus \{(0,0)\}$, we deduce $u(2\sigma, 0) = 0$. Hence, we can apply the fundamental theorem of calculus and derive

$$
\begin{aligned}
|F(u(2\sigma - r_*, r_*))| &= \left| \int_0^{r_*} f(u)\big|_{K_\sigma} \left( -\partial_t u + \partial_r u \right)\big|_{K_\sigma} dr \right| \\
&\leq \left( \int_0^{r_*} \left\{ \frac{f^2(u)|_{K_\sigma}}{r^2} \right\} r\, dr \right)^{1/2} \\
&\quad \cdot \left( \int_0^{r_*} \left\{ (-\partial_t u + \partial_r u)^2 |_{K_\sigma} \right\} r\, dr \right)^{1/2} \\
&\leq \frac{2}{k} \mathcal{E}_{\mathrm{ang}}^{1/2}(\sigma) \mathcal{E}_{\mathrm{rad}}^{1/2}(\sigma) \leq \frac{1}{k} \mathcal{E}(\sigma),
\end{aligned}
\qquad (6.53)
$$

for all $r_* \in [0, \sigma]$, which proves (6.50).

An almost identical argument produces

$$\left| F(u(r_*, r_*)) - F(u\big|_{\mathcal{C}_{[0,r_*]}}(0,0)) \right| \leq \frac{2}{k} \mathcal{F}_{\mathrm{ang}}^{1/2}(r_*) \mathcal{F}_{\mathrm{rad}}^{1/2}(r_*) \leq \frac{1}{k} \mathcal{F}(r_*). \qquad (6.54)$$

To finish the argument for (6.51), it suffices to prove that $u\big|_{\mathcal{C}_{[0,r_*]}}(0,0) = 0$. Since $\mathcal{F}(r_*)$ is finite, we must have that $f(u\big|_{\mathcal{C}_{[0,r_*]}}(0,0)) = 0$, which implies that $u\big|_{C_{[0,r_*]}}(0,0)$ is either 0 or $U_r$. For the latter case, using (6.49), we pick $\epsilon$ sufficiently small such that

$$F(U_r) - \frac{1}{k}\mathcal{E}(T) > \epsilon. \tag{6.55}$$

Furthermore, due to (6.45), we can find $r$ satisfying $\mathcal{F}(r) < k\epsilon$. On one hand, if we rely on (6.54), then we infer

$$|F(u(r,r)) - F(U_r)| \leq \frac{1}{k}\mathcal{F}(r) < \epsilon. \tag{6.56}$$

On the other hand, based on $(r,r) \in K_r$ and (6.50), we obtain

$$|F(u(r,r))| \leq \sup_{K_r} F(u) \leq (1/k)\mathcal{E}(T) < F(U_r),$$

which, coupled with (6.55), yields

$$F(U_r) - F(u(r,r)) > \epsilon.$$

Obviously, this is in contradiction to (6.56) and, hence, the proof of the lemma is concluded. $\qquad\square$

**Remark 6.3.** We can further refine (6.52) (i.e., the bound on the size of $u$) via the following line of reasoning. We choose $(t_*, r_*) \in K_\sigma$ such that $|u(t_*, r_*)| = \sup_{K_\sigma} |u|$ and, using the second line in (6.53), we deduce

$$k|F(u(t_*, r_*))| \leq \frac{1}{2} \int\limits_{0 \leq r \leq r_*} \left\{ (\partial_t u - \partial_r u)^2 + \frac{k^2}{r^2} f^2(u) \right\}\bigg|_{K_\sigma} r\,dr.$$

Similarly, one derives

$$k|F(u(t_*, r_*)) - F(u(\sigma, \sigma))| \leq \frac{1}{2} \int\limits_{r_* \leq r \leq \sigma} \left\{ (\partial_t u - \partial_r u)^2 + \frac{k^2}{r^2} f^2(u) \right\}\bigg|_{K_\sigma} r\,dr,$$

and, by adding it to the previous estimate, we infer

$$|F(u(t_*, r_*))| \leq \frac{1}{2}\left\{ |F(u(\sigma, \sigma))| + \frac{1}{k}\mathcal{E}(\sigma) \right\}.$$

If we rely now on (6.51) and (6.45), then we obtain the refined bound

$$\sup_{K_\sigma} |u| \leq U_*(\sigma) + o(\sigma), \tag{6.57}$$

where $U_*(\sigma)$ is uniquely defined by

$$F(U_*(\sigma)) \stackrel{\text{def}}{=} \frac{1}{2k}\mathcal{E}(\sigma). \tag{6.58}$$

### 6.2.3 Conditional regularity

After these initial energy arguments, we now turn to Grillakis's main conditional regularity result, which requires one more assumption on $f$.

**Theorem 6.1.** *Under the assumptions of Lemma 6.2, if $f$ additionally satisfies*

$$f(u)\big(f(u) + f'(u)u\big) > 0, \qquad \forall u \in (0, U_*(T)], \qquad (6.59)$$

*where $U_*$ is the one defined by (6.58), then $u$ is regular in the set $D_T$.*

**Remark 6.4.** This theorem claims that the origin cannot be a blow-up point for $u$. In order to see an example of what the assumptions actually imply, let us consider the case of the target manifold being $\mathbb{S}^2$, i.e., $f(u) = \sin(u)$, and $k = 1$. Hence, $U_r = \pi$, $F(u) = 1 - \cos(u)$, and the energy threshold in (6.49) is $kF(U_r) = 2$. In fact, the bound (6.49) is optimal, in the sense that the next chapter on collapsing solutions shows that there exist equivariant wave maps that blow up in finite time, whose initial energy[3] is just above 2. Furthermore, for this case, we derive from (6.58) that $U_*(T) < \pi/2$ and the assumption (6.59) reads as

$$\sin(u)(\sin(u) + \cos(u)u) > 0, \qquad \forall u \in (0, U_*(T)],$$

which is obviously true.

The proof for this theorem combines two key a priori estimates with standard regularity theory based on Strichartz bounds. Given the well-established nature of the latter, for which we provide nevertheless some final comments, our focus will be on the arguments for the a priori estimates.

#### 6.2.3.1 Christodoulou-Zadeh exterior bounds

In [29], Christodoulou and Zadeh made the very important observation that the energy cannot concentrate[4] in any region of the type $\{\lambda_0 t \le r \le t\}$, with $\lambda_0 \in (0, 1)$. Here, we will build off this fact and show that, for fixed $\lambda_0 \in (0, 1)$ and $\epsilon > 0$, there exists $\sigma > 0$ sufficiently small such that

$$\sup_{D_{\lambda_0, \sigma}} |u| < \epsilon, \qquad (6.60)$$

---

[3]The energy used in this section is one half of the energy associated to collapsing solutions in Chapter 7.

[4]This remark is relevant because it points to the fact that energy may concentrate only near the central line $r = 0$, in a very narrow region, which is, of course, in agreement with blow-up results.

where

$$D_{\lambda_0,\sigma} \overset{\text{def}}{=} \left\{ (t,r) \mid t + r \leq 2\sigma, \ 0 < \lambda_0 t \leq r \leq t \right\}. \tag{6.61}$$

This is one of the two a priori estimates we need in the proof of Theorem 6.1.

We start by introducing

$$w \overset{\text{def}}{=} r^{1/2}u, \qquad \partial_+ \overset{\text{def}}{=} \partial_t + \partial_r, \qquad \partial_- \overset{\text{def}}{=} \partial_t - \partial_r, \tag{6.62}$$

and, due to the following elementary fact,

$$\partial_r^2 \left( r^{1/2}u \right) = r^{1/2}\Delta_r u - \frac{1}{4r^{3/2}}u,$$

we can write the original equation (6.1) for $u$ in the alternative form

$$\partial_+\partial_- w + \frac{g(u)}{r^2}w = 0. \tag{6.63}$$

In the above,

$$g(u) \overset{\text{def}}{=} k^2 \frac{f(u)f'(u)}{u} - \frac{1}{4},$$

and the assumptions made on $f$ easily imply that $g$ is a bounded function.

The way we are going to prove (6.60) is by integrating (6.63) over line segments contained in regions of the type described by (6.61). For this purpose, we proceed to set up the necessary terminology and notation. If $(t_0, r_0) \in D_{\lambda_0, T}$, let us denote the characteristic lines passing through $(t_0, r_0)$ by

$$l_+(t_0, r_0) \overset{\text{def}}{=} \left\{ (t,r) \mid t = t_0 + \sigma, \ r = r_0 + \sigma \right\}, \tag{6.64}$$

$$l_-(t_0, r_0) \overset{\text{def}}{=} \left\{ (t,r) \mid t = t_0 - \tau, \ r = r_0 + \tau \right\}, \tag{6.65}$$

and also define

$$\tau_0 \overset{\text{def}}{=} \frac{t_0 - r_0}{2}, \qquad \sigma_0 \overset{\text{def}}{=} \frac{t_0 + r_0}{2}.$$

We are going to pay particular attention to the line segment $l_-^{\lambda_0}(\sigma_0)$, which is the intersection of $l_-(t_0, r_0)$ and $D_{\lambda_0, T}$. It can be parametrized as follows

$$l_-^{\lambda_0}(\sigma_0) \overset{\text{def}}{=} \left\{ (t,r) \mid t = \sigma_0 + \tau, \ r = \sigma_0 - \tau, \ 0 \leq \tau \leq \frac{1 - \lambda_0}{1 + \lambda_0}\sigma_0 \right\}, \tag{6.66}$$

and, associated to it, we define two quantities:

$$m_{\lambda_0}(\sigma_0) \overset{\text{def}}{=} \sup_{l_-^{\lambda_0}(\sigma_0)} |\partial_- w|, \tag{6.67}$$

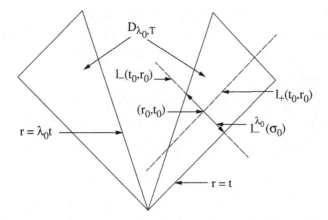

Fig. 6.2   The line segment $l^{\lambda_0}_-(\sigma_0)$ in the region $D_{\lambda_0,T}$.

$$\mathcal{E}_{\lambda_0}(\sigma_0) \overset{\text{def}}{=} \frac{1}{2} \int\limits_{l^{\lambda_0}_-(\sigma_0)} \left\{ (\partial_t u - \partial_r u)^2 + \frac{k^2}{r^2} f^2(u) \right\} r\,dr. \qquad (6.68)$$

In the next lemma, we are able to show a certain type of control over the quantity $m_{\lambda_0}(\sigma_0)$, which ultimately leads to both the desired a priori bound (6.60) and the smallness of $\mathcal{E}_{\lambda_0}(\sigma_0)$ close to the origin.

**Lemma 6.3.** *Under the assumptions of Lemma 6.2, for a fixed* $\lambda_0 \in (0,1)$, *the following estimates hold true:*

$$m_{\lambda_0}(\sigma_0) \le C(\lambda_0) \frac{\mathcal{F}^{1/2}_{\text{ang}}(\sigma_1)}{\sigma_0^{1/2}} + m_{\lambda_0}(\sigma_1), \qquad (6.69)$$

$$\sup_{l^{\lambda_0}_-(\sigma_0)} |u| \le C(\lambda_0) \left( |u(\sigma_0,\sigma_0)| + \mathcal{F}^{1/2}_{\text{ang}}(\sigma_1) + m_{\lambda_0}(\sigma_1)\sigma_0^{1/2} \right), \qquad (6.70)$$

$$\mathcal{E}_{\lambda_0}(\sigma_0) \le C(\lambda_0) \left( u^2(\sigma_0,\sigma_0) + \mathcal{F}_{\text{ang}}(\sigma_1) + m^2_{\lambda_0}(\sigma_1)\sigma_0 \right), \qquad (6.71)$$

*for all* $0 < \sigma_0 < \sigma_1 \le T$, *where* $C = C(\lambda_0)$ *is a constant depending strictly on* $\lambda_0$.

*As a consequence, if* $\epsilon > 0$ *is arbitrary, then there exists* $\sigma_2$ *sufficiently small such that*

$$\sup_{D_{\lambda_0,\sigma_2}} |u| \le C(\lambda_0)\epsilon, \qquad (6.72)$$

$$\sup_{\sigma \in (0,\sigma_2]} \mathcal{E}_{\lambda_0}(\sigma) \le C(\lambda_0)\epsilon. \qquad (6.73)$$

**Proof.** We start by integrating (6.63) over the line segment uniting the points $(t_0, r_0)$ and $(\sigma_1 + \tau_0, \sigma_1 - \tau_0) \in K_{\sigma_1}$, which is parametrized by

$$\{(t_0 + \sigma, r_0 + \sigma) \mid 0 \leq \sigma \leq \sigma_1 - \sigma_0\}.$$

As we recognize that

$$\partial_\sigma \{h(t_0 + \sigma, r_0 + \sigma)\} = (\partial_+ h)(t_0 + \sigma, r_0 + \sigma),$$

we can write

$$(\partial_- w)(t_0, r_0) = \int_0^{\sigma_1 - \sigma_0} \left\{ \frac{(g(u)w)(t_0 + \sigma, r_0 + \sigma)}{(r_0 + \sigma)^2} \right\} d\sigma$$
$$+ (\partial_- w)(\sigma_1 + \tau_0, \sigma_1 - \tau_0).$$

Secondly, for fixed $\sigma \in [0, \sigma_1 - \sigma_0]$, we simply integrate $\partial_- w$ over the line segment connecting $(t_0 + \sigma, r_0 + \sigma)$ to $(\sigma_0 + \sigma, \sigma_0 + \sigma) \in \mathcal{C}_{[\sigma_0, \sigma_1]}$ that has the parametrization

$$\{(t_0 + \sigma - \tau, r_0 + \sigma + \tau) \mid 0 \leq \tau \leq \tau_0\}.$$

Due to

$$\partial_\tau \{h(t_0 + \sigma - \tau, r_0 + \sigma + \tau)\} = -(\partial_- h)(t_0 + \sigma - \tau, r_0 + \sigma + \tau),$$

we infer

$$w(t_0 + \sigma, r_0 + \sigma) = \int_0^{\tau_0} (\partial_- w)(t_0 + \sigma - \tau, r_0 + \sigma + \tau) d\tau$$
$$+ w(\sigma_0 + \sigma, \sigma_0 + \sigma)$$

If we substitute this equation into the previous one, take absolute values, and use the fact that $g$ is bounded, then we deduce

$$\left| (\partial_- w)(t_0, r_0) \right| \leq \left| (\partial_- w)(\sigma_1 + \tau_0, \sigma_1 - \tau_0) \right|$$
$$+ C \int_0^{\sigma_1 - \sigma_0} \frac{|w(\sigma_0 + \sigma, \sigma_0 + \sigma)|}{(r_0 + \sigma)^2} d\sigma \qquad (6.74)$$
$$+ C \int_0^{\sigma_1 - \sigma_0} \left\{ \int_0^{\tau_0} \frac{|(\partial_- w)(t_0 + \sigma - \tau, r_0 + \sigma + \tau)|}{(r_0 + \sigma)^2} d\tau \right\} d\sigma.$$

Next, we work on the first integral on the right-hand side and, recalling that $w = r^{1/2} u$, an application of the Cauchy-Schwarz inequality yields

$$\int_0^{\sigma_1 - \sigma_0} \frac{|w(\sigma_0 + \sigma, \sigma_0 + \sigma)|}{(r_0 + \sigma)^2} d\sigma \leq \left( \int_{\mathcal{C}_{[\sigma_0, \sigma_1]}} \left\{ \frac{u^2}{r^2} \right\} r dr \right)^{1/2}$$
$$\cdot \left( \int_0^{\sigma_1 - \sigma_0} \frac{(\sigma_0 + \sigma)^2}{(r_0 + \sigma)^4} d\sigma \right)^{1/2}.$$

For the integral on $\mathcal{C}_{[\sigma_0,\sigma_1]}$, we argue that a foliation of $D_{[\sigma_0,\sigma_1]}$ by $K_\sigma$, with $\sigma \in [\sigma_0, \sigma_1]$, and (6.52) imply

$$\sup_{D_{[\sigma_0,\sigma_1]}} |u| = \sup_{\sigma \in [\sigma_0,\sigma_1]} \sup_{K_\sigma} |u| \leq \sup_{\sigma \in [\sigma_0,\sigma_1]} U(\sigma) = U(\sigma_1) < U_r,$$

and, based on the assumptions on $f$, there exists $C > 0$ such that

$$|u| \leq C|f(u)| \quad \text{on} \quad D_{[\sigma_0,\sigma_1]}. \tag{6.75}$$

Hence,

$$\int_{\mathcal{C}_{[\sigma_0,\sigma_1]}} \left\{ \frac{u^2}{r^2} \right\} r dr \leq C \mathcal{F}_{\text{ang}}(\sigma_1).$$

For the second integral term above, due to $t_0 + r_0 = 2\sigma_0$ and $\lambda_0 t_0 \leq r_0 \leq t_0$, we have

$$\frac{2\lambda_0}{1 + \lambda_0} \sigma_0 \leq r_0 \leq \sigma_0,$$

and, as a consequence,

$$\int_0^{\sigma_1 - \sigma_0} \frac{(\sigma_0 + \sigma)^2}{(r_0 + \sigma)^4} d\sigma \leq C(\lambda_0) \int_0^{\sigma_1 - \sigma_0} \frac{1}{(r_0 + \sigma)^2} d\sigma \leq \frac{C(\lambda_0)}{\sigma_0}.$$

Therefore, by combining the last three integral bounds, we derive

$$\int_0^{\sigma_1 - \sigma_0} \frac{|w(\sigma_0 + \sigma, \sigma_0 + \sigma)|}{(r_0 + \sigma)^2} d\sigma \leq C(\lambda_0) \frac{\mathcal{F}_{\text{ang}}^{1/2}(\sigma_1)}{\sigma_0^{1/2}}. \tag{6.76}$$

Using this estimate and the definition (6.67) for $m_{\lambda_0}$ in the context of (6.74), we obtain

$$m_{\lambda_0}(\sigma_0) \leq m_{\lambda_0}(\sigma_1) + C(\lambda_0) \frac{\mathcal{F}_{\text{ang}}^{1/2}(\sigma_1)}{\sigma_0^{1/2}}$$

$$+ \int_0^{\sigma_1 - \sigma_0} \frac{\tau_0}{(r_0 + \sigma)^2} m_{\lambda_0}(\sigma_0 + \sigma) d\sigma. \tag{6.77}$$

Then, the desired bound, (6.69), follows by virtue of Grönwall's inequality, as this can be applied due to

$$\int_0^{T - \sigma_0} \frac{\tau_0}{(r_0 + \sigma)^2} d\sigma \leq \frac{\tau_0}{r_0} \leq C(\lambda_0).$$

In arguing for (6.70), we start by integrating $\partial_- w$ over portions of $l_-^{\lambda_0}(\sigma_0)$ and, taking advantage of (6.66), we infer

$$\sup_{l_-^{\lambda_0}(\sigma_0)} |w| \leq |w(\sigma_0, \sigma_0)| + \frac{1 - \lambda_0}{1 + \lambda_0} \sigma_0 \, m_{\lambda_0}(\sigma_0).$$

Moreover, it is easy to notice that

$$\min_{l_-^{\lambda_0}(\sigma_0)} r = \frac{2\lambda_0}{1+\lambda_0}\sigma_0.$$

Thus, due to $w = r^{1/2}u$, we have

$$\sup_{l_-^{\lambda_0}(\sigma_0)} |u| \le C(\lambda_0)\left(|u(\sigma_0,\sigma_0)| + \sigma_0^{1/2}m_{\lambda_0}(\sigma_0)\right),$$

and (6.70) follows immediately by invoking (6.69).

In what concerns (6.71), using the assumptions on $f$ and (6.52), we first deduce that

$$|f(u)| \le C|u| \quad \text{on } l_-^{\lambda_0}(\sigma_0).$$

Moreover, the parametrization (6.66) and the fact that $w = r^{1/2}u$ imply

$$\int_{l_-^{\lambda_0}(\sigma_0)} \{(\partial_t u - \partial_r u)^2\}\, r\, dr \le C \le \int_{l_-^{\lambda_0}(\sigma_0)} (\partial_- w)^2\, d\tau.$$

Hence, by combining the last two estimates with (6.68), we deduce

$$\mathcal{E}_{\lambda_0}(\sigma_0) \le C\left(\int_{l_-^{\lambda_0}(\sigma_0)} (\partial_- w)^2\, d\tau + \int_{l_-^{\lambda_0}(\sigma_0)} \frac{u^2}{r}\, dr\right).$$

The bound we are looking for, (6.71), is then obtained from the previous estimate by also factoring in (6.69) and (6.70).

Finally, due to (6.46) and (6.52), for a fixed $\epsilon > 0$, we can first choose $0 < \sigma_1 \le T$ such that $\mathcal{F}_{ang}(\sigma_1) < \epsilon^2$ and then $0 < \sigma_2 < \sigma_1$ satisfying

$$\sup_{\mathcal{C}_{[0,\sigma_2]}} |u| + m_{\lambda_0}(\sigma_1)\sigma_2^{1/2} < \epsilon.$$

It follows that both (6.72) and (6.73) hold true as consequences of (6.70) and (6.71), respectively.                                    □

### 6.2.3.2   *The Morawetz-type estimate*

Here, we follow up on previous computations done for the vector field prescribed by (6.12) and we derive from them an important integral identity. This serves as one of the main ingredients in proving a certain Morawetz-type estimate, which, coupled with the Christodoulou-Zadeh bounds, helps us in showing that $u$ is regular at the origin.

We proceed by integrating (6.20) over the region $D_{[\sigma_0,\sigma_1]}$, with $0 < \sigma_0 < \sigma_1 \leq T$, and, based on (6.34)-(6.36), we infer

$$
\int_{K_{\sigma_1}} \left\{ P^0 + \frac{x_j}{r} P^j \right\} r\,dr + \int_{D_{[\sigma_0,\sigma_1]}} \{S\}\, r\,dr\,dt
$$
$$
= \int_{C_{[\sigma_0,\sigma_1]}} \left\{ P^0 - \frac{x_j}{r} P^j \right\} r\,dr + \int_{K_{\sigma_0}} \left\{ P^0 + \frac{x_j}{r} P^j \right\} r\,dr. \tag{6.78}
$$

Using the formulas (6.22) and (6.23), we deduce

$$
P^0 + \frac{x_j}{r} P^j = \frac{1-\lambda^{1-\alpha}}{2}(\partial_t u - \partial_r u)^2 + \frac{1+\lambda^{1-\alpha}}{2}\frac{k^2}{r^2}f^2(u)
$$
$$
+ (\partial_t - \partial_r)\left\{ \frac{bu^2}{4} \right\} + (-\partial_t b + \partial_r b)\frac{u^2}{2},
$$

$$
P^0 - \frac{x_j}{r} P^j = \frac{1+\lambda^{1-\alpha}}{2}(\partial_t u + \partial_r u)^2 + \frac{1-\lambda^{1-\alpha}}{2}\frac{k^2}{r^2}f^2(u)
$$
$$
+ (\partial_t + \partial_r)\left\{ \frac{bu^2}{4} \right\} - (\partial_t b + \partial_r b)\frac{u^2}{2}.
$$

We can integrate by parts a couple of the above terms, according to

$$
\int_{K_\sigma} \left\{ (\partial_t - \partial_r)\frac{bu^2}{4} \right\} r\,dr = \int_{K_\sigma} \left\{ \frac{bu^2}{4r} \right\} r\,dr - \left\{ \frac{rbu^2}{4} \right\}\Big|_{r=\sigma}^{r=0},
$$
$$
\int_{C_{[\sigma_0,\sigma_1]}} \left\{ (\partial_t + \partial_r)\frac{bu^2}{4} \right\} r\,dr = - \int_{C_{[\sigma_0,\sigma_1]}} \left\{ \frac{bu^2}{4r} \right\} r\,dr + \left\{ \frac{rbu^2}{4} \right\}\Big|_{r=\sigma_0}^{r=\sigma_1}.
$$

Hence, taking into account the formulas (6.14) and (6.18) and that $\lambda = 1$ on $C_{[\sigma_0,\sigma_1]}$, we can rewrite (6.78) as

$$
\mathcal{E}^{\mathrm{id}}(\sigma_1) + \int_{D_{[\sigma_0,\sigma_1]}} \{S\}\, r\,dr\,dt = \mathcal{E}^{\mathrm{id}}(\sigma_0) + \mathcal{F}^{\mathrm{id}}(\sigma_0,\sigma_1), \tag{6.79}
$$

where

$$
\mathcal{E}^{\mathrm{id}}(\sigma) \overset{\mathrm{def}}{=} \int_{K_\sigma} \left\{ \frac{1-\lambda^{1-\alpha}}{2}(\partial_t u - \partial_r u)^2 + \frac{1+\lambda^{1-\alpha}}{2}\frac{k^2}{r^2}f^2(u) + \frac{\tilde{b}}{4r^2}u^2 \right\} r\,dr,
$$

$$
\mathcal{F}^{\mathrm{id}}(\sigma_0,\sigma_1) \overset{\mathrm{def}}{=} \int_{C_{[\sigma_0,\sigma_1]}} \left\{ (\partial_t u + \partial_r u)^2 + \frac{1}{4r^2}u^2 \right\} r\,dr,
$$

and

$$\tilde{b} \stackrel{\text{def}}{=} (\alpha - 2\alpha^2)\lambda^{1-\alpha} + (1 - \alpha)(\lambda + 2\lambda^2 + 2\alpha\lambda^{2-\alpha}).$$

One notices immediately that $\tilde{b}$ is uniformly bounded on $D_{[\sigma_0,\sigma_1]}$ and, moreover, is nonnegative if $0 \le \alpha \le 1/2$. If we combine this fact with (6.75) and (6.38)-(6.43), then we obtain

$$\left|\mathcal{E}^{\text{id}}(\sigma)\right| \le C\mathcal{E}(\sigma) \quad \text{and} \quad \mathcal{F}^{\text{id}}(\sigma_0,\sigma_1) \le C\mathcal{F}(\sigma_0,\sigma_1), \tag{6.80}$$

for some universal constant $C$. Therefore, due to (6.44) and (6.79), we derive

$$\left| \int_{D_{[\sigma_0,\sigma_1]}} \{S\} \, r dr dt \right| \le C\mathcal{E}(\sigma_1). \tag{6.81}$$

Our next goal is to show that $S$ is nonnegative on $D_{[\sigma_0,\sigma_1]}$ if we choose both $\sigma_1$ and $\alpha$ to be sufficiently small. Based on (6.24), (6.27), and (6.28), this would follow if we can prove the same claim about $S^{(1)}$. In fact, we argue that, for $\sigma_1$ and $\alpha$ small enough, there exists $\delta > 0$ such that

$$S^{(1)} \ge \frac{\delta}{tr^2}f^2(u), \quad \text{on} \quad D_{\sigma_1}. \tag{6.82}$$

For this purpose, we let $\lambda_0 \in (0,1)$ be arbitrary and we split $D_{\sigma_1}$ according to

$$D_{\sigma_1} = (D_{\sigma_1} \setminus D_{\lambda_0,\sigma_1}) \cup (D_{\sigma_1} \cap D_{\lambda_0,\sigma_1}).$$

We analyze the interior region first, for which we discuss initially the subregion where the size of $u$ is small enough. Due to $f(0) = 0$ and $f'(0) = 1$, we can find $\delta_1 > 0$ such that

$$f(u)f'(u)u \ge \delta_1 f^2(u),$$

and, as a result, we claim (6.82) directly from (6.26) by choosing $\alpha$ to be sufficiently small. On the other hand, in a subregion where for a certain $\overline{u} > 0$ we have $|u| \ge \overline{u}$, we can apply (6.59) and derive the existence of a $\delta_2 > 0$ satisfying

$$\frac{f(u) + f'(u)u}{f^2(u)} \ge \delta_2.$$

Hence, we obtain yet again (6.82) if we rely on $\lambda \le \lambda_0$ (which holds as we are in $D_{\sigma_1} \setminus D_{\lambda_0,\sigma_1}$), choose $\lambda_0$ small enough, and, eventually, readjust $\alpha$ to be sufficiently small. Finally, in the exterior region, by picking $\sigma_1$ small enough, we can apply the Christodoulou-Zadeh bound (6.72) and claim that we are in the regime when the size of $u$ is small and (6.82) follows as above.

Thus, we can now work under the assumption that $S$ is nonnegative and we are justified in taking $\sigma_0 \to 0$ in (6.81) to deduce

$$\int_{D_{\sigma_1}} \{S\} r\, dr\, dt \leq C\mathcal{E}(\sigma_1), \tag{6.83}$$

when both $\sigma_1$ and $\alpha$ are sufficiently small. As we shall see, this estimate provides a certain type of weak decay for portions of the energy. It implies

$$\int_{D_{\sigma_1}} \{S^{(0)} + S^{(1)}\} r\, dr\, dt < \infty.$$

Next, we foliate $D_{\sigma_1}$ with $K_\sigma$ ($\sigma \in [0, \sigma_1]$) and notice that $\sigma \leq t \leq 2\sigma$ on $K_\sigma$. Hence, using the previous bound, (6.28), and (6.82), we infer

$$\int_0^{\sigma_1} \frac{1}{\sigma} \left\{ \int_{K_\sigma} \left\{ (1 - \lambda^{1-\alpha})(\partial_t u - \partial_r u)^2 + \frac{\delta}{r^2} f^2(u) \right\} r\, dr \right\} d\sigma < \infty,$$

which forces the existence of a sequence $\sigma^{(n)} \to 0$ such that

$$\lim_{n \to \infty} \int_{K_{\sigma^{(n)}}} \left\{ (1 - \lambda^{1-\alpha})(\partial_t u - \partial_r u)^2 + \frac{\delta}{r^2} f^2(u) \right\} r\, dr = 0.$$

By looking at the formula for $\mathcal{E}^{\mathrm{id}}$ and relying on (6.75) and the properties for $\tilde{b}$, the previous limit allows us to write

$$\lim_{n \to \infty} \mathcal{E}^{\mathrm{id}}(\sigma^{(n)}) = 0.$$

As a consequence, we can revisit (6.79) and use $\sigma^{(n)}$ for $\sigma_0$, followed by taking $n \to \infty$. The end result is the identity

$$\mathcal{E}^{\mathrm{id}}(\sigma_1) + \int_{D_{\sigma_1}} \{S\} r\, dr\, dt = \lim_{n \to \infty} \mathcal{F}^{\mathrm{id}}(\sigma^{(n)}, \sigma_1),$$

which, coupled with (6.80) and (6.45), yields

$$\lim_{\sigma \to 0} \mathcal{E}^{\mathrm{id}}(\sigma) + \int_{D_\sigma} \{S\} r\, dr\, dt = 0.$$

As both $\mathcal{E}^{\mathrm{id}}$ and $S$ are nonnegative, we derive $\lim_{\sigma_1 \to 0} \mathcal{E}^{\mathrm{id}}(\sigma_1) = 0$. It is easy to see that, for $0 \leq \alpha \leq 1/2$, we have

$$\mathcal{E}^{\mathrm{id}}(\sigma) \geq \mathcal{E}_{\mathrm{ang}}(\sigma).$$

This implies

$$\lim_{\sigma \to 0} \mathcal{E}_{\mathrm{ang}}(\sigma) = 0$$

and, due to (6.50), we conclude that $u$ is continuous at the origin. At this point, we claim that continuity implies regularity of the solutions.

One further refinement for the estimates already proven comes from the fact that, with the information obtained so far, we can choose $\alpha = 1$ and run the previous scheme again to deduce

$$\lim_{\sigma \to 0} \int_{D_\sigma} \{S\}\, r dr dt = 0.$$

In this case, $S^{(0)} = S^{(2)} = 0$ while

$$S^{(1)} = \frac{k^2}{2r^3} \left( f(u)f'(u)u + f^2(u) - \frac{1}{2k^2} u^2 \right).$$

Based on the smallness of $u$ near the tip of the characteristic cone and the assumptions on $f$, we finally derive

$$\lim_{\sigma \to 0} \int_{D_\sigma} \left\{ \frac{f^2(u)}{r^3} \right\} r dr t = 0. \tag{6.84}$$

### 6.2.3.3  *Final remarks on obtaining regularity*

We would like to finish with some comments regarding the general method[5] used in proving regularity. For our problem, this involves combining the estimate (6.84) with local Strichartz bounds. These are obtained by multiplying the solution with a cutoff function in space and time, whose gradient is timelike (i.e., the level sets of the cutoff function are spacelike). Such a Strichartz estimate is an $L^6 - L^{6/5}$ inequality, which is applied to the original equation (6.1) differentiated in time. The crucial fact is that the smallness of the integral in (6.84) allows one to balance the terms in the local Strichartz bound. Since these types of argument are by now standard in the literature, we leave the clarification of the above heuristics for the diligent reader.

---

[5]The essence of this method is that, if a certain portion of the energy does not concentrate, then one argues for global regularity by relying on estimates for the linear equation in appropriate spaces. Figuring out what these spaces are is sometimes a challenging problem.

## 6.3 Small data global well-posedness for Skyrme and Skyrme-like models

The common theme for the three arguments leading to the results of this section is a contraction mapping approach, which is done in function spaces motivated by scaling heuristics. Moreover, in the analyses for the Skyrme and Faddeev problems, a prominent role is played by Tataru-type radial space, which are used in handling null form nonlinearities.

### 6.3.1 *Preliminaries*

We start by further specializing the assumptions made at the beginning of the chapter to the case when

$$N = \mathbb{S}^n \quad \text{and} \quad \Omega(\omega) = \omega.$$

For the Adkins-Nappi model, the physically consistent "equivariant" hypothesis on the gauge field $A : \mathbb{R}^{3+1} \to \mathbb{R}^4$ is[6]

$$A(t, r, \omega) = (V(t, r), 0, 0, 0).$$

Using this ansatz for the actions describing the three theories and setting $\Lambda = 1$ in (1.7) and $c = 1$ in (1.13), careful computations show that the resulting variational equations are[7]

$$
\left(1 + \frac{2\sin^2 u}{r^2}\right)(u_{tt} - u_{rr}) - \frac{2}{r}u_r
$$
$$
+ \frac{\sin 2u}{r^2}\left(1 + u_t^2 - u_r^2 + \frac{\sin^2 u}{r^2}\right) = 0 \tag{6.85}
$$

for the Skyrme model,

$$
\left(1 + \frac{\sin^2 u}{r^2}\right)(u_{tt} - u_{rr}) - \left(1 - \frac{\sin^2 u}{r^2}\right)\frac{u_r}{r}
$$
$$
+ \frac{\sin 2u}{2r^2}\left(u_t^2 - u_r^2 + 1\right) = 0 \tag{6.86}
$$

for the 2 + 1-dimensional Faddeev model, and

$$
\begin{cases}
u_{tt} - u_{rr} - \frac{2}{r}u_r + \frac{\sin 2u}{r^2} + \frac{(u - \sin u \cos u)(1 - \cos 2u)}{r^4} = 0 \\
\partial_r(r^2 V_r) + 2\sin^2 u \, u_r = 0
\end{cases} \tag{6.87}
$$

---

[6]This is to say that the only relevant components of the associated electromagnetic tensor are the electrical ones (i.e., $F_{0i}$ with $1 \leq i \leq 3$).

[7]In this context, we switched the derivative notation from $\partial_t u$ and $\partial_r u$ to $u_t$ and $u_r$, respectively, as it is more user-friendly.

for the Adkins-Nappi model. In the last system, the two equations decouple and so, for all three models, the only relevant equation is the one for the azimuthal angle $u$.

An important role in the analysis is played by the respective conserved energies, which are given by

$$\mathcal{E}[u](t) = \int_0^\infty \left[ \left(1 + \frac{2\sin^2 u}{r^2}\right) \frac{u_t^2 + u_r^2}{2} + \frac{\sin^2 u}{r^2} + \frac{\sin^4 u}{2r^4} \right] r^2 dr \quad (6.88)$$

for the Skyrme model,

$$\mathcal{E}[u](t) = \int_0^\infty \left[ \left(1 + \frac{\sin^2 u}{r^2}\right) \frac{u_t^2 + u_r^2}{2} + \frac{\sin^2 u}{2r^2} \right] r \, dr \quad (6.89)$$

for the Faddeev model, and

$$\mathcal{E}[u](t) = \int_0^\infty \left[ \frac{u_t^2 + u_r^2}{2} + \frac{\sin^2 u}{r^2} + \frac{(u - \sin u \cos u)^2}{2r^4} \right] r^2 dr \quad (6.90)$$

for the Adkins-Nappi model.

### 6.3.1.1  *Small energy heuristics*

If we want to work with finite energy maps, then it is easy to see that we need to assume

$$\sin u(t, 0) = \sin u(t, \infty) = 0.$$

Under the equivariant assumption, direct calculations based on (1.3) and (1.4) yield

$$Q(\phi)(t) = \frac{\cos u(t, 0) - \cos u(t, \infty)}{2} \quad (6.91)$$

and

$$Q(\phi)(t) = \frac{(2u(t, \infty) - \sin 2u(t, \infty)) - (2u(t, 0) - \sin 2u(t, 0))}{2\pi} \quad (6.92)$$

for $n = 2$ and $n = 3$, respectively. All of the results discussed in this section concern maps with zero charge, i.e.,

$$u(t, 0) = u(t, \infty) = 0.$$

In fact, as the next lemma shows, for all the three models, a small energy regime is possible only for maps like the previous ones.

**Lemma 6.4.** *If $u$ is the azimuthal angle associated with an equivariant ansatz for the Skyrme, $2+1$-dimensional Faddeev, or Adkins-Nappi models, satsifying $u(t, 0) = 0$, then the following a priori bound holds:*

$$\|u\|_{L^\infty_{t,x}} \leq C(\mathcal{E}[u]), \quad (6.93)$$

*with $C(a) \to 0$ as $a \to 0$.*

**Proof.** Consider the functionals

$$I_1(z) = 2z - \sin 2z,$$

$$I_2(z) = \int_0^z |\sin w| \, dw,$$

and

$$I_3(z) = z^2 - \sin^2 z,$$

for the Skyrme, $2 + 1$-dimensional Faddeev, and Adkins-Nappi models, respectively. It is easy to see that, for all $1 \le k \le 3$,

$$I_k(0) = 0, \qquad |I_k(z)| > 0 \, (z \neq 0), \qquad \lim_{|z| \to \infty} |I_k(z)| = \infty.$$

Hence, the bound (6.93) is true if we prove

$$|I_k(u(t, r))| \lesssim \mathcal{E}[u](t)$$

for each respective model. But this is immediate by an application of the fundamental theorem of calculus and of the Cauchy-Schwarz inequality. We only show the argument for $I_1$, as the other two follow along the same lines.

$$|I_1(u(t, r))| = |I_1(u(t, r)) - I_1(u(t, 0))|$$

$$= \left| \int_0^r 4 \sin^2 u(t, s) \, u_r(t, s) \, ds \right|$$

$$\lesssim \left( \int_0^r \sin^2 u(t, s) \, u_r^2(t, s) \, ds \right)^{1/2} \left( \int_0^r \sin^2 u(t, s) \, ds \right)^{1/2}$$

$$\lesssim \mathcal{E}[u](t).$$

$\square$

The previous result is also useful in formulating scaling heuristics for the three problems as, unlike the wave map system, none of these equations (i.e., (6.85)-(6.87)) is scale-invariant. However, if we work in a small energy regime and rely on (6.93), then we can formally write scale-invariant approximations for all of them:

$$\left( 1 + \frac{2u^2}{r^2} \right) (u_{tt} - u_{rr}) - \frac{2}{r} u_r + \frac{2u}{r^2} \left( 1 + u_t^2 - u_r^2 + \frac{u^2}{r^2} \right) = 0$$

is the approximation for the Skyrme model,

$$\left( 1 + \frac{u^2}{r^2} \right) (u_{tt} - u_{rr}) - \left( 1 - \frac{u^2}{r^2} \right) \frac{u_r}{r} + \frac{u}{r^2} \left( u_t^2 - u_r^2 + 1 \right) = 0$$

is the approximation for the Faddeev model, and

$$u_{tt} - u_{rr} - \frac{2}{r}u_r + \frac{2u}{r^2} + \frac{u^5}{r^4} = 0$$

is the approximation for the Adkins-Nappi model. For the first two approximations, the scaling invariance is given by

$$u_\lambda(t,r) = \lambda\, u\left(\frac{t}{\lambda}, \frac{r}{\lambda}\right),$$

whereas

$$u_\lambda(t,r) = \lambda^{1/2}\, u\left(\frac{t}{\lambda}, \frac{r}{\lambda}\right)$$

corresponds to the Adkins-Nappi approximation. Arguing as in Chapter 3, these heuristics suggest that the critical homogeneous Sobolev space is: $\dot{H}^{5/2}(\mathbb{R}^3)$ for (6.85), $\dot{H}^2(\mathbb{R}^2)$ for (6.86), and $\dot{H}^2(\mathbb{R}^3)$ for (6.87). Thus, one expects to be able to prove small data global regularity results in function spaces similar to the previous ones and this is precisely what the authors achieved in [51] and [52].

### 6.3.1.2 *The energy-supercritical character of the problems*

Next, we would like to explain that all these equations are supercritical with respect to their corresponding energies (i.e., (6.88)-(6.90)), in the sense that the minimal Sobolev regularity needed to control these energies is less than the one predicted above by scaling. Thus, we expect the energy to be of limited use when proving regularity results for these problems, making the analysis more challenging. Specifically, we claim that the Skyrme energy (6.88) is at $\dot{H}^{7/4} \cap \dot{H}^1(\mathbb{R}^3)$-regularity level, the Faddeev energy (6.89) is at $\dot{H}^{3/2} \cap \dot{H}^1(\mathbb{R}^2)$-regularity level, whereas the Adkins-Nappi energy (6.90) is at $\dot{H}^{5/3} \cap \dot{H}^1(\mathbb{R}^3)$-regularity level. We are going to argue for the last two and leave the discussion for the Skyrme energy to the interested reader. Given that the arguments are entirely spatial in nature, we suppress the dependance of $u$ on time.

For the Faddeev model, our main focus will be on the $\sin^2 u \cdot u_r^2/r^2$ term, as it is not hard to see that the components $u_t^2$, $u_r^2$, and $\sin^2 u/r^2$ of the energy are all at $\dot{H}^1(\mathbb{R}^2)$-regularity level. We start by using the Hardy-type inequality (2.16) to infer

$$\int_0^\infty \frac{\sin^2 u \cdot u_r^2}{r^2}\, r\, dr \lesssim \left\|\frac{u_r}{r^{1/2}}\right\|_{L^2(\mathbb{R}^2)}^2 \left\|\frac{\sin u}{r^{1/2}}\right\|_{L^\infty(\mathbb{R}^2)}^2$$

$$\lesssim \|u\|_{\dot{H}^{3/2}(\mathbb{R}^2)}^2 \left\|\frac{\sin u}{r^{1/2}}\right\|_{L^\infty(\mathbb{R}^2)}^2.$$

The radial Sobolev estimate (2.13) implies

$$\frac{|\sin u|}{r^{1/2}} \lesssim \frac{1}{r^{1/2+\epsilon}} \|u\|_{\dot{H}^{1-\epsilon}(\mathbb{R}^2)} \lesssim \|u\|_{\dot{H}^{1-\epsilon}(\mathbb{R}^2)},$$

if $0 < \epsilon < 1/2$ and $r \geq 1$. Thus, we are left to investigate what happens with the right-hand side when $r < 1$. For that, we take advantage of $u(0) = 0$ and, based on the fundamental theorem of calculus, Hölder's inequality, and Sobolev embeddings, we derive

$$\frac{|\sin u(r)|}{r^{1/2}} \leq \frac{|u(r)|}{r^{1/2}} \leq \int_0^r \frac{|u_r(s)|}{s^{1/2}}\, ds \lesssim \|u_r\|_{L^p(\mathbb{R}^2)} \lesssim \|u\|_{\dot{H}^{2-2/p}(\mathbb{R}^2)},$$

if $p > 4$. Together with the previous bound, this estimate yields

$$\frac{|\sin u|}{r^{1/2}} \lesssim \|u\|_{\dot{H}^{1-\epsilon} \cap \dot{H}^{3/2+\epsilon}(\mathbb{R}^2)},$$

and, as a result,

$$\int_0^\infty \frac{\sin^2 u \cdot u_r^2}{r^2} r\, dr \lesssim \|u\|_{\dot{H}^{1-\epsilon} \cap \dot{H}^{3/2+\epsilon}(\mathbb{R}^2)}^4,$$

which concludes the analysis.

We now turn to the Adkins-Nappi model, for which the relevant energy term is $(u - \sin u \cos u)^2 / r^4$. We first notice that

$$|u - \sin u \cos u| \lesssim \min\{|u|, |u^3|\}$$

and, applying the radial Sobolev inequality (2.13), we deduce

$$\int_1^\infty \frac{(u - \sin u \cos u)^2}{r^4} r^2\, dr \lesssim \int_1^\infty \frac{u^2}{r^2}\, dr \lesssim \int_1^\infty \frac{1}{r^3}\, dr \cdot \|u\|_{\dot{H}^1(\mathbb{R}^3)}^2.$$

For the domain of integration $\{r < 1\}$, we argue as in the case of the Faddeev model. Using the fundamental theorem of calculus, Hölder's inequality, and Sobolev embeddings, we obtain

$$\frac{|u(r)|}{r^{1/6+\epsilon}} \leq \int_0^r \frac{|u_r(s)|}{s^{1/6+\epsilon}}\, ds \lesssim \|u_r\|_{L^p(\mathbb{R}^3)} \lesssim \|u\|_{\dot{H}^{5/2-3/p}(\mathbb{R}^3)},$$

if $\epsilon > 0$ is sufficiently small and $p > 18/(5 - 6\epsilon)$. Hence, it follows that

$$\int_0^1 \frac{(u - \sin u \cos u)^2}{r^4} r^2\, dr \lesssim \int_0^1 \frac{u^6}{r^2}\, dr \lesssim \int_0^1 \frac{1}{r^{1-6\epsilon}}\, dr \cdot \|u\|_{\dot{H}^{5/3+\epsilon}(\mathbb{R}^3)}^6,$$

which, together with the previous integral bound, finishes the discussion about the supercriticality of the Adkins-Nappi problem with respect to the energy.

### 6.3.1.3    *A reformulation step and statement of the main results*

A brief look at the three differential equations describing our models reveals that the Skyrme and Faddeev equations are quasilinear in nature, as the coefficients of the leading order terms depend on the unknown function itself, whereas the Adkins-Nappi equation contains highly singular terms like $(u - \sin u \cos u)(1 - \cos 2u)/r^4$. This tells us that investigating these equations in their current form is a very challenging task.

In order to make the analysis more tractable, we perform the classical substitution[8] $u = rv$, which has the role of transforming all three equations into semilinear wave problems at the expense of increasing the number of spatial dimensions. Concretely, the counterparts of (6.85)-(6.87) for the function $v$ are

$$
\begin{aligned}
\Box_{5+1}v = {}&h_{11}(r,u)\,v^3 + h_{12}(r,u)\,v^5 \\
&+ h_{13}(r,u)\,v^3 v_r + h_{14}(r,u)\,v(v_t^2 - v_r^2),
\end{aligned}
\tag{6.94}
$$

$$
\begin{aligned}
\Box_{4+1}v = {}&h_{21}(r,u)\,v^3 + h_{22}(r,u)\,v^5 \\
&+ h_{23}(r,u)\,v^3 v_r + h_{24}(r,u)\,v(v_t^2 - v_r^2),
\end{aligned}
\tag{6.95}
$$

$$
\Box_{5+1}v = h_{31}(u)\,v^3 + h_{32}(u)\,v^5,
\tag{6.96}
$$

where

$$
\Box_{n+1} \overset{\text{def}}{=} -\partial_t^2 + \partial_r^2 + \frac{n-1}{r}\,\partial_r
$$

is the radial wave operator on $\mathbb{R}^{n+1}$ and the coefficients of the nonlinearities are given by

$$
\begin{aligned}
h_{11}(r,u) &= \frac{\sin 2u - 2u}{u^3\,\Phi_1(r,u)}, & h_{12}(r,u) &= \frac{\sin 2u(\sin^2 u - u^2)}{u^5\,\Phi_1(r,u)}, \\[2mm]
h_{13}(r,u) &= \frac{4\sin u(\sin u - u\cos u)}{u^3\,\Phi_1(r,u)}, & h_{14}(r,u) &= \frac{\sin 2u}{u\,\Phi_1(r,u)},
\end{aligned}
\tag{6.97}
$$

$$
\begin{aligned}
h_{21}(r,u) &= \frac{\sin 2u - 2u}{2u^3\,\Phi_2(r,u)}, & h_{22}(r,u) &= \frac{\sin u(\sin u - u\cos u)}{u^4\,\Phi_2(r,u)}, \\[2mm]
h_{23}(r,u) &= \frac{2\sin u(\sin u - u\cos u)}{u^3\,\Phi_2(r,u)}, & h_{24}(r,u) &= \frac{\sin 2u}{2u\,\Phi_2(r,u)},
\end{aligned}
\tag{6.98}
$$

$$
h_{31}(u) = \frac{\sin 2u - 2u}{u^3}, \qquad h_{32}(u) = \frac{(u - \sin u \cos u)(1 - \cos 2u)}{u^5},
\tag{6.99}
$$

---

[8]This is also used in proving regularity for equivariant wave maps.

with

$$\Phi_1(r,u) = 1 + \frac{2\sin^2 u}{r^2} \quad \text{and} \quad \Phi_2(r,u) = 1 + \frac{\sin^2 u}{r^2}. \quad (6.100)$$

We can easily recognize the similarities between the respective coefficients for the three equations, e.g.,

$$\Phi_1(r,u)h_{11}(r,u) = 2\Phi_2(r,u)h_{11}(r,u) = h_{31}(u),$$

$$\Phi_1(r,u)h_{13}(r,u) = 2\Phi_2(r,u)h_{23}(r,u) \overset{\text{def}}{=} \tilde{h}_3(u),$$

$$\Phi_1(r,u)h_{14}(r,u) = 2\Phi_2(r,u)h_{24}(r,u) \overset{\text{def}}{=} \tilde{h}_4(u),$$

which is further proof of the intimate relations between these models. Using also the notation

$$\Phi_1(r,u)h_{12}(r,u) \overset{\text{def}}{=} \tilde{h}_{12}(u), \qquad \Phi_2(r,u)h_{22}(r,u) \overset{\text{def}}{=} \tilde{h}_{22}(u),$$

direct computations based on Maclaurin series yield the following useful result.

**Lemma 6.5.** *The functions $h_{31}$, $h_{32}$, $\tilde{h}_{12}$, $\tilde{h}_{22}$, $\tilde{h}_3$, and $\tilde{h}_4$ are all analytic on $\mathbb{R}$ and satisfy the decay estimates*

$$|h_{31}| + \langle u\rangle|\partial_u^{1+j}h_{31}| + |\partial_u^j\tilde{h}_3| \lesssim \langle u\rangle^{-2}, \quad (6.101)$$

$$|\partial_u^j\tilde{h}_{12}| + |\partial_u^j\tilde{h}_{22}| \lesssim \langle u\rangle^{-3}, \quad (6.102)$$

$$|\partial_u^j h_{32}| \lesssim \langle u\rangle^{-4}, \qquad |\partial_u^j\tilde{h}_4| \lesssim \langle u\rangle^{-1}, \quad (6.103)$$

*where $j \geq 0$ is an integer and $\langle u\rangle = (1+u^2)^{1/2}$.*

We can now state the main results, which address the small data global well-posedness and scattering for (6.94)-(6.96) and which validate the previous scaling heuristics.

**Theorem 6.2.** *There exists $\delta > 0$ such that for any radial initial data $(v(0,r), \partial_t v(0,r))$ decaying as $r \to \infty$ and satisfying*

$$\|\partial v(0,\cdot)\|_{\dot{B}_{2,1}^{3/2}\cap L^2(\mathbb{R}^5)} \leq \delta,$$

*the equation (6.94) admits a unique global solution $v$ satisfying*

$$\partial v \in C(\mathbb{R}; \dot{B}_{2,1}^{3/2}\cap L^2(\mathbb{R}^5)) \cap L^2(\mathbb{R}; \dot{B}_{4,1}^{3/4}\cap \dot{B}_{4,2}^{-3/4}(\mathbb{R}^5)).$$

*Moreover, for some $v_\pm$ solving the free wave equation,*

$$\lim_{t\to\pm\infty}\|\partial(v-v_\pm)(t)\|_{\dot{B}_{2,1}^{3/2}\cap L^2(\mathbb{R}^5)} = 0.$$

**Theorem 6.3.** *There exists $\delta > 0$ such that for any radial initial data $(v(0,r), \partial_t v(0,r))$ decaying as $r \to \infty$ and satisfying*

$$\|\partial v(0,\cdot)\|_{\dot{B}^1_{2,1} \cap \dot{B}^0_{2,1}(\mathbb{R}^4)} \le \delta,$$

*the equation (6.95) admits a unique global solution $v$ satisfying*

$$\partial v \in C(\mathbb{R}; \dot{B}^1_{2,1} \cap \dot{B}^0_{2,1}(\mathbb{R}^4)) \cap L^2(\mathbb{R}; \dot{B}^{1/6}_{6,1} \cap \dot{B}^{-5/6}_{6,1}(\mathbb{R}^4)).$$

*Moreover, for some $v_\pm$ solving the free wave equation,*

$$\lim_{t \to \pm\infty} \|\partial(v - v_\pm)(t)\|_{\dot{B}^1_{2,1} \cap \dot{B}^0_{2,1}(\mathbb{R}^4)} = 0.$$

**Theorem 6.4.** *There exists $\delta > 0$ such that for any radial initial data $(v(0,r), \partial_t v(0,r))$ decaying as $r \to \infty$ and satisfying*

$$\|\partial v(0,\cdot)\|_{\dot{H}^1 \cap L^2(\mathbb{R}^5)} \le \delta,$$

*the equation (6.96) admits a unique global solution $v$ satisfying*

$$\partial v \in C(\mathbb{R}; \dot{H}^1 \cap L^2(\mathbb{R}^5)) \cap L^2(\mathbb{R}; \dot{B}^{1/4}_{4,2} \cap \dot{B}^{-3/4}_{4,2}(\mathbb{R}^5))$$

*and for some $v_\pm$ solving the free wave equation,*

$$\lim_{t \to \pm\infty} \|\partial(v - v_\pm)(t)\|_{\dot{H}^1 \cap L^2(\mathbb{R}^5)} = 0.$$

In connection to these statements, we make the following comment.

**Remark 6.5.** The above theorems can be turned into similar results for the original equations in $u$ by relying on

$$\|u\|_{\dot{H}^s(\mathbb{R}^n)} \simeq \|v\|_{\dot{H}^s(\mathbb{R}^{n+2})}, \qquad \|r^{2/p-1} u\|_{\dot{B}^s_{p,q}(\mathbb{R}^n)} \simeq \|v\|_{\dot{B}^s_{p,q}(\mathbb{R}^{n+2})}.$$

The presence of two distinct regularities in the formulation of these theorems is due to the fact that all these equations are not scale-invariant and it was the authors' preference to work with homogeneous spaces. Finally, the scattering results are directly obtained as a byproduct of the fixed-point method used in proving global well-posedness.

### 6.3.1.4 *Method of proof*

In proving the three theorems, we rely on the contraction mapping frame-work developed in Chapter 3. Accordingly, we write the corresponding equations (6.94)-(6.96) in the generic integral form

$$v = S(v_0, v_1) + \Box^{-1}(N(v)),$$

where $S = S(v_0, v_1)$ is the homogeneous solution operator

$$\Box S = 0, \qquad S(0) = v_0, \qquad S_t(0) = v_1,$$

$\Box^{-1} = \Box^{-1} H$ is the inhomogeneous solution operator

$$\Box\left(\Box^{-1}H\right) = H, \qquad \left(\Box^{-1}H\right)(0) = \left(\Box^{-1}H\right)_t(0) = 0,$$

and $N = N(v)$ is the respective right-hand side of each equation. This reduces the proof of the theorems to finding a data space $D$ and a solution space $X$ for each problem, such that the following estimates are true:

$$\|S(v_0, v_1)\|_X \lesssim \|(v_0, v_1)\|_D, \qquad \|w\|_{L_t^\infty D} \lesssim \|w\|_X, \qquad (6.104)$$

$$\|\Box^{-1}(N(w_1) - N(w_2))\|_X \lesssim (\|w_1\|_X + \|w_2\|_X)\|w_1 - w_2\|_X, \qquad (6.105)$$

where the last bound holds for $\|v_1\|_X$ and $\|v_2\|_X$ sufficiently small.

The data spaces are described by:

- for the Skyrme problem,

$$\|(v_0, v_1)\|_D = \|v_0\|_{\dot{B}_{2,1}^{5/2} \cap \dot{H}^1(\mathbb{R}^5)} + \|v_1\|_{\dot{B}_{2,1}^{3/2} \cap L^2(\mathbb{R}^5)};$$

- for the Faddeev problem,

$$\|(v_0, v_1)\|_D = \|v_0\|_{\dot{B}_{2,1}^2 \cap \dot{B}_{2,1}^1(\mathbb{R}^4)} + \|v_1\|_{\dot{B}_{2,1}^1 \cap \dot{B}_{2,1}^0(\mathbb{R}^4)};$$

- for the Adkins-Nappi problem,

$$\|(v_0, v_1)\|_D = \|v_0\|_{\dot{H}^2 \cap \dot{H}^1(\mathbb{R}^5)} + \|v_1\|_{\dot{H}^1 \cap L^2(\mathbb{R}^5)}.$$

In what concerns the solution spaces, for the Skyrme and Faddeev problems, we first introduce the function space $|\nabla|F$, which is defined with the help of Tataru's $F$-spaces (see Definition 2.4 in Chapter 2) by

$$\|v\|_{|\nabla|F} \overset{\text{def}}{=} \sum_{\lambda \in 2^{\mathbb{Z}}} \lambda^{n/2-1} \|A_\lambda(\nabla)v\|_{F_\lambda}. \qquad (6.106)$$

With its help, we construct the solution space

$$Z \overset{\text{def}}{=} \{v \in \mathcal{S}'(\mathbb{R}^{n+1}) | \ v = v(t, x) = v(t, r), \ \|v\|_{F \cap |\nabla|F} < \infty\}, \qquad (6.107)$$

and choose its $5 + 1$-dimensional and $4 + 1$-dimensional versions for the Skyrme and Faddeev problems, respectively. Finally, for the Adkins-Nappi problem, we pick the solution space $X$ to be the closure of $\mathcal{S}'(\mathbb{R}^{n+1})$ under the norm

$$\|v\|_X = \|\partial v\|_{L^\infty(\dot{H}^1 \cap L^2)(\mathbb{R} \times \mathbb{R}^5)} + \|\partial v\|_{L^2\left(\dot{B}_{4,2}^{1/4} \cap \dot{B}_{4,2}^{-3/4}\right)(\mathbb{R} \times \mathbb{R}^5)}. \quad (6.108)$$

If we turn our attention now to the estimates that need to be proven in connection to these data and solution spaces, then we first notice that the homogeneous bounds (6.104) follow relatively easily from the properties of Tataru's spaces for (6.94) and (6.95) and from Strichartz estimates for (6.96). The most involved part of the argument is the proof of the nonlinear bound (6.105), which will be our main task for the remainder of this section. Based on the analytic properties in Lemma 6.5 for the nonlinear coefficients appearing in $N(v)$ and the polynomial structure of $N(v)$, we will show that, for all equations and for $\|v\|_X$ sufficiently small,

$$\|\Box^{-1}(N(v))\|_X \lesssim \|v\|_X^3. \quad (6.109)$$

It will be clear then that the nonlinear estimate (6.105) is derived by the same argument and ingredients, for which the detail will be omitted.

In proving (6.109), we have two main steps. First, we analyze the pure power nonlinearities (i.e., cubic, quartic, and quintic), for which we only use Hölder-type estimates and the compatibility of Strichartz inequalities with Tataru's spaces. Following this, for the Skyrme and Faddeev problems, we are left to investigate the last nonlinear term, for which we rely on its null structure

$$v_t^2 - v_r^2 = -\frac{\Box v^2}{2} + v \Box v. \quad (6.110)$$

As we shall see, in proving (6.109) for this nonlinearity, it is enough to show that

$$\|r\, v_1 v_2\|_Z \lesssim \|v_1\|_Z \|v_2\|_Z.$$

Unfortunately, the argument for this bilinear bound works only for $n \geq 5$, thus finishing the analysis for the Skyrme problem. For $n = 4$ (the case of the Faddeev model), we demonstrate instead the trilinear estimate

$$\|r^2\, v_1 v_2 v_3\|_Z \lesssim \|v_1\|_Z \|v_2\|_Z \|v_3\|_Z, \quad (6.111)$$

which is enough to claim (6.109) on the basis of the special structure for the coefficient $h_{24}(r, u)$.

In presenting the argument for (6.109), we start by showing it first for the Adkins-Nappi problem. Then we proceed to proving it for the pure power nonlinearities of the Faddeev problem. Finally, we conclude the section with the proof of the above bilinear and trilinear bounds and with the argument justifying their sufficiency in claiming (6.109) for the null-form terms. The motivation for the choices and order in this presentation has to do with the fact that the coefficients for the Adkins-Nappi nonlinearities are simpler than the ones for the Skyrme and Faddeev nonlinearities, as the presence of $\Phi_1$ and $\Phi_2$ complicates the analysis quite a bit. Furthermore, the analyses for the pure power nonlinearities of the Skyrme and Faddeev problems are very similar and this is why we choose to present only one of them. We believe the careful reader can then make easy adjustments and complete the analysis for the other problem on his own.

### 6.3.2 *Main analysis for the Adkins-Nappi problem*

Here, we begin by applying the inhomogeneous Strichartz estimate (2.52) in the context of the norm (6.108) to deduce

$$\|\Box^{-1}(N(v))\|_X \lesssim \|N(v)\|_{L^1(\dot{H}^1 \cap L^2)(\mathbb{R} \times \mathbb{R}^5)}. \tag{6.112}$$

Next, we investigate the $L^1 L^2$ component of the right-hand side and, with the help of the $\mathbb{R}^5$-Sobolev embeddings[9]

$$\dot{B}_{4,2}^{1/4} \subset L^5, \quad \dot{H}^2 \subset \dot{B}_{4,2}^{3/4} \subset L^{10},$$

we derive

$$\|v^3\|_{L^1 L^2} \lesssim \|v\|_{L^\infty L^{10}} \|v\|_{L^2 L^5}^2 \lesssim \|v\|_{L^\infty \dot{H}^2} \|v\|_{L^2 \dot{B}_{4,2}^{1/4}}^2 \lesssim \|v\|_X^3,$$

$$\|v^5\|_{L^1 L^2} \lesssim \|v\|_{L^\infty L^{10}}^3 \|v\|_{L^2 L^{10}}^2 \lesssim \|v\|_{L^\infty \dot{H}^2}^3 \|v\|_{L^2 \dot{B}_{4,2}^{3/4}}^2 \lesssim \|v\|_X^5.$$

If we couple these estimates with the uniform bounds from Lemma 6.5 for the coefficients $h_{31}$ and $h_{32}$, then we obtain that

$$\|N(v)\|_{L^1 L^2(\mathbb{R} \times \mathbb{R}^5)} \lesssim \|v\|_X^3, \tag{6.113}$$

for $\|v\|_X$ sufficiently small.

Following this, we work on the $L^1 \dot{H}^1$ component of the right-hand side in (6.112), for which we first infer that

$$(h_{32}(u)v^5)_r = v^4 v_r [5 h_{32}(u) + u h'_{32}(u)] + v^4 \frac{v}{r} [u h'_{32}(u)].$$

---

[9]These are consequences of the original embeddings (2.6)-(2.7).

Using the uniform bound (6.103) and the Hardy-type inequality (2.15), we deduce

$$\|h_{32}(u)v^5\|_{\dot{H}^1} \lesssim \|v^4\|_{L^5}(\|v_r\|_{L^{10/3}} + \|v/r\|_{L^{10/3}}) \lesssim \|v\|_{L^{20}}^4\|v_r\|_{L^{10/3}}.$$

Next, we rely on the Sobolev embeddings

$$\dot{H}^2 \subset \dot{H}^{1,10/3} \subset \dot{B}_{4,\infty}^{3/4}, \qquad \dot{B}_{4,1}^1 \subset L^{20},$$

and the real interpolation[10]

$$\left(\dot{B}_{4,\infty}^{3/4}, \dot{B}_{4,\infty}^{5/4}\right)_{1/2,1} = \dot{B}_{4,1}^1,$$

to derive

$$\|v\|_{L^{20}} \lesssim \|v_r\|_{L^{10/3}}^{1/2}\|v\|_{\dot{B}_{4,\infty}^{5/4}}^{1/2},$$

and, subsequently,

$$\|h_{32}(u)v^5\|_{L^1\dot{H}^1} \lesssim \|v_r\|_{L^\infty L^{10/3}}^3\|v\|_{L^2\dot{B}_{4,\infty}^{5/4}}^2 \lesssim \|v\|_{L^\infty\dot{H}^2}^3\|v\|_{L^2\dot{B}_{4,2}^{5/4}}^2 \lesssim \|v\|_X^5.$$

For the cubic nonlinearity, we proceed similarly to obtain first

$$\|h_{31}(u)v^3\|_{\dot{H}^1} \lesssim \|v^2\|_{L^4}\|v_r\|_{L^4} \lesssim \|v\|_{L^{10}}\|v\|_{L^{20/3}}\|v_r\|_{L^4}.$$

After another round of Sobolev embeddings (e.g., $\dot{H}^{7/4} \subset L^{20/3}$), it follows that

$$\|h_{31}(u)v^3\|_{L^1\dot{H}^1} \lesssim \|v\|_{L^2\dot{B}_{4,2}^{3/4}}\|v\|_{L^\infty\dot{H}^{7/4}}\|v\|_{L^2\dot{B}_{4,2}^1} \lesssim \|v\|_X^3.$$

Thus, one has

$$\|N(v)\|_{L^1\dot{H}^1(\mathbb{R}\times\mathbb{R}^5)} \lesssim \|v\|_X^3,$$

for $\|v\|_X$ sufficiently small, which, together with (6.113) and (6.112), finishes the argument.

### 6.3.3 *The analysis for the pure power nonlinearities in the Faddeev problem*

This subsection is devoted to the proof of

$$\|\Box^{-1}(\tilde{N}(v))\|_X \lesssim \|v\|_X^3,$$

where

$$\tilde{N}(v) = h_{21}(r,u)\,v^3 + h_{22}(r,u)\,v^5 + h_{23}(r,u)\,v^3v_r$$

---

[10]This is a particular case of the more general interpolation relation (2.9) in Chapter 2.

and $X$ is the $4+1$-dimensional version of the function space $Z$ defined by (6.107). This will follow if we show that

$$\|\tilde{N}(v)\|_{L^1\left(\dot{B}_{2,1}^1 \cap \dot{B}_{2,1}^0\right)(\mathbb{R}\times\mathbb{R}^4)}$$
$$\lesssim \|\partial v\|^3_{L^\infty\left(\dot{B}_{2,1}^1 \cap \dot{B}_{2,1}^0\right)(\mathbb{R}\times\mathbb{R}^4)} + \|\partial v\|^3_{L^2\left(\dot{B}_{6,1}^{1/6} \cap \dot{B}_{6,1}^{-5/6}\right)(\mathbb{R}\times\mathbb{R}^4)}. \tag{6.114}$$

Indeed, the Strichartz-type estimate (2.129) implies

$$\|v\|_{\tilde{X}} := \|\partial v\|_{L^\infty\left(\dot{B}_{2,1}^1 \cap \dot{B}_{2,1}^0\right)(\mathbb{R}\times\mathbb{R}^4)} + \|\partial v\|_{L^2\left(\dot{B}_{6,1}^{1/6} \cap \dot{B}_{6,1}^{-5/6}\right)(\mathbb{R}\times\mathbb{R}^4)}$$

$$\lesssim \|v\|_{F\cap|\nabla|F} = \|v\|_X,$$

while, as a consequence of the inhomogeneous bound (2.125), we deduce

$$\|\Box^{-1}H\|_X \lesssim \|H\|_{L^1\left(\dot{B}_{2,1}^1 \cap \dot{B}_{2,1}^0\right)(\mathbb{R}\times\mathbb{R}^4)}.$$

In proving (6.114), we will work with product estimates in Besov spaces, Sobolev-type embeddings, and interpolation relations. The main difficulties in the analysis are created by $\Phi_2$, given by (6.100), which we approach by writing

$$\frac{1}{\Phi_2(r,u)} = \sum_{k=0}^{\infty}(-V^2)^k, \qquad V \overset{\text{def}}{=} \frac{\sin u}{r} = \frac{\sin rv}{r}, \tag{6.115}$$

when the size of $V$ is sufficiently small. This decomposition implies that for estimating $\tilde{N}(v)$ we will be facing generic terms of the type

$$h_{31}(u)\,v^3\,V^{2k}, \quad \tilde{h}_{22}(u)\,v^5\,V^{2k}, \quad \text{or} \quad \tilde{h}_3(u)\,v^3 v_r\,V^{2k},$$

which motivates the choice for the aforementioned analytic tools.

We start the argument for (6.114) by applying the Besov product estimate (2.12) and the Sobolev-type embedding $\dot{B}_{2,1}^2(\mathbb{R}^4) \subset L^\infty(\mathbb{R}^4)$ to infer that the space

$$Y \overset{\text{def}}{=} \dot{B}_{2,1}^2 \cap \dot{B}_{2,1}^1(\mathbb{R}^4)$$

is an algebra. Moreover, a deeper analysis, which also uses radial Sobolev inequalities and which can be found by the reader in Lemma 3 of [52], shows that if $w_1$ and $w_2$ are a pair of radial functions, then

$$\|r\,w_1\,w_2\|_Y \lesssim \|w_1\|_Y\,\|w_2\|_Y.$$

As it turns out, this bound is the main tool needed to control the terms in the above geometric series. In point of fact, a direct argument based on induction yields the estimate

$$\|V^{2k}\|_Y \lesssim C^k\|v\|_Y^{2k}, \tag{6.116}$$

which holds for all radial functions $v$ with $\|v\|_Y \leq 1$ and all integers $k \geq 1$, with $C > 0$ being a constant independent of $v$ and $k$.

### 6.3.3.1 *Estimating the cubic nonlinearity*

For this nonlinearity, we prove that

$$\left\| h_{21}(r, u) v^3 \right\|_{L^1 \left( \dot{B}^1_{2,1} \cap \dot{B}^0_{2,1} \right)(\mathbb{R} \times \mathbb{R}^4)} \lesssim \| v \|^3_{\tilde{X}}, \tag{6.117}$$

if $\| v \|_{\tilde{X}} \ll 1$. Invoking again the embedding $\dot{B}^2_{2,1} \subset L^\infty$, we deduce

$$\| V \|_{L^\infty_{t,x}} \leq \| v \|_{L^\infty_{t,x}} \lesssim \| v \|_{L^\infty \dot{B}^2_{2,1}} \lesssim \| v \|_{\tilde{X}} \ll 1,$$

which allows us to use the expansion (6.115) and write

$$h_{21}(r, u) \, v^3 = \frac{1}{2} \sum_{k=0}^\infty h_{31}(u) \, v^3 \, (-V^2)^k.$$

This reduces the argument for (6.117) to obtaining relevant bounds for the general term of the previous series.

We work first on the $L^1 \dot{B}^1_{2,1}$ norm, for which an application of the Besov product estimate (2.12) and (6.116) implies, for $k \geq 1$,

$$\begin{aligned}
\left\| h_{31}(u) \, v^3 \, V^{2k} \right\|_{L^1 \dot{B}^1_{2,1}} \\
\lesssim \left\| h_{31}(u) \, v^3 \right\|_{L^1 \dot{B}^1_{2,1}} \left\| V^{2k} \right\|_{L^\infty_{t,x}} \\
+ \left\| h_{31}(u) \, v^3 \right\|_{L^1 L^\infty} \left\| V^{2k} \right\|_{L^\infty \dot{B}^1_{2,1}} \\
\lesssim C^k \| v \|^{2k}_{\tilde{X}} \left( \left\| h_{31}(u) \, v^3 \right\|_{L^1 \dot{B}^1_{2,1}} + \left\| h_{31}(u) \, v^3 \right\|_{L^1 L^\infty} \right).
\end{aligned} \tag{6.118}$$

For the $L^1 L^\infty$ norm, the decay estimate (6.101), the embedding $\dot{B}^{2/3}_{6,1} \subset L^\infty$, and the interpolation relation

$$\left( \dot{B}^{7/6}_{6,1}, \dot{B}^{1/6}_{6,1} \right)_{\frac{1}{2}, 1} = \dot{B}^{2/3}_{6,1}$$

jointly yield

$$\begin{aligned}
\left\| h_{31}(u) v^3 \right\|_{L^1 L^\infty} \lesssim \| v^3 \|_{L^1 L^\infty} \lesssim \| v \|_{L^\infty_{t,x}} \| v \|^2_{L^2 L^\infty} \\
\lesssim \| v \|_{\tilde{X}} \| v \|^2_{L^2 \dot{B}^{2/3}_{6,1}} \lesssim \| v \|_{\tilde{X}} \| v \|_{L^2 \dot{B}^{7/6}_{6,1}} \| v \|_{L^2 \dot{B}^{1/6}_{6,1}} \tag{6.119} \\
\lesssim \| v \|^3_{\tilde{X}}.
\end{aligned}$$

In what concerns the $L^1 \dot{B}^1_{2,1}$ norm of $h_{31}(u) \, v^3$, the argument is more involved and it starts by using the interpolation relation

$$\left( \dot{H}^2, L^2 \right)_{\frac{1}{2}, 1} = \dot{B}^1_{2,1}$$

to infer

$$\left\| h_{31}(u) \, v^3 \right\|_{L^1 \dot{B}^1_{2,1}} \lesssim \left\| h_{31}(u) \, v^3 \right\|^{1/2}_{L^1 \dot{H}^2} \left\| h_{31}(u) \, v^3 \right\|^{1/2}_{L^1 L^2}. \tag{6.120}$$

Based on (6.101) and the Besov embeddings $\dot{B}_{6,1}^{1/6} \subset L^8$ and $\dot{B}_{2,1}^1 \subset L^4$, we derive

$$\left\| h_{31}(u)\,v^3 \right\|_{L^1 L^2} \lesssim \|v^3\|_{L^1 L^2} \lesssim \|v\|_{L^2 L^8}^2 \|v\|_{L^\infty L^4}$$

$$\lesssim \|v\|_{L^2 \dot{B}_{6,1}^{1/6}}^2 \|v\|_{L^\infty \dot{B}_{2,1}^1} \lesssim \|v\|_X^3.$$

For the $L^1 \dot{H}^2$ norm of $h_{31}(u)v^3$, we take advantage of

$$\left\| h_{31}(u)\,v^3 \right\|_{L^1 \dot{H}^2} \simeq \left\| \partial_r^2 \left( h_{31}(u)v^3 \right) \right\|_{L^1 L^2}, \tag{6.121}$$

and apply the decay estimate (6.101) to derive

$$\left| \partial_r^2 \left( h_{31}(u)\,v^3 \right) \right| \lesssim \left| v^2\,v_{rr} \right| + \left| v\,v_r^2 \right| + \left| v^3\,v_r \right| + \left| v^5 \right|.$$

Each of the terms on the right-hand side of the last inequality can be estimated in $L^1 L^2$ through straightforward Besov embeddings, e.g.,

$$\|v^3 v_r\|_{L^1 L^2} \lesssim \|v\|_{L^\infty L^\infty} \|v\|_{L^2 L^\infty}^2 \|v_r\|_{L^\infty L^2} \lesssim \|v\|_X^3 \|\partial v\|_{L^\infty \dot{B}_{2,1}^0} \lesssim \|v\|_X^4.$$

The end result is the bound

$$\left\| \partial_r^2 \left( h_{31}(u)v^3 \right) \right\|_{L^1 L^2} \lesssim \|v\|_X^3,$$

when $\|v\|_X$ is sufficiently small. If we couple this estimate with (6.120) and (6.121), then we obtain

$$\left\| h_{31}(u)\,v^3 \right\|_{L^1 \dot{B}_{2,1}^1} \lesssim \|v\|_X^3.$$

On the account of (6.118) and (6.119), this finishes the proof of

$$\left\| h_{21}(r,u)v^3 \right\|_{L^1 \dot{B}_{2,1}^1 (\mathbb{R} \times \mathbb{R}^4)} \lesssim \|v\|_X^3.$$

All which is left to argue for claiming (6.117) is that

$$\left\| h_{21}(r,u)v^3 \right\|_{L^1 \dot{B}_{2,1}^0 (\mathbb{R} \times \mathbb{R}^4)} \lesssim \|v\|_X^3 \tag{6.122}$$

holds when $\|v\|_X \ll 1$. For this purpose, we start by using the Besov product estimate (2.12) and the embedding $\dot{B}_{4/3,1}^1 \subset \dot{B}_{2,1}^0$ to deduce

$$\left\| h_{31}(u)\,v^3\,V^{2k} \right\|_{L^1 \dot{B}_{2,1}^0} \lesssim \left\| h_{31}(u)\,v^3 \right\|_{L^1 \dot{B}_{2,1}^1} \left\| V^{2k} \right\|_{L^\infty L^4}$$

$$+ \left\| h_{31}(u)\,v^3 \right\|_{L^1 L^4} \left\| V^{2k} \right\|_{L^\infty \dot{B}_{2,1}^1}.$$

The first and fourth norms on the right-hand side have already been estimated in the $L^1 \dot{B}_{2,1}^1$ analysis. For the second and third norms we rely again on Besov embeddings involving $L^4$ and $L^8$, respectively, to derive

$$\left\| V^{2k} \right\|_{L^\infty L^4} \lesssim \|V\|_{L_{t,x}^\infty}^{2k-1} \|V\|_{L^\infty L^4} \lesssim \|v\|_{L_{t,x}^\infty}^{2k-1} \|v\|_{L^\infty \dot{B}_{2,1}^1} \lesssim \|v\|_X^{2k}$$

and

$$\left\| h_{31}(u)v^3 \right\|_{L^1 L^4} \lesssim \|v\|_{L_{t,x}^\infty} \|v\|_{L^2 L^8}^2 \lesssim \|v\|_X^3.$$

Following this, one can check immediately that this is all that is needed to obtain (6.122) and thus conclude the analysis for the cubic nonlinearity.

### 6.3.3.2   *Estimating the quintic nonlinearity*

For the quintic nonlinearity, we are able to prove

$$\|h_{22}(r,u)\, v^5\|_{L^1(\dot{B}^1_{2,1} \cap \dot{B}^0_{2,1})(\mathbb{R}\times\mathbb{R}^4)} \lesssim \|v\|_{\tilde{X}}^5. \tag{6.123}$$

The argument has many similarities with the one for the cubic nonlinearity, as the extra $v^2$ present here in most expressions can be bounded directly in $L^\infty_{t,x}$, which is controlled by $\|v\|_{\tilde{X}}$. Moreover, the main coefficients of the cubic and quintic nonlinearities (i.e., $h_{31}(u)$ and $\tilde{h}_{22}(u)$, respectively) verify almost identical decay estimates according to Lemma 6.5.

### 6.3.3.3   *Estimating the quartic nonlinearity*

Here, the goal is to show that

$$\|h_{23}(r,u)\, v^3 v_r\|_{L^1(\dot{B}^1_{2,1} \cap \dot{B}^0_{2,1})(\mathbb{R}\times\mathbb{R}^4)} \lesssim \|v\|_{\tilde{X}}^4 \tag{6.124}$$

holds when $\|v\|_{\tilde{X}} \ll 1$. As in the case of the cubic nonlinearity, we can reduce the argument to proving relevant bounds for the generic term

$$\tilde{h}_3(u)\, v^3 v_r\, V^{2k},$$

where $k \geq 1$ is an integer.

We work first on the $L^1 \dot{B}^1_{2,1}$ norm and we proceed as before by applying (2.12) to infer

$$\left\|\tilde{h}_3(u)\, v^3 v_r\, V^{2k}\right\|_{L^1 \dot{B}^1_{2,1}} \lesssim \left\|\tilde{h}_3(u)\, v^3 v_r\right\|_{L^1 \dot{B}^1_{2,1}} \left\|V^{2k}\right\|_{L^\infty_{t,x}}$$
$$+ \left\|\tilde{h}_3(u)\, v^3 v_r\right\|_{L^1 L^4} \left\|V^{2k}\right\|_{L^\infty \dot{B}^1_{4,1}}.$$

The second norm has already been analyzed. Based on the embeddings $\dot{B}^2_{2,1} \subset \dot{B}^1_{4,1}$ and $\dot{B}^{1/6}_{6,1} \subset L^8$, together with the power estimate (6.116), it follows that

$$\left\|V^{2k}\right\|_{L^\infty \dot{B}^1_{4,1}} \lesssim \left\|V^{2k}\right\|_{L^\infty \dot{B}^2_{2,1}} \lesssim C^k \|v\|_{\tilde{X}}^{2k}$$

and

$$\left\|\tilde{h}_3(u)\, v^3 v_r\right\|_{L^1 L^4} \lesssim \|v\|_{L^\infty_{t,x}}^2 \|v\|_{L^2 L^8} \|v_r\|_{L^2 L^8} \lesssim \|v\|_{\tilde{X}}^4.$$

Thus, we are left to find an applicable bound for the $L^1 \dot{B}^1_{2,1}$ norm of $\tilde{h}_3(u)\, v^3 v_r$, which is the more intricate part of the proof for (6.124). This is because an interpolation approach like the one used for the cubic nonlinearity introduces a $v_{rrr}$ term for which we do not have good estimates.

Yet again, we rely on (2.12) and Besov embeddings to deduce

$$\left\| \tilde{h}_3(u)\, v^3 v_r \right\|_{L^1 \dot{B}^1_{2,1}}$$

$$\lesssim \left\| \tilde{h}_3(u)\, v^3 \right\|_{L^1 L^\infty} \|v_r\|_{L^\infty \dot{B}^1_{2,1}} + \left\| \tilde{h}_3(u)\, v^3 \right\|_{L^1 \dot{B}^1_{4,1}} \|v_r\|_{L^\infty L^4}$$

$$\lesssim \|v\|_{\tilde{X}}^4 + \left\| \tilde{h}_3(u)\, v^3 \right\|_{L^1 \dot{B}^1_{4,1}} \|v\|_{\tilde{X}}.$$

Next, we use the interpolation relations

$$\dot{B}^1_{4,1} = \left( \dot{H}^{5/4,4}, \dot{H}^{3/4,4} \right)_{1/2,1} = \left( \left( \dot{H}^2, \dot{H}^{1,6} \right)_{[3/4]}, \left( L^2, \dot{H}^{1,6} \right)_{[3/4]} \right)_{1/2,1}$$

to derive the inequality

$$\|f\|_{L^1 \dot{B}^1_{4,1}} \lesssim \|f\|_{L^1 \dot{H}^2}^{1/8} \|f\|_{L^1 L^2}^{1/8} \|f\|_{L^1 \dot{H}^{1,6}}^{3/4}.$$

The first two norms on the right are treated for $\tilde{h}_3(u)\, v^3$ in the same way as for $h_{31}(u)\, v^3$, since $\tilde{h}_3$ and $h_{31}$ have similar decay rates according to (6.101).

Hence, it suffices to analyze the $L^1 \dot{H}^{1,6}$ norm, for which (6.101) implies

$$\left\| \tilde{h}_3(u)\, v^3 \right\|_{L^1 \dot{H}^{1,6}} \simeq \left\| \partial_r(\tilde{h}_3(u)\, v^3) \right\|_{L^1 L^6} \lesssim \|v^2 v_r\|_{L^1 L^6} + \|v^4\|_{L^1 L^6}.$$

Working with previous estimates, we obtain

$$\|v^2 v_r\|_{L^1 L^6} \lesssim \|v\|_{L^\infty_{t,x}} \|v\|_{L^2 L^{24}} \|v_r\|_{L^2 L^8} \lesssim \|v\|_{\tilde{X}}^3$$

and

$$\|v^4\|_{L^1 L^6} \lesssim \|v\|_{L^\infty_{t,x}} \|v^3\|_{L^1 L^6} \lesssim \|v\|_{\tilde{X}}^4.$$

If we put together all the bounds obtained so far in connection to the $L^1 \dot{B}^1_{2,1}$ norm of $\tilde{h}_3(u)\, v^3 v_r\, V^{2k}$, then we conclude that

$$\left\| \tilde{h}_3(u)\, v^3 v_r\, V^{2k} \right\|_{L^1 \dot{B}^1_{2,1}} \lesssim \|v\|_{\tilde{X}}^{2k+4}$$

holds for $\|v\|_{\tilde{X}}$ sufficiently small. This is a serviceable estimate for proving (6.124).

Finally, we investigate the $L^1 \dot{B}^0_{2,1}$ norm of $\tilde{h}_3(u)\, v^3 v_r\, V^{2k}$, for which an application of the Besov product estimate (2.12) yields

$$\left\| \tilde{h}_3(u)\, v^3 v_r\, V^{2k} \right\|_{L^1 \dot{B}^0_{2,1}} \lesssim \left\| \tilde{h}_3(u)\, v^3 v_r \right\|_{L^1 \dot{B}^1_{2,1}} \|V^{2k}\|_{L^\infty L^4}$$

$$+ \left\| \tilde{h}_3(u)\, v^3 v_r \right\|_{L^1 L^4} \|V^{2k}\|_{L^\infty \dot{B}^1_{2,1}}.$$

However, we claim that the careful reader can check that all the norms appearing on the right-hand side have already been analyzed and they provide favorable estimates leading to

$$\left\| \tilde{h}_3(u)\, v^3 v_r\, V^{2k} \right\|_{L^1 \dot{B}^0_{2,1}} \lesssim \|v\|_{\tilde{X}}^{2k+4}.$$

This finishes the argument for (6.124) and, thus, ends the discussion for the pure power nonlinearities of the Faddeev problem.

### 6.3.4 The analysis for the null-form nonlinearity in the Skyrme and Faddeev problems

In this subsection, we prove the nonlinear estimate

$$\|\Box^{-1}\left(h_{i4}(r,u)\,v(v_t^2 - v_r^2)\right)\|_Z \lesssim \|v\|_Z^3, \qquad 1 \leq i \leq 2, \qquad (6.125)$$

for $\|v\|_Z$ sufficiently small, where $Z$ is the appropriate[11] function space defined by (6.107). This will allow us to conclude the argument for the inequality (6.109) in the cases of the Skyrme and Faddeev problems, which also finishes the proof of their global well-posedness results.

As one expects from the definition of $Z$, the main tools used in showing (6.125) are linear and multilinear bounds involving Tataru's spaces. For the convenience of the reader, we recall here from Chapter 2 the following estimates:

$$\|\Box u\|_{\Box F} \lesssim \|u\|_F, \qquad (6.126)$$

$$\|\Box^{-1} u\|_F \lesssim \|u\|_{\Box F}, \qquad (6.127)$$

$$\|u\,v\|_F \lesssim \|u\|_F \|v\|_F, \qquad (6.128)$$

$$\|u\,v\|_{\Box F} \lesssim \|u\|_F \|v\|_{\Box F}. \qquad (6.129)$$

The first two hold for any spatial dimension $n \geq 2$, while the last two require $n \geq 4$. These will be supplemented by four new inequalities, which are included in the next lemma and for which we introduce the notation

$$\|v\|_{|\nabla|\Box F} \stackrel{\text{def}}{=} \sum_{\lambda \in 2^{\mathbb{Z}}} \lambda^{n/2-1} \|A_\lambda(\nabla)v\|_{\Box F_\lambda} \quad \text{and} \quad \Box Z \stackrel{\text{def}}{=} \Box F \cap |\nabla|\Box F.$$

**Lemma 6.6.** *i) For general functions, the bilinear estimates*

$$\|u\,v\|_{|\nabla|F} \lesssim \|u\|_{|\nabla|F} \|v\|_F, \qquad (6.130)$$

$$\|u\,v\|_{|\nabla|\Box F} \lesssim \|u\|_{|\nabla|F} \|v\|_{\Box F} \qquad (6.131)$$

*are true if $n \geq 4$.*

*ii) For radial functions, the trilinear bound*

$$\|r^2 uvw\|_Z \lesssim \|u\|_Z \|v\|_Z \|w\|_Z \qquad (6.132)$$

*holds for $n \geq 4$. If $n \geq 5$, then the stronger[12] bilinear bound*

$$\|ruv\|_Z \lesssim \|u\|_Z \|v\|_Z \qquad (6.133)$$

*is true.*

---

[11] $Z$ is $5+1$-dimensional for $i = 1$ (Skyrme problem) and $4+1$-dimensional for $i = 2$ (Faddeev problem).

[12] It is immediate to argue that (6.133) implies (6.132).

Before proving this lemma, let us convince the reader that we now have all the ingredients to demonstrate (6.125). First, if one revisits the arguments in Chapter 2 for (6.126) and (6.127), then it is trivial to recognize that they also provide the bounds

$$\|\Box u\|_{|\nabla|\Box F} \lesssim \|u\|_{|\nabla|F}, \qquad \|\Box^{-1} u\|_{|\nabla|F} \lesssim \|u\|_{|\nabla|\Box F}.$$

Together with (6.126) and (6.127), these imply

$$\|\Box u\|_{\Box Z} \lesssim \|u\|_Z, \qquad \|\Box^{-1} u\|_Z \lesssim \|u\|_{\Box Z}.$$

Secondly, the bilinear bounds (6.128), (6.129), (6.130), and (6.131) easily yield

$$\|u\, v\|_Z \lesssim \|u\|_Z \|v\|_Z, \qquad \|u\, v\|_{\Box Z} \lesssim \|u\|_Z \|v\|_{\Box Z}.$$

Next, we rely on these estimates for $Z$ and $\Box Z$ and on the null structure (6.110) of $v_t^2 - v_r^2$ to infer

$$\begin{aligned}
\|\Box^{-1} & \left( h_{i4}(r, u)\, v(v_t^2 - v_r^2) \right)\|_Z \\
& \lesssim \|h_{i4}(r, u)\, v(v_t^2 - v_r^2)\|_{\Box Z} \\
& \lesssim \|h_{i4}(r, u)\, v\|_Z \|v_t^2 - v_r^2\|_{\Box Z} \\
& \lesssim \|h_{i4}(r, u)\, v\|_Z \left( \|\Box v^2\|_{\Box Z} + \|v \Box v\|_{\Box Z} \right) \\
& \lesssim \|h_{i4}(r, u)\, v\|_Z \|v\|_Z^2.
\end{aligned} \tag{6.134}$$

Finally, the trilinear bound (6.132) leads by induction to

$$\|u_1 u_2 \cdots u_{2j+1}\|_{rZ} \lesssim C^j \|u_1\|_{rZ} \|u_2\|_{rZ} \cdots \|u_{2j+1}\|_{rZ}, \qquad \forall j \geq 0,$$

where $\|w\|_{rZ} := \|w/r\|_Z$. As a consequence, for any fixed $\alpha \in \mathbb{R}$,

$$\begin{aligned}
\|\sin \alpha u\|_{rZ} & \lesssim \sum_{j=0}^{\infty} \frac{|\alpha|^{2j+1}}{(2j+1)!} \|u^{2j+1}\|_{rZ} \\
& \lesssim \sum_{j=0}^{\infty} \frac{C^j |\alpha|^{2j+1}}{(2j+1)!} \|u\|_{rZ}^{2j+1} \\
& = \sum_{j=0}^{\infty} \frac{C^j |\alpha|^{2j+1}}{(2j+1)!} \|v\|_Z^{2j+1} \lesssim \|v\|_Z
\end{aligned}$$

holds when $\|v\|_Z \ll 1$. Using the decomposition

$$i\, h_{i4}(r, u)\, v = \frac{\sin 2u}{r} \sum_{k=0}^{\infty} (-1)^k \left( \frac{\sin u}{r} \right)^{2k},$$

and the fact that $Z$ is an algebra, we deduce on the basis of the last estimate that

$$\|h_{i4}(r,u)\,v\|_Z \lesssim \|\sin 2u\|_{rZ} \sum_{k=0}^{\infty} C^k \|\sin u\|_{rZ}^{2k} \lesssim \|v\|_Z. \qquad (6.135)$$

Jointly with (6.134), this bound implies (6.125).

To conclude this subsection, we present the argument for the four new multilinear estimates (6.130)-(6.133).

**Proof of Lemma 6.6.** The approach for proving these bounds is in the same spirit with the one used in Chapter 2 for showing (6.128) and (6.129). As all the function spaces involved here use $l^1$ summability over the dyadic decomposition given by $(A_\lambda(\nabla))_{\lambda \in 2^{\mathbb{Z}}}$, it is enough to prove these estimates for single dyadic pieces. To simplify the notation, we denote $u_\lambda := A_\lambda(\nabla)u$, $v_\mu := A_\mu(\nabla)u$, and $w_\nu := A_\nu(\nabla)w$.

In what concerns (6.130) and (6.131), we first notice that the inequalities for the cases when either $\lambda \gg \mu$ or $\lambda \ll \mu$ follow directly as consequences of (6.128) and (6.129). For example, in any of those instances, the definition (6.106) of $|\nabla|F$ and (6.128) imply

$$\|u_\lambda v_\mu\|_{|\nabla|F} \simeq \max\{\lambda,\mu\}^{-1}\|u_\lambda v_\mu\|_F \lesssim \max\{\lambda,\mu\}^{-1}\|u_\lambda\|_F\|v_\mu\|_F$$
$$\lesssim \lambda^{-1}\|u_\lambda\|_F\|v_\mu\|_F \simeq \|u_\lambda\|_{|\nabla|F}\|v_\mu\|_F.$$

Therefore, we are left to investigate only the $\lambda \sim \mu$ scenario, for which we are going to abuse the notation and write $u_\lambda v_\lambda$ in place of $u_\lambda v_\mu$. The support of its spacetime Fourier transform is in the region $|\tau| + |\xi| \lesssim \lambda$. Together with the Strichartz-type estimates (2.129) for Tataru's spaces, this fact leads to

$$\|u_\lambda v_\lambda\|_{|\nabla|F} \simeq \sum_{\lambda_1 \lesssim \lambda} \lambda_1^{n/2-1}\|A_{\lambda_1}(u_\lambda v_\lambda)\|_{F_{\lambda_1}}$$
$$\lesssim \sum_{\lambda_1 \lesssim \lambda} \lambda_1^{n/2-1}\|A_{\lambda_1}(u_\lambda v_\lambda)\|_{X_{\lambda_1}^{1/2}}$$
$$\lesssim \sum_{\lambda_2 \leq \lambda_1 \lesssim \lambda} \lambda_1^{n/2-1}\lambda_2^{1/2}\|A_{\lambda_1}B_{\lambda_2}(u_\lambda v_\lambda)\|_{L_{t,x}^2}$$
$$\lesssim \lambda^{(n-1)/2}\|u_\lambda v_\lambda)\|_{L_{t,x}^2}$$
$$\lesssim \lambda^{(n-1)/2}\|u_\lambda\|_{L^\infty L^2}\|v_\lambda\|_{L^2 L^\infty}$$
$$\lesssim \lambda^{(n-1)/2}\|u_\lambda\|_{F_\lambda}\lambda^{(n-1)/2}\|v_\lambda\|_{F_\lambda}$$
$$\simeq \|u_\lambda\|_{|\nabla|F}\|v_\lambda\|_F,$$

which finishes the proof of (6.130). The argument for (6.131) is very similar, with the only extra ingredient being the embedding $\Box F_\lambda \subset \lambda^{3/2} L^2_{t,x}$, and we leave it to the interested reader.

Now, we start the argument for the multilinear bounds (6.132) and (6.133) by deriving weighted Strichartz-type estimates for the dyadic pieces $u_\lambda$ with radial symmetry. A standard stationary phase approach yields

$$\|r^{(n-1)/2} S_\lambda(\nabla_x)\varphi\|_{L^\infty} \lesssim \lambda^{1/2}\|\varphi\|_{L^2},$$

which holds for any radial $\varphi = \varphi(x) = \check{\varphi}(r) \in L^2_x$ with $S_\lambda = S_\lambda(\nabla_x)$ being a Littlewood-Paley projector at frequency $\lambda$ in spatial frequency. If we interpolate the previous bound consecutively with the well-known estimates

$$\|S_\lambda(\nabla_x)\varphi\|_{L^\infty} \lesssim \|\varphi\|_{L^\infty} \quad \text{and} \quad \|S_\lambda(\nabla_x)\varphi\|_{L^p} \lesssim \|\varphi\|_{L^p},\ 2 \le p \le \infty,$$

then we derive that

$$\|r^{(n-1)(1/p-1/q)} S_\lambda(\nabla_x)\varphi\|_{L^q} \lesssim \lambda^{1/p-1/q}\|\varphi\|_{L^p}$$

is true for all $2 \le p \le q \le \infty$. One more interpolation between this bound and the Bernstein-type inequality

$$\|S_\lambda(\nabla_x)\varphi\|_{L^q} \lesssim \lambda^{n(1/p-1/q)}\|\varphi\|_{L^p}$$

produces the estimate

$$\|r^{\alpha(1/p-1/q)} S_\lambda(\nabla_x)\varphi\|_{L^q} \lesssim \lambda^{(n-\alpha)(1/p-1/q)}\|\varphi\|_{L^p},$$

which holds for all radial functions $\varphi \in L^p$, $0 \le \alpha \le n-1$, and $2 \le p \le q \le \infty$. Next, we combine this bound with the Strichartz-type estimates (2.129) for Tataru's spaces and, as a result, we infer that, for wave-admissible triples $(q,p,n)$ with $n \ge 4$ and $u = u(t,r)$, the following inequality is true:

$$\|r^{(n-1)/p} u_\lambda\|_{L^q L^\infty} \lesssim \lambda^{n/2-1/q-(n-1)/p}\|u_\lambda\|_{F_\lambda}.$$

In particular, this implies

$$\|r^\alpha u_\lambda\|_{L^\infty_{t,x}} \lesssim \lambda^{n/2-\alpha}\|u_\lambda\|_{F_\lambda}, \qquad 0 \le \alpha \le (n-1)/2, \qquad (6.136)$$

$$\|r^\alpha u_\lambda\|_{L^2 L^\infty} \lesssim \lambda^{(n-1)/2-\alpha}\|u_\lambda\|_{F_\lambda}, \qquad 0 \le \alpha \le (n-3)/2, \qquad (6.137)$$

which are the weighted Strichartz-type estimates we were after and which serve as crucial ingredients for the actual proof of (6.132) and (6.133). We notice that $r^{-1/2}$ is the best decay for the $L^2 L^\infty$ norm when $n = 4$, whereas one has $r^{-1}$ decay for the same norm if $n \ge 5$. This is why, with this approach, we can prove the bilinear estimate (6.133) only in spatial dimensions higher than 4.

As explained in the beginning of the proof, the two multilinear bounds will follow from their dyadic versions, i.e.,

$$\|r^2 u_\lambda v_\mu w_\nu\|_Z \lesssim (\lambda^2 + \lambda)(\mu^2 + \mu)(\nu^2 + \nu)\|u_\lambda\|_{F_\lambda}\|v_\mu\|_{F_\mu}\|w_\nu\|_{F_\nu} \quad (6.138)$$

and

$$\|r\, u_\lambda v_\mu\|_Z \lesssim (\lambda^{n/2} + \lambda^{(n-2)/2})(\mu^{n/2} + \mu^{(n-2)/2})\|u_\lambda\|_{F_\lambda}\|v_\mu\|_{F_\mu}, \quad (6.139)$$

where we can assume by symmetry that $\lambda \geq \mu \geq \nu$ and $\lambda \geq \mu$, respectively. In both cases, we use the operator $\tilde{B}_\mu = \tilde{B}_\mu(\nabla)$ to decompose $u_\lambda$ according to

$$u_\lambda^{<\mu} \stackrel{\text{def}}{=} \tilde{B}_\mu(\nabla)u_\lambda, \qquad u_\lambda^{>\mu} \stackrel{\text{def}}{=} u_\lambda - u_\lambda^{<\mu}.$$

Corollary 2.4 provides the estimates

$$\|u_\lambda^{<\mu}\|_{F_\lambda} \lesssim \|u_\lambda\|_{F_\lambda}, \quad \|u_\lambda^{<\mu}\|_{\Box F_\lambda} \lesssim \|u_\lambda\|_{\Box F_\lambda}, \quad \|u_\lambda^{>\mu}\|_{L^1 L^2} \lesssim \mu^{-1}\|u_\lambda\|_{Y_\lambda}, \quad (6.140)$$

which hold uniformly for $\mu \leq \lambda$. Together with (6.136) and (6.137), these yield

$$\|r^2 u_\lambda^{<\mu} v_\mu\, w_\nu\|_{L^2_{t,x}} \lesssim \|u_\lambda^{<\mu}\|_{L^\infty L^2}\|r^{3/2}v_\mu\|_{L^\infty_{t,x}}\|r^{1/2}w_\nu\|_{L^2 L^\infty}$$

$$\lesssim \mu^{1/2}\nu\,\|u_\lambda\|_{F_\lambda}\,\|v_\mu\|_{F_\mu}\,\|w_\nu\|_{F_\nu}$$

when $n = 4$, and

$$\|r\, u_\lambda^{<\mu} v_\mu\|_{L^2_{t,x}} \lesssim \|u_\lambda^{<\mu}\|_{L^\infty L^2}\|r\, v_\mu\|_{L^2 L^\infty} \lesssim \mu^{(n-3)/2}\|u_\lambda\|_{F_\lambda}\|v_\mu\|_{F_\mu}$$

when $n \geq 5$. Given that the support of the spacetime Fourier transform for both $r^2 u_\lambda^{<\mu}v_\mu w_\nu$ and $r u_\lambda^{<\mu}v_\mu$ lies in the region $\|\tau| - |\xi\| \lesssim \mu$, it follows that

$$\|r^2 u_\lambda^{<\mu} v_\mu\, w_\nu\|_{X_\lambda^{1/2}} \lesssim \mu\,\nu\,\|u_\lambda\|_{F_\lambda}\,\|v_\mu\|_{F_\mu}\,\|w_\nu\|_{F_\nu} \quad (6.141)$$

and

$$\|r\, u_\lambda^{<\mu} v_\mu\|_{X_\lambda^{1/2}} \lesssim \mu^{(n-2)/2}\,\|u_\lambda\|_{F_\lambda}\,\|v_\mu\|_{F_\mu}, \quad (6.142)$$

are true for $n = 4$ and $n \geq 5$, respectively.

We are left to investigate the terms $r^2 u_\lambda^{>\mu}v_\mu w_\nu$ and $r u_\lambda^{>\mu}v_\mu$, for which the analysis is clearly relevant only if $\lambda \gg \mu$. Based on the definition (2.118) of $F_\lambda$, we can write

$$u_\lambda^{>\mu} = \sum_{16\mu < d \leq \lambda} u_\lambda^d + u_\lambda^0,$$

such that the support of the spacetime Fourier transform for $u_\lambda^d$ is a subset of the domain $|\tau^2 - |\xi|^2| \sim d\lambda$ and

$$\|u_\lambda^{>\mu}\|_{F_\lambda} \sim \sum_d \|u_\lambda^d\|_{X_\lambda^{1/2}} + \|u_\lambda^0\|_{Y_\lambda}.$$

If we apply (6.136) and (6.140), then we obtain

$$\left\| r^2 \sum_d u_\lambda^d v_\mu w_\nu \right\|_{X_\lambda^{1/2}} \lesssim \sum_d d^{1/2} \|r^2 u_\lambda^d v_\mu w_\nu\|_{L_{t,x}^2}$$

$$\lesssim \|r v_\mu\|_{L_{t,x}^\infty} \|r w_\nu\|_{L_{t,x}^\infty} \sum_d d^{1/2} \|u_\lambda^d\|_{L_{t,x}^2} \quad (6.143)$$

$$\lesssim \mu\nu \|u_\lambda\|_{F_\lambda} \|v_\mu\|_{F_\mu} \|w_\nu\|_{F_\nu}$$

and

$$\left\| r \sum_d u_\lambda^d v_\mu \right\|_{X_\lambda^{1/2}} \lesssim \sum_d d^{1/2} \|r u_\lambda^d v_\mu\|_{L_{t,x}^2}$$

$$\lesssim \|r v_\mu\|_{L_{t,x}^\infty} \sum_d d^{1/2} \|u_\lambda^d\|_{L_{t,x}^2} \quad (6.144)$$

$$\lesssim \mu^{(n-2)/2} \|u_\lambda\|_{F_\lambda} \|v_\mu\|_{F_\mu}.$$

For the $Y_\lambda$ component of our two terms, due to $Y_\lambda \subset L^\infty L^2$, (6.136), and (6.140), we derive

$$\|r^2 u_\lambda^0 v_\mu w_\nu\|_{Y_\lambda} = \|r^2 u_\lambda^0 v_\mu w_\nu\|_{L^\infty L^2} + \lambda^{-1} \|\Box(r^2 u_\lambda^0 v_\mu w_\nu)\|_{L^1 L^2}$$

$$\lesssim \|u_\lambda^0\|_{L^\infty L^2} \|r^2 v_\mu w_\nu\|_{L_{t,x}^\infty}$$

$$+ \lambda^{-1} \left( \|r^2 v_\mu w_\nu \Box u_\lambda^0\|_{L^1 L^2} + \|[\Box, r^2 v_\mu w_\nu] u_\lambda^0\|_{L^1 L^2} \right)$$

$$\lesssim \mu\nu \left( \|u_\lambda^0\|_{L^\infty L^2} + \lambda^{-1} \|\Box u_\lambda^0\|_{L^1 L^2} \right) \|v_\mu\|_{F_\mu} \|w_\nu\|_{F_\nu}$$

$$+ \mu \|r^2 v_\mu w_\nu\|_{L_{t,x}^\infty} \|u_\lambda^0\|_{L^1 L^2}$$

$$\lesssim \mu\nu \|u_\lambda^0\|_{Y_\lambda} \|v_\mu\|_{F_\mu} \|w_\nu\|_{F_\nu}$$

$$\lesssim \mu\nu \|u_\lambda\|_{F_\lambda} \|v_\mu\|_{F_\mu} \|w_\nu\|_{F_\nu}$$

and

$$\|r u_\lambda^0 v_\mu\|_{Y_\lambda} = \|r u_\lambda^0 v_\mu\|_{L^\infty L^2} + \lambda^{-1} \|\Box(r u_\lambda^0 v_\mu)\|_{L^1 L^2}$$

$$\lesssim \|u_\lambda^0\|_{L^\infty L^2} \|r v_\mu\|_{L_{t,x}^\infty}$$

$$+ \lambda^{-1} \left( \|r v_\mu \Box u_\lambda^0\|_{L^1 L^2} + \|[\Box, r v_\mu] u_\lambda^0\|_{L^1 L^2} \right)$$

$$\lesssim \mu^{(n-2)/2} \left( \|u_\lambda^0\|_{L^\infty L^2} + \lambda^{-1} \|\Box u_\lambda^0\|_{L^1 L^2} \right) \|v_\mu\|_{F_\mu}$$

$$+ \mu \|r v_\mu\|_{L_{t,x}^\infty} \|u_\lambda^0\|_{L^1 L^2}$$

$$\lesssim \mu^{(n-2)/2} \|u_\lambda^0\|_{Y_\lambda} \|v_\mu\|_{F_\mu}$$

$$\lesssim \mu^{(n-2)/2} \|u_\lambda\|_{F_\lambda} \|v_\mu\|_{F_\mu}.$$

Combining the last two bounds with (6.141)-(6.144), we conclude that

$$\|r^2 u_\lambda v_\mu w_\nu\|_Z \lesssim (\lambda^2 + \lambda)\mu\,\nu \|u_\lambda\|_{F_\lambda} \|v_\mu\|_{F_\mu} \|w_\nu\|_{F_\nu}$$

holds if $n = 4$ and $\lambda \geq \mu \geq \nu$, and

$$\|r\, u_\lambda v_\mu\|_Z \lesssim (\lambda^{n/2} + \lambda^{(n-2)/2})(\mu^{n/2} + \mu^{(n-2)/2}) \|u_\lambda\|_{F_\lambda} \|v_\mu\|_{F_\mu}$$

is true if $n \geq 5$ and $\lambda \geq \mu$. These obviously imply the desired estimates, (6.138) and (6.139), respectively.

$\square$

## 6.4 Large data global regularity for the $2 + 1$-dimensional Faddeev model

In this section, we present an abridged version of Creek's large data result [31] for the 2+1-dimensional equivariant Faddeev problem, which is inspired by Li's argument [108] for the similar setting of the Skyrme model. It addresses the global-in-time evolution for the variational equation (6.86) with boundary conditions

$$u(t, 0) = 0 \quad \text{and} \quad u(t, \infty) = \pi,$$

which corresponds, according to (6.91), to Faddeev maps with unit charge.

### 6.4.1 *Statement of the main result and initial reductions*

Creek prefers to work in Li's framework and, for that matter, switches from $u$ to $\pi - u$. This changes the boundary conditions to

$$u(t, 0) = \pi \quad \text{and} \quad u(t, \infty) = 0, \tag{6.145}$$

but it does not alter the differential equation satisfied by $u$, which is written in the form

$$\Box_{2+1} u = F(u), \tag{6.146}$$

where $\Box_{2+1}$ is the previously defined radial wave operator on $\mathbb{R}^{2+1}$,

$$F(u) = 2\left(1 - \frac{1}{\Phi_2(r, u)}\right)\frac{u_r}{r} + \frac{(u_t^2 - u_r^2 + 1)\sin 2u}{2r^2 \Phi_2(r, u)}, \tag{6.147}$$

and $\Phi_2$ is given by (6.100). The following is the main result in [31].

**Theorem 6.5.** *Let $(u_0, u_1)$ be radial initial data with $u_0(0) = \pi$, $u_0(\infty) = u_1(0) = u_1(\infty) = 0$, and*

$$(u_0, u_1) \in H^s \times H^{s-1}(\mathbb{R}^2), \quad s \geq 4.$$

*Then there exists a global radial solution u to the Cauchy problem associated to (6.146) with $(u(0), u_t(0)) = (u_0, u_1)$, which satisfies the boundary conditions (6.145) and*

$$u \in C([0, T], H^s(\mathbb{R}^2)) \cap C^1([0, T], H^{s-1}(\mathbb{R}^2)), \qquad \forall T > 0.$$

In proving this theorem, the first step is similar to the one employed for the small data results of the previous section and it has to to do with performing the substitution

$$u(t, r) = r \, v(t, r) + \varphi(r), \qquad (6.148)$$

with $\varphi : \mathbb{R}_+ \to \mathbb{R}$ being a smooth, decreasing function, verifying $\varphi \equiv \pi$ on $[0, 1]$ and $\varphi \equiv 0$ on $[2, \infty)$. For what follows, we will also need to introduce a finer version of $\varphi$, denoted by $\varphi_{<1}$, which shares the same smoothness and monotonicity with $\varphi$, but satisfies now $\varphi_{<1} \equiv 1$ on $[0, 1/2]$ and $\varphi_{<1} \equiv 0$ on $[1, \infty)$. Furthermore, we write $\varphi_{>1}$ to label the function $1 - \varphi_{<1}$. These notational conventions and the ones described by (6.98) allow us to derive the following nonlinear wave equation for $v$:

$$
\begin{aligned}
\Box_{4+1} v \\
&= -\frac{v}{r^2} - \frac{\Delta \varphi}{r} + \varphi_{<1} \frac{F(rv + \pi)}{r} + \varphi_{>1} \frac{F(rv + \varphi)}{r} \\
&= -\frac{\Delta \varphi}{r} + \varphi_{>1} \left( \frac{F(rv + \varphi)}{r} - \frac{v}{r^2} \right) + \varphi_{<1} \big( h_{21}(r, rv) \, v^3 \\
&\quad + h_{22}(r, rv) \, v^5 + 2 \, h_{22}(r, rv) \, r v^4 v_r + h_{24}(r, rv) \, v(v_t^2 - v_r^2) \big),
\end{aligned}
\qquad (6.149)
$$

where

$$\Delta \varphi = \varphi_{rr} + \frac{\varphi_r}{r}.$$

One cannot help but notice the stark similarities between the right-hand sides of this equation and the one written as (6.95). The only slight difference is that we work here with the nonlinearity $r v^4 v_r$ instead of $v^3 v_r$ and, as a result, we need to adjust the coefficient.

The motivation for the substitution (6.148) is that Theorem 6.5 comes as a direct consequence of the subsequent result concerning (6.149).

**Theorem 6.6.** *Let $(v_0, v_1)$ be radial initial data with*

$$(v_0, v_1) \in H^s \times H^{s-1}(\mathbb{R}^4), \quad s \geq 4.$$

*Then there exists a global radial solution v to the Cauchy problem associated to (6.149) with $(v(0), v_t(0)) = (v_0, v_1)$, which satisfies*

$$v \in C([0, T], H^s(\mathbb{R}^4)) \cap C^1([0, T], H^{s-1}(\mathbb{R}^4)), \qquad \forall T > 0.$$

Indeed, if $(u_0, u_1)$ is an initial data satisfying the hypotheses of Theorem 6.5, then one can check directly that

$$(v_0, v_1) \overset{\text{def}}{=} \left( \frac{u_0 - \varphi}{r}, \frac{u_1}{r} \right)$$

is an allowable initial data for the last theorem, which yields a global solution $v$ for (6.149). Reversing the derivation of (6.149) from (6.146), we deduce that $u$ given by (6.148) solves the initial value problem comprised of (6.146), $u(0) = u_0$, and $u_t(0) = u_1$. Moreover, one easily obtains the required Sobolev regularity for $u$. It remains to verify that $u$ satisfies the boundary conditions (6.145). For this purpose, we fix $T > 0$ and argue that the regularity of $v$ implies, on the basis of a Sobolev embedding, that $v \in L_{t,x}^\infty([0, T] \times \mathbb{R}^4)$. This yields $u(t, 0) = \varphi(0) = \pi$, for all $0 \leq t \leq T$. On the other hand, using the radial Sobolev estimate (2.14) in the context of $v(t) \in H^1(\mathbb{R}^4)$, we derive that $|v(t, r)| \lesssim r^{-3/2}$. This implies $|u(t, r)| \lesssim r^{-1/2}$ if $r \geq 2$, which obviously leads to $u(t, \infty) = 0$.

We are left to prove Theorem 6.6, for which the approach is to obtain global solutions from local ones, which satisfy a certain continuation criterion. This can be made precise through the following variation of a classical result (e.g., Theorem 6.4.11 in Hörmander [73]), adapted to our specific problem.

**Theorem 6.7.** *Let $s \geq 4$ be an integer and consider radial initial data $(v_0, v_1)$ with*

$$(v_0, v_1) \in H^s \times H^{s-1}(\mathbb{R}^4).$$

*Then there exist $T > 0$ and a local radial solution $v$ on $[0, T)$ to the Cauchy problem associated to (6.149) with $(v(0), v_t(0)) = (v_0, v_1)$, which satisfies*

$$v \in C([0, T), H^s(\mathbb{R}^4)) \cap C^1([0, T), H^{s-1}(\mathbb{R}^4)).$$

*Furthermore, this solution can be continued past $T < \infty$ if*

$$\|(1 + r)(|v| + |\nabla_{t,x} v|)\|_{L_{t,x}^\infty([0,T) \times \mathbb{R}^4)} < \infty. \tag{6.150}$$

With this result in hand, the proof of Theorem 6.6 is reduced to arguing that:

*For any $0 < T < \infty$, a radial solution $v$ on $[0, T)$ to (6.149) with $(v(0), v_t(0)) \in H^4 \times H^3(\mathbb{R}^4)$ satisfies (6.150).*

The strategy for demonstrating this claim is somewhat indirect, in the sense that, using $v$, we first construct an auxiliary function $\Phi$ according to

$$
\Phi(t,r) = \frac{1}{r} \left[ \int_\pi^{u(t,r)} \left( 1 + \frac{\sin^2 y}{r^2} \right)^{1/2} dy \right.
$$
$$
\left. + \varphi_{>1}(r) \int_0^\pi \left( 1 + \frac{\sin^2 y}{r^2} \right)^{-3/2} dy \right], \tag{6.151}
$$

where $u$ and $\varphi_{>1}$ have the previous meaning. Secondly, using energy-type and Strichartz estimates, we prove that

$$
\sum_{i=0}^4 \| \partial_t^i \Phi \|_{L^\infty H^{4-i}([0,T] \times \mathbb{R}^4)} < \infty. \tag{6.152}
$$

Finally, based on this bound, Sobolev embeddings, and radial Sobolev inequalities, we infer that $v$ satisfies (6.150).

### 6.4.2 Motivation for the construction of the auxiliary function

The functions to be analyzed are solutions of the semilinear equation (6.149). Even if the right-hand side of this equation does not exhibit any essential singularities (i.e., terms involving negative powers of $r$ supported in domains where $r = 0$ is allowed), its structure is quite intricate and one can make the argument that it cannot be easily investigated using energy-type or Strichartz bounds. Thus, a natural question is whether one can simplify (6.149) even further or relate $v$ to an auxiliary function that satisfies a more friendly equation.

We choose to pursue the latter and, for that matter, we start back at (6.146), the equation for $u$. For an arbitrary function $\Psi : \mathbb{R}_+ \times \mathbb{R}$, $\Psi = \Psi(s,y)$, we want to first write a wave equation satisfied by

$$
\Psi_1(t,r) \stackrel{\text{def}}{=} \Psi(r, u(t,r)).
$$

A direct computation that takes into account (6.146) yields

$$
\Box_{2+1} \Psi_1 = 2 \left[ \left( 1 - \frac{1}{\Phi_2(r,u)} \right) \frac{\Psi_y(r,u)}{r} + \Psi_{sy}(r,u) \right] u_r
$$
$$
+ \left( \Psi_{yy}(r,u) - \frac{\sin 2u}{2r^2 \Phi_2(r,u)} \Psi_y(r,u) \right) (u_r^2 - u_t^2)
$$
$$
+ \Psi_{ss}(r,u) + \frac{1}{r} \Psi_s(r,u) + \frac{\sin 2u}{2r^2 \Phi_2(r,u)} \Psi_y(r,u),
$$

and we would like to pick $\Psi$ such that the terms on the right-hand side involving derivatives of $u$ disappear entirely. This can be done if $\Psi$ satisfies the differential equations

$$\left(1 - \frac{1}{\Phi_2(s,y)}\right)\frac{\Psi_y(s,y)}{s} + \Psi_{sy}(s,y) = 0$$

and

$$\Psi_{yy}(s,y) - \frac{\sin 2y}{2s^2\Phi_2(s,y)}\Psi_y(s,y) = 0.$$

The first one can be easily integrated with respect to $s$ and produces the profile

$$\Psi_y(s,y) = h(y)\left(1 + \frac{\sin^2 y}{s^2}\right)^{1/2},$$

which plugged into the second equation implies that the function $h$ is constant. We choose to work with $h \equiv 1$ and, thus, deduce

$$\Psi(r, u(t,r)) = \int_{y_0}^{u(t,r)}\left(1 + \frac{\sin^2 y}{r^2}\right)^{1/2} dy + \Psi(r, y_0).$$

If we rely now on the boundary condition (6.145), then it follows that we can go forward with

$$\Psi_1(t,r) = \int_{\pi}^{u(t,r)}\left(1 + \frac{\sin^2 y}{r^2}\right)^{1/2} dy, \qquad (6.153)$$

whose corresponding wave equation can be written in the form

$$\Box_{2+1}\Psi_1 = \left(\frac{1}{r^2} - 1\right)\Psi_1 + \int_{\pi}^{u}\left(1 + \frac{\sin^2 y}{r^2}\right)^{-3/2} dy. \qquad (6.154)$$

Next, we want to treat the $\Psi_1/r^2$ singularity in this equation. For this purpose, we introduce

$$\Psi_2(t,r) \overset{\text{def}}{=} \frac{1}{r}\Psi_1(t,r) \qquad (6.155)$$

and we lift the spatial dimensions in the radial wave operator from 2 to 4. As a result, we derive

$$\Box_{4+1}\Psi_2 = -\Psi_2 + \frac{1}{r}\int_{\pi}^{u}\left(1 + \frac{\sin^2 y}{r^2}\right)^{-3/2} dy.$$

Recalling the relation (6.148) and the definitions of $\varphi$, $\varphi_{<1}$, and $\varphi_{<1}$, it follows that we can recast the previous equation as

$$
\Box_{4+1}\Psi_2 = -\Psi_2 + \varphi_{<1} \int_0^v \left(1 + \frac{\sin^2 ry}{r^2}\right)^{-3/2} dy
$$

$$
+ \frac{\varphi_{>1}}{r} \int_\pi^u \left(1 + \frac{\sin^2 y}{r^2}\right)^{-3/2} dy,
$$

(6.156)

whose right-hand side does not present any singularities. However, even though we fixed the equation to be investigated, a new issue arises with the function $\Psi_2$ itself. At a basic level, we would like to have $\Psi_2 \in L^\infty L^2([0,T) \times \mathbb{R}^4)$. Using the conservation of energy (6.89) for solutions to (6.146), one obtains that $u \in L^\infty \dot{H}^1([0,T) \times \mathbb{R}^2)$, which, combined with the radial Sobolev inequality (2.13), yields the decay estimate

$$
|u(t,r)| \lesssim \frac{1}{r^{1/2-\epsilon}}.
$$

Based on the formulas (6.153) and (6.155), it follows that

$$
|\Psi_2(t,r)| \simeq \frac{1}{r}
$$

holds uniformly in $t$ if $r$ is sufficiently large. This is obviously incompatible with the above integrability condition that we require of $\Psi_2$.

In order to fix this shortcoming of $\Psi_2$, we perform one final conversion and consider

$$
\Phi(t,r) \overset{\text{def}}{=} \Psi_2(t,r) + \frac{\varphi_{>1}(r)}{r} \int_0^\pi \left(1 + \frac{\sin^2 y}{r^2}\right)^{-3/2} dy.
$$

(6.157)

Several integral manipulations, which include a change of variable, lead to

$$
\Phi = \int_0^v \left(1 + \frac{\sin^2(rw + \varphi)}{r^2}\right)^{1/2} dw + \frac{\varphi_{>1}}{r} \int_0^\varphi \left(1 + \frac{\sin^2 y}{r^2}\right)^{1/2} dy
$$

$$
- \frac{\varphi_{>1}}{r} \int_0^\pi \left(1 + \frac{\sin^2 y}{r^2}\right)^{-3/2} \left(2\frac{\sin^2 y}{r^2} + \frac{\sin^4 y}{r^4}\right) dy.
$$

The second integral term is time-independent and supported in the region $\{1/2 \leq r \leq 2\}$, due to the presence of both $\varphi_{>1}$ and $\varphi$. Hence, it can be written as $\varphi_{\sim 1} = \varphi_{\sim 1}(r)$, which denotes a generic smooth, bounded function, with support in the previous region. The third integral term is also time-independent and supported in the region $\{r \geq 1/2\}$, which implies that

$$
1 + \frac{\sin^2 y}{r^2} \simeq 1
$$

holds uniformly for $0 \leq y \leq \pi$. As a consequence, we can write it in the form $\varphi_{\geq 1/2}/r^3$, where $\varphi_{\geq 1/2} = \varphi_{\geq 1/2}(r)$ is a generic smooth, bounded function, with support in the domain $\{r \geq 1/2\}$. It follows that we can absorb the second term into the third one and thus deduce

$$\Phi = \int_0^v \left(1 + \frac{\sin^2(rw + \varphi)}{r^2}\right)^{1/2} dw + \frac{\varphi_{\geq 1/2}}{r^3}. \qquad (6.158)$$

This formula makes the case for the deep connection between $\Phi$ and $v$, as the latter is the only time-dependent function appearing on the right-hand side.

Next, we want to derive a wave equation satisfied by $\Phi$ and, for this purpose, we first combine (6.156) and (6.157) to infer that[13]

$$\Box_{4+1}\Psi_2 = -\Phi + \int_0^v \left(1 + \frac{\sin^2(rw + \varphi)}{r^2}\right)^{-3/2} dw$$

$$+ \frac{\varphi_{>1}}{r} \int_0^\varphi \left(1 + \frac{\sin^2 y}{r^2}\right)^{-3/2} dy$$

$$= -\Phi + \int_0^v \left(1 + \frac{\sin^2(rw + \varphi)}{r^2}\right)^{-3/2} dw + \varphi_{\sim 1}.$$

Secondly, using again (6.157) and the previous notational conventions, we have

$$\Phi - \Psi_2 = \frac{\varphi_{\geq 1/2}}{r},$$

while a tedious, but direct computation shows that

$$\Box_{4+1}\left(\frac{\varphi_{\geq 1/2}}{r}\right) = \frac{\varphi_{\geq 1/2}}{r^3}.$$

If we put together all these facts, then we arrive at

$$\Box_{4+1}\Phi = -\Phi + \int_0^v \left(1 + \frac{\sin^2(rw + \varphi)}{r^2}\right)^{-3/2} dw + \frac{\varphi_{\geq 1/2}}{r^3}, \qquad (6.159)$$

which argues yet again for the strong interdependence between $\Phi$ and $v$. This concludes the motivation and derivation of the auxiliary function $\Phi$.

---

[13]To streamline the exposition, we keep the same notation for cutoff functions localized in a certain region.

### 6.4.3 *Analysis of* $\Phi$ *based on energy-type arguments*

We remind the reader that our next goal is to show that $\Phi$ satisfies the estimate (6.152). Here, we take the first step and, using energy-type arguments, we show that[14]

$$\|\Phi\|_{L^\infty H^1([0,T)\times\mathbb{R}^4)} + \|\partial_t \Phi\|_{L^\infty L^2([0,T)\times\mathbb{R}^4)} < \infty. \qquad (6.160)$$

Employing the formula (6.158), we derive that

$$\partial_t \Phi = \left(1 + \frac{\sin^2 u}{r^2}\right)^{1/2} \frac{u_t}{r},$$

which implies

$$\|\partial_t \Phi\|_{L^\infty L^2} \simeq \left\|\left(1 + \frac{\sin^2 u}{r^2}\right)^{1/2} u_t\right\|_{L^\infty L^2([0,T)\times\mathbb{R}^2)}.$$

If we rely on the energy formula (6.89), the conservation of energy for solutions to (6.146), and $(u(0), u_t(0)) \in H^4 \times H^3(\mathbb{R}^2)$, then we deduce that

$$\left\|\left(1 + \frac{\sin^2 u}{r^2}\right)^{1/2} u_t\right\|_{L^\infty L^2([0,T)\times\mathbb{R}^2)} \lesssim \mathcal{E}[u](0) < \infty.$$

Hence, the second term in (6.160) is finite.

We take advantage of this information and prove next that $\Phi \in L^\infty L^2$. As $T$ is finite, the mean value theorem implies

$$\|\Phi\|_{L^\infty L^2} \leq \|\Phi(0)\|_{L^2(\mathbb{R}^4)} + T\|\partial_t \Phi\|_{L^\infty L^2},$$

which reduces the argument to showing $\Phi(0) \in L^2(\mathbb{R}^4)$. It is easy to see that if $r \geq 1$, then

$$1 + \frac{\sin^2(rw + \varphi)}{r^2} \simeq 1,$$

while if $r < 1$, then

$$1 + \frac{\sin^2(rw + \varphi)}{r^2} = 1 + \frac{\sin^2(rw)}{r^2} \leq 1 + w^2.$$

Thus, on the basis of (6.158), we infer that

$$|\Phi(0,r)| \lesssim |v(0,r)| + |v(0,r)|^2 + \frac{|\varphi_{\geq 1/2}|}{r^3},$$

which further implies

$$\|\Phi(0)\|_{L^2(\mathbb{R}^4)} \lesssim \|v(0)\|_{L^2(\mathbb{R}^4)} + \|v(0)\|_{L^4(\mathbb{R}^4)}^2 + 1 \lesssim \left(1 + \|v(0)\|_{H^1(\mathbb{R}^4)}\right)^2 < \infty,$$

---

[14]For ease of notation, we will assume henceforth that all the norms are computed over the domain $[0, T) \times \mathbb{R}^4$ unless otherwise specified.

due to Sobolev embeddings.

In order to finish the proof of (6.160), we are left to argue[15] that $\nabla_x \Phi \in L^\infty L^2$. As $\Phi$ is a radial function, we have $|\nabla_x \Phi|^2 = (\partial_r \Phi)^2$ and, hence, we can directly focus on $\partial_r \Phi$. In the spirit of general energy-based arguments, we start by multiplying the equation (6.159) by $\partial_t \Phi$ and then integrating the outcome with respect to the spatial variables using Gauss's theorem. As a result, we obtain

$$\frac{1}{2} \partial_t \left\{ \int_{\mathbb{R}^4} (\partial_t \Phi)^2 + (\partial_r \Phi)^2 \right\}$$
$$= \int_{\mathbb{R}^4} \partial_t \Phi \left[ \Phi - \int_0^v \left( 1 + \frac{\sin^2(rw + \varphi)}{r^2} \right)^{-3/2} dw - \frac{\varphi_{\geq 1/2}}{r^3} \right].$$

For one of the terms on the right-hand side, we rely on (6.158) to derive

$$\left| \int_0^v \left( 1 + \frac{\sin^2(rw + \varphi)}{r^2} \right)^{-3/2} dw \right| \leq \left| \int_0^v \left( 1 + \frac{\sin^2(rw + \varphi)}{r^2} \right)^{1/2} dw \right|$$
$$\leq |\Phi| + \frac{|\varphi_{\geq 1/2}|}{r^3}.$$

Consequently, if we apply the Cauchy-Schwarz inequality, then we deduce that

$$\left\| \int_{\mathbb{R}^4} \partial_t \Phi \left[ \Phi - \int_0^v \left( 1 + \frac{\sin^2(rw + \varphi)}{r^2} \right)^{-3/2} dw - \frac{\varphi_{\geq 1/2}}{r^3} \right] \right\|_{L^\infty([0,T))}$$
$$\lesssim \|\partial_t \Phi\|_{L^\infty L^2} (1 + \|\Phi\|_{L^\infty L^2}).$$

Coupling this estimate with the last integral identity and the facts proven so far about $\partial_t \Phi$ and $\Phi$, it follows that proving $\partial_r \Phi \in L^\infty L^2$ reduces to showing that $\partial_r \Phi(0) \in L^2(\mathbb{R}^4)$.

To achieve this, we use (6.158) one more time to infer

$$\partial_r \Phi(0) = \left( 1 + \frac{\sin^2 u(0)}{r^2} \right)^{1/2} v_r(0)$$
$$+ \frac{1}{2} \int_0^{v(0)} \left( 1 + \frac{\sin^2(rw + \varphi)}{r^2} \right)^{-1/2} \left[ \frac{-2\sin^2(rw + \varphi)}{r^3} \right.$$
$$\left. + \frac{\sin 2(rw + \varphi) \cdot (w + \varphi_r)}{r^2} \right] dw$$
$$+ \frac{\varphi_{\geq 1/2}}{r^3}.$$

---

[15]This is the part in the main argument where Creek's approach for the Faddeev model is more direct than Li's for the Skyrme model. While Creek strictly uses energy-type techniques, Li needs to rely also on a sharp Hardy inequality.

The last term is easily seen to be in $L^2(\mathbb{R}^4)$, while for the first one, using the boundary conditions (6.145), we derive that

$$\frac{\sin^2 u(0,r)}{r^2} = \frac{\sin^2(u(0,r) - u(0,0))}{r^2} \leq \|u_r(0)\|_{L^\infty(\mathbb{R}^2)}^2 \tag{6.161}$$

$$\lesssim \|u(0)\|_{H^{2+\epsilon}(\mathbb{R}^2)}^2.$$

As $v(0) \in H^4(\mathbb{R}^4)$, it follows that the first term belongs to $L^2(\mathbb{R}^4)$ too.

We finally come to discussing the integral term for which we analyze separately the cases $r \geq 1$ and $r < 1$. When $r \geq 1$, we estimate the integrand directly by

$$\left(1 + \frac{\sin^2(rw + \varphi)}{r^2}\right)^{-1/2} \left| \frac{-2\sin^2(rw + \varphi)}{r^3} + \frac{\sin 2(rw + \varphi) \cdot (w + \varphi_r)}{r^2} \right|$$

$$\lesssim \frac{1 + |w|}{r^2},$$

which implies that the absolute value of the integral term can be bounded from above by a multiple of

$$\frac{|v(0)| + |v(0)|^2}{r^2}.$$

However, we already know from the analysis of $\|\Phi(0)\|_{L^2(\mathbb{R}^4)}$ that this function lies in the space $L^2(\{x \in \mathbb{R}^4; |x| \geq 1\})$. When $r < 1$, we have $\varphi \equiv \pi$ and the integrand takes the form

$$\left(1 + \frac{\sin^2 rw}{r^2}\right)^{-1/2} \frac{-2\sin^2 rw + \sin 2rw \cdot rw}{r^3}. \tag{6.162}$$

Using Maclaurin series, we can estimate its absolute value from above by $rw^4$ and, hence, the norm of the integral term in the space $L^2(\{x \in \mathbb{R}^4; |x| < 1\})$ is dominated by $\|v(0)\|_{L^{10}(\mathbb{R}^4)}^5$. Based on Sobolev embeddings, we conclude that the regularity of $v(0)$ is more than enough to control this quantity, which proves it to be finite.

In conclusion, we obtain $\partial_r \Phi(0) \in L^2(\mathbb{R}^4)$, which finishes the proof of (6.160).

## 6.4.4 *Analysis of $\Phi$ based on Strichartz estimates*

In this part of the argument, the analysis builds on the one done previously using energy-type techniques and takes advantage of radial Strichartz estimates for the linear wave equation to finally conclude (6.152). There are three steps in this analysis, each one corresponding to showing

$$\sum_{i=0}^{s} \|\partial_t^i \Phi\|_{L^\infty \dot{H}^{s-i}} < \infty \tag{6.163}$$

for $s = 2$, $3$, and $4$, respectively. It is clear that these bounds, when combined with (6.160), yield (6.152).

The three steps are proven in succession, starting with $s = 2$ and ending with $s = 4$. Each one significantly uses information obtained in the previous steps, with the argument for (6.163) when $s = 2$ relying on facts proven in connection to (6.160). The approaches for these steps are very similar, in the sense that, for a specific $s$:

- first, one differentiates the equation (6.159) $s - 1$ times with respect to the time variable;
- next, by mainly applying Strichartz estimates, the regularity of the initial data $(v(0), v_t(0))$, and the connection between $\Phi$ and $v$ through (6.158), it can be shown that

$$\|\partial_t^s \Phi\|_{L^\infty L^2} + \|\partial_t^{s-1} \Phi\|_{L^\infty \dot{H}^1} < \infty; \qquad (6.164)$$

- finally, by further using the wave equation derived at the previous step (i.e., for $s - 1$) and the connecting formula (6.158), one deduces the finiteness for the remaining norms in (6.163).

For pedagogical reasons, we present here only the first step (i.e., the argument for (6.163) when $s = 2$), which is the most transparent of all three. We believe the reader can then follow without much difficulty the argument in [31] for the remaining two steps.

As mentioned above, we start by differentiating both (6.159) and (6.158) with respect to $t$ and thus deduce

$$
\begin{aligned}
\Box_{4+1} \partial_t \Phi &= -\partial_t \Phi + \left(1 + \frac{\sin^2 u}{r^2}\right)^{-3/2} v_t, \\
\partial_t \Phi &= \left(1 + \frac{\sin^2 u}{r^2}\right)^{1/2} v_t,
\end{aligned}
\qquad (6.165)
$$

which lead to the convenient equation

$$\Box_{4+1} \partial_t \Phi = \left[-1 + \left(1 + \frac{\sin^2 u}{r^2}\right)^{-2}\right] \partial_t \Phi. \qquad (6.166)$$

For this equation, we apply the inhomogeneous version of the radial Strichartz estimates in Theorem 2.6 to derive

$$
\begin{aligned}
&\|\partial_t \Phi\|_{L^p L^{\frac{4p}{p-1}}} + \|\partial_t^2 \Phi\|_{L^\infty L^2} + \|\partial_t \Phi\|_{L^\infty \dot{H}^1} \\
&\qquad \lesssim \|\partial_t^2 \Phi(0)\|_{L^2(\mathbb{R}^4)} + \|\partial_t \Phi(0)\|_{\dot{H}^1(\mathbb{R}^4)} + \|\Box_{4+1} \partial_t \Phi\|_{L^1 L^2},
\end{aligned}
\qquad (6.167)
$$

for all $2 \le p \le \infty$. Our goal will be to show that the right-hand side is finite, which implies, among others, that (6.164) holds for $s = 2$. One may wonder at this point why we need to use Strichartz estimates to control $\|\partial_t \Phi\|_{L^p L^{\frac{4p}{p-1}}}$ and not rely strictly on energy-type bounds to obtain (6.163). This is definitely not clear in the first step, but as one proceeds to compute $\Box_{4+1} \partial_t^s \Phi$ for $s \ge 2$, it is quite evident that, in order to estimate a quantity like $\|\Box_{4+1} \partial_t^s \Phi\|_{L^1 L^2}$, the control of the previous Strichartz norms is a key ingredient.

In analyzing the right-hand side of (6.167), the easiest term to deal with is the last one. Due to (6.166) and (6.160), we obtain

$$\|\Box_{4+1} \partial_t \Phi\|_{L^1 L^2} \lesssim \|\partial_t \Phi\|_{L^1 L^2} \lesssim T \|\partial_t \Phi\|_{L^\infty L^2} < \infty. \tag{6.168}$$

Next, we work on $\|\partial_t \Phi(0)\|_{\dot{H}^1(\mathbb{R}^4)}$, for which one has first that

$$\|\partial_t \Phi(0)\|_{\dot{H}^1(\mathbb{R}^4)} \simeq \|\partial_{tr}^2 \Phi(0)\|_{L^2(\mathbb{R}^4)}.$$

A direct computation based on (6.165) shows that

$$\partial_{tr}^2 \Phi = \left(1 + \frac{\sin^2 u}{r^2}\right)^{1/2} v_{tr} + \frac{1}{2} \left(1 + \frac{\sin^2 u}{r^2}\right)^{-1/2} v_t \left(\frac{\sin 2u}{r^2} u_r - \frac{2\sin^2 u}{r^3}\right).$$

Using (6.161), Sobolev embeddings, and the regularity of the initial data for both (6.146) and (6.149), we infer

$$\left\| \left(1 + \frac{\sin^2 u(0)}{r^2}\right)^{1/2} \right\|_{L^\infty(\mathbb{R}^4)} \lesssim 1 + \|u(0)\|_{H^{2+\epsilon}(\mathbb{R}^2)} < \infty, \tag{6.169}$$

$$\|v_t(0)\|_{L^\infty(\mathbb{R}^4)} \lesssim \|v_t(0)\|_{H^{2+\epsilon}(\mathbb{R}^4)} < \infty, \tag{6.170}$$

$$\|v_{tr}(0)\|_{L^2(\mathbb{R}^4)} \lesssim \|v_t(0)\|_{H^1(\mathbb{R}^4)} < \infty. \tag{6.171}$$

For the remaining term in the expression of $\partial_{tr}^2 \Phi$, we discuss separately the regimes $r \le 1$ and $r > 1$. If $r \le 1$, then arguing as for (6.162) we deduce

$$\left| \frac{\sin 2u}{r^2} u_r - \frac{2\sin^2 u}{r^3} \right| \le \left| \frac{\sin 2rv \cdot rv - 2\sin^2 rv}{r^3} \right| + \left| \frac{\sin 2rv}{r} v_r \right| \lesssim rv^4 + |vv_r|.$$

For $r > 1$, we can write the coarse bound

$$\left| \frac{\sin 2u}{r^2} u_r - \frac{2\sin^2 u}{r^3} \right| \le \frac{|v| + |\varphi_r|}{r^2} + \frac{|v_r|}{r} + \frac{1}{r^3}.$$

Combining these two estimates, we easily derive

$$\left\| \frac{\sin 2u(0)}{r^2} u_r(0) - \frac{2\sin^2 u(0)}{r^3} \right\|_{L^2(\mathbb{R}^4)}$$

$$\lesssim 1 + \|\varphi_r\|_{L^2(\mathbb{R}^4)} + \|v_r(0)\|_{L^2(\mathbb{R}^4)}(1 + \|v(0)\|_{L^\infty(\mathbb{R}^4)})$$

$$+ \|v(0)\|_{L^2(\mathbb{R}^4)}(1 + \|v(0)\|_{L^\infty(\mathbb{R}^4)}^3)$$

$$\lesssim 1 + \|\varphi_r\|_{L^2(\mathbb{R}^4)} + \|v(0)\|_{H^1(\mathbb{R}^4)}(1 + \|v(0)\|_{H^{2+\epsilon}(\mathbb{R}^4)})^3 < \infty.$$

The joint conclusion of this bound and (6.169)-(6.171) is then

$$\|\partial_t \Phi(0)\|_{\dot{H}^1(\mathbb{R}^4)} < \infty. \tag{6.172}$$

We are left to show the finiteness of $\|\partial_t^2 \Phi(0)\|_{L^2(\mathbb{R}^4)}$, for which we rely on (6.165) to easily calculate

$$\partial_t^2 \Phi = \left(1 + \frac{\sin^2 u}{r^2}\right)^{1/2} v_{tt} + \frac{1}{2}\left(1 + \frac{\sin^2 u}{r^2}\right)^{-1/2} \frac{\sin 2u}{r} v_t^2.$$

In what concerns the second term, we can write

$$\left(1 + \frac{\sin^2 u}{r^2}\right)^{-1/2} \frac{|\sin 2u|}{r} \lesssim 1,$$

which implies

$$\left\|\frac{1}{2}\left(1 + \frac{\sin^2 u(0)}{r^2}\right)^{-1/2} \frac{\sin 2u(0)}{r} v_t^2(0)\right\|_{L^2(\mathbb{R}^4)}$$

$$\lesssim \|v_t(0)\|_{L^\infty(\mathbb{R}^4)} \|v_t(0)\|_{L^2(\mathbb{R}^4)}$$

$$\lesssim \|v_t(0)\|_{H^{2+\epsilon}(\mathbb{R}^4)} \|v_t(0)\|_{L^2(\mathbb{R}^4)} < \infty.$$

If we also factor in (6.169), then the desired conclusion follows by proving the finiteness of $\|v_{tt}(0)\|_{L^2(\mathbb{R}^4)}$.

For this purpose, we use the wave equation (6.149) and argue that

$$\|v_{tt}(0)\|_{L^2(\mathbb{R}^4)} \leq \|\Delta v(0)\|_{L^2(\mathbb{R}^4)} + \|\Box_{4+1} v(0)\|_{L^2(\mathbb{R}^4)}$$

$$\leq \|v(0)\|_{H^2(\mathbb{R}^4)} + \|\Box_{4+1} v(0)\|_{L^2(\mathbb{R}^4)}.$$

In proving the finiteness of the norm associated to the d'Alembertian, one uses a direct, term-by-term investigation, from which we selected the parts corresponding to two representative terms. The first one is $\varphi_{>1} F(u)/r$ and, taking advantage of (6.147), we can estimate it pointwise by

$$\left|\varphi_{>1}\frac{F(u)}{r}\right| \lesssim |\varphi_{>1}|\left(\frac{|u_r|}{r^2} + \frac{|u_r|^2 + |u_t|^2 + 1}{r^3}\right).$$

Hence,

$$\left\|\varphi_{>1}\frac{F(u)(0)}{r}\right\|_{L^2(\mathbb{R}^4)} \lesssim 1 + \|u_t(0)\|_{L^\infty(\mathbb{R}^2)}\|u_t(0)\|_{L^2(\mathbb{R}^2)}$$

$$+ (1 + \|u_r(0)\|_{L^\infty(\mathbb{R}^2)})\|u_r(0)\|_{L^2(\mathbb{R}^2)}$$

$$\lesssim 1 + \|u_t(0)\|_{H^{1+\epsilon}(\mathbb{R}^2)}\|u_t(0)\|_{L^2(\mathbb{R}^2)}$$

$$+ (1 + \|u(0)\|_{H^{2+\epsilon}(\mathbb{R}^2)})\|u(0)\|_{H^1(\mathbb{R}^2)} < \infty.$$

The second term is $\varphi_{<1}\, h_{24}(r, rv)v(v_t^2 - v_r^2)$ and, using the inequality (6.103) for $\tilde{h}_4(u) = 2\Psi_2(r, u)h_{24}(r, u)$, we derive

$$\left|\varphi_{<1}\, h_{24}(r, rv)v(v_t^2 - v_r^2)\right| \lesssim |v|(|v_t|^2 + |v_r|^2).$$

Accordingly, we obtain

$$\left\|\varphi_{<1}\, h_{24}(r, rv)(0)v(0)(v_t^2(0) - v_r^2(0))\right\|_{L^2(\mathbb{R}^4)}$$
$$\lesssim \|v(0)\|_{L^\infty(\mathbb{R}^4)}\left(\|v_t(0)\|_{L^\infty(\mathbb{R}^4)}\|v_t(0)\|_{L^2(\mathbb{R}^4)}\right.$$
$$\left. + \|v_r(0)\|_{L^\infty(\mathbb{R}^4)}\|v_r(0)\|_{L^2(\mathbb{R}^4)}\right)$$
$$\lesssim \|v(0)\|_{H^{2+\epsilon}(\mathbb{R}^4)}\left(\|v_t(0)\|_{H^{2+\epsilon}(\mathbb{R}^4)}\|v_t(0)\|_{L^2(\mathbb{R}^4)}\right.$$
$$\left. + \|v(0)\|_{H^{3+\epsilon}(\mathbb{R}^4)}\|v(0)\|_{H^1(\mathbb{R}^4)}\right) < \infty,$$

which finishes the discussion for the finiteness of $\|\partial_t^2\Phi(0)\|_{L^2(\mathbb{R}^4)}$. Together with (6.167), (6.168), and (6.172), this proves

$$\|\partial_t^2\Phi\|_{L^\infty L^2} + \|\partial_t\Phi\|_{L^\infty \dot{H}^1} < \infty.$$

Therefore, to finally claim (6.163), we are left to show

$$\|\Phi\|_{L^\infty \dot{H}^2} < \infty,$$

for which it is obviously enough to argue for the finiteness of $\|\Delta\Phi\|_{L^\infty L^2}$. First, we use the wave equation (6.159) satisfied by $\Phi$ to infer

$$|\Delta\Phi| \leq |\partial_t^2\Phi| + |\Phi| + |v| + \frac{|\varphi_{\geq 1/2}|}{r^3}.$$

Secondly, based on (6.158), we deduce

$$|v| \leq |\Phi| + \frac{|\varphi_{\geq 1/2}|}{r^3}. \tag{6.173}$$

As a consequence, it follows that

$$\|\Delta\Phi\|_{L^\infty L^2} \lesssim \|\partial_t^2\Phi\|_{L^\infty L^2} + \|\Phi\|_{L^\infty L^2} + \left\|\frac{|\varphi_{\geq 1/2}|}{r^3}\right\|_{L^2(\mathbb{R}^4)},$$

and the previous analysis concludes the argument.

### 6.4.5 *Confirmation of the continuation criterion*

We now have all the necessary tools to finish the proof of Theorem 6.5, which was reduced to showing that $v$ satisfies the continuation criterion given by (6.150). If we invoke classical Sobolev embeddings, then we derive

$$\|\Phi\|_{L^\infty_{t,x}} + \|\nabla_{t,x}\Phi\|_{L^\infty_{t,x}} \lesssim \|\Phi\|_{L^\infty H^4} + \|\partial_t\Phi\|_{L^\infty H^3}.$$

On the other hand, using the radial Sobolev inequality (2.13), we infer that

$$\|r\Phi\|_{L_{t,x}^{\infty}} + \|r\nabla_{t,x}\Phi\|_{L_{t,x}^{\infty}} \lesssim \|\Phi\|_{L^{\infty}H^2} + \|\partial_t\Phi\|_{L^{\infty}H^1}.$$

Thus, on the basis of (6.152), we obtain that

$$\|(1+r)(|\Phi| + |\nabla_{t,x}\Phi|)\|_{L_{t,x}^{\infty}} < \infty. \tag{6.174}$$

Now, the goal is to exploit the connection between $\Phi$ and $v$ and argue that the previous estimate implies (6.150). Due to (6.165) and (6.173), it follows immediately that

$$(1+r)(|v| + |v_t|) \lesssim (1+r)(|\Phi| + |\partial_t\Phi|) + \frac{|\varphi_{\geq 1/2}|}{r^2}.$$

Next, an analysis very similar to the one done for $\partial_r\Phi(0)$ yields the pointwise bounds:

$$|\nabla_x v| \lesssim |\nabla_x \Phi| + r|v|^5 + \frac{|\varphi_{\geq 1/2}|}{r^3}, \qquad \text{if } r \leq 1,$$

and

$$|\nabla_x v| \lesssim |\nabla_x \Phi| + \frac{|v| + |v|^2}{r^2} + \frac{|\varphi_{\geq 1/2}|}{r^3}, \qquad \text{if } r > 1.$$

It is very clear then that combining the last three estimates with (6.174) produces (6.150), thus finishing the whole argument.

# Chapter 7

# The question of collapse for wave maps

In this chapter, we finally get to investigate the possibility of collapsing solutions for the wave map problem. By the general term "collapse", we mean the occurrence of either finite time blow-up or finite time loss of regularity for a certain solution. We will concentrate on recent advances for this issue in the context of the wave map problem with the base manifold being $\mathbb{R}^{2+1}$ and $\mathbb{S}^2$ as its target. The regularity theory developed before, especially the analysis of the large data problem in Chapter 4, reveals that the wave map problem can experience collapse only if two conditions are met:

- the problem admits stationary solutions (i.e., harmonic maps);
- there exists concentration of energy strictly inside a characteristic cone.

In fact, we will see that what we need is stationary solutions which are invariant under spatial dilations. The dilation invariance is a feature of the associated elliptic problem, which is conformally invariant in two spatial dimensions, but is not a property of the evolution equation. This is an indication that collapsing solutions may have quite delicate dynamic behavior.

As energy concentration is crucial to collapsing solutions, we will consider the phenomenon of energy concentration inside the characteristic cone as the hallmark of collapse. A rigorous analysis of collapsing solutions has been carried out only in the restricted equivariant case, leaving open the problem for other relevant settings. We will see that this investigation demonstrates that the collapsing solutions are stable under small equivariant perturbations, but it does not say anything about the stability under general, non-equivariant perturbations. Nevertheless, the analysis we present here answers an important question raised for the wave map problem.

The issue of collapse for the nonlinear sigma model has a long history. We recall from Chapter 1 that, at the classical level, this field theory was proposed as a particle model and its static solutions represent elementary particles. Furthermore, from the same chapter, we remember the heuristic argument based on dilation and Derrick's result asserting that harmonic maps might be dynamically unstable. For wave maps having the domain manifold $\mathbb{R}^{n+1}$, with $n \geq 3$, we have the pioneering results of Shatah [143], Shatah-Tavilhdar-Zadeh [148], and Cazenave-Shatah-Tavilhdar-Zadeh [25] demonstrating the existence of self-similar collapsing solutions. In what concerns the $2 + 1$-dimensional problem, it was conjectured that solutions collapse if their energy is sufficiently large (i.e., above the threshold energy of a harmonic map), with progress being slow up until the breakthroughs made by Rodnianski-Sterbenz [138] and Krieger-Schlag-Tataru [97]. However, these results had strong motivation in the numerical evidence obtained by Bizoń-Chmaj-Tabor [18], the heuristic analysis of Ovchinnikov-Sigal [128], and the instability result on harmonic maps of Côte [30]. The most complete up-to-date $2 + 1$-dimensional result, covering the important case of collapsing wave maps with unit charge, is due to Raphaël-Rodnianski [136], whose work in turn was inspired by contributions of Perelman [129] and Merle-Raphaël [115–119] toward understanding the formation of singularities for critical nonlinear Schrödinger equations.

In this chapter, our goal is to provide a novel approach to the collapse part of Raphaël-Rodnianski's result. For pedagogical reasons, we do not cover their derivation of sharp blow-up profiles, which is more technically involved, and we refer the reader for more details on it to the original paper and also to the work of Rodnianski-Sterbenz [138]. In proving the collapse, our strategy is similar to the one used by Raphaël-Rodnianski: we deduce a pair of ingenious differential inequalities, which imply together the existence of a "trapping regime" driving the solution toward collapse. However, we achieve this by relying on a new setting, which makes the structure of the problem and the estimates more transparent.

The organization of the material is as follows. First, we provide a precise setup of the problem, recall the equivariant reduction which leads to a single semilinear wave equation, and discuss the relevant static solutions serving as prototypes of blow-up profiles. The latter will include comments on the Belavin-Polyakov analysis [14] of two-dimensional harmonic maps. Following this, we present the complex version of the associated Hodge system and its equivariant reductions, which will be our basic set of equations.

This is a system of first-order equations coupled with a set of commutation relations between certain operators. Next, we derive a second-order energy identity and the Pohozaev conformal identity, which are crucial ingredients in our analysis. With all these prerequisites covered, we focus subsequently on deriving various operator and norm estimates, as well as the differential inequalities needed to demonstrate the collapse. This is the most technical part of our argument. In the final section, we use only the differential inequalities to conclude the proof, thus making the case for this part to be read independently in order to see where all the technical material leads to.

## 7.1 Introduction

The main object of study for our analysis is the equivariant case for the 2+1-dimensional wave map problem with $\mathbb{S}^2$ as the target manifold. Writing the metric on $\mathbb{S}^2$ in standard spherical coordinates,

$$dR^2 + \sin^2(R)d\Theta^2, \tag{7.1}$$

the wave map system[1] reads as

$$\Box R + 2\sin(R)\cos(R)\left\{\nabla_\mu \Theta \nabla^\mu \Theta\right\} = 0,$$
$$-\nabla_\mu \left\{\sin^2(R)\nabla^\mu \Theta\right\} = 0.$$

Using now the polar coordinates $(t, r, \theta)$ on $\mathbb{R}^{2+1}$ and making the equivariant ansatz

$$(R, \Theta) = (R(t, r), k\theta), \qquad k \in \mathbb{Z}, \tag{7.2}$$

we see that the second equation is automatically satisfied, whereas the first one reduces to

$$\Box_{t,r} R + \frac{k^2 \sin(2R)}{2r^2} = 0, \qquad \Box_{t,r} = \partial_t^2 - \partial_r^2 - \frac{\partial_r}{r}. \tag{7.3}$$

The goal of our argument is to construct collapsing solutions for this equation. Given the radial nature of the problem, if we prescribe data at $t = 0$, i.e.,

$$R(0, r) = R_0(r), \qquad R_{,t}(0, r) = R_1(r), \tag{7.4}$$

and solve the equation forward in time, then a potential singularity point is of the type $(T, 0)$, where $T > 0$ is some future time. As we can imagine,

---

[1]One may have considered a target manifold with the metric $h = dR^2 + f^2(R)d\Theta^2$, with $f(0) = f(R_*) = 0$ and $f > 0$ in $(0, R_*)$. However, this complicates matters and does not lead to any valuable insight.

$\theta$ disappears completely from further considerations; however, an implicit integration with respect to this variable still survives in all the integrals to be considered henceforth, contributing a factor of $2\pi$. To streamline the computational aspect of the presentation, we scale out this factor. Furthermore, we will also write $\Box$ for $\Box_{t,r}$, unless we really want to emphasize the radial feature of our problem.

Following this, we briefly discuss the conservation of energy for (7.3), since it plays a crucial role in the process of collapse. If we define the energy and momentum densities by

$$e(R) \stackrel{\text{def}}{=} R_{,t}^2 + R_{,r}^2 + \frac{k^2 \sin^2(R)}{r^2},$$

$$p(R) \stackrel{\text{def}}{=} 2R_{,t}R_{,r},$$

then we have the identity

$$\partial_t\{r\,e(R)\} - \partial_r\{r\,p(R)\} = 0,$$

which implies that the evolution of the equation preserves the total energy

$$E_{\text{tot}}(R, R_{,t}) \stackrel{\text{def}}{=} \int_0^\infty \{e(R)\}\,r\,dr.$$

On the other hand, the previous differential identity also leads to local energy conservation. An integration over the region $\{(t,r)\,|\,r \leq T-t,\, t_1 \leq t \leq t_2\}$, with $0 < t_1 < t_2 < T$, yields

$$\int_{D_{t_1}} \{e(R)\}\,r\,dr + \int_{\partial C[t_1,t_2]} \left\{(R_{,t} + R_{,r})^2 + \frac{k^2 \sin^2(R)}{r^2}\right\}r\,dr = \int_{D_{t_2}} \{e(R)\}\,r\,dr,$$

where the domains for the integrals are defined by

$$D_t \stackrel{\text{def}}{=} \{(t,r)\,|\,t = \text{fixed},\, r \leq T - t\}$$

$$\partial C[t_1, t_2] \stackrel{\text{def}}{=} \{(t,r)\,|\,r = T - t,\, t_1 \leq t \leq t_2 < T\}.$$

Thus, the map $t \mapsto \int_{D_t}\{e(R)\}r\,dr$ is decreasing on the interval $[0,T)$ and we know that a necessary condition for collapse at the point $(T,0)$ is

$$\lim_{t \to T} \int_{D_t} \{e(R)\}\,r\,dr \geq 4|k|.$$

In fact, the *quantization conjecture* states that we should have equality in the limit and the expectation is that the collapsing solution converges in some sense to a harmonic map. This energy concentration condition follows from the argument in Chapter 4, where it was shown that a solution whose energy is below the critical level $4|k|$ remains regular for all times. Moreover, the general regularity theory there claims that if there is no energy concentration inside a cone narrower than the characteristic one, then the solution is regular. Here, we see the complement of this scenario, namely collapsing solutions for which the main part of the energy concentrates in a narrow region near the central timeline $\{r = 0\}$, while the remainder (also called excess energy) is radiated outside the characteristic cone.

The other essential ingredient for collapse is the existence of stationary/static solutions with finite energy and we know that (7.3) admits such solutions. The following describes a simple way to construct them. What we try to do is minimize the energy of a static solution $R = Q(r)$, which can be rewritten as

$$\int_0^\infty \left\{ Q_{,r}^2 + k^2 \frac{\sin^2(Q)}{r^2} \right\} r\,dr$$

$$= \int_0^\infty \left( Q_{,r} \mp k \frac{\sin(Q)}{r} \right)^2 r\,dr \pm 2k \int_0^\infty Q_{,r} \sin(Q)\,dr$$

$$= \int_0^\infty \left( Q_{,r} \mp k \frac{\sin(Q)}{r} \right)^2 r\,dr \mp 2k \left( \cos Q(\infty) - \cos Q(0) \right).$$

In the above, the choice of sign is such that the last term is equal to $4|k|$. This computation indicates that minimizers of the energy should solve the ordinary differential equation

$$\frac{dQ}{dr} = \pm k \frac{\sin(Q)}{r},$$

which can be easily integrated using separation of variables. The end result[2] is the family of static solutions

$$Q_\lambda^{(k)}(r) = 2 \arctan\left( (\lambda r)^k \right),$$

where $k \in \mathbb{Z}$ and $\lambda$ is a free real parameter. The latter points to the invariance of solutions under scaling, with the energy being also independent of $\lambda$. If we introduce $Q(r) := 2 \arctan(r)$ and[3]

$$Q^{(k)}(r) \stackrel{\text{def}}{=} Q(r^k), \qquad k \in \mathbb{Z}, \tag{7.5}$$

---

[2]This can also be quantified as the identity

$$\Gamma_{\mathbb{S}^2}(Q) \mp k \Gamma_{\mathbb{R}^2}(r) = \text{constant},$$

with $\Gamma_M$ denoting the Green's function of the manifold $M$.

[3]It is enough to consider the case $k \geq 1$, due to the identity $Q^{(k)}(r) + Q^{(-k)}(r) = \pi$.

then our previous family of static maps can be written concisely as $Q_\lambda^{(k)}(r) := Q^{(k)}(\lambda r)$, where we use the subscript $\lambda$ to conveniently denote dilated functions. The energy associated with these maps is

$$\int_0^\infty \left\{ (Q_{,r}^{(k)})^2 + k^2 \frac{\sin^2(Q^{(k)})}{r^2} \right\} r\,dr = 4|k|,$$

which corroborates the computation done before.

Next, the idea is to look for rescaled solutions of the type $Q^{(k)}(r/\lambda(t))$, where $\lambda = \lambda(t)$ is some function satisfying $\lambda(t) \to 0$ as $t \to T$. Since the energy needs to concentrate strictly inside the characteristic cone, it is tempting to investigate first self-similar solutions, i.e., solutions of the form $\widetilde{Q}(r/(T-t))$. In what follows, we prove that such solutions exist; however, they have infinite energy and, thus, do not serve our purpose. We recall the hyperbolic coordinates used in parametrizing the characteristic cone $\{r \leq T - t\}$,

$$T - t = \rho \cosh(\chi), \qquad r = \rho \sinh(\chi), \qquad \theta = \theta,$$

where $\rho := \sqrt{(T-t)^2 - r^2}$ and $\chi := \tanh^{-1}(r/(T-t))$. The metric on the base manifold becomes

$$d\rho^2 + \rho^2 \left( d\chi^2 + \sinh^2(\chi) d\theta^2 \right),$$

with the corresponding radial d'Alembertian being given by

$$\Box_{\rho,\chi} = \partial_\rho^2 + \frac{2}{\rho}\partial_\rho - \frac{1}{\rho^2 \sinh(\chi)} \partial_\chi \left\{ \sinh(\chi)\partial_\chi \right\}.$$

Thus, the equivariant equation (7.3) can be restated in these coordinates as

$$R_{,\rho\rho} + \frac{2}{\rho}R_{,\rho} - \frac{1}{\sinh(\chi)}\partial_\chi \left( \sinh(\chi)R_{,\chi} \right) + \frac{k^2 \sin(2R)}{2\sinh^2(\chi)} = 0,$$

where self-similar solutions are described by $R = \widetilde{Q}(\chi)$ (i.e., $R$ is independent of $\rho$, which plays the role of the time variable). In the same spirit with the computations done for harmonic maps, if we try to minimize the energy of such a function, which is

$$\int_0^\infty \left\{ \widetilde{Q}_{,\chi}^2 + \frac{k^2 \sin^2(\widetilde{Q})}{\sinh^2(\chi)} \right\} \sinh(\chi)d\chi$$

$$= \int_0^\infty \left( \widetilde{Q}_{,\chi} \mp \frac{k\sin(\widetilde{Q})}{\sinh(\chi)} \right)^2 \sinh(\chi)d\chi \pm 2k \int_0^\infty \sin(\widetilde{Q})\widetilde{Q}_\chi d\chi,$$

then we are led to solve the ordinary differential equation

$$\widetilde{Q}_{,\chi} = \frac{k\sin(\widetilde{Q})}{\sinh(\chi)}.$$

One easily obtains the family of solutions

$$\widetilde{Q}^{(k)}(\chi) = 2\arctan\left(\lambda^k \tanh^k(\chi/2)\right) = 2\arctan\left((\lambda p)^k\right) = Q\left((\lambda p)^k\right),$$

where $k \in \mathbb{Z}$ and

$$p \overset{\text{def}}{=} \tanh(\chi/2) = \frac{r}{T - t + \sqrt{(T-t)^2 - r^2}}.$$

As previously anticipated, based on

$$p_{,t} = \frac{p}{\rho}, \qquad p_{,r} = \frac{((T-t)/r)p}{\rho},$$

a straightforward calculation shows that these solutions have infinite energy and, hence, they are not acceptable. Nevertheless, they serve as prototypes for the collapsing solutions we construct later.

The large data result of Sterbenz-Tataru discussed in Chapter 4 tells us that, for collapse to occur, energy concentration must happen inside a cone of the form $r = (T - t)\delta(t)$, with $\delta(t) \to 0$ as $t \to T$. Hence, we can further refine the profile $Q^{(k)}(r/\lambda(t))$ of potential collapsing solutions to include $\lambda(t)/(T - t) \to 0$ as $t \to T$. In what follows, we write $\tau_k = \tau_k(t)$ for the rescaled time function $\lambda(t)$ corresponding to a specific value of the parameter $k \in \mathbb{Z}$. The main result of Raphaël-Rodnianski [136], whose collapse part is discussed for the remainder of this chapter, addresses exactly the above profiles and shows that they are stable blow-up solutions under small equivariant perturbations. Here is its precise formulation.

**Theorem 7.1.** ([136]) *For a fixed integer $k \geq 1$, there exists an open set $\mathcal{O}$ of initial data $(R_0, R_1)$ for which the corresponding Cauchy problem (7.3)-(7.4) admits a collapsing solution $R^{(k)} = R^{(k)}(t, r)$ on the time interval $[0, T)$, with $T = T(R_0, R_1)$. Moreover, this solution satisfies*

$$\lim_{t \to T} \left\| R^{(k)}(t, \tau_k(t)y) - Q^{(k)}(y) \right\|_{H^1_{\text{loc},y}} = 0, \qquad \text{as} \qquad t \to T \qquad (7.6)$$

*where the rescaled time function $\tau_k = \tau_k(t)$ has the asymptotic behavior*

$$\tau_k(t) \sim (T - t)a_k(t) \qquad \text{as} \quad t \to T,$$

*with*

$$a_1(t) = e^{-\sqrt{|\log(T-t)|} + O(1)} \qquad \text{if} \ \ k = 1,$$

$$a_k(t) = \frac{c_k + o(1)}{|\log(T - t)|^{\frac{1}{2k-2}}} \qquad \text{if} \ \ k \geq 2,$$

*while the speed $b_k := -d\tau_k/dt$ behaves like*

$$b_k(t) \sim a_k(t) \to 0 \qquad as \quad t \to T. \tag{7.7}$$

*In addition, we can find a pair of radial functions $(U, V)$ such that*

$$\lim_{t \to T} \left\| \left( R^{(k)}(t, r) - Q^{(k)}\left(\frac{r}{\tau_k(t)}\right) - U(r), \ \partial_t R(t, r) - V(r) \right) \right\|_H = 0, \tag{7.8}$$

*where the $H$-norm is defined by*

$$\|(U, V)\|_H^2 \stackrel{\text{def}}{=} \int_0^\infty \left\{ V^2 + (\partial_r U)^2 + \frac{U^2}{r^2} \right\} r \, dr. \tag{7.9}$$

*Finally, the energy is quantized in the sense that*

$$E_{\text{tot}}(R^{(k)}, \partial_t R^{(k)}) = E_{\text{tot}}(Q^{(k)}, 0) + E_{\text{tot}}(U, V)$$

*with*

$$E_{\text{tot}}(Q^{(k)}, 0) = 4|k|, \qquad E_{\text{tot}}(U, V) \ll 1.$$

**Remark 7.1.** The article by Raphaël-Rodnianski also covers the equivariant case of the $SO(4)$ Yang-Mills problem

$$F_{\alpha\beta} = 2\partial_{[\alpha}, A_{\beta]} + [A_\alpha, A_\beta],$$
$$\partial_\beta F^{\alpha\beta} + [A_\beta, F^{\alpha\beta}] = 0,$$

with $0 \le \alpha, \beta \le 3$. The equivariant ansatz

$$A_\alpha^{ij} = 2\delta_\alpha^{[i} x^{j]} \frac{1 - u(t, r)}{r^2}$$

reduces the previous system to the semilinear wave equation

$$\Box_{t,r} u - \frac{2u(1 - u^2)}{r^2} = 0.$$

The similarity with (7.3) is striking and the ideas in the blow-up analysis are much the same, if not somewhat simpler.

Let us provide now some historical remarks on the problem that makes the subject of the previous theorem. The first rigorous analysis of collapse appeared in works by Rodnianski-Sterbenz [138] and, independently, Krieger-Schlag-Tataru [97]. The former deals with the case $k \ge 4$ and constructs collapsing solutions of the type

$$R(t, r) = (Q + \epsilon)\left(t, \frac{r}{\tau(t)}\right), \tag{7.10}$$

with $U = U(t,r) := \epsilon\,(t, r/\tau(t))$ satisfying

$$\|U, \partial_t U\|_{\dot{H}^1 \times L^2(\mathbb{R}^2)} \ll 1$$

and the rescaled time function behaving as

$$\lim_{t \to T} \tau(t) = 0, \qquad \tau(t) \geq \frac{T - t}{|\log(T - t)|^{1/4}}.$$

On the other hand, the result of Krieger-Schlag-Tataru addresses the challenging case $k = 1$ and builds solutions having the same profile as above, but with

$$\tau(t) = (T - t)^\nu, \qquad \nu > \frac{3}{2}.$$

It is not clear at the moment if there is a relation between the two results. Even if there is numerical evidence indicating that the blow-up regime obtained by Krieger-Schlag-Tataru is unstable, a precise explanation is still missing.

Next, we want to comment on some related open questions, which have to do with the alternative derivation of static solutions presented in Subsection 1.2.9 (also see Belavin-Polyakov [14]). As a reminder, by using the stereographic projection

$$w = \tan\left(R/2\right)e^{i\Theta} \tag{7.11}$$

and complexifying $\mathbb{R}^{2+1}$ according to

$$z \stackrel{\text{def}}{=} x^1 + ix^2, \qquad (t, x^1, x^2) \mapsto (t, \bar{z}, z),$$

the wave map system becomes

$$\left(\partial_t^2 - 4\partial_{\bar{z}}\partial_z\right)w + \frac{2\overline{w}}{1 + |w|^2}\left(4\partial_{\bar{z}}w\partial_z w - (\partial_t w)^2\right) = 0. \tag{7.12}$$

It is clear that $w = f(z)$ and $w = \overline{f(z)}$, where $f$ is an arbitrary analytic function, are static solutions to this equation and their energy is given by

$$E(f) = \int_{\mathbb{C}} \left\{ \frac{4|f'(z)|^2}{(1 + |f(z)|^2)^2} \right\} dm(z), \qquad dm(z) = \frac{1}{2i}d\bar{z} \wedge dz.$$

Here, we want to focus on the restricted family of static, equivariant solutions

$$w^{(k)}(z) = z^k, \qquad k \in \mathbb{Z},$$

which are the projection of the solutions found earlier, namely $Q^{(k)}(r)e^{ik\theta}$. Using the invariance of the problem under dilations, we deduce that $(z/\lambda)^k$, with $\lambda$ being a nonzero, real parameter, is also a static, equivariant solution. The next idea is to look for equivariant wave maps of the form

$$\left(\frac{z}{\lambda(t)}\right)^k + u(t,r)e^{ik\theta}$$

and investigate the dynamics of the rescaled time function $\lambda = \lambda(t)$. Theorem 7.1 can be interpreted as a stability result for blow-up profiles of the above type, under a smallness assumption for $u$. A first, related open problem is to look for general solutions of the form

$$\left(\frac{z}{\lambda(t)}\right)^k + u(t,z)$$

and formulate heuristics for the dynamics of $\lambda(t)$, together with an investigation of stability for potential blow-ups. A more difficult problem is to research the same issues, but with the monomial $z^k$ replaced by an expression of the type

$$c(t)\prod_{j=1}^{k}(z - z_j(t)).$$

This makes sense, as we know that the polynomial

$$w = c\prod_{j=1}^{k}(z - z_j), \qquad c, z_j \in \mathbb{C},$$

is a static solution for (7.12), whose energy is $4|k|$.

Following this, let us discuss the connection between our problem and the question of blow-up for the cubic focusing nonlinear Schrödinger equation, whose analysis inspired in part the work of Raphaël-Rodnianski. This equation reads as

$$i\psi_{,t} - \Delta\psi - |\psi|^2\psi = 0, \qquad \psi = \psi(t, \mathbf{x}), \qquad (t, \mathbf{x}) \in \mathbb{R} \times \mathbb{R}^2,$$

and the corresponding conserved energy is given by

$$\int_{\mathbb{R}^2}\left\{|\nabla\psi(t)|^2 - \frac{1}{2}|\psi(t)|^4\right\}d\mathbf{x}.$$

Due to the energy being indefinite, one can picture a possible blow-up picture in which the overall energy stays constant, but its two terms blow-up separately. In fact, the situation is quite subtle, with Weinstein [187] observing that collapse requires for $\|\psi\|_{L^2}$ to be above a certain threshold and constructing special blow-up solutions. On the other hand, Glassey [58] gave an earlier, very short argument for collapse based on a set of differential inequalities. Sharp blow-up profiles for radially-symmetric solutions were conjectured and checked numerically by Landman-LeMesurier-Papanicolaou-Sulem-Sulem [98] and LeMesurier-Papanicolaou-Sulem-Sulem [104–106], using a delicate asymptotic analysis inspired by work of Talanov-Vlasov [166]. After many years, a rigorous argument certifying these profiles was finally developed by Perelman [129] and Merle-Raphaël [115–119], proving that, near the blow-up time $T$, the solution behaves like $\psi(r/\lambda(t))$, with the rescaled time function satisfying the asymptotics

$$\lambda(t) \sim \sqrt{\frac{2\pi(T-t)}{\log|\log(T-t)|}} \qquad \text{as} \quad t \to T.$$

Contrasting this Schrödinger equation, the energy of our problem is positive-definite and the possible blow-up mechanism is harder to understand. The only feature that points to collapse for the wave map problem is the dilation invariance of the static problem, which gives the possibility of energy concentration. As we will see, even our modest goal of proving finite time collapse (and not analyzing sharp blow-up profiles) is quite challenging.

In finishing this introduction, it is perhaps useful to describe in a nutshell what we try to achieve in the following sections. The idea is simple, yet the implementation is not so simple. We consider trial blow-up profiles of the form $Q(r/\tau(t))$, with $\tau = \tau(t) \geq 0$ being a dynamic time function. For collapsing solutions we need to have $\tau(T) = 0$ for some $T > 0$, and the goal is to establish a differential inequality of the type

$$\tau|\ddot{\tau}| \leq \delta(\dot{\tau})^2, \qquad \text{with} \quad \delta \ll 1.$$

We choose $\tau(0) > 0$ and $\dot{\tau}(0) < 0$ (we may pick instead $-\dot{\tau}(0)/\tau(0) > 0$). Due to the differential inequality, we infer that

$$\frac{d}{dt}\left(\frac{-\dot{\tau}}{\tau^{2\delta}}\right) = \frac{2\delta(\dot{\tau})^2 - \tau\ddot{\tau}}{\tau^{2\delta+1}} \geq \frac{\delta(\dot{\tau})^2}{\tau^{2\delta+1}} \geq 0$$

holds as long as $\tau(t) > 0$. Hence, we deduce that

$$\frac{-\dot{\tau}(t)}{\tau^{2\delta}(t)} \geq \frac{-\dot{\tau}(0)}{\tau^{2\delta}(0)} > 0,$$

which implies, after integration over an interval $[0, T]$,

$$\tau^{1-2\delta}(T) \leq \tau^{1-2\delta}(0) - CT.$$

Obviously, this estimate tells us that $t \mapsto \tau(t)$ cannot be positive for all times.

## 7.2   The associated Hodge system in complex coordinates

Our approach to proving collapse is to work with the associated Hodge system (instead of the wave equation (7.3)), which we claim allows for certain aspects of the evolution to become more transparent. We recall from Subsection 1.2.7 that the Hodge system is the first-order system

$$\nabla_\mu^A \psi^\mu = 0, \qquad \nabla_\mu^A \psi_\nu = \nabla_\nu^A \psi_\mu,$$

$$\nabla_\mu A_\nu - \nabla_\nu A_\mu = \frac{K}{2i}\left(\psi_\mu \overline{\psi_\nu} - \overline{\psi_\mu}\psi_\nu\right), \qquad (7.13)$$

where $0 \le \mu, \nu \le 2$,

$$\psi_\nu \stackrel{\text{def}}{=} \sqrt{h}\,\nabla_\nu w, \qquad K \stackrel{\text{def}}{=} \frac{-2}{h}\cdot\frac{\partial^2 \log(h)}{\partial w \partial \overline{w}},$$

$$A_\mu \stackrel{\text{def}}{=} \frac{1}{2i}\left(\partial_w \log(h)\nabla_\mu w - \partial_{\overline{w}} \log(h)\nabla_\mu \overline{w}\right).$$

In our setting,

$$h(w) = \frac{4}{(1+|w|^2)^2}$$

and easy calculations yield

$$\psi_\nu = \frac{2\nabla_\nu w}{1+|w|^2}, \qquad A_\mu = \frac{1}{i(1+|w|^2)}\left(w\nabla_\mu \overline{w} - \overline{w}\nabla_\mu w\right), \qquad K = 1.$$

We wish to write the system (7.13) in complex coordinates, and, for this purpose, we introduce the notation $(t, z) := (x^0, x^1 + ix^2)$ and

$$\partial_{\overline{z}} \stackrel{\text{def}}{=} \frac{1}{2}(\partial_1 + i\partial_2), \qquad \partial_z \stackrel{\text{def}}{=} \frac{1}{2}(\partial_1 - i\partial_2).$$

Furthermore, we can find the pair of potentials $(\omega, v)$ such that

$$A_1 = \partial_2 \omega + \partial_1 v, \qquad A_2 = -\partial_1 \omega + \partial_2 v, \qquad (7.14)$$

which implies

$$\frac{1}{2}(A_1 + iA_2) = -i\omega_{,\overline{z}} + v_{,\overline{z}}. \qquad (7.15)$$

Using these conventions, the derivatives associated to the gauge field $A$ become

$$\mathcal{G} \stackrel{\text{def}}{=} \partial_{\overline{z}} + \omega_{,\overline{z}} + iv_{,\overline{z}}, \quad \mathcal{G}^* \stackrel{\text{def}}{=} -\partial_z + \omega_{,z} - iv_{,z}, \quad \partial_t^A \stackrel{\text{def}}{=} \partial_t + iA_0. \quad (7.16)$$

On the other hand, for the complex-valued field $\psi$ we construct

$$\psi_{\overline{z}} \stackrel{\text{def}}{=} \frac{1}{2}(\psi_1 + i\psi_2), \qquad \psi_z \stackrel{\text{def}}{=} \frac{1}{2}(\psi_1 - i\psi_2), \qquad \psi_0 \stackrel{\text{def}}{=} \psi_0, \qquad (7.17)$$

while for $G_{\mu\nu} := (1/2i)\big(\psi_\mu \overline{\psi_\nu} - \overline{\psi_\mu}\psi_\nu\big)$ we define

$$B_0 \stackrel{\text{def}}{=} -\frac{1}{2}G_{12} = \frac{1}{2}\big(|\psi_z|^2 - |\psi_{\bar z}|^2\big), \tag{7.18}$$

$$B_1 \stackrel{\text{def}}{=} \frac{i}{2}\big(G_{01} + i\,G_{02}\big) = \frac{1}{2}\big(\psi_0 \overline{\psi_z} - \overline{\psi_0}\psi_{\bar z}\big). \tag{7.19}$$

We notice that $B_0$ is real-valued and $B_1$ is complex-valued, and we warn the reader that the indices in (7.17) do not indicate differentiation, whereas $\psi_{\bar z}$ is not the complex conjugate of $\psi_z$.

With these prerequisites, the Hodge system (7.13) can be written in the form

$$\mathcal{G}\psi_z = \frac{1}{4}\partial_t^A \psi_0, \qquad \mathcal{G}^*\psi_{\bar z} = -\frac{1}{4}\partial_t^A \psi_0, \tag{7.20}$$

$$\partial_t^A \psi_{\bar z} = \mathcal{G}\psi_0, \qquad -\partial_t^A \psi_z = \mathcal{G}^*\psi_0, \tag{7.21}$$

$$[\mathcal{G}, \mathcal{G}^*] = K B_0, \qquad [\partial_t^A, \mathcal{G}] = K B_1, \qquad [\partial_t^A, \mathcal{G}^*] = K\overline{B_1}. \tag{7.22}$$

In relation to comments made in Chapter 1, the first set of equations describes the evolution of the system, while the remaining equations represent compatibility conditions. For later use, the commutation relations can also be restated as

$$\partial_{\bar z}\big\{\partial_t(\omega + iv) - iA_0\big\} = K B_1, \tag{7.23}$$

$$\partial_z\big\{\partial_t(\omega - iv) + iA_0\big\} = K\overline{B_1}, \tag{7.24}$$

$$\partial_{\bar z}\partial_z\omega = \frac{K}{2}B_0. \tag{7.25}$$

We also observe that $B_0$ and $B_1$ are connected through the equation

$$\partial_t B_0 - \partial_z B_1 - \partial_{\bar z}\overline{B_1} = 0,$$

which implies the conservation of the quantity

$$\deg(\psi) = \int_{\mathbb{C}} \{B_0(t,z)\}\, dm(z).$$

From the system (7.20)-(7.22), we can easily derive the wave equations

$$\Big\{\big(\partial_t^A\big)^2 + 4\mathcal{G}^*\mathcal{G}\Big\}\psi_z = -K\overline{B_1}\psi_0, \qquad \Big\{\big(\partial_t^A\big)^2 + 4\mathcal{G}\mathcal{G}^*\Big\}\psi_{\bar z} = K B_1\psi_0,$$

$$\Big\{\big(\partial_t^A\big)^2 + 4\mathcal{G}\mathcal{G}^*\Big\}\psi_0 = 4K B_1\psi_z, \qquad \Big\{\big(\partial_t^A\big)^2 + 4\mathcal{G}^*\mathcal{G}\Big\}\psi_0 = -4K\overline{B_1}\psi_{\bar z}.$$

One may ask at this moment what the point is with all these computations. The answer is that certain structures emerge in this setting, which can be exploited in the study of the evolution problem. To be precise, we will take advantage of these structures in order to derive energy-type estimates for the equivariant case. Another advantage of this framework is that it is definitely easier to work with a first-order system, rather than with the original second-order equation (7.3).

In concluding this section, let us notice that there is an elegant way in which one can write the previous complex Hodge system. A straightforward calculation reveals

$$
\mathbb{G} \overset{\text{def}}{=} \begin{pmatrix} e^{-\omega-iv} & 0 \\ 0 & e^{\omega-iv} \end{pmatrix} \begin{pmatrix} 0 & \partial_{\bar{z}} \\ -\partial_z & 0 \end{pmatrix} \begin{pmatrix} e^{-\omega+iv} & 0 \\ 0 & e^{\omega+iv} \end{pmatrix} = \begin{pmatrix} 0 & \mathcal{G} \\ \mathcal{G}^* & 0 \end{pmatrix},
$$

which can be used then to cast the complex Hodge system in the matrix form

$$
\begin{pmatrix} \frac{i}{2}\partial_t^A & \mathcal{G} \\ \mathcal{G}^* & \frac{i}{2}\partial_t^A \end{pmatrix} \begin{pmatrix} -\frac{i}{2}\psi_0 & \psi_{\bar{z}} \\ -\psi_z & -\frac{i}{2}\psi_0 \end{pmatrix} = 0.
$$

## 7.3 The equivariant reduction

Our next task is to use the equivariant ansatz and, thus, express the Hodge system (7.20)-(7.22) in this setting. It is helpful first to separate equivariant maps into "sectors/fibers" according to the scheme

$$
\mathbb{F}_n \overset{\text{def}}{=} \left\{ w : \mathbb{R}^{2+1} \to \mathbb{C} \,\middle|\, w = u(t,r)e^{in\theta},\, u \in \mathbb{R} \right\}, \qquad n \in \mathbb{Z}. \qquad (7.26)
$$

In polar coordinates, we have

$$
\partial_{\bar{z}} = \frac{e^{i\theta}}{2}\left( \partial_r + \frac{i}{r}\partial_\theta \right), \qquad\qquad \partial_z = \frac{e^{-i\theta}}{2}\left( \partial_r - \frac{i}{r}\partial_\theta \right),
$$

and, subsequently, if $w \in \mathbb{F}_n$, then

$$
\psi_{\bar{z}} = \frac{2w_{,\bar{z}}}{1+|w|^2} = \frac{u_{,r} - (n/r)u}{1+u^2}e^{i(n+1)\theta} \in \mathbb{F}_{n+1}, \qquad (7.27)
$$

$$
\psi_z = \frac{2w_{,z}}{1+|w|^2} = \frac{u_{,r} + (n/r)u}{1+u^2}e^{i(n-1)\theta} \in \mathbb{F}_{n-1}, \qquad (7.28)
$$

$$
\psi_0 = \frac{2\partial_t w}{1+|w|^2} = \frac{2u_{,t}}{1+u^2}e^{in\theta} \in \mathbb{F}_n. \qquad (7.29)
$$

For the operators $\mathcal{G}$ and $\mathcal{G}^*$, one deduces

$$
\mathcal{G} = \frac{e^{i\theta}}{2}\left( \partial_r + \frac{i}{r}\partial_\theta + \omega_{,r} + iv_{,r} \right), \qquad \mathcal{G} : \mathbb{F}_n \mapsto \mathbb{F}_{n+1},
$$

$$
\mathcal{G}^* = \frac{e^{-i\theta}}{2}\left( -\partial_r + \frac{i}{r}\partial_\theta + \omega_{,r} - iv_{,r} \right), \qquad \mathcal{G}^* : \mathbb{F}_{n+1} \mapsto \mathbb{F}_n,
$$

and we set

$$\mathcal{L} \overset{\text{def}}{=} 2e^{-i\theta}\mathcal{G}, \qquad\qquad \mathcal{L}^{\dagger} \overset{\text{def}}{=} 2e^{i\theta}\mathcal{G}^{*}.$$

Since we consider equivariant solutions, we want to isolate the radial part of these operators by defining their restriction on the above fibers,

$$\mathcal{L}\big|_{\mathbb{F}_n} = \partial_r - \frac{n}{r} + \omega_{,r} + iv_{,r} \overset{\text{def}}{=} \mathcal{L}_n, \qquad (7.30)$$

$$\mathcal{L}^{\dagger}\big|_{\mathbb{F}_n} = -\partial_r - \frac{n}{r} + \omega_{,r} - iv_{,r} \overset{\text{def}}{=} \mathcal{L}_{n-1}^{*}. \qquad (7.31)$$

The last notation is consistent with the easily-checked fact that it indeed represents the adjoint of $\mathcal{L}_{n-1}$ with respect to real-valued inner product

$$\langle f|g \rangle \overset{\text{def}}{=} \int_0^{\infty} \{fg\}\, r\,dr. \qquad (7.32)$$

In a similar spirit, we isolate the radial parts of $\psi_{\bar{z}}$, $\psi_z$, and $\psi_0$, respectively. For this purpose, we recall the original target coordinates $(R, \Theta)$ and that, for an equivariant map $w \in \mathbb{F}_k$, one has

$$w = \tan(R/2)\, e^{i\Theta} = u(t, r)e^{ik\theta}.$$

Based on (7.27)-(7.29), we can then define:

$$\phi_{\uparrow}(R) \overset{\text{def}}{=} \frac{2(u_{,r} - (k/r)u)}{1 + u^2} = R_{,r} - \frac{k\sin(R)}{r}, \qquad (7.33)$$

$$\phi_{\downarrow}(R) \overset{\text{def}}{=} \frac{2(u_{,r} + (k/r)u)}{1 + u^2} = R_{,r} + \frac{k\sin(R)}{r}, \qquad (7.34)$$

$$\phi_0(R) \overset{\text{def}}{=} \frac{2u_{,t}}{1 + u^2} = R_{,t}. \qquad (7.35)$$

These notations are inspired by the fact that they correspond to the radial part of $\psi_{\bar{z}}$, $\psi_z$, and $\psi_0$ in the fibers $\mathbb{F}_{k+1}$, $\mathbb{F}_{k-1}$, and $\mathbb{F}_k$, respectively.

Next, we make the choice of working in the temporal gauge, i.e., $A_0 = 0$, and we show that we can also assume $v = 0$. This is achieved as follows. In the equivariant setting, with the help of (7.18) and (7.25), we derive that

$$\partial_r \left\{ r\omega_{,r} + \frac{2k}{1 + u^2} \right\} = 0. \qquad (7.36)$$

We want to integrate this equation such that $r\omega_{,r}$ is regular at $r = 0$ and, knowing that $R(t, 0) = 0$ when $k > 0$ and $R(t, 0) = \pi$ when $k < 0$, we pick

$$r\omega_{,r} = \frac{-2k}{1 + u^2} = -k(1 + \cos(R)), \qquad \text{if } k < 0, \qquad (7.37)$$

$$r\omega_{,r} = \frac{2ku^2}{1 + u^2} = k(1 - \cos(R)), \qquad \text{if } k > 0. \qquad (7.38)$$

These can be written jointly as

$$r\omega_{,r} = k\big(\operatorname{sgn}(k) - \cos(R)\big).$$

Following this, we use (7.19) to obtain

$$B_1 = \frac{ke^{i\theta}}{r}\partial_t\left\{\frac{-1}{1+u^2}\right\},$$

which implies, due to the gauge choice and (7.23), that

$$\omega_{,tr} + iv_{,tr} = \frac{2k}{r}\partial_t\left\{\frac{-1}{1+u^2}\right\}.$$

If we rely on (7.36), then we get $v_{,tr} = 0$ and, thus, we can work forward with $v = 0$.

**Remark 7.2.** Based on the computations done so far, we can write our basic operators $\mathcal{L}_n$ and $\mathcal{L}_n^*$ as

$$\mathcal{L}_n = \partial_r - \frac{n}{r} + \frac{k(\operatorname{sgn}(k) - \cos(R))}{r}, \tag{7.39}$$

$$\mathcal{L}_n^* = -\partial_r - \frac{n+1}{r} + \frac{k(\operatorname{sgn}(k) - \cos(R))}{r}. \tag{7.40}$$

Moreover, if we define the weight $w_k := e^\omega$ (to emphasize the dependance of $\omega$ on $k$), then we can express them in the form of weighted derivatives:

$$\mathcal{L}_n = \frac{r^n}{w_k(t,r)}\partial_r\left\{\frac{w_k(t,r)}{r^n}\cdot\right\}, \quad \mathcal{L}_n^* = -\frac{w_k(t,r)}{r^{n+1}}\partial_r\left\{\frac{r^{n+1}}{w_k(t,r)}\cdot\right\}. \tag{7.41}$$

With all these prerequisites completed, we can now restate the Hodge system (7.20)-(7.22) in the equivariant setting as

$$\mathcal{L}_{k-1}\phi_\downarrow = \partial_t\phi_0, \qquad \mathcal{L}_k^*\phi_\uparrow = -\partial_t\phi_0, \tag{7.42}$$

$$\mathcal{L}_{k-1}^*\phi_0 = -\partial_t\phi_\downarrow, \qquad \mathcal{L}_k\phi_0 = \partial_t\phi_\uparrow, \tag{7.43}$$

$$[\partial_t, \mathcal{L}] = [\partial_t, \mathcal{L}^*] = \frac{1}{2}\phi_0(\phi_\downarrow - \phi_\uparrow), \tag{7.44}$$

$$\mathcal{L}_k^*\mathcal{L}_k - \mathcal{L}_{k-1}\mathcal{L}_{k-1}^* = -\frac{1}{2}(\phi_\downarrow^2 - \phi_\uparrow^2). \tag{7.45}$$

The way we are going to use the commutation relations is by restricting $\mathcal{L}$ and $\mathcal{L}^*$ to a specific fiber and thus deduce

$$[\partial_t, \mathcal{L}_n] = [\partial_t, \mathcal{L}_n^*] = \frac{1}{2}\phi_0(\phi_\downarrow - \phi_\uparrow), \qquad \forall n \in \mathbb{Z}. \tag{7.46}$$

The values of interest for $n$ are $k-1$ and $k$. Henceforth, the previous system of equations will be our basic object of study and we intend to construct solutions $(\phi_\uparrow, \phi_\downarrow, \phi_0) \in \mathbb{R}^3$, whose formula in terms of the original solution $R$ of (7.3) is given by

$$\phi_\uparrow(R) \stackrel{\text{def}}{=} \partial_r R - \frac{k\sin(R)}{r}, \tag{7.47}$$

$$\phi_\downarrow(R) \stackrel{\text{def}}{=} \partial_r R + \frac{k\sin(R)}{r}, \tag{7.48}$$

$$\phi_0(R) \stackrel{\text{def}}{=} \partial_t R. \tag{7.49}$$

**Remark 7.3.** It is now important to connect the harmonic maps (i.e., $R = Q(r)$) with the previous Hodge system. For such a map, one has $\phi_0 = 0$ and either $\phi_\downarrow = 0$ or $\phi_\uparrow = 0$. In our case, we pick $\phi_\uparrow = 0$, which leads by integration to $Q(r) = 2\arctan(r^k)$ and, subsequently, to $w_k(r) = 1 + r^{2k}$ and

$$\phi_\downarrow(Q(r)) = \frac{2k\sin(Q(r))}{r} = \frac{4kr^{k-1}}{1 + r^{2k}}.$$

One can of course dilate this solution by an arbitrary parameter $\lambda$ to create the family of solutions

$$\phi_\downarrow(Q_\lambda) = \frac{2k\sin(Q(\lambda r))}{r}.$$

In constructing collapsing solutions, the idea is to turn $\lambda$ into a dynamic parameter, meaning to choose $\lambda = 1/\tau(t)$ and work with $Q_{\tau(t)}(r) := Q(r/\tau(t))$. This profile is obviously no longer static and one expects that the collapsing solution looks like $R = Q_\tau + U$, with $U$ being "small" in an appropriate sense:

$$\phi_0(R) \approx 0, \qquad \phi_\uparrow(R) \approx 0, \qquad \phi_\downarrow(R) \approx \phi_\downarrow(Q_\tau).$$

This intuition would have to be justified then by estimating $\phi_0(R)$, $\phi_\uparrow(R)$, and $\phi_\downarrow(R) - \phi_\downarrow(Q_\tau)$ in some suitable norms.

## 7.4 Differential identities and applications

In this section, we focus on deriving differential identities, which will be useful in obtaining the main energy estimates in the argument. First, we obtain an identity for the so-called "second-order energy", and then we work on deducing a conformal-type identity and other conservation laws.

### 7.4.1    *The second-order energy*

Starting from the Hodge system (7.42)-(7.45), we can write wave equations for the field $(\phi_\downarrow, \phi_\uparrow, \phi_0)$. For example, using (7.42)-(7.43), it follows that

$$
\begin{aligned}
\left\{\partial_t^2 + \mathcal{L}_k\mathcal{L}_k^*\right\}\phi_\uparrow &= \partial_t\mathcal{L}_k\phi_0 - \mathcal{L}_k\partial_t\phi_0 = [\partial_t, \mathcal{L}_k]\phi_0 \\
&= \frac{1}{2}\phi_0^2(\phi_\downarrow - \phi_\uparrow).
\end{aligned}
\tag{7.50}
$$

Similarly, one deduces

$$
\left\{\partial_t^2 + \mathcal{L}_{k-1}^*\mathcal{L}_{k-1}\right\}\phi_\downarrow = -[\partial_t, \mathcal{L}_{k-1}^*]\phi_0,
\tag{7.51}
$$

$$
\left\{\partial_t^2 + \mathcal{L}_k^*\mathcal{L}_k\right\}\phi_0 = -[\partial_t, \mathcal{L}_k^*]\phi_\uparrow,
\tag{7.52}
$$

$$
\left\{\partial_t^2 + \mathcal{L}_{k-1}\mathcal{L}_{k-1}^*\right\}\phi_0 = [\partial_t, \mathcal{L}_{k-1}]\phi_\downarrow.
\tag{7.53}
$$

It is tempting to derive energy-type estimates from the wave equations above and we will do just that. For this purpose, we introduce:

$$
e(\phi_\uparrow) \stackrel{\text{def}}{=} \frac{1}{2}\left\{(\partial_t\phi_\uparrow)^2 + (\mathcal{L}_k^*\phi_\uparrow)^2\right\}, \qquad p(\phi_\uparrow) \stackrel{\text{def}}{=} \partial_t\phi_\uparrow \cdot \mathcal{L}_k^*\phi_\uparrow,
\tag{7.54}
$$

$$
e(\phi_\downarrow) \stackrel{\text{def}}{=} \frac{1}{2}\left\{(\partial_t\phi_\downarrow)^2 + (\mathcal{L}_{k-1}\phi_\downarrow)^2\right\}, \qquad p(\phi_\downarrow) \stackrel{\text{def}}{=} -\partial_t\phi_\downarrow \cdot \mathcal{L}_{k-1}\phi_\downarrow.
\tag{7.55}
$$

If we multiply (7.50) by $r\partial_t\phi_\uparrow$, then the second term on the left-hand side can be written as

$$
\begin{aligned}
\mathcal{L}_k\mathcal{L}_k^*\phi_\uparrow \cdot r\partial_t\phi_\uparrow &= \frac{r^k}{w_k}\partial_r\left(\frac{w_k}{r^k}\mathcal{L}_k^*\phi_\uparrow\right) \cdot r\partial_t\phi_\uparrow \\
&= \partial_r\left\{r\,\mathcal{L}_k^*\phi_\uparrow \cdot \partial_t\phi_\uparrow\right\} - r\,\mathcal{L}_k^*\phi_\uparrow \cdot \frac{w_k}{r^{k+1}}\partial_r\left\{\frac{r^{k+1}}{w_k}\partial_t\phi_\uparrow\right\} \\
&= \partial_r\left\{rp(\phi_\uparrow)\right\} + r\,\mathcal{L}_k^*\phi_\uparrow \cdot \mathcal{L}_k^*\partial_t\phi_\uparrow \\
&= \partial_r\left\{rp(\phi_\uparrow)\right\} + \partial_t\left\{\frac{r}{2}\left(\mathcal{L}_k^*\phi_\uparrow\right)^2\right\} + r\mathcal{L}_k^*\phi_\uparrow \cdot [\mathcal{L}_k^*, \partial_t]\phi_\uparrow.
\end{aligned}
$$

Using now the commutation formula (7.46) and (7.42), we derive the following second-order[4] energy identity:

$$
\begin{aligned}
\partial_t\left\{re(\phi_\uparrow)\right\} + \partial_r\left\{rp(\phi_\uparrow)\right\} &= \frac{r}{2}\phi_0(\phi_\downarrow - \phi_\uparrow)(\phi_0\,\partial_t\phi_\uparrow + \phi_\uparrow\,\mathcal{L}_k^*\phi_\uparrow) \\
&= \frac{r}{2}\phi_0(\phi_\downarrow - \phi_\uparrow)(\phi_0\,\partial_t\phi_\uparrow - \phi_\uparrow\,\partial_t\phi_0).
\end{aligned}
\tag{7.56}
$$

**Remark 7.4.** In a similar spirit, one can work with (7.51) to infer

$$
\partial_t\left\{re(\phi_\downarrow)\right\} - \partial_r\left\{rp(\phi_\downarrow)\right\} = \frac{r}{2}\phi_0(\phi_\downarrow - \phi_\uparrow)\left(-\phi_0\,\partial_t\phi_\downarrow + \phi_\downarrow\,\partial_t\phi_0\right).
$$

---

[4]We use the "second-order energy" terminology because the energy densities in (7.54)-(7.55) are at the level of second-order derivatives of $R$, according to (7.47)-(7.48).

If we denote

$$\phi_- \stackrel{\text{def}}{=} \phi_\downarrow - \phi_\uparrow = \frac{2k\sin(R)}{r},$$

then, by adding the last two differential identities, we obtain

$$\partial_t \left\{ r\left(e(\phi_\uparrow) + e(\phi_\downarrow)\right) \right\} + \partial_r \left\{ r\left(p(\phi_\uparrow) - p(\phi_\downarrow)\right) \right\} = \frac{r}{4}\left(\phi_-^2\, \partial_t \phi_0^2 - \phi_0^2\, \partial_t \phi_-^2\right).$$

In fact, for the main argument, we will not use any of the last three differential identities, but rather a slight modification of (7.56), which is deduced by further working on its right-hand side. On the basis of (7.47)-(7.49), we derive

$$\partial_t\left(\phi_\downarrow - \phi_\uparrow\right) = \partial_t \left\{ \frac{2k\sin(R)}{r} \right\} = \frac{2k\cos(R)}{r}\, \partial_t R = \frac{2k\cos(R)}{r}\, \phi_0.$$

If we factor in also (7.42), then we obtain

$$\phi_0\left(\phi_\downarrow - \phi_\uparrow\right)\phi_0\, \partial_t \phi_\uparrow$$
$$= \partial_t \left\{ \phi_0^2\left(\phi_\downarrow - \phi_\uparrow\right)\phi_\uparrow \right\} - 2\phi_0\left(\phi_\downarrow - \phi_\uparrow\right)\phi_\uparrow\, \partial_t \phi_0 - \phi_0^2 \phi_\uparrow\, \partial_t\left(\phi_\downarrow - \phi_\uparrow\right)$$
$$= \partial_t \left\{ \phi_0^2\left(\phi_\downarrow - \phi_\uparrow\right)\phi_\uparrow \right\} + 2\phi_0\left(\phi_\downarrow - \phi_\uparrow\right)\phi_\uparrow\, \mathcal{L}_k^* \phi_\uparrow - \phi_0^3 \frac{\phi_\uparrow}{r}\, 2k\cos(R).$$

Combining this identity with (7.56), it follows that

$$\partial_t \left\{ r\left[ e(\phi_\uparrow) - \frac{1}{2}\left(\phi_\downarrow - \phi_\uparrow\right)\phi_0^2\phi_\uparrow \right] \right\} + \partial_r \left\{ rp(\phi_\uparrow) \right\}$$
$$= r\left[ \frac{3\phi_0}{2}\left(\phi_\downarrow - \phi_\uparrow\right)\phi_\uparrow\, \mathcal{L}_k^* \phi_\uparrow - \phi_0^3 \phi_\uparrow \frac{k\cos(R)}{r} \right].$$
(7.57)

By integrating this relation with respect to the radial variable, we infer

$$\frac{d}{dt}\left\{ \widehat{\mathbb{D}}_\uparrow(t) + \widehat{N}(t) \right\} = F(t),$$
(7.58)

where

$$\widehat{\mathbb{D}}_\uparrow(t) \stackrel{\text{def}}{=} \int_0^\infty \left\{ e(\phi_\uparrow) \right\} r\, dr,$$
(7.59)

$$\widehat{N}(t) \stackrel{\text{def}}{=} \int_0^\infty \left\{ -\frac{1}{2}\phi_0^2\left(\phi_\downarrow - \phi_\uparrow\right)\phi_\uparrow \right\} r\, dr,$$
(7.60)

$$F(t) \stackrel{\text{def}}{=} \int_0^\infty \left\{ \frac{3\phi_0}{2}\left(\phi_\downarrow - \phi_\uparrow\right)\phi_\uparrow\left(\mathcal{L}_k^* \phi_\uparrow\right) - \phi_0^3 \phi_\uparrow \frac{k\cos(R)}{r} \right\} r\, dr,$$
(7.61)

and we can think of $\widehat{\mathbb{D}}_\uparrow$ as the "energy" associated to $\phi_\uparrow$. One of the main parts of the argument will be devoted to estimating this energy.

### 7.4.2    *Conformal identity and other conservation laws*

Even if the wave map problem is not conformally-invariant, we can still obtain conformal-type identities as follows. Using the formulas (7.39)-(7.40), it is easy to check that $r\mathcal{L}_{k-1} = \mathcal{L}_k r$, which, due to (7.42), implies

$$\mathcal{L}_k\left(r\phi_\downarrow\right) = \partial_r\left\{r\phi_0\right\}.$$

Coupling this equation with $\mathcal{L}_k^*\phi_\uparrow = -\partial_t\phi_0$ (i.e., the second equation in (7.42)), we deduce

$$r\phi_\uparrow\mathcal{L}_k\left(r\phi_\downarrow\right) - r^2\phi_\downarrow\,\mathcal{L}_k^*\phi_\uparrow = \partial_t\left\{r^2\phi_0(\phi_\uparrow + \phi_\downarrow)\right\} - r^2\phi_0\,\partial_t\left\{\phi_\uparrow + \phi_\downarrow\right\}.$$

On the other hand, if we rely on (7.39)-(7.40) and (7.47)-(7.49), then we derive

$$r\phi_\uparrow\mathcal{L}_k\left(r\phi_\downarrow\right) - r^2\phi_\downarrow\,\mathcal{L}_k^*\phi_\uparrow = \partial_r\left\{r^2\phi_\uparrow\phi_\downarrow\right\}, \quad \partial_t\left\{\phi_\uparrow + \phi_\downarrow\right\} = 2\partial_r\phi_0.$$

Putting together these last three relations, we finally arrive at the conformal identity

$$\partial_t\left\{r^2\phi_0(\phi_\downarrow + \phi_\uparrow)\right\} - \partial_r\left\{r^2\left(\phi_0^2 + \phi_\uparrow\phi_\downarrow\right)\right\} + 2r\phi_0^2 = 0. \qquad (7.62)$$

In what concerns energy conservation, let us notice first that (7.42)-(7.43) yield

$$\left(\mathcal{L}_{k-1}\phi_\downarrow - \partial_t\phi_0\right)\phi_0 - \left(\mathcal{L}_{k-1}^*\phi_0 + \partial_t\phi_\downarrow\right)\phi_\downarrow = 0,$$

$$\left(\mathcal{L}_k^*\phi_\uparrow + \partial_t\phi_0\right)\phi_0 - \left(\mathcal{L}_k\phi_0 - \partial_t\phi_\uparrow\right)\phi_\uparrow = 0,$$

which can be easily rewritten as

$$\partial_t\left\{\frac{r}{2}\left(\phi_0^2 + \phi_\downarrow^2\right)\right\} - \partial_r\left\{r\phi_0\phi_\downarrow\right\} = 0, \qquad (7.63)$$

$$\partial_t\left\{\frac{r}{2}\left(\phi_0^2 + \phi_\uparrow^2\right)\right\} - \partial_r\left\{r\phi_0\phi_\uparrow\right\} = 0. \qquad (7.64)$$

Adding these two relations, we also obtain the overall energy differential identity, i.e.,

$$\partial_t\left\{r\left[\phi_0^2 + \frac{1}{2}\left(\phi_\uparrow^2 + \phi_\downarrow^2\right)\right]\right\} - \partial_r\left\{r\phi_0(\phi_\downarrow + \phi_\uparrow)\right\} = 0. \qquad (7.65)$$

We will use the notation

$$e_\uparrow(R) \stackrel{\text{def}}{=} \frac{1}{2}\left(\phi_0^2 + \phi_\uparrow^2\right), \qquad\qquad \mathbb{E}_\uparrow \stackrel{\text{def}}{=} \int_0^\infty \{e_\uparrow(R)\}r\,dr, \qquad (7.66)$$

$$e_\downarrow(R) \stackrel{\text{def}}{=} \frac{1}{2}\left(\phi_0^2 + \phi_\downarrow^2\right), \qquad\qquad \mathbb{E}_\downarrow \stackrel{\text{def}}{=} \int_0^\infty \{e_\downarrow(R)\}r\,dr, \qquad (7.67)$$

$$e(R) \stackrel{\text{def}}{=} \phi_0^2 + \frac{1}{2}\left(\phi_\uparrow^2 + \phi_\downarrow^2\right), \qquad \mathbb{E} \stackrel{\text{def}}{=} \int_0^\infty \{e(R)\}r\,dr, \qquad (7.68)$$

for the energy densities and their respective energies. If we integrate (7.63)-(7.65) with respect to the radial variable, it follows that the three energies are conserved by the evolution and we will use this fact for the stability argument in the next section.

## 7.5 Initial setup, key energy facts, and stability issues

Here, we follow up on the ideas presented in Section 7.1 concerning the construction of collapsing solutions to (7.3), for which one can assume, without loss of generality, that $k \geq 1$. We recall that, for a fixed $k$, the intuition is to start with the harmonic map

$$Q^{(k)}(r) = 2 \arctan(r^k)$$

and look for a finite-energy solution $R = R(t, r)$ behaving near a time $T > 0$ like $Q^{(k)}(r/\tau_k(t))$, where $\tau_k = \tau_k(t)$ is a dynamic time function satisfying $\tau_k(t) \to 0$ as $t \to T$.

### 7.5.1 *Notational conventions and initial considerations*

As our analysis treats in a unitary way all the values of $k \geq 1$, we are going to abuse the notation and write $Q$ for $Q^{(k)}$, keeping in mind, however, its dependance on $k$. If one defines the scaled versions $Q_\lambda(r) := Q(r/\lambda)$ (which are also static solutions to (7.3)), then, by sending $\lambda \to 0$, these functions will concentrate near the origin $r = 0$. Furthermore, due to (7.42)-(7.43), it is easy to see that they satisfy the following identities:

$$\phi_\uparrow(Q_\lambda) = \phi_0(Q_\lambda) = 0, \qquad \mathcal{L}_{k-1}\phi_\downarrow(Q_\lambda) = 0, \qquad \partial_t\phi_\downarrow(Q_\lambda) = 0. \quad (7.69)$$

Next, we substitute for $\lambda$ the dynamic time function $\tau = \tau(t)$ and introduce

$$Q_\tau = Q_\tau(t, r) \overset{\text{def}}{=} Q(r/\tau(t)) \quad (7.70)$$

which is, of course, no longer a solution to the wave map equation. Our goal will be to try and write solutions to (7.3) in the form

$$R(t, r) = (Q_\tau + U)(t, r), \quad (7.71)$$

with $U$ being a small perturbation. As we will see, this is particularly challenging in the case $k = 1$, because $\partial_t R$ might not belong to $L^2$ and, thus, the solution would not have finite energy.

In order to accommodate this issue, we bring in a smooth, decreasing function $\chi : \mathbb{R}_+ \to \mathbb{R}$, verifying $\chi \equiv 1$ on $[0, 1]$ and $\chi \equiv 0$ on $[2, \infty)$, and denote, for an arbitrary $M_1 \gg 1$,

$$\chi_{M_1} = \chi_{M_1}(y) \overset{\text{def}}{=} \chi(y/M_1).$$

Instead of (7.71), we look for solutions of the type[5]

$$R = \chi_{M_1}(r/\tau(t))Q_\tau(t, r) + \big(1 - \chi_{M_1}(r/\tau(t))\big)\pi + U(t, r), \quad (7.72)$$

---

[5] We need the cutoff $\chi_{M_1}$ to ensure that the initial data have finite energy for $k = 1$.

where we kept in mind that $Q(0) = 0$ and $\lim_{r \to \infty} Q(r) = \pi$. However, we will also use the expression

$$R(t,r) = (Q_\tau + U_1)(t,r), \tag{7.73}$$

with

$$U_1 \overset{\text{def}}{=} U(t,r) + \left(1 - \chi_{M_1}(r/\tau(t))\right)\left(\pi - Q_\tau(t,r)\right). \tag{7.74}$$

At this point, the dynamical scaling $\tau = \tau(t)$ is unknown and, in order to fix it, we impose a "gauge" condition on the perturbation $U$. This is the orthogonality condition

$$\langle U | \chi_M \sin(Q_\tau) \rangle = 0, \tag{7.75}$$

where the parameter $M$ is chosen such that $M_1 \gg M \gg 1$ and the inner product is the one defined by (7.32). It follows immediately that one also has

$$\langle U_1 | \chi_M \sin(Q_\tau) \rangle = 0. \tag{7.76}$$

After these considerations, we introduce further notational conventions, which help streamline the rest of this exposition in this section. These are

$$b(t) \overset{\text{def}}{=} -\dot\tau(t), \tag{7.77}$$

$$ds \overset{\text{def}}{=} \frac{dt}{\tau(t)}, \tag{7.78}$$

$$y \overset{\text{def}}{=} \frac{r}{\tau(t)}, \tag{7.79}$$

representing the speed, the rescaled time variable, and the scaled space variable, respectively. We will also write $f'$ for $df/ds$ and rely, in particular, on the following expression for the acceleration,

$$b'(s) = \tau(t)\dot b(t). \tag{7.80}$$

In what concerns the variable $y$, because of (7.70), we have $Q_\tau(t,r) = Q(y)$. Henceforth, if we want to emphasize the dependence of the previous quantity on $r$, then we use the $Q_\tau$ notation. Otherwise, we work with the $Q$ notation.

We turn next to the operators $\mathcal{L}_n$ and $\mathcal{L}_n^*$, which, for a solution to (7.3), are given by the formulas (7.39) and (7.40), respectively (also, recall (7.41)). Now, the goal is to write versions of these operators adapted to $Q_\tau$ and they are as follows:

$$\begin{aligned}
\widehat{\mathcal{L}}_n &\overset{\text{def}}{=} \partial_r - \frac{n}{r} + \frac{k\left(1 - \cos(Q_\tau)\right)}{r} \\
&= \frac{r^n}{\widehat{w}_k(r/\tau(t))} \partial_r \left\{ \frac{\widehat{w}_k(r/\tau(t))}{r^n} \, \cdot \right\}
\end{aligned} \tag{7.81}$$

and

$$\widehat{\mathcal{L}}_n^* \overset{\text{def}}{=} -\partial_r - \frac{n+1}{r} + \frac{k\big(1-\cos(Q_\tau)\big)}{r}$$
$$= -\frac{\widehat{w}_k(r/\tau(t))}{r^{n+1}}\partial_r\left\{\frac{r^{n+1}}{\widehat{w}_k(r/\tau(t))}\cdot\right\}. \tag{7.82}$$

In the above, $\widehat{w}_k(y) := 1 + y^{2k}$ and a direct computation yields

$$\cos(Q(y)) = \frac{1-y^{2k}}{1+y^{2k}}.$$

The relationship between the pairs of operators $(\mathcal{L}_n, \mathcal{L}_n^*)$ and $(\widehat{\mathcal{L}}_n, \widehat{\mathcal{L}}_n^*)$ is given by

$$\mathcal{L}_n = \widehat{\mathcal{L}}_n + \frac{V_k(U_1)}{r}, \qquad \mathcal{L}_n^* = \widehat{\mathcal{L}}_n^* + \frac{V_k(U_1)}{r}, \tag{7.83}$$

where the potential $V_k(U_1)$ has the expression

$$V_k(U_1) \overset{\text{def}}{=} k\left[\cos(Q_\tau)\big(1-\cos(U_1)\big) + \sin(Q_\tau)\sin(U_1)\right]. \tag{7.84}$$

It is important to remark at this point that $V_k(U_1)$ is small if $U_1$ is small.

Following these considerations, we focus on $\phi_\uparrow(R)$, $\phi_\downarrow(R)$, and $\phi_0(R)$, and we decompose them according to[6]

$$\phi_\uparrow(R) = \phi_\uparrow(Q_\tau) + \widehat{\mathcal{L}}_k U_1 + \frac{kf_k(U_1)}{r}, \tag{7.85}$$

$$\phi_\downarrow(R) = \phi_\downarrow(Q_\tau) - \widehat{\mathcal{L}}_{k-1}^* U_1 - \frac{kf_k(U_1)}{r}, \tag{7.86}$$

$$\phi_0(R) = \chi_M \phi_0(Q_\tau) + W, \tag{7.87}$$

with

$$f_k(U_1) \overset{\text{def}}{=} \sin(Q_\tau)\big(1-\cos(U_1)\big) + \cos(Q_\tau)\big(U_1 - \sin(U_1)\big), \tag{7.88}$$

$$W \overset{\text{def}}{=} \phi_0(U_1) + \big(1 - \chi_M\big)\phi_0(Q_\tau). \tag{7.89}$$

For $f_k(U_1)$, it will be useful to write it as

$$f_k(U_1) = \sin(Q_\tau)\frac{U_1^2}{2} + \widetilde{f}_k(U_1), \tag{7.90}$$

where $\widetilde{f}_k(U_1) \sim U_1^3 + O(U_1^4)$. Furthermore, in connection to (7.85)-(7.87), simple calculations based on (7.70) yield

$$\phi_\uparrow(Q_\tau) = 0, \qquad \phi_\downarrow(Q_\tau) = \frac{2k\sin(Q_\tau)}{r}, \qquad \phi_0(Q_\tau) = \frac{kb}{\tau}\sin(Q_\tau). \tag{7.91}$$

---

[6]To simplify the notation, from this point onward, we will write $(\phi_\uparrow, \phi_\downarrow, \phi_0)$ to strictly denote $(\phi_\uparrow(R), \phi_\downarrow(R), \phi_0(R))$.

### 7.5.2　Energy formulas

In our construction of collapsing solutions, we will be working with energies which are slightly above the one of a harmonic map. This means that we choose $\mathbb{E} = 4k + \epsilon^2$, where $\epsilon^2$, also called the "excess energy", is a small parameter at our disposal; in particular,

$$\mathbb{E}_\uparrow \stackrel{\text{def}}{=} \epsilon_1^2, \qquad \mathbb{E}_\downarrow \stackrel{\text{def}}{=} 4k + \epsilon_2^2, \qquad \epsilon_1^2 + \epsilon_2^2 = \epsilon^2. \qquad (7.92)$$

A simple motivation for this choice is offered by the following calculation,

$$\int_0^\infty \left\{ \frac{1}{2} \phi_\downarrow^2 (Q_\tau) \right\} r\, dr = \int_0^\infty \frac{8k^2 y^{2k-1} dy}{(1 + y^{2k})^2} \, dy = 4k, \qquad (7.93)$$

which shows that the "bulk" of the energy is carried by $\phi_\downarrow(Q_\tau)$. To further justify (7.92), we focus here on writing formulas for $\mathbb{E}_\uparrow$ and $\mathbb{E}_\downarrow - 4k$ in terms of quantities which can be proven later to be small[7], for a suitable choice of initial data.

As a first step, we square the equations (7.85)-(7.87) and, based on (7.90)-(7.91), we deduce

$$\phi_0^2 = \chi_M^2 \phi_0^2 (Q_\tau) + W^2 + 2\chi_M \phi_0(Q_\tau) W, \qquad (7.94)$$

$$\phi_\uparrow^2 = \left( \widehat{\mathcal{L}}_k U_1 \right)^2 + \frac{k^2 f_k^2(U_1)}{r^2} + 2 \frac{k f_k(U_1)}{r} \widehat{\mathcal{L}}_k U_1, \qquad (7.95)$$

$$\begin{aligned}
\phi_\downarrow^2 = \phi_\downarrow^2(Q_\tau) &+ \left( \widehat{\mathcal{L}}_{k-1}^* U_1 \right)^2 + \frac{k^2 f_k^2(U_1)}{r^2} \\
&- 2\phi_\downarrow(Q_\tau) \widehat{\mathcal{L}}_{k-1}^* U_1 + 2 \frac{k f_k(U_1)}{r} \widehat{\mathcal{L}}_{k-1}^* U_1 \\
&- 2\phi_\downarrow(Q_\tau) \left\{ \frac{1}{4} \phi_\downarrow(Q_\tau) U_1^2 + \frac{k \widetilde{f}_k(U_1)}{r} \right\}.
\end{aligned} \qquad (7.96)$$

Due to (7.66)-(7.67), our next goal is to integrate the above expressions in order to derive useful formulas for $\mathbb{E}_\uparrow$ and $\mathbb{E}_\downarrow - 4k$.

**Lemma 7.1.** *The following formula holds for $\mathbb{E}_\uparrow$:*

$$\begin{aligned}
\mathbb{E}_\uparrow = \frac{1}{2} \bigg\{ D_k(M)\, b^2 &+ \| \widehat{\mathcal{L}}_k U_1 \|_{L^2(rdr)}^2 + \left\| \frac{k f_k(U_1)}{r} \right\|_{L^2(rdr)}^2 \\
&+ \| W \|_{L^2(rdr)}^2 \bigg\} + \left\langle \frac{k f_k(U_1)}{r} \middle| \widehat{\mathcal{L}}_k U_1 \right\rangle + \left( -A_k(\widetilde{U}_1) + J_k \right) b^2,
\end{aligned} \qquad (7.97)$$

---

[7] As we will see, the crucial idea in proving the smallness for these quantities is that the orthogonality condition (7.76) allows us to control the size of $U_1$.

*where $D_k(M)$, $\tilde{U}_1$, $A_k(\tilde{U}_1)$[8], and $J_k$ are given by*

$$D_k(M) \overset{\text{def}}{=} k^2 \|\chi_M \sin(Q)\|^2_{L^2(ydy)} \sim \begin{cases} \log(M) & if \quad k = 1, \\ C & if \quad k > 1, \end{cases} \qquad (7.98)$$

$$\tilde{U}_1 = \tilde{U}_1(t, y) \overset{\text{def}}{=} U_1(t, \tau(t)y), \qquad (7.99)$$

$$A_k(\tilde{U}_1) \overset{\text{def}}{=} k\langle \tilde{U}_1 | [y\partial_y \chi_M + k\chi_M(\cos(Q) + 1)] \sin(Q)\rangle_{ydy}, \qquad (7.100)$$

$$J_k \overset{\text{def}}{=} k^2 \langle (1 - \chi_M) \sin(Q) | \chi_M \sin(Q)\rangle_{ydy}. \qquad (7.101)$$

**Proof.** On the account of (7.94)-(7.95), the formula (7.97) is proved if we show that

$$\|\chi_M \phi_0(Q_\tau)\|^2_{L^2(rdr)} = D_k(M)b^2$$

and

$$\langle W | \chi_M \phi_0(Q_\tau)\rangle = \left(-A_k(\tilde{U}_1) + J_k\right) b^2.$$

The former is immediate as the elementary property

$$\langle \cdot | \cdot \rangle = \tau^2 \langle \cdot | \cdot \rangle_{ydy}$$

and the formula for $\phi_0(Q_\tau)$ in (7.91) imply

$$\|\chi_M \phi_0(Q_\tau)\|^2_{L^2(rdr)} = b^2 k^2 \|\chi_M \sin(Q)\|^2_{L^2(ydy)} = b^2 D_k(M).$$

Moreover, the asymptotics for $D_k(M)$ in (7.98) can be easily justified using $\sin(Q) = 2y^k/(1 + y^{2k})$.

For the latter, we use the same two facts as above coupled with the definition (7.89) and the orthogonality condition (7.76) to infer

$$\langle W | \chi_M \phi_0(Q_\tau)\rangle$$

$$= \left\langle \partial_t U_1 + \frac{kb}{\tau}(1 - \chi_M) \sin(Q_\tau) \Big| \frac{kb}{\tau} \chi_M \sin(Q_\tau)\right\rangle$$

$$= \frac{kb}{\tau} \left[\partial_t \{\langle U_1 | \chi_M \sin(Q_\tau)\rangle\} - \langle U_1 | \partial_t \{\chi_M \sin(Q_\tau)\}\rangle\right]$$

$$+ \frac{k^2 b^2}{\tau^2} \langle (1 - \chi_M) \sin(Q_\tau) | \chi_M \sin(Q_\tau)\rangle$$

$$= -kb^2 \langle \tilde{U}_1 | [y\partial_y \chi_M + k\chi_M \cos(Q)] \sin(Q)\rangle_{ydy} + J_k b^2$$

$$= \left(-A_k(\tilde{U}_1) + J_k\right) b^2.$$

$\square$

---

[8]Due to the orthogonality condition (7.76), it is clear that one can replace 1 in the formula for $A_k(\tilde{U}_1)$ with an arbitrary real number. The reason we choose 1 is that it provides a subtle cancellation later in the derivation of a differential equation satisfied by $b$.

We proceed next to prove a corresponding result for $\mathbb{E}_\downarrow - 4k$, for which one argues in a similar way. We do not use this result in our analysis, but we state it here for completeness.

**Lemma 7.2.** *The following formula holds for $\mathbb{E}_\downarrow - 4k$:*

$$\mathbb{E}_\downarrow - 4k = \frac{1}{2}\left\{ D_k(M)b^2 + \|\widehat{\mathcal{L}}_k U_1\|_{L^2(rdr)}^2 + \left\|\frac{kf_k(U_1)}{r}\right\|_{L^2(rdr)}^2 \right.$$

$$+ \|W\|_{L^2(rdr)}^2 \Bigg\} + \left(-A_k(\widetilde{U}_1) + J_k\right)b^2$$

$$- \left\langle \frac{kf_k(U_1)}{r} \middle| \widehat{\mathcal{L}}_k U_1 \right\rangle - 2\left\langle \frac{kf_k(U_1)}{r} \middle| \frac{k\cos(Q_\tau)U_1}{r} \right\rangle \qquad (7.102)$$

$$- \left\langle \phi_\downarrow(Q_\tau) \middle| \frac{k\widetilde{f}_k(U_1)}{r} \right\rangle.$$

**Proof.** Due to (7.96), (7.93), and (7.97), it is easy to see that the above formula is true if we show that

$$\|\widehat{\mathcal{L}}_{k-1}^* U_1\|_{L^2(rdr)}^2 - \frac{1}{2}\|\phi_\downarrow(Q_\tau)U_1\|_{L^2(rdr)}^2 = \|\widehat{\mathcal{L}}_k U_1\|_{L^2(rdr)}^2,$$

$$\left\langle \phi_\downarrow(Q_\tau) \middle| \widehat{\mathcal{L}}_{k-1}^* U_1 \right\rangle = 0,$$

$$\left\langle \frac{kf_k(U_1)}{r} \middle| \widehat{\mathcal{L}}_{k-1}^* U_1 \right\rangle = -\left\langle \frac{kf_k(U_1)}{r} \middle| \widehat{\mathcal{L}}_k U_1 + 2\frac{k\cos(Q_\tau)U_1}{r} \right\rangle.$$

For the second equality, an integration by parts and (7.69) imply

$$\left\langle \phi_\downarrow(Q_\tau) \middle| \widehat{\mathcal{L}}_{k-1}^* U_1 \right\rangle = \left\langle \widehat{\mathcal{L}}_{k-1}\phi_\downarrow(Q_\tau) \middle| U_1 \right\rangle = 0.$$

The last equality can be justified on the basis of (7.81) and (7.82) as follows:

$$\widehat{\mathcal{L}}_{k-1}^* = -\partial_r - \frac{k\cos(Q_\tau)}{r} = -\widehat{\mathcal{L}}_k - 2\frac{k\cos(Q_\tau)}{r}.$$

Finally, for the first equality, we use again (7.81) and (7.82), together with an integration by parts, to infer that

$$\|\widehat{\mathcal{L}}_{k-1}^* U_1\|_{L^2(rdr)}^2 - \frac{1}{2}\|\phi_\downarrow(Q_\tau)U_1\|_{L^2(rdr)}^2$$

$$= \|\widehat{\mathcal{L}}_k U_1\|_{L^2(rdr)}^2 + 2k\left(\left\langle 2\partial_r U_1 \cdot U_1 \middle| \frac{\cos(Q_\tau)}{r} \right\rangle - \left\langle U_1^2 \middle| \frac{k\sin^2(Q_\tau)}{r^2} \right\rangle\right)$$

$$= \|\widehat{\mathcal{L}}_k U_1\|_{L^2(rdr)}^2 + 2k\left(\left\langle \partial_r\{U_1^2\} \middle| \frac{\cos(Q_\tau)}{r} \right\rangle - \left\langle U_1^2 \middle| \frac{k\sin^2(Q_\tau)}{r^2} \right\rangle\right)$$

$$= \|\widehat{\mathcal{L}}_k U_1\|_{L^2(rdr)}^2 - 2k\left\langle U_1^2 \middle| \frac{\partial_r\{\cos(Q_\tau)\}}{r} + \frac{k\sin^2(Q_\tau)}{r^2} \right\rangle$$

$$= \|\widehat{\mathcal{L}}_k U_1\|_{L^2(rdr)}^2.$$

This finishes the proof of this lemma. $\qquad\qquad\square$

### 7.5.3   A differential equation for the speed

In the main argument proving collapse, we need to complement the information about energy with precise details concerning the way the speed $b = b(t)$, given by (7.77), varies. In particular, with the help of the orthogonality condition (7.23), we obtain a useful differential equation satisfied by this function, which is the object of the next lemma.

**Lemma 7.3.** *For the acceleration $b' = b'(s)$ described by (7.80), the following formula holds:*

$$b' \left( D_{k,1}(M) - \frac{1}{k} A_k(\widetilde{U}_1) \right)$$

$$= b^2 \left\{ D_{k,2}(M) + \left( 2 + \frac{1}{k} \right) A_k(\widetilde{U}_1) + B_k(\widetilde{U}_1) \right\} \qquad (7.103)$$

$$+ 2\frac{b\tau}{k} A_k(\widetilde{\partial_t U}_1) - Z^{\mathrm{bulk}}(\phi_\uparrow) - Z^{\mathrm{bdry}}(\phi_\uparrow),$$

*where the terms appearing above are given by*

$$D_{k,1}(M) \stackrel{\mathrm{def}}{=} k \big\langle \sin(Q) \big| \chi_M \sin(Q) \big\rangle_{ydy}, \qquad (7.104)$$

$$D_{k,2}(M) \stackrel{\mathrm{def}}{=} \frac{k}{2} \big\langle \sin(Q) \big| y \partial_y \chi_M \sin(Q) \big\rangle_{ydy}, \qquad (7.105)$$

$$Z^{\mathrm{bulk}}(\phi_\uparrow) \stackrel{\mathrm{def}}{=} \left\langle \frac{V_k(U_1)}{r} \phi_\uparrow \Big| \chi_M \sin(Q_\tau) \right\rangle, \qquad (7.106)$$

$$Z^{\mathrm{bdry}}(\phi_\uparrow) \stackrel{\mathrm{def}}{=} \left\langle \phi_\uparrow \Big| y \partial_y \chi_M \frac{\sin(Q_\tau)}{r} \right\rangle, \qquad (7.107)$$

$$B_k(\widetilde{U}_1) \stackrel{\mathrm{def}}{=} \big\langle \widetilde{U}_1 \big| [ y\partial_y(y\partial_y \chi_M) + 2ky\partial_y \chi_M \cos(Q) \\ + k^2 \chi_M \cos(2Q) ] \sin(Q) \big\rangle_{ydy}, \qquad (7.108)$$

*with $A_k(\widetilde{U}_1)$ being previously defined in (7.100) and $A_k(\widetilde{\partial_t U}_1)$ coming from the same formula by substituting $\partial_t U_1$ for $U_1$.*

**Proof.** The starting point for the derivation of the above formula is the identity

$$\phi_0 = \phi_0(Q_\tau) + \partial_t U_1,$$

which results from the decomposition $R = Q_\tau + U_1$. This implies that

$$\big\langle \partial_t \phi_0 | \chi_M \sin(Q_\tau) \big\rangle = \big\langle \partial_t \phi_0(Q_\tau) | \chi_M \sin(Q_\tau) \big\rangle \\ + \big\langle \partial_t^2 U_1 | \chi_M \sin(Q_\tau) \big\rangle. \qquad (7.109)$$

For the left-hand side, we can use (7.42) and (7.83) to rewrite it as

$$\langle \partial_t \phi_0 | \chi_M \sin(Q_\tau) \rangle$$
$$= -\langle \mathcal{L}_k^* \phi_\uparrow | \chi_M \sin(Q_\tau) \rangle$$
$$= -\langle \widehat{\mathcal{L}}_k^* \phi_\uparrow | \chi_M \sin(Q_\tau) \rangle - \left\langle \frac{V_k(U_1)}{r} \phi_\uparrow \middle| \chi_M \sin(Q_\tau) \right\rangle$$
$$= -\langle \phi_\uparrow | \widehat{\mathcal{L}}_k \left( \chi_M \sin(Q_\tau) \right) \rangle - \left\langle \frac{V_k(U_1)}{r} \phi_\uparrow \middle| \chi_M \sin(Q_\tau) \right\rangle.$$

A direct computation based on the easily-checked fact $\widehat{\mathcal{L}}_k \sin(Q_\tau) = 0$ shows that

$$\widehat{\mathcal{L}}_k \left( \chi_M \sin(Q_\tau) \right) = \partial_r \chi_M \, \sin(Q_\tau) = y \partial_y \chi_M \frac{\sin(Q_\tau)}{r}.$$

Hence, according to (7.106) and (7.107), we deduce that

$$\langle \partial_t \phi_0 | \chi_M \sin(Q_\tau) \rangle = -Z^{\text{bulk}}(\phi_\uparrow) - Z^{\text{bdry}}(\phi_\uparrow). \tag{7.110}$$

For the right-hand side of (7.109), the product rule of differentiation yields

$$\langle \partial_t \phi_0(Q_\tau) | \chi_M \sin(Q_\tau) \rangle = \partial_t \left\{ \langle \phi_0(Q_\tau) | \chi_M \sin(Q_\tau) \rangle \right\}$$
$$- \langle \phi_0(Q_\tau) | \partial_t \left( \chi_M \sin(Q_\tau) \right) \rangle, \tag{7.111}$$

$$\langle \partial_t^2 U_1 | \chi_M \sin(Q_\tau) \rangle = \partial_t^2 \left\{ \langle U_1 | \chi_M \sin(Q_\tau) \rangle \right\}$$
$$- 2\langle \partial_t U_1 | \partial_t \left( \chi_M \sin(Q_\tau) \right) \rangle \tag{7.112}$$
$$- \langle U_1 | \partial_t^2 \left( \chi_M \sin(Q_\tau) \right) \rangle.$$

We take aside each of these equations and work further on their right-hand sides. For the first one, due to (7.91) and (7.104), we have

$$\langle \phi_0(Q_\tau) | \chi_M \sin(Q_\tau) \rangle = kb\tau \langle \sin(Q) | \chi_M \sin(Q) \rangle_{ydy} = b\tau D_{k,1}(M),$$

which, jointly with (7.77) and (7.80), yields

$$\partial_t \left\{ \langle \phi_0(Q_\tau) | \chi_M \sin(Q_\tau) \rangle \right\} = (\dot{b}\tau + b\dot{\tau}) D_{k,1}(M)$$
$$= (b' - b^2) D_{k,1}(M). \tag{7.113}$$

For the second term involving $\phi_0(Q_\tau)$, we are using (7.91) and integration by parts to derive

$$\langle \phi_0(Q_\tau) | \partial_t (\chi_M \sin(Q_\tau)) \rangle$$

$$= \frac{kb^2}{\tau^2} \langle \sin(Q_\tau) | y \partial_y (\chi_M \sin(Q_\tau)) \rangle$$

$$= kb^2 \langle \sin(Q) | y \partial_y (\chi_M \sin(Q)) \rangle_{ydy}$$

$$= kb^2 \int_0^\infty \{ y^2 \partial_y (\chi_M \sin(Q)) \sin(Q) \} \, dy$$

$$= kb^2 \int_0^\infty \left\{ y^2 \left[ \partial_y \left( \frac{1}{2} \chi_M \sin^2(Q) \right) + \frac{1}{2} \partial_y \chi_M \, \sin^2(Q) \right] \right\} \, dy$$

$$= kb^2 \int_0^\infty \left\{ -\chi_M \sin^2(Q) + \frac{1}{2} y \partial_y \chi_M \, \sin^2(Q) \right\} \, y \, dy$$

$$= kb^2 \left( -\langle \sin(Q) | \chi_M \sin(Q) \rangle_{ydy} + \frac{1}{2} \langle \sin(Q) | y \partial_y \chi_M \sin(Q) \rangle_{ydy} \right)$$

$$= b^2 \left( -D_{k,1}(M) + D_{k,2}(M) \right).$$

$$(7.114)$$

This is precisely the moment where we can witness the subtle cancellation mentioned previously. It has to do with the presence of $-b^2 D_{k,1}(M)$ in both (7.113) and (7.114), which, as a consequence, implies the vanishing of this term from (7.111). Later, this is a crucial fact in arguing that the coefficient of $b^2$ in (7.103) can be made small.

Next, we focus on the terms appearing on the right-hand side of (7.112). The first one is zero, due to the orthogonality condition (7.76). For the second and third terms, we are going to compute the time derivatives of $\chi_M \sin(Q_\tau)$ by taking advantage of the straightforward identity

$$\partial_t \sin(Q_\tau) = \frac{kb}{\tau} \cos(Q_\tau) \sin(Q_\tau).$$

As such, we can infer

$$\partial_t (\chi_M \sin(Q_\tau)) = \frac{b}{\tau} [y \partial_y \chi_M + k \chi_M \cos(Q_\tau)] \sin(Q_\tau)$$

$$\partial_t^2 (\chi_M \sin(Q_\tau)) = \frac{b' + b^2}{\tau^2} [y \partial_y \chi_M + k \chi_M \cos(Q_\tau)] \sin(Q_\tau)$$

$$+ \frac{b^2}{\tau^2} [y \partial_y (y \partial_y \chi_M) + 2ky \partial_y \chi_M \cos(Q_\tau)$$

$$+ k^2 \chi_M \cos(2Q_\tau)] \sin(Q_\tau).$$

Moreover, relying on (7.76) and (7.100), we infer that

$$\langle \partial_t U_1 | \chi_M \sin(Q_\tau) \rangle = -\langle U_1 | \partial_t (\chi_M \sin(Q_\tau)) \rangle = -\frac{b\tau}{k} A_k(\widetilde{U}_1).$$

Using the last three identities, it follows that

$$\langle \partial_t U_1 | \partial_t (\chi_M \sin(Q_\tau)) \rangle = \frac{b\tau}{k} A_k(\widetilde{\partial_t U_1}) + b^2 A_k(\widetilde{U}_1), \tag{7.115}$$

$$\langle U_1 | \partial_t^2 (\chi_M \sin(Q_\tau)) \rangle = \frac{b' + b^2}{k} A_k(\widetilde{U}_1) + b^2 B_k(\widetilde{U}_1). \tag{7.116}$$

The formula to be proved is then the combined consequence of the previous two relations, (7.109), (7.110), (7.111), (7.112), (7.113), and (7.114). □

### 7.5.4    *Preliminary assumptions on the initial data and the stability argument*

Our next goal is to take advantage of the orthogonality condition (7.76) and the energy formula (7.97) and show that, under minimal assumptions on the initial data (e.g., (7.92)), the main components of $E_\uparrow$ satisfy important stability-type estimates. These can be used then to have good control on the size of $U_1$.

For this purpose, we start by proving a key technical lemma, which applies to both $U$ and $U_1$.

**Lemma 7.4.** *Let* $u = u(t,r)$ *be a function satisfying*

$$u(t,0) = \lim_{r \to \infty} u(t,r) = 0$$

*and the orthogonality condition* $\langle u | \chi_M \sin(Q_\tau) \rangle = 0$. *If one uses the operator notation* $-\mathcal{D}_k^* := \partial_r + k/r$, *then the following fixed-time estimate hold:*

$$k^{1/2} \|u\|_{L^\infty} + \|\partial_r u\|_{L^2(rdr)} + k \left\| \frac{u}{r} \right\|_{L^2(rdr)} \leq C \|\mathcal{D}_k^* u\|_{L^2(rdr)} \tag{7.117}$$

$$\leq C_k(M) \|\widehat{\mathcal{L}}_k u\|_{L^2(rdr)},$$

*where the constant* $C_k(M)$ *can be chosen such that*

$$C_k(M) \sim \begin{cases} M/\log(M) & \text{if } k = 1, \\ \log(M) & \text{if } k = 2, \\ C & \text{if } k > 2. \end{cases} \tag{7.118}$$

**Proof.** First, we make some preliminary observations that have to do with switching between the coordinates $r$ and $y = r/\tau(t)$. If we introduce the notation $\widetilde{u} = \widetilde{u}(t,y) := u(t,\tau(t)y)$ and define the differential operator

$$\widehat{\mathcal{L}}_{y,k} \stackrel{\text{def}}{=} \frac{y^k}{\widehat{w}_k(y)} \partial_y \left\{ \frac{\widehat{w}_k(y)}{y^k} \cdot \right\},$$

which is applicable to functions $v = v(t, y)$, then simple computations yield

$$\widehat{\mathcal{L}}_k u = \frac{1}{\tau(t)} \widehat{\mathcal{L}}_{y,k} \widetilde{u} \quad \text{and} \quad \int_a^b \left( \widehat{\mathcal{L}}_k u \right)^2 r dr = \int_{\frac{a}{\tau(t)}}^{\frac{b}{\tau(t)}} \left( \widehat{\mathcal{L}}_{y,k} \widetilde{u} \right)^2 y dy.$$

We recall that $\widehat{w}_k(y) = 1 + y^{2k}$ and

$$\widehat{\mathcal{L}}_k = \frac{r^k}{\widehat{w}_k(r/\tau(t))} \partial_r \left\{ \frac{\widehat{w}_k(r/\tau(t))}{r^k} \cdot \right\} = \partial_r - \frac{k \cos(Q_\tau)}{r}.$$

Using the definition of $\widehat{\mathcal{L}}_{y,k}$ and the fundamental theorem of calculus, we deduce

$$\widetilde{u}(t, y) = \frac{y^k}{w_k(y)} \int_1^y \left\{ \frac{w_k(p)}{p^k} \widehat{\mathcal{L}}_{p,k} \widetilde{u} \right\} dp + \widetilde{u}(t, 1) \frac{2y^k}{1 + y^{2k}} \tag{7.119}$$

$$\stackrel{\text{def}}{=} \widetilde{u}^\perp + \widetilde{u}(t, 1) \, \widetilde{u}_{\text{hom}}.$$

The reasoning behind the above notation has to do with the easily verified fact that $\widehat{\mathcal{L}}_k u_{\text{hom}} = 0$ and we caution the reader that $u^\perp$ is not necessarily orthogonal to the kernel of $\widehat{\mathcal{L}}_k$, being just a convenient notation.

Next, we focus on deriving a uniform bound in $y$ for $\widetilde{u}^\perp(t, y)$. An application of the Cauchy-Schwarz inequality directly on its formula, coupled with a straightforward calculation, reveals that

$$\sup_y |\widetilde{u}^\perp(t, y)|$$

$$\leq \sup_y \left[ \frac{y^k}{1 + y^{2k}} \left| \int_1^y \left\{ \frac{(1 + p^{2k})^2}{p^{2k+1}} \right\} dp \right|^{1/2} \left| \int_1^y \left( \widehat{\mathcal{L}}_{p,k} \widetilde{u} \right)^2 p dp \right|^{1/2} \right]$$

$$\leq \frac{C}{k^{1/2}} \| \widehat{\mathcal{L}}_k u \|_{L^2(rdr)}.$$

Following this, we recognize first that $\widetilde{u}_{\text{hom}} = \sin(Q)$. If we rely on the orthogonality condition in the hypothesis and we take the inner product of the equation (7.119) with $\chi_M \sin(Q)$, then it leads us to

$$|\widetilde{u}(t, 1)| = \frac{|\langle \widetilde{u}^\perp | \chi_M \sin(Q) \rangle_{ydy}|}{\langle \sin(Q) | \chi_M \sin(Q) \rangle_{ydy}} \leq \frac{\| u^\perp \|_{L^\infty} \| \chi_M \sin(Q) \|_{L^1(ydy)}}{\| \chi_M \sin^2(Q) \|_{L^1(ydy)}}.$$

We choose $C_k(M)$ to be a suitable multiple of

$$\frac{\| \chi_M \sin(Q) \|_{L^1(ydy)}}{\| \chi_M \sin^2(Q) \|_{L^1(ydy)}},$$

for which a tedious, yet direct computation yields the asymptotics in (7.118). Putting together all the uniform estimates obtained so far, we obtain the overall fixed-time bound

$$\| u \|_{L^\infty} \leq \frac{C_k(M)}{k^{1/2}} \| \widehat{\mathcal{L}}_k u \|_{L^2(rdr)}. \tag{7.120}$$

Now, we turn to proving the $L^2$ estimates in (7.117). For this purpose, we use the formulas for $\widehat{\mathcal{L}}_k$ and $\mathcal{D}_k^*$ and integration by parts to deduce

$$\|\widehat{\mathcal{L}}_k u\|_{L^2(rdr)}^2 = \int_0^\infty \left\{ (\partial_r u)^2 + \frac{k^2 \cos(2Q_\tau) u^2}{r^2} \right\} rdr,$$

$$\|\mathcal{D}_k^* u\|_{L^2(rdr)}^2 = \int_0^\infty \left\{ (\partial_r u)^2 + \frac{k^2 u^2}{r^2} \right\} rdr.$$

which imply, due to (7.120),

$$\|\mathcal{D}_k^* u\|_{L^2(rdr)}^2 \leq \|\widehat{\mathcal{L}}_k u\|_{L^2(rdr)}^2 + 2\|u\|_{L^\infty}^2 \int_0^\infty \left\{ \frac{k^2 \sin^2(Q_\tau)}{r^2} \right\} rdr$$

$$= \|\widehat{\mathcal{L}}_k u\|_{L^2(rdr)}^2 + 4k\|u\|_{L^\infty}^2$$

$$\leq C_k^2(M) \|\widehat{\mathcal{L}}_k u\|_{L^2(rdr)}^2.$$

To conclude the proof of (7.117), we rely one more time on integration by parts to derive

$$k\, u^2(t, r) = \int_0^r \{2ku\, \partial_s u\}\, ds \leq \int_0^r \left\{ (\partial_s u)^2 + \frac{k^2 u^2}{s^2} \right\} sds \leq \|\mathcal{D}_k^* u\|_{L^2(rdr)}^2,$$

thus deriving

$$k^{1/2} \|u\|_{L^\infty} \leq \|\mathcal{D}_k^* u\|_{L^2(rdr)}.$$

$\square$

Following this, we make some preliminary assumptions on the initial data and calibrate the constants involved in the argument. Specifically, in addition to (7.92), we choose

$$\|\widehat{\mathcal{L}}_k U(0)\|_{L^2(rdr)} \leq C\epsilon$$

and pick $M_1 \gg M \gg 1$ such that

$$M_1 \geq C\epsilon^{-1} \quad \text{and} \quad C_k(M) \leq C\epsilon^{-1/4}.$$

A simple consequence of the definitions (7.74) and (7.81) is the fixed-time estimate

$$\|\widehat{\mathcal{L}}_k U_1\|_{L^2(rdr)} \leq \|\widehat{\mathcal{L}}_k U\|_{L^2(rdr)} + \frac{C}{M_1},$$

which, based on the previous choices for $U(0)$ and $M_1$, implies

$$\|\widehat{\mathcal{L}}_k U_1(0)\|_{L^2(rdr)} \leq C\epsilon.$$

This is the extent to which $M_1$ is involved in our analysis.

Now, we have all the ingredients to state and prove the main result of this subsection.

**Theorem 7.2.** *Under the assumptions*
$$\mathbb{E}_\uparrow < \epsilon^2, \qquad C_k(M) \le C\epsilon^{-1/4}, \qquad \|\widehat{\mathcal{L}}_k U_1(0)\|_{L^2(rdr)} \le C\epsilon, \qquad (7.121)$$
*if $\epsilon$ is sufficiently small and $[0, T)$ is time interval of existence for the solution to (7.3) given by (7.72), then one has the a priori bound*
$$\sup_{t \in [0,T)} \left\{ |b(t)| + \|\widehat{\mathcal{L}}_k U_1(t)\|_{L^2(rdr)} + \|W(t)\|_{L^2(rdr)} \right\} \le C\epsilon. \qquad (7.122)$$
*Furthermore, the following estimate involving $U_1$ holds:*
$$\sup_{t \in [0,T)} \left\{ \|U_1(t)\|_{L^\infty} + \left\| \frac{U_1(t,r)}{r} \right\|_{L^2(rdr)} \right\} \le C\epsilon^{3/4}. \qquad (7.123)$$

**Proof.** In proving (7.122), we are going to rely on the energy formula (7.97), for which a number of terms are estimated using Lemma 7.4. Based on (7.100) and (7.117), we deduce, after performing a simple integration,
$$|A_k(\tilde{U}_1)|$$
$$\le k \, \|[y\partial_y \chi_M + k\chi_M(\cos(Q) + 1)] \sin(Q)\|_{L^1(ydy)} \|U_1\|_{L^\infty} \qquad (7.124)$$
$$\le \tilde{C}_k(M) C_k(M) \|\widehat{\mathcal{L}}_k U_1\|_{L^2(rdr)},$$
where $\tilde{C}_k(M)$ can be chosen to satisfy
$$\tilde{C}_k(M) \sim \begin{cases} M & \text{if } k = 1, \\ \log(M) & \text{if } k = 2, \\ C & \text{if } k > 2. \end{cases}$$
Another direct integration yields
$$|J_k| \le \frac{C}{M^{k-1}}.$$
Next, we bound terms in (7.97) involving $f_k(U_1)$ and, due to (7.88), we infer
$$|f_k(U_1)| \le C \left( |\sin(Q_\tau)| + |\cos(Q_\tau)| \right) U_1^2.$$
If we factor in (7.117), this implies
$$\left\| \frac{k f_k(U_1)}{r} \right\|_{L^2(rdr)}$$
$$\le C \left[ \left\| \frac{k \sin(Q_\tau)}{r} \right\|_{L^2(rdr)} \|U_1\|_{L^\infty}^2 + \left\| \frac{k \cos(Q_\tau) U_1)}{r} \right\|_{L^2(rdr)} \|U_1\|_{L^\infty} \right]$$
$$= C \left[ \left\| \frac{k \sin(Q_\tau)}{r} \right\|_{L^2(rdr)} \|U_1\|_{L^\infty}^2 + \left\| \partial_r U_1 - \widehat{\mathcal{L}}_k U_1 \right\|_{L^2(rdr)} \|U_1\|_{L^\infty} \right]$$
$$\le C_k^2(M) \|\widehat{\mathcal{L}}_k U_1\|_{L^2(rdr)}^2.$$

Moreover, an application of the elementary inequality $|yz| \le y^2/4 + z^2$ gives

$$|\langle k f_k(U_1)/r | \mathcal{L}_k U_1 \rangle| \le \frac{1}{4} \|\widehat{\mathcal{L}}_k U_1\|^2_{L^2(rdr)} + \|k f_k(U_1)/r\|^2_{L^2(rdr)}.$$

Collecting all these estimates and using them in the context of (7.97), we find that

$$\frac{1}{4}\left(1 - C_k^2(M)\|\widehat{\mathcal{L}}_k U_1(t)\|_{L^2(rdr)}\right) D_k(M) b^2(t) + \frac{1}{2}\|W(t)\|^2_{L^2(rdr)}$$

$$+ \frac{1}{4}\|\widehat{\mathcal{L}}_k U_1(t)\|^2_{L^2(rdr)} - C_k^4(M)\|\widehat{\mathcal{L}}_k U_1(t)\|^4_{L^2(rdr)}$$

$$\le \left(\frac{1}{2}D_k(M) - \widetilde{C}_k(M)C_k(M)\|\widehat{\mathcal{L}}_k U_1(t)\|_{L^2(rdr)} - \frac{C}{M^{k-1}}\right) b^2(t)$$

$$+ \frac{1}{2}\|W(t)\|^2_{L^2(rdr)} + \frac{1}{4}\|\widehat{\mathcal{L}}_k U_1(t)\|^2_{L^2(rdr)} - \frac{1}{2}\left\|\frac{k f_k(U_1)(t)}{r}\right\|^2_{L^2(rdr)}$$

$$\le \mathbb{E}_\uparrow < \epsilon^2.$$

Next, we employ a continuity argument to show that

$$\sup_{t \in [0,T)} \|\widehat{\mathcal{L}}_k U_1(t)\|_{L^2(rdr)} \le C\epsilon. \tag{7.125}$$

For this purpose, we define the polynomial

$$P(X) \overset{\text{def}}{=} C_k^4(M)X^4 - \frac{X^2}{4} + \epsilon^2$$

and notice that $P(X) > 0$ if $|X| < \epsilon/2$, whereas the assumption on $C_k(M)$ implies $P(10\epsilon) < 0$, if $\epsilon$ is sufficiently small. Considering now the continuous map

$$t \in [0,T) \mapsto f(t) \overset{\text{def}}{=} P\left(\|\widehat{\mathcal{L}}_k U_1(t)\|_{L^2(rdr)}\right),$$

it follows that the assumption on $\|\widehat{\mathcal{L}}_k U_1(0)\|_{L^2(rdr)}$ ensures $f(0) > 0$ and the above bound involving $b$, $W$, and $U_1$ can be restated as

$$f(t) > \frac{1}{4}\left(1 - C_k^2(M)\|\widehat{\mathcal{L}}_k U_1(t)\|_{L^2(rdr)}\right) D_k(M) b^2(t) + \frac{1}{2}\|W(t)\|^2_{L^2(rdr)},$$

for all $t \in [0,T)$. We claim that these lead to

$$f(t) \ge 0, \qquad \forall t \in [0,T).$$

The alternative is the existence of $t_0 \in (0,T)$ such that $f(t_0) < 0$, which forces

$$C_k^2(M)\|\widehat{\mathcal{L}}_k U_1(t_0)\|_{L^2(rdr)} < 1$$

and, consequently,

$$f(t_0) \geq \frac{3}{4}\|\widehat{\mathcal{L}}_k U_1(t_0)\|_{L^2(rdr)}^2 + \epsilon^2 > 0,$$

thus, a contradiction. Hence, coupling all these findings, we obtain that (7.125) holds for $\epsilon$ sufficiently small.

A direct consequence of this bound is that $C_k^2(M)\|\widehat{\mathcal{L}}_k U_1(t)\|_{L^2(rdr)} \leq C\epsilon^{1/2}$. Using this estimate in the context of the first inequality including $b$ and $W$, we deduce

$$\sup_{t \in [0,T)} \left\{|b(t)| + \|W(t)\|_{L^2(rdr)}\right\} \leq C\epsilon.$$

This concludes the argument for (7.122). In what concerns (7.123), we argue that it holds as a result of coupling (7.117) and (7.122). $\qquad\square$

**Remark 7.5.** Instead of using the calibration $C_k(M) \leq \epsilon^{-1/4}$ for all values of $k$, one could be more diligent and use the specific asymptotics in (7.118) to derive that (7.123) is true with $\epsilon^{3/4}$ replaced by

$$\epsilon_k \sim \begin{cases} \epsilon^{3/4} & \text{if } k = 1, \\ \epsilon/\log(1/\epsilon) & \text{if } k = 2, \\ \epsilon & \text{if } k > 2. \end{cases}$$

However, because we want to simplify the argument and, thus, avoid separating the cases $k = 1$, $k = 2$, and $k > 2$, we will go ahead with (7.123), which is enough for proving the collapse.

**Remark 7.6.** A similar analysis can be done for certain terms in (7.102), as we know that $0 \leq \mathbb{E}_\downarrow - 4k < \epsilon^2$, and we leave this as an exercise for the interested reader.

**Remark 7.7.** We will make frequent use of (7.122) and (7.123) in the following three sections, which are highly technical in nature. The impatient reader may look ahead at Section 7.8 in order to see what our final goal is.

## 7.6 Operator estimates and comparison estimates

In this section, we plan to develop the estimates needed to analyze the second-order energy $\widehat{\mathbb{D}}_\uparrow(t)$, given by (7.59), which is a crucial element in the argument for collapse. According to (7.59), (7.54), and (7.43), we can write

$$\widehat{\mathbb{D}}_\uparrow(t) = \int_0^\infty \frac{1}{2}\left\{\left(\mathcal{L}_k \phi_0\right)^2 + \left(\mathcal{L}_k^* \phi_\uparrow\right)^2\right\} rdr. \tag{7.126}$$

Thus, one goal is to derive practical a priori bounds involving the operators $\mathcal{L}_k$ and $\mathcal{L}_k^*$. Moreover, we will also prove estimates for the version of these operators defined with the help of $Q_\tau$, i.e., $\widehat{\mathcal{L}}_k$ and $\widehat{\mathcal{L}}_k^*$. As it turns out, these bounds are quite straightforward when $k > 1$; however, the case $k = 1$ is special and we need to treat it with extra care.

We recall that we operate under the assumption that $\mathbb{E}_\uparrow < \epsilon^2$, which, based on (7.66), implies $\|\phi_\uparrow\|_{L^2(rdr)} \leq C\epsilon$. For our first bound, we use this fact to show that $\mathcal{L}_k^*$ has "nice" repulsive properties.

**Lemma 7.5.** *If $k \geq 1$, $\|\phi_\uparrow\|_{L^2(rdr)} \leq C\epsilon$, and $\epsilon$ is sufficiently small, then $\|\mathcal{L}_k^*\phi_\uparrow\|_{L^2(rdr)}$ can be bounded from below in the following manner:*

$$
\begin{aligned}
\|\mathcal{L}_k^*\phi_\uparrow\|_{L^2(rdr)}^2 \geq \int_0^\infty & \left\{ \left(1 - C\|\phi_\uparrow\|_{L^2(rdr)}^2\right)\left(\partial_r\phi_\uparrow\right)^2 \right. \\
& \left. + \frac{(k\cos R + 1)^2}{r^2}\phi_\uparrow^2 + \frac{1}{4}\phi_\downarrow^2\phi_\uparrow^2 \right\}rdr \qquad (7.127) \\
\geq \left(1 - C\epsilon^2\right)&\|\phi_\uparrow\|_{\dot{H}^1(rdr)}^2 + \frac{1}{4}\|\phi_\downarrow\phi_\uparrow\|_{L^2(rdr)}^2.
\end{aligned}
$$

**Proof.** We start by using the formula (7.40) to infer

$$
\left(\mathcal{L}_k^*\phi_\uparrow\right)^2 = \left(\partial_r\phi_\uparrow\right)^2 + \frac{(k\cos R + 1)^2}{r^2}\phi_\uparrow^2 + 2\frac{k\cos R + 1}{r}\partial_r\phi_\uparrow\,\phi_\uparrow.
$$

Integrating by parts the last term and relying on

$$
-k\,\partial_r\{\cos R\} = k\sin(R)\partial_r R = \frac{1}{4}\left(\phi_\downarrow^2 - \phi_\uparrow^2\right),
$$

we deduce

$$
\begin{aligned}
\|\mathcal{L}_k^*\phi_\uparrow\|_{L^2(rdr)}^2 = \int_0^\infty & \left\{ \left(\partial_r\phi_\uparrow\right)^2 + \frac{(k\cos R + 1)^2}{r^2}\phi_\uparrow^2 \right. \\
& \left. + \frac{1}{4}\left(\phi_\downarrow^2 - \phi_\uparrow^2\right)\phi_\uparrow^2 \right\}rdr. \qquad (7.128)
\end{aligned}
$$

Next, we take advantage of the Sobolev embedding $\dot{H}^{1/2}(\mathbb{R}^2) \subset L^4(\mathbb{R}^2)$ and the interpolation relation

$$
\left(\dot{H}^1, L^2\right)_{\frac{1}{2},2} = \dot{H}^{1/2}
$$

to obtain

$$
\|\phi_\uparrow\|_{L^4}^4 \leq C\|\phi_\uparrow\|_{L^2}^2\|\phi_\uparrow\|_{\dot{H}^1}^2. \qquad (7.129)
$$

Together with $\|\phi_\uparrow\|_{L^2(rdr)} \leq C\epsilon$, this allows us to absorb the negative term in (7.128) in the first one and thus deduce (7.127). $\qquad\square$

**Remark 7.8.** It is worth noticing that the above estimate also holds in the non-equivariant setting, as the only assumption is that $\|\phi_\uparrow\|_{L^2(rdr)}$ is small, which follows from the choice of initial data and energy conservation.

Next, we focus on proving a priori bounds for the operators $\widehat{\mathcal{L}}^*_k$, defined by (7.82), for $k > 1$.

**Lemma 7.6.** *If $k > 1$ and $u = u(r)$ is a function satisfying the boundary conditions*

$$u(0) = \lim_{r \to \infty} u(r) = 0,$$

*then the following estimates hold:*

$$\|u\|_{L^\infty} \leq \frac{1}{\sqrt{k-1}} \|\widehat{\mathcal{L}}^*_k u\|_{L^2(rdr)}, \tag{7.130}$$

$$\left\|\frac{u}{r}\right\|_{L^2(rdr)} \leq \frac{1}{k-1} \|\widehat{\mathcal{L}}^*_k u\|_{L^2(rdr)}. \tag{7.131}$$

**Proof.** Similar to the previous argument, we square the formula for $\widehat{\mathcal{L}}^*_k u$ and integrate by parts with the joint help of

$$\partial_r \{\cos(Q_\tau)\} = -\frac{k \sin^2(Q_\tau)}{r}$$

and the boundary conditions to produce

$$\begin{aligned}
\|\widehat{\mathcal{L}}^*_k u\|^2_{L^2(rdr)} \\
= \int_0^\infty \left\{ (\partial_r u)^2 + \frac{(1 + k\cos(Q_\tau))^2}{r^2} u^2 + \frac{k^2 \sin^2(Q_\tau)}{r^2} u^2 \right\} rdr \\
= \int_0^\infty \left\{ (\partial_r u)^2 + \frac{k^2 + 2k\cos(Q_\tau) + 1}{r^2} u^2 \right\} rdr \\
\geq \int_0^\infty \left\{ (\partial_r u)^2 + \frac{(k-1)^2}{r^2} u^2 \right\} rdr.
\end{aligned} \tag{7.132}$$

This implies immediately (7.131).

For (7.130), we use the fundamental theorem of calculus, $u(0) = 0$, and the elementary inequality $2ab \leq a^2 + b^2$ to derive

$$\begin{aligned}
(k-1)u^2(r_*) = \int_0^{r_*} \{2(k-1)u\partial_r u\} dr \\
\leq \int_0^{r_*} \left\{ (\partial_r u)^2 + \frac{(k-1)^2}{r^2} u^2 \right\} rdr \\
\leq \|\widehat{\mathcal{L}}^*_k u\|^2_{L^2(rdr)}.
\end{aligned}$$

This completes the proof of the lemma. $\qquad\square$

Following this, we prove that, under our working assumptions, we can compare $\|\mathcal{L}_k^* u\|_{L^2(rdr)}$ to $\|\widehat{\mathcal{L}}_k^* u\|_{L^2(rdr)}$ for $k > 1$.

**Lemma 7.7.** *Let $k > 1$ and assume $\|U_1\|_{L^\infty} \leq C\epsilon^{3/4}$ (see (7.123)). If $u = u(r)$ is a function as in the previous lemma and $\epsilon$ is sufficiently small, then*

$$\|\mathcal{L}_k^* u\|_{L^2(rdr)} \geq C\|\widehat{\mathcal{L}}_k^* u\|_{L^2(rdr)} \geq \frac{C}{2}\|\phi_\downarrow(Q_\tau)u\|_{L^2(rdr)}. \tag{7.133}$$

**Proof.** We recall that $\phi_\downarrow(Q_\tau) = 2k\sin(Q_\tau)/r$ and, based on (7.132), we deduce

$$\|\widehat{\mathcal{L}}_k^* u\|_{L^2(rdr)} \geq \frac{1}{2}\|\phi_\downarrow(Q_\tau)u\|_{L^2(rdr)}. \tag{7.134}$$

For the first comparison estimate, according to (7.83) and (7.84), we have

$$\mathcal{L}_k^* = \widehat{\mathcal{L}}_k^* + \frac{V_k(U_1)}{r},$$

where

$$V_k(U_1) = k\left[\cos(Q_\tau)\bigl(1 - \cos(U_1)\bigr) + \sin(Q_\tau)\sin(U_1)\right].$$

Hence,

$$\|\mathcal{L}_k^* u\|_{L^2(rdr)}^2 = \|\widehat{\mathcal{L}}_k^* u\|_{L^2(rdr)}^2 + \left\|\frac{V_k(U_1)u}{r}\right\|_{L^2(rdr)}^2 + 2\left\langle \widehat{\mathcal{L}}_k^* u \,\Big|\, \frac{V_k(U_1)u}{r}\right\rangle.$$

In proving the desired estimate, the goal is to absorb the last term on the right-hand side into the first one. Using the Cauchy-Schwarz inequality, the hypothesis on $\|U_1\|_{L^\infty}$, and (7.134), we derive

$$\left|\left\langle \widehat{\mathcal{L}}_k^* u \,\Big|\, \frac{V_k(U_1)u}{r}\right\rangle\right| \leq Ck\|\widehat{\mathcal{L}}_k^* u\|_{L^2(rdr)}\left\{\|U_1\|_{L^\infty}^2\left\|\frac{u}{r}\right\|_{L^2(rdr)}\right.$$

$$\left. + \|U_1\|_{L^\infty}\|\phi_\downarrow(Q_\tau)u\|_{L^2(rdr)}\right\}$$

$$\leq C\epsilon^{3/4}\|\widehat{\mathcal{L}}_k^* u\|_{L^2(rdr)}^2,$$

which finishes the argument by picking $\epsilon$ to be sufficiently small. $\square$

Now, we address the counterparts of Lemmas 7.6 and 7.7 for the delicate case when $k = 1$. First, one notices that the operator $\widehat{\mathcal{L}}_1^*$ is not as strongly repulsive as $\widehat{\mathcal{L}}_k^*$ when $k > 1$. This can be seen by a close inspection of

(7.132), which reveals that $\|\widehat{\mathcal{L}}_1^* u\|_{L^2(rdr)}$ no longer controls $\|u/r\|_{L^2(rdr)}$. Instead, we have a result which is slightly weaker than Lemma 7.6.

**Lemma 7.8.** *If* $\|\phi_\uparrow\|_{L^2(rdr)} \leq C\epsilon$, $\|U_1\|_{L^\infty} \leq C\epsilon^{3/4}$, *and* $\epsilon$ *is sufficiently small, then the following estimate holds:*

$$\|\mathcal{L}_1^* \phi_\uparrow\|_{L^2(rdr)} + \|\widehat{\mathcal{L}}_1^* \phi_\uparrow\|_{L^2(rdr)} \geq C \left\| \frac{U_1 \phi_\uparrow}{r} \right\|_{L^2(rdr)}. \tag{7.135}$$

**Proof.** Our initial goal is to prove that

$$\|\mathcal{L}_1^* \phi_\uparrow\|_{L^2(rdr)}^2 \geq C \int_0^\infty \left\{ \frac{1 + \cos(R)}{r^2} \phi_\uparrow^2 \right\} r dr, \tag{7.136}$$

$$\|\widehat{\mathcal{L}}_1^* \phi_\uparrow\|_{L^2(rdr)}^2 \geq C \int_0^\infty \left\{ \frac{1 + \cos(Q_\tau)}{r^2} \phi_\uparrow^2 \right\} r dr. \tag{7.137}$$

The second bound is immediate, as one can just read it off of (7.132).

For the first one, we start by rewriting (7.127) for $k = 1$, i.e.,

$$\|\mathcal{L}_1^* \phi_\uparrow\|_{L^2(rdr)}^2$$
$$\geq \int_0^\infty \left\{ (1 - C\epsilon^2)(\partial_r \phi_\uparrow)^2 + \frac{(1 + \cos R)^2}{r^2} \phi_\uparrow^2 + \frac{1}{4} \phi_\downarrow^2 \phi_\uparrow^2 \right\} r dr. \tag{7.138}$$

Next, we take advantage of

$$\phi_\downarrow = \frac{2\sin(R)}{r} + \phi_\uparrow$$

to infer that

$$\int_0^\infty \left\{ \frac{1}{4} \phi_\downarrow^2 \phi_\uparrow^2 \right\} r dr = \int_0^\infty \left\{ \frac{\sin^2(R)}{r^2} \phi_\uparrow^2 - \frac{1}{4} \phi_\uparrow^4 + \frac{1}{2} \phi_\downarrow \phi_\uparrow^3 \right\} r dr. \tag{7.139}$$

For the last two integrands in the above formula, we use (7.129), the hypothesis, and Hölder's inequality to deduce

$$\|\phi_\uparrow\|_{L^4(rdr)} \leq C\epsilon^{1/2} \|\phi_\uparrow\|_{\dot{H}^1(rdr)}^{1/2}$$

and, subsequently,

$$\|\phi_\downarrow \phi_\uparrow^3\|_{L^1(rdr)} \leq \|\phi_\downarrow \phi_\uparrow\|_{L^2(rdr)} \|\phi_\uparrow\|_{L^4(rdr)}^2 \leq C\epsilon \|\phi_\downarrow \phi_\uparrow\|_{L^2(rdr)} \|\phi_\uparrow\|_{\dot{H}^1(rdr)}.$$

Hence, combining these two bounds with (7.138) and (7.139), we derive the estimate

$$\|\mathcal{L}_1^* \phi_\uparrow\|_{L^2(rdr)}^2 + C\epsilon \|\phi_\downarrow \phi_\uparrow\|_{L^2(rdr)}^2$$
$$\geq \int_0^\infty \left\{ (1 - C\epsilon)(\partial_r \phi_\uparrow)^2 + \frac{2(1 + \cos(R))}{r^2} \phi_\uparrow^2 \right\} r dr.$$

Then (7.136) follows if we also factor in from (7.138) that

$$\|\phi_\downarrow \phi_\uparrow\|^2_{L^2(rdr)} \leq 4\|\mathcal{L}^*_1 \phi_\uparrow\|^2_{L^2(rdr)}.$$

Next, the claim is that (7.136) and (7.137) are sufficient to prove (7.135). Indeed, for the region of integration

$$\left\{ |U_1| y = \frac{|U_1| r}{\tau} \leq 1 \right\},$$

we have

$$1 + \cos(Q_\tau) = \frac{2}{1+y^2} \geq U_1^2$$

and, as a consequence of (7.137), we obtain

$$\|\widehat{\mathcal{L}}^*_1 \phi_\uparrow\|^2_{L^2(rdr)} \geq C \int\limits_{|U_1|y \leq 1} \left\{ \frac{U_1^2 \phi_\uparrow^2}{r^2} \right\} r dr. \tag{7.140}$$

For the complementary case, $\{|U_1| y > 1\}$, it follows that $y > |U_1|^{-1} \geq C\epsilon^{-3/4} \gg 1$. Hence,

$$|\pi - R| = |\pi - Q_\tau - U_1| = 2\left| \arctan\left(\frac{1}{y}\right) - \frac{U_1}{2} \right| \ll 1,$$

which implies

$$1 + \cos(R) = 2\sin^2\left(\frac{\pi - R}{2}\right) \geq C\left| \arctan\left(\frac{1}{y}\right) - \frac{U_1}{2} \right|^2.$$

Similarly, we deduce

$$1 + \cos(Q_\tau) \geq C \arctan^2\left(\frac{1}{y}\right).$$

If we couple these two bounds with (7.136) and (7.137), then we derive

$$\|\mathcal{L}^*_1 \phi_\uparrow\|^2_{L^2(rdr)} + \|\widehat{\mathcal{L}}^*_1 \phi_\uparrow\|^2_{L^2(rdr)} \geq C \int\limits_{|U_1|y > 1} \left\{ \frac{U_1^2 \phi_\uparrow^2}{r^2} \right\} r dr. \tag{7.141}$$

Together with (7.140), this estimate proves (7.135), thus finalizing the argument. $\square$

With the previous result in hand, we have the tools to compare $\|\mathcal{L}^*_1 \phi_\uparrow\|_{L^2(rdr)}$ to $\|\widehat{\mathcal{L}}^*_1 \phi_\uparrow\|_{L^2(rdr)}$, a fact which is addressed by the following lemma.

**Lemma 7.9.** *Under the assumptions of Lemma 7.8, the comparison estimate*

$$\|\mathcal{L}^*_1 \phi_\uparrow\|_{L^2(rdr)} \geq C\|\widehat{\mathcal{L}}^*_1 \phi_\uparrow\|_{L^2(rdr)} \geq \frac{C}{2}\|\phi_\downarrow(Q_\tau)\phi_\uparrow\|_{L^2(rdr)} \tag{7.142}$$

*is true.*

**Proof.** The argument is quite similar to the one used for showing (7.133). In fact, we claim that

$$\|\widehat{\mathcal{L}}_1^* \phi_\uparrow\|_{L^2(rdr)} \geq \frac{1}{2}\|\phi_\downarrow(Q_\tau)\phi_\uparrow\|_{L^2(rdr)} \tag{7.143}$$

is covered by (7.133), as a close inspection of the argument for

$$\|\widehat{\mathcal{L}}_k^* u\|_{L^2(rdr)} \geq \frac{1}{2}\|\phi_\downarrow(Q_\tau)u\|_{L^2(rdr)}$$

reveals that it does not rely in any way on $k > 1$.

As before, in proving the first estimate in (7.142), we proceed by writing

$$\|\mathcal{L}_1^* \phi_\uparrow\|_{L^2(rdr)}^2 = \|\widehat{\mathcal{L}}_1^* \phi_\uparrow\|_{L^2(rdr)}^2 + \left\|\frac{V_1(U_1)\phi_\uparrow}{r}\right\|_{L^2(rdr)}^2 + 2\left\langle \widehat{\mathcal{L}}_1^* \phi_\uparrow \middle| \frac{V_1(U_1)\phi_\uparrow}{r}\right\rangle$$

and the goal is still the derivation of a favorable bound for the last term. If we rely on the formula for $V_1$ (see (7.84)), the fact that $\phi_\downarrow(Q_\tau) = 2\sin(Q_\tau)/r$, the smallness of $U_1$, (7.135), and (7.143), then we directly deduce that

$$\left|\left\langle \widehat{\mathcal{L}}_1^* \phi_\uparrow \middle| \frac{V_1(U_1)\phi_\uparrow}{r}\right\rangle\right|$$

$$\leq C\|\widehat{\mathcal{L}}_1^* \phi_\uparrow\|_{L^2(rdr)}\|U_1\|_{L^\infty}\left(\left\|\frac{U_1\phi_\uparrow}{r}\right\|_{L^2(rdr)} + \|\phi_\downarrow(Q_\tau)\phi_\uparrow\|_{L^2(rdr)}\right)$$

$$\leq C\epsilon^{3/4}\|\widehat{\mathcal{L}}_1^* \phi_\uparrow\|_{L^2(rdr)}\left(\|\mathcal{L}_1^* \phi_\uparrow\|_{L^2(rdr)} + \|\widehat{\mathcal{L}}_1^* \phi_\uparrow\|_{L^2(rdr)}\right).$$

Applying this estimate in the context of the previous identity involving $\|\mathcal{L}_1^* \phi_\uparrow\|_{L^2(rdr)}$ and $\|\widehat{\mathcal{L}}_1^* \phi_\uparrow\|_{L^2(rdr)}$ concludes the proof of the lemma. □

At the beginning of this section, we made the case for the need of deriving useful a priori bounds for both operators $\mathcal{L}_k$ and $\mathcal{L}_k^*$. So far, we exclusively addressed the latter. Now, we turn our attention to $\mathcal{L}_k$ and, in particular, $\mathcal{L}_k\phi_0$, which appears as a component of the second-order energy according to (7.126). As $\mathcal{L}_k$ has nontrivial kernel, we need to perform a special decomposition for $\phi_0$ and then rely in a crucial way on the orthogonality condition (7.76).

**Lemma 7.10.** *Let $k \geq 1$ and assume that*

$$\|\phi_\downarrow\|_{L^2(rdr)} \leq C \tag{7.144}$$

*and*

$$\|U_1\|_{L^\infty} + \|U_1/r\|_{L^2(rdr)} \leq C\epsilon^{3/4}. \tag{7.145}$$

*For the decomposition*

$$\phi_0 = \phi_0(Q_\tau) + \partial_t U_1,$$

*the following bound is true:*

$$\|\partial_t U_1\|_{L^\infty} \leq C_k(M) \left( \|\mathcal{L}_k \phi_0\|_{L^2(rdr)} + \epsilon^{3/4} \frac{b}{\tau} \right). \tag{7.146}$$

*Furthermore, we can write $\phi_0$ as*

$$\phi_0 = \phi_0^\perp + l_k \, \phi_0(Q_\tau) \tag{7.147}$$

*where the functions $\phi_0^\perp = \phi_0^\perp(t,r)$ and $l_k = l_k(t,r)$ satisfy the estimates*

$$\|\phi_0^\perp\|_{L^\infty} + \left\| \frac{\phi_0^\perp}{r} \right\|_{L^2(rdr)} \leq C \|\mathcal{L}_k \phi_0\|_{L^2(rdr)}, \tag{7.148}$$

$$\|l_k\|_{L^\infty} \leq C_k(M) \left( \frac{\tau}{b} \|\mathcal{L}_k \phi_0\|_{L^2(rdr)} + \epsilon^{3/4} \right) + C. \tag{7.149}$$

**Proof.** The argument has many similarities with the one for Lemma 7.4, where it is preferable to work in the $(t,y)$ coordinates than in the $(t,r)$ ones. We remind the reader that for a function $u = u(t,r)$, we adopted the notational convention $\tilde{u} = \tilde{u}(t,y) = u(t, \tau(t)y)$. We also recall from (7.39) and (7.41) that the operator $\mathcal{L}_k$, acting on $u = u(t,r)$, has the form

$$\mathcal{L}_k = \partial_r - \frac{k \cos(R)}{r} = \frac{r^k}{w_k(t,r)} \partial_r \left\{ \frac{w_k(t,r)}{r^k} \cdot \right\}, \tag{7.150}$$

where

$$w_k(t,r) = \exp \left\{ k \int_0^r \left\{ \frac{1 - \cos(R(t,p))}{p} \right\} dp \right\}.$$

As in Lemma 7.4, we introduce a version of $\mathcal{L}_k$ adapted to the $(t,y)$ coordinates, i.e.,

$$\mathcal{L}_{y,k} \overset{\text{def}}{=} \frac{y^k}{\widetilde{w}_k(t,y)} \partial_y \left\{ \frac{\widetilde{w}_k(t,y)}{y^k} \cdot \right\}, \tag{7.151}$$

for which we have

$$\mathcal{L}_{y,k} \tilde{u} = \tau \mathcal{L}_k u$$

and, subsequently,

$$\|\mathcal{L}_{y,k} \tilde{u}\|_{L^2(ydy)} = \|\mathcal{L}_k u\|_{L^2(rdr)}.$$

Next, we use $R = Q_\tau + U_1$ to infer

$$\widetilde{w}_k(t,y) = \exp\left\{ k \int_0^y \left\{ \frac{1 - \cos(\widetilde{R}(t,p))}{p} \right\} dp \right\}$$

$$= \widehat{w}_k(y) \exp\left\{ k \int_0^y \left\{ \frac{\cos(Q)(1 - \cos(\widetilde{U}_1))}{p} \right\} dp \right\} \qquad (7.152)$$

$$\cdot \exp\left\{ k \int_0^y \left\{ \frac{\sin(Q)\sin(\widetilde{U}_1)}{p} \right\} dp \right\},$$

where, as we know from before,

$$\widehat{w}_k(y) = \exp\left\{ k \int_0^y \left\{ \frac{1 - \cos(Q)}{p} \right\} dp \right\} = 1 + y^{2k}.$$

If we rely on (7.145), then we can estimate the two exponents as follows:

$$\left| \int_0^y \left\{ \frac{\cos(Q)(1 - \cos(\widetilde{U}_1))}{p} \right\} dp \right| \le C \int_0^y \left\{ \frac{\widetilde{U}_1^2}{p} \right\} dp \le C \left\| \frac{U_1}{r} \right\|_{L^2(rdr)}$$

$$\le C\epsilon^{3/2},$$

$$\left| \int_0^y \left\{ \frac{\sin(Q)\sin(\widetilde{U}_1)}{p} \right\} dp \right|$$

$$\le \left( \int_0^y \left\{ \frac{\sin^2(Q)}{p} \right\} dp \right)^{1/2} \left( \int_0^y \left\{ \frac{\sin^2(\widetilde{U}_1)}{p} \right\} dp \right)^{1/2}$$

$$\le C \left( \int_0^\infty \left\{ \frac{p}{(1 + p^2)^2} \right\} dp \right)^{1/2} \left( \int_0^y \left\{ \frac{\widetilde{U}_1^2}{p} \right\} dp \right)^{1/2}$$

$$\le C\epsilon^{3/4}.$$

These bounds imply that we can write

$$\widetilde{w}_k(t,y) = \widehat{w}_k(y) m_k(t,y), \qquad (7.153)$$

with

$$\|m_k - 1\|_{L^\infty} \le C\epsilon^{3/4}. \qquad (7.154)$$

Now, we can deduce by integration from the formula (7.151) that

$$\widetilde{\phi}_0 = \frac{y^k}{\widetilde{w}_k(t,y)} \left[ \int_1^y \left\{ \frac{\widetilde{w}_k(t,p)}{p^k} (\mathcal{L}_{p,k}\widetilde{\phi}_0) \right\} dp + \widetilde{w}_k(t,1)\widetilde{\phi}_0(t,1) \right]$$

$$\stackrel{\text{def}}{=} \widetilde{\phi}_0^\perp + (1 + L)\frac{\phi_0(Q)}{m_k},$$

where

$$\widetilde{\phi}_0^\perp \overset{\text{def}}{=} \frac{y^k}{\widetilde{w}_k(t,y)} \int_1^y \left\{ \frac{\widetilde{w}_k(t,p)}{p^k} \left( \mathcal{L}_{p,k} \widetilde{\phi}_0 \right) \right\} dp$$

and

$$1 + L \overset{\text{def}}{=} \frac{\tau}{2kb} \widetilde{w}_k(t,1) \widetilde{\phi}_0(t,1).$$

The motivation for these definitions is given by (7.91), i.e.,

$$\phi_0(Q) = \frac{kb}{\tau} \sin(Q), \tag{7.155}$$

which easily implies

$$\mathcal{L}_{y,k} \left\{ \frac{\phi_0(Q)}{m_k} \right\} = 0. \tag{7.156}$$

Hence, to achieve the decomposition (7.147), we take $l_k = l_k(t,r)$ to satisfy

$$\widetilde{l}_k(t,y) \overset{\text{def}}{=} \frac{1+L}{m_k}. \tag{7.157}$$

On the other hand, we have

$$\widetilde{\phi}_0 = \widetilde{\phi}_0(R) = \phi_0(Q) + \widetilde{\partial_t U_1},$$

which yields

$$\widetilde{\partial_t U_1} = \widetilde{\phi}_0^\perp + \frac{1 - m_k + L}{m_k} \phi_0(Q). \tag{7.158}$$

At this point, we have all the prerequisites for proving (7.149) and (7.148).

We start by invoking the formulas for $\widetilde{\phi}_0^\perp$ and $\widetilde{w}_k$, the Cauchy-Schwarz inequality, and (7.154), to argue that

$$\|\widetilde{\phi}_0^\perp\|_{L^\infty}$$

$$\leq \sup_y \left\{ \frac{y^k}{\widetilde{w}_k(t,y)} \left( \int_1^y \left\{ \frac{\widetilde{w}_k^2(t,p)}{p^{2k+1}} \right\} dp \right)^{1/2} \right\} \|\mathcal{L}_{y,k} \widetilde{\phi}_0\|_{L^2(ydy)}$$

$$\leq C \sup_y \left\{ \frac{y^k}{\widehat{w}_k(y)} \left( \int_1^y \left\{ \frac{\widehat{w}_k^2(p)}{p^{2k+1}} \right\} dp \right)^{1/2} \right\} \|\mathcal{L}_k \phi_0\|_{L^2(rdr)} \tag{7.159}$$

$$\leq C \|\mathcal{L}_k \phi_0\|_{L^2(rdr)}.$$

This proves half of (7.148).

For the other half, due to (7.156), we notice that

$$\mathcal{L}_k \phi_0 = \mathcal{L}_k \phi_0^\perp.$$

A direct computation based on (7.150) shows that

$$\|\mathcal{L}_k\phi_0^\perp\|_{L^2(rdr)}^2$$

$$= \int_0^\infty \left\{ (\partial_r\phi_0^\perp)^2 + \frac{k^2\cos^2(R)}{r^2}(\phi_0^\perp)^2 + \frac{1}{4}(\phi_\uparrow^2 - \phi_\downarrow^2)(\phi_0^\perp)^2 \right\} rdr$$

$$\geq \int_0^\infty \left\{ \frac{k^2\cos^2(R)}{r^2}(\phi_0^\perp)^2 \right\} rdr - \frac{1}{4}\|\phi_\downarrow\|_{L^2(rdr)}^2\|\phi_0^\perp\|_{L^\infty}^2.$$

If we rely on (7.159) and (7.144), then we derive

$$\int_0^\infty \left\{ \frac{k^2\cos^2(R)}{r^2}(\phi_0^\perp)^2 \right\} rdr \leq C\|\mathcal{L}_k\phi_0^\perp\|_{L^2(rdr)}^2.$$

Furthermore, from $R = Q_\tau + U_1$ and (7.145), we deduce

$$\left| \cos(R) - \frac{1 - (r/\tau)^{2k}}{1 + (r/\tau)^{2k}} \right| = |\cos(R) - \cos(Q_\tau)| \leq C\epsilon^{3/4}.$$

Thus, for $r/\tau \notin [1/2, 2]$, we have $|\cos(R)| \geq C > 0$ and, subsequently,

$$\int_{\frac{r}{\tau}\notin[\frac{1}{2},2]} \left\{ \frac{(\phi_0^\perp)^2}{r^2} \right\} rdr \leq C\|\mathcal{L}_k\phi_0^\perp\|_{L^2(rdr)}^2.$$

For the complimentary case (i.e., $1/2 \leq r/\tau \leq 2$) we directly obtain

$$\int_{\frac{1}{2}\leq\frac{r}{\tau}\leq 2} \left\{ \frac{(\phi_0^\perp)^2}{r^2} \right\} rdr \leq C\|\phi_0^\perp\|_{L^\infty}^2 \leq C\|\mathcal{L}_k\phi_0^\perp\|_{L^2(rdr)}^2,$$

which finishes the argument for the other half of (7.148).

Finally, we prove (7.149) and, with its help, also show that (7.146) holds. This is the moment in the argument where we use the orthogonality condition (7.76), in the same way it was relied upon for Lemma 7.1. Thus, we start by deriving

$$\tau^2\langle\widetilde{\partial_t U_1}|\chi_M\sin(Q)\rangle_{ydy} = \langle\partial_t U_1|\chi_M\sin(Q_\tau)\rangle$$

$$= -\langle U_1|\partial_t(\chi_M\sin(Q_\tau))\rangle \qquad (7.160)$$

$$= -\frac{b\tau}{k}A_k(\widetilde{U}_1),$$

where $A_k(\widetilde{U}_1)$ is defined by (7.100). Next, we use (7.158) to deduce

$$\langle\widetilde{\partial_t U_1}|\chi_M\sin(Q)\rangle_{ydy} = \langle\widetilde{\phi_0}^\perp|\chi_M\sin(Q)\rangle_{ydy}$$

$$+ \left\langle \frac{1 - m_k}{m_k}\phi_0(Q)\Big|\chi_M\sin(Q) \right\rangle_{ydy} \qquad (7.161)$$

$$+ L\left\langle \frac{\phi_0(Q)}{m_k}\Big|\chi_M\sin(Q) \right\rangle_{ydy}$$

and the goal is to obtain from this identity a favorable estimate for $L$.

For the left-hand side, we employ (7.160), (7.124), and (7.145) to infer

$$\left| \langle \widetilde{\partial_t U_1} | \chi_M \sin(Q) \rangle_{ydy} \right| = \frac{b}{k\tau} \left| A_k(\tilde{U}_1) \right| \leq \tilde{C}_k(M) \frac{b}{\tau} \|U_1\|_{L^\infty}$$
$$\leq \tilde{C}_k(M) \epsilon^{3/4} \frac{b}{\tau}.$$

Next, the first term on the right-hand side of the previous identity can be easily bounded as

$$\left| \langle \widetilde{\phi_0^\perp} | \chi_M \sin(Q) \rangle_{ydy} \right| \leq \|\phi_0^\perp\|_{L^\infty} \|\chi_M \sin(Q)\|_{L^1(ydy)}$$
$$\leq \tilde{C}_k(M) \|\mathcal{L}_k \phi_0\|_{L^2(rdr)},$$

by applying (7.159) and performing a simple integration. For the terms involving $m_k$, due to (7.154), we make the observation that

$$\left\langle \frac{1}{m_k} \sin(Q) \middle| \chi_M \sin(Q) \right\rangle_{ydy} \sim D_k(M),$$

where $D_k(M)$ is given by (7.98). Hence, based on (7.155) and (7.154), we have

$$\left\langle \frac{1}{m_k} \phi_0(Q) \middle| \chi_M \sin(Q) \right\rangle_{ydy} \sim D_k(M) \frac{b}{\tau},$$
$$\left| \left\langle \frac{1 - m_k}{m_k} \phi_0(Q) \middle| \chi_M \sin(Q) \right\rangle_{ydy} \right| \leq C D_k(M) \epsilon^{3/4} \frac{b}{\tau}.$$

Coupling all these facts about the terms in (7.161), we finally obtain

$$|L| \leq (C + C_k(M)) \epsilon^{3/4} + C_k(M) \frac{\tau}{b} \|\mathcal{L}_k \phi_0\|_{L^2(rdr)}$$
$$\leq C_k(M) \left( \frac{\tau}{b} \|\mathcal{L}_k \phi_0\|_{L^2(rdr)} + \epsilon^{3/4} \right),$$

which, according to (7.157), implies (7.149). Finally, using (7.158), (7.159), and the previous estimate, it follows that

$$\|\partial_t U_1\|_{L^\infty} \leq \|\phi_0^\perp\|_{L^\infty} + C_k(M) \left( \frac{\tau}{b} \|\mathcal{L}_k \phi_0\|_{L^2(rdr)} + \epsilon^{3/4} \right) \frac{b}{\tau}$$
$$\leq C_k(M) \left( \|\mathcal{L}_k \phi_0\|_{L^2(rdr)} + \epsilon^{3/4} \frac{b}{\tau} \right),$$

thus proving (7.146). This finishes the whole argument for the lemma. $\square$

We conclude this section with one more estimate concerning $\mathcal{L}_1^* \phi_\uparrow$, which complements the ones derived in Lemmas 7.5, 7.8, and 7.9. This is facilitated by the new information we have on the weight $w_1$, obtained in the previous result.

**Lemma 7.11.** *If we assume that* $\|U_1/r\|_{L^2(rdr)} \leq C\epsilon^{3/4}$, *then the following bound,*

$$\sup_{\frac{r}{\tau} \leq M} |\phi_\uparrow| \leq C(\log(M))^{1/2} \|\mathcal{L}_1^* \phi_\uparrow\|_{L^2(rdr)}, \tag{7.162}$$

*is true.*

**Proof.** We remind the reader that, according to (7.41),

$$\mathcal{L}_1^* = -\frac{w_1(t,r)}{r^2} \partial_r \left\{ \frac{r^2}{w_1(t,r)} \cdot \right\},$$

and, introducing as before,

$$\mathcal{L}_{y,1}^* \overset{\text{def}}{=} -\frac{\widetilde{w}_1(t,y)}{y^2} \partial_y \left\{ \frac{y^2}{\widetilde{w}_1(t,y)} \cdot \right\}, \tag{7.163}$$

we have

$$\mathcal{L}_{y,1}^* \tilde{u} = \tau \mathcal{L}_1^* u, \qquad \|\mathcal{L}_{y,1}^* \tilde{u}\|_{L^2(ydy)} = \|\mathcal{L}_1^* u\|_{L^2(rdr)}.$$

Using (7.153) and (7.154), we know that

$$\widetilde{w}_1(t,y) = (1+y^2)m_1(t,y), \qquad \|m_1 - 1\|_{L^\infty} \leq C\epsilon^{3/4}.$$

If we rely on the finiteness of $\mathbb{E}_\uparrow$, defined by (7.66), then we deduce that

$$\lim_{r \to 0} \phi_\uparrow(t,r)\, r = 0,$$

which allows us to integrate the formula (7.163) on the interval $[0,y]$ and thus derive

$$\widetilde{\phi}_\uparrow = -\frac{\widetilde{w}_1(t,y)}{y^2} \int_0^y \left\{ \frac{p^2}{\widetilde{w}_1(t,p)} \mathcal{L}_{p,1}^* \widetilde{\phi}_\uparrow \right\} dp.$$

Applying the Cauchy-Schwarz inequality coupled with the previous facts on $\widetilde{w}_1$, we obtain

$$\sup_{y \leq M} |\widetilde{\phi}_\uparrow| \leq \sup_{y \leq M} \left\{ \frac{\widetilde{w}_1(t,y)}{y^2} \left( \int_0^y \left\{ \frac{p^3}{\widetilde{w}_1^2(t,p)} \right\} dp \right)^{1/2} \right\} \|\mathcal{L}_{y,1}^* \widetilde{\phi}_\uparrow\|_{L^2(ydy)}$$

$$\leq C(\log(M))^{1/2} \|\mathcal{L}_1^* \phi_\uparrow\|_{L^2(rdr)}.$$

This proves the desired estimate by switching back from the $y$ coordinate to the radial one. $\qquad \square$

## 7.7 Estimating the second-order energy and the acceleration

One of the central objects in our collapse analysis is the second-order energy $\widehat{\mathbb{D}}_\uparrow = \widehat{\mathbb{D}}_\uparrow(t)$, introduced by (7.59) and for which (7.126) provides an alternative formula. It satisfies the differential identity (7.58), which also involves two other functions of time, $\widehat{N} = \widehat{N}(t)$ and $F = F(t)$. For technical reasons, which lead to simplification of certain considerations, we choose to scale the definitions of $\widehat{\mathbb{D}}_\uparrow$ and $\widehat{N}$ (i.e., (7.59) and (7.60), respectively) by a factor of $\tau^2(t)$ and hence work with

$$\mathbb{D}_\uparrow(t) \overset{\text{def}}{=} \frac{\tau^2(t)}{2} \int_0^\infty \left\{ \left( \mathcal{L}_k \phi_0 \right)^2 + \left( \mathcal{L}_k^* \phi_\uparrow \right)^2 \right\} r\, dr, \tag{7.164}$$

$$N(t) \overset{\text{def}}{=} -\frac{\tau^2(t)}{2} \int_0^\infty \left\{ (\phi_\downarrow - \phi_\uparrow) \phi_0^2 \phi_\uparrow \right\} r\, dr, \tag{7.165}$$

$$F(t) \overset{\text{def}}{=} \int_0^\infty \left\{ \frac{3\phi_0}{2} (\phi_\downarrow - \phi_\uparrow) \phi_\uparrow \, \mathcal{L}_k^* \phi_\uparrow - \phi_0^3 \phi_\uparrow \frac{k \cos(R)}{r} \right\} r\, dr, \tag{7.166}$$

and the differential identity

$$\frac{d}{dt} \left\{ \frac{\mathbb{D}_\uparrow(t) + N(t)}{\tau^2(t)} \right\} = F(t). \tag{7.167}$$

The first goal of this section is to turn this identity into a differential inequality for the second-order energy, which is one of the crucial ingredients in proving the collapse. This is the content of the next theorem.

**Theorem 7.3.** *Under the assumptions*

$$\|\phi_\uparrow\|_{L^2(rdr)} \le C\epsilon, \qquad \|\phi_\downarrow\|_{L^2(rdr)} \le C, \qquad C_k(M) \le C\epsilon^{-1/4},$$

$$\|U_1\|_{L^\infty} + \|U_1/r\|_{L^2(rdr)} \le C\epsilon^{3/4},$$

*if $\epsilon$ is sufficiently small and*

$$I \overset{\text{def}}{=} F + \frac{3b}{4\tau} \|\phi_\downarrow(Q_\tau)\phi_\uparrow\|_{L^2(rdr)}^2, \tag{7.168}$$

*then*

$$|I| \le \frac{Cb^3 \mathbb{D}_\uparrow^{1/2} + C_k(M)\epsilon^{3/4} b \mathbb{D}_\uparrow + C_k(M) \mathbb{D}_\uparrow^{3/2} + C_k^3(M) \mathbb{D}_\uparrow^2}{\tau^3}. \tag{7.169}$$

*Furthermore, the integral given by (7.165) can be estimated as follows:*

$$|N| \le C \left( b^2 \mathbb{D}_\uparrow^{1/2} + \epsilon \mathbb{D}_\uparrow \right) + C_k^2(M) \mathbb{D}_\uparrow^{3/2}. \tag{7.170}$$

**Proof.** We start the argument by rewriting the first integrand in the formula (7.166) for $F$ in a form convenient to deriving (7.168) and (7.169). For this purpose, we recall that, associated to $R = Q_\tau + U_1$ and due to (7.47)-(7.49), (7.91), and (7.83), we have the following decompositions:

$$\phi_0 = \phi_0(Q_\tau) + \partial_t U_1 = \frac{kb}{\tau}\sin(Q_\tau) + \partial_t U_1,$$

$$\phi_\downarrow - \phi_\uparrow = \frac{2k\sin(R)}{r} = \phi_\downarrow(Q_\tau) + \frac{2kg_k(U_1)}{r},$$

$$\mathcal{L}_k^* = \widehat{\mathcal{L}}_k^* + \frac{V_k(U_1)}{r},$$

where

$$g_k(U_1) \overset{\text{def}}{=} \sin(Q_\tau)(\cos(U_1) - 1) + \cos(Q_\tau)\sin(U_1)$$

and $V_k(U_1)$ is given by (7.84). Accordingly, we deduce that

$$\phi_0(\phi_\downarrow - \phi_\uparrow)\phi_\uparrow \mathcal{L}_k^*\phi_\uparrow$$

$$= \phi_0(Q_\tau)(\phi_\downarrow - \phi_\uparrow)\phi_\uparrow \mathcal{L}_k^*\phi_\uparrow + \partial_t U_1(\phi_\downarrow - \phi_\uparrow)\phi_\uparrow \mathcal{L}_k^*\phi_\uparrow$$

$$= \phi_0(Q_\tau)\phi_\downarrow(Q_\tau)\phi_\uparrow \mathcal{L}_k^*\phi_\uparrow + \phi_0(Q_\tau)\frac{2kg_k(U_1)}{r}\phi_\uparrow \mathcal{L}_k^*\phi_\uparrow$$

$$+ \partial_t U_1(\phi_\downarrow - \phi_\uparrow)\phi_\uparrow \mathcal{L}_k^*\phi_\uparrow$$

$$= \phi_0(Q_\tau)\phi_\downarrow(Q_\tau)\phi_\uparrow \widehat{\mathcal{L}}_k^*\phi_\uparrow + \phi_0(Q_\tau)\phi_\downarrow(Q_\tau)\phi_\uparrow\frac{V_k(U_1)\phi_\uparrow}{r}$$

$$+ \phi_0(Q_\tau)\frac{2kg_k(U_1)}{r}\phi_\uparrow \mathcal{L}_k^*\phi_\uparrow + \partial_t U_1(\phi_\downarrow - \phi_\uparrow)\phi_\uparrow \mathcal{L}_k^*\phi_\uparrow.$$

In relation to the previous identity, we introduce the following integral terms,

$$I_0 \overset{\text{def}}{=} \frac{3}{2}\int_0^\infty \left\{\phi_0(Q_\tau)\phi_\downarrow(Q_\tau)\phi_\uparrow \widehat{\mathcal{L}}_k^*\phi_\uparrow\right\} r\,dr, \tag{7.171}$$

$$I_1 \overset{\text{def}}{=} \frac{3}{2}\int_0^\infty \left\{\phi_0(Q_\tau)\phi_\downarrow(Q_\tau)\phi_\uparrow^2\frac{V_k(U_1)}{r}\right\} r\,dr, \tag{7.172}$$

$$I_2 \overset{\text{def}}{=} 3\int_0^\infty \left\{\phi_0(Q_\tau)\frac{kg_k(U_1)}{r}\phi_\uparrow \mathcal{L}_k^*\phi_\uparrow\right\} r\,dr, \tag{7.173}$$

$$I_3 \overset{\text{def}}{=} \frac{3}{2}\int_0^\infty \left\{\partial_t U_1(\phi_\downarrow - \phi_\uparrow)\phi_\uparrow \mathcal{L}_k^*\phi_\uparrow\right\} r\,dr, \tag{7.174}$$

for which one obviously has

$$\frac{3}{2}\int_0^\infty \left\{\phi_0(\phi_\downarrow - \phi_\uparrow)\phi_\uparrow \mathcal{L}_k^*\phi_\uparrow\right\} r\,dr = I_0 + I_1 + I_2 + I_3. \tag{7.175}$$

We proceed by computing a closed-form expression for $I_0$, while the rest of the integral terms are estimated using the results of the previous section.

For $I_0$, we use the formulas (7.91) and (7.82) for $\phi_0(Q_\tau)$, $\phi_\downarrow(Q_\tau)$, and $\widehat{\mathcal{L}}_k^*$, respectively, to infer that

$$\phi_0(Q_\tau)\phi_\downarrow(Q_\tau)\phi_\uparrow \widehat{\mathcal{L}}_k^*\phi_\uparrow = \frac{b}{\tau}\frac{k^2\sin^2(Q_\tau)}{r}\left\{\widehat{\mathcal{L}}_k^*\phi_\uparrow^2 - \frac{1+k\cos(Q_\tau)}{r}\phi_\uparrow^2\right\}.$$

Next, we rely on

$$\langle u|\widehat{\mathcal{L}}_k^*v\rangle = \langle \widehat{\mathcal{L}}_k u|v\rangle,$$

the formula (7.81) for $\widehat{\mathcal{L}}_k$, and straightforward calculations to derive

$$I_0 = \frac{3b}{2\tau}\int_0^\infty \left\{\left[\widehat{\mathcal{L}}_k\left(\frac{k^2\sin^2(Q_\tau)}{r}\right)\right.\right.$$
$$\left.\left. - \frac{k^2\sin^2(Q_\tau)(1+k\cos(Q_\tau))}{r^2}\right]\phi_\uparrow^2\right\}r\,dr \qquad (7.176)$$
$$= -\frac{3b}{\tau}\int_0^\infty \left\{\frac{k^2\sin^2(Q_\tau)}{r^2}\phi_\uparrow^2\right\}r\,dr = -\frac{3b}{4\tau}\|\phi_\downarrow(Q_\tau)\phi_\uparrow\|_{L^2(rdr)}^2.$$

The main point of this derivation is that $I_0$ is negative provided $b > 0$, which turns out to be the case.

For $I_1$, we argue first, based on the bounds (7.133) and (7.142) and the formula (7.164) for $\mathbb{D}_\uparrow$, that

$$\|\phi_\downarrow(Q_\tau)\phi_\uparrow\|_{L^2(rdr)} \le C\|\mathcal{L}_k^*\phi_\uparrow\|_{L^2(rdr)} \le C\frac{\mathbb{D}_\uparrow^{1/2}}{\tau}. \qquad (7.177)$$

On the other hand, using (7.91), we have

$$\phi_0(Q_\tau)\phi_\downarrow(Q_\tau) = \frac{br}{2\tau}\phi_\downarrow^2(Q_\tau),$$

while $\|U_1\|_{L^\infty} \le C\epsilon^{3/4}$ implies

$$\|V_k(U_1)\|_{L^\infty} \le C\|U_1\|_{L^\infty} \le C\epsilon^{3/4}.$$

If we couple the last three facts, then we obtain

$$|I_1| \le C\frac{b}{\tau}\int_0^\infty \left\{\phi_\downarrow^2(Q_\tau)\phi_\uparrow^2|V_k(U_1)|\right\}r\,dr$$
$$\le C\frac{b}{\tau}\|\phi_\downarrow(Q_\tau)\phi_\uparrow\|_{L^2(rdr)}^2\|V_k(U_1)\|_{L^\infty} \qquad (7.178)$$
$$\le \frac{C\epsilon^{3/4}b\mathbb{D}_\uparrow}{\tau^3}.$$

For $I_2$, the argument is very similar to the one for $I_1$, with

$$\phi_0(Q_\tau)\frac{kg_k(U_1)}{r}\phi_\uparrow \mathcal{L}_k^*\phi_\uparrow = \frac{kb}{2\tau}g_k(U_1)\phi_\downarrow(Q_\tau)\phi_\uparrow \mathcal{L}_k^*\phi_\uparrow$$

and

$$\|g_k(U_1)\|_{L^\infty} \leq C\|U_1\|_{L^\infty} \leq C\epsilon^{3/4}.$$

Factoring in also (7.177), we deduce that

$$\begin{aligned}
|I_2| &\leq C\frac{b}{\tau}\|g_k(U_1)\|_{L^\infty}\|\phi_\downarrow(Q_\tau)\phi_\uparrow\|_{L^2(rdr)}\|\mathcal{L}_k^*\phi_\uparrow\|_{L^2(rdr)} \\
&\leq \frac{C\epsilon^{3/4}b\,\mathbb{D}_\uparrow}{\tau^3}.
\end{aligned} \tag{7.179}$$

For the last integral term, $I_3$, an application of the Cauchy-Schwarz inequality yields

$$|I_3| \leq \|\partial_t U_1\|_{L^\infty}\big(\|\phi_\downarrow\phi_\uparrow\|_{L^2(rdr)} + \|\phi_\uparrow\|_{L^4(rdr)}^2\big)\|\mathcal{L}_k^*\phi_\uparrow\|_{L^2(rdr)}.$$

The first factor on the right-hand side is estimated using (7.146) and the easily derived fact from (7.164),

$$\|\mathcal{L}_k\phi_0\|_{L^2(rdr)} \leq C\frac{\mathbb{D}_\uparrow^{1/2}}{\tau}. \tag{7.180}$$

It follows that

$$\|\partial_t U_1\|_{L^\infty} \leq C_k(M)\left(\|\mathcal{L}_k\phi_0\|_{L^2(rdr)} + \frac{\epsilon^{3/4}b}{\tau}\right) \leq \frac{C_k(M)\big(\mathbb{D}_\uparrow^{1/2} + \epsilon^{3/4}b\big)}{\tau}.$$

For the second factor, we rely on $\|\phi_\uparrow\|_{L^2(rdr)} \leq C\epsilon$ and the estimates (7.127) and (7.129) to infer

$$\begin{aligned}
\|\phi_\downarrow\phi_\uparrow\|_{L^2(rdr)} &+ \|\phi_\uparrow\|_{L^4(rdr)}^2 \\
&\leq C\left(\|\mathcal{L}_k^*\phi_\uparrow\|_{L^2(rdr)} + \|\phi_\uparrow\|_{L^2(rdr)}\|\phi_\uparrow\|_{\dot{H}^1(rdr)}\right) \\
&\leq C\|\mathcal{L}_k^*\phi_\uparrow\|_{L^2(rdr)} \\
&\leq C\frac{\mathbb{D}_\uparrow^{1/2}}{\tau}.
\end{aligned} \tag{7.181}$$

Thus, collecting all these facts, we obtain

$$|I_3| \leq \frac{C_k(M)\big(\mathbb{D}_\uparrow^{1/2} + \epsilon^{3/4}b\big)\mathbb{D}_\uparrow}{\tau^3}. \tag{7.182}$$

Following these considerations, we now turn to the second integrand in the formula (7.166) for $F$. First, we employ the decomposition (7.147) for $\phi_0$ and argue that

$$|\phi_0^3| \leq C\big(|\phi_0^\perp|^3 + |l_k|^3|\phi_0(Q_\tau)|^3\big),$$

which implies

$$\left| \int_0^\infty \left\{ \phi_0^3 \phi_\uparrow \frac{k \cos(R)}{r} \right\} r \, dr \right|$$

$$\leq C \left( \int_0^\infty \left\{ \frac{|\phi_0^\perp|^3 |\phi_\uparrow|}{r} \right\} r \, dr + \int_0^\infty \left\{ \frac{|l_k|^3 |\phi_0(Q_\tau)|^3 |\phi_\uparrow|}{r} \right\} r \, dr \right). \tag{7.183}$$

For the first integral on the right-hand side, we use $\|\phi_\uparrow\|_{L^2(rdr)} \leq C\epsilon$, (7.148), and (7.180) to deduce

$$\int_0^\infty \left\{ \frac{|\phi_0^\perp|^3 |\phi_\uparrow|}{r} \right\} r \, dr \leq \|\phi_0^\perp\|_{L^\infty}^2 \|\phi_0^\perp / r\|_{L^2(rdr)} \|\phi_\uparrow\|_{L^2(rdr)}$$

$$\leq C\epsilon \|\mathcal{L}_k \phi_0\|_{L^2(rdr)}^3 \tag{7.184}$$

$$\leq \frac{C\epsilon \mathbb{D}_\uparrow^{3/2}}{\tau^3}.$$

For the second integral, due to (7.91), we notice first that

$$\phi_0^3(Q_\tau) = \frac{k^2 b^3 r}{2\tau^3} \sin^2(Q_\tau) \phi_\downarrow(Q_\tau).$$

Moreover, a simple computation reveals that

$$\| \sin^2(Q_\tau) \|_{L^2(rdr)} = C\tau. \tag{7.185}$$

Hence, taking advantage also of (7.149), (7.177), and (7.180), we derive

$$\int_0^\infty \left\{ \frac{|l_k|^3 |\phi_0(Q_\tau)|^3 |\phi_\uparrow|}{r} \right\} r \, dr$$

$$\leq C \frac{b^3}{\tau^3} \|l_k\|_{L^\infty}^3 \| \sin^2(Q_\tau) \|_{L^2(rdr)} \|\phi_\downarrow(Q_\tau) \phi_\uparrow\|_{L^2(rdr)}$$

$$\leq C \frac{b^3}{\tau^3} \left[ C_k^3(M) \left( \frac{\tau^3}{b^3} \|\mathcal{L}_k \phi_0\|_{L^2(rdr)}^3 + \epsilon^{9/4} \right) + C \right] \tau \frac{\mathbb{D}_\uparrow^{1/2}}{\tau} \tag{7.186}$$

$$\leq C \frac{b^3}{\tau^3} \mathbb{D}_\uparrow^{1/2} \left[ C_k^3(M) \left( \frac{\mathbb{D}_\uparrow^{3/2}}{b^3} + \epsilon^{9/4} \right) + C \right]$$

$$\leq \frac{C_k^3(M) \mathbb{D}_\uparrow^2 + \left( C_k^3(M) \epsilon^{9/4} + C \right) b^3 \mathbb{D}_\uparrow^{1/2}}{\tau^3}.$$

If we combine now (7.166), (7.175), (7.176), (7.178), (7.179), (7.182), (7.183), (7.184), (7.186), and the obvious bound

$$0 < C_1 \leq C_k(M) \leq C\epsilon^{-1/4}, \tag{7.187}$$

then we arrive at

$$\left| F + \frac{3b}{4\tau} \|\phi_\downarrow(Q_\tau)\phi_\uparrow\|_{L^2(rdr)}^2 \right|$$

$$\leq \frac{Cb^3\mathbb{D}_\uparrow^{1/2} + C_k(M)\epsilon^{3/4}b\,\mathbb{D}_\uparrow + C_k(M)\mathbb{D}_\uparrow^{3/2} + C_k^3(M)\mathbb{D}_\uparrow^2}{\tau^3},$$

thus proving (7.168) and (7.169).

All that is left to finish the proof of this theorem is estimating $N(t)$, for which the argument is quite similar to the one for the second integrand in $F(t)$. We rely again on the decomposition (7.147) to claim that

$$\phi_0^2 \leq 2\left[(\phi_0^\perp)^2 + l_k^2\phi_0^2(Q_\tau)\right],$$

which, jointly with $\phi_\downarrow - \phi_\uparrow = 2k\sin(R)/r$, yields

$$\left|\frac{N}{\tau^2}\right| \leq C\left(\int_0^\infty \left\{\frac{(\phi_0^\perp)^2 |\phi_\uparrow|}{r}\right\} rdr + \int_0^\infty \left\{l_k^2\phi_0^2(Q_\tau)\,|(\phi_\downarrow - \phi_\uparrow)\phi_\uparrow|\right\} rdr\right).$$

The first integral is easily estimated with the help of (7.148), (7.180), and $\|\phi_\uparrow\|_{L^2(rdr)} \leq C\epsilon$ by

$$\int_0^\infty \left\{\frac{(\phi_0^\perp)^2 |\phi_\uparrow|}{r}\right\} rdr \leq \|\phi_0^\perp\|_{L^\infty}\|\phi_0^\perp/r\|_{L^2(rdr)}\|\phi_\uparrow\|_{L^2(rdr)}$$

$$\leq C\epsilon\|\mathcal{L}\phi_0\|_{L^2(rdr)}^2 \tag{7.188}$$

$$\leq \frac{C\epsilon\mathbb{D}_\uparrow}{\tau^2}.$$

For the second integral, we use (7.149), (7.91), (7.185), (7.181), (7.180), and (7.187) to deduce

$$\int_0^\infty \left\{l_k^2\phi_0^2(Q_\tau)\,|(\phi_\downarrow - \phi_\uparrow)\phi_\uparrow|\right\} rdr$$

$$\leq C\frac{b^2}{\tau^2}\|l_k\|_{L^\infty}^2\|\sin^2(Q_\tau)\|_{L^2(rdr)}\left(\|\phi_\downarrow\phi_\uparrow\|_{L^2(rdr)} + \|\phi_\uparrow\|_{L^4(rdr)}^2\right)$$

$$\leq C\frac{b^2}{\tau^2}\left[C_k^2(M)\left(\frac{\tau^2}{b^2}\|\mathcal{L}_k\phi_0\|_{L^2(rdr)}^2 + \epsilon^{3/2}\right) + C\right]\tau\frac{\mathbb{D}_\uparrow^{1/2}}{\tau} \tag{7.189}$$

$$\leq C\frac{b^2}{\tau^2}\mathbb{D}_\uparrow^{1/2}\left(C_k^2(M)\frac{\mathbb{D}_\uparrow}{b^2} + C\right)$$

$$\leq \frac{C_k^2(M)\mathbb{D}_\uparrow^{3/2} + Cb^2\mathbb{D}_\uparrow^{1/2}}{\tau^2}.$$

Putting together the last three estimates, it follows that (7.170) holds true and, in this way, we conclude the whole argument. $\qquad\square$

Apart from the previous estimates concerning the second-order energy, another key fact in proving the collapse is having precise information on the speed $b = b(t)$ and the associated acceleration $b' = b'(s)$. For that matter, we use the formula (7.103) proved in Lemma 7.3 that relates the two functions and turn it into a useful differential inequality.

**Theorem 7.4.** *Under the assumptions of Theorem 7.3, the following a priori bound holds:*

$$
|b'| \leq \left( \frac{M^{-2k+2}}{D_k(M)} + C_k^2(M)\epsilon^{3/4} + C_k(M)\|U_1\|_{L^\infty} \right) b^2
$$
$$
+ \left( \frac{Z_k(M)}{D_k(M)} + \frac{\|U_1\|_{L^\infty}}{D_k^{1/2}(M)} + C_k^2(M)b \right) \mathbb{D}^{1/2},
$$
(7.190)

*where $D_k(M)$ is defined by (7.98) and $Z_k(M)$ can be chosen such that*

$$
Z_k(M) \sim \begin{cases} (\log(M))^{1/2} & \text{if } k = 1, \\ M^{-k+1} & \text{if } k > 1. \end{cases}
$$
(7.191)

**Proof.** The first step in the argument is to derive estimates and asymptotics for the coefficients of $b'$, $b^2$, and $b\tau$, appearing in (7.103). Based on (7.98) and (7.104), it is easy to claim that

$$
D_{k,1}(M) \sim D_k(M) \sim \begin{cases} \log(M) & \text{if } k = 1, \\ C & \text{if } k > 1. \end{cases}
$$

As $y \mapsto y\partial_y \chi_M$ is a uniformly bounded function in terms of $M$, which is supported on the compact interval $[M, 2M]$, a simple integration from the formula (7.105) reveals that

$$
|D_{k,2}(M)| \leq CM^{-2k+2}.
$$

We recall from (7.124) that

$$
\left| A_k(\widetilde{U}_1) \right| \leq \widetilde{C}_k(M)\|U_1\|_{L^\infty},
$$

where

$$
\widetilde{C}_k(M) \sim \begin{cases} M & \text{if } k = 1, \\ \log(M) & \text{if } k = 2, \\ C & \text{if } k > 2. \end{cases}
$$

Moreover, a careful inspection of the proof for the previous bound and the observation that $A_k(\partial_t U_1)$ is simply obtained by substituting $\partial_t U_1$ for $U_1$ in $A_k(\widetilde{U}_1)$ allows us to infer

$$
\left| B_k(\widetilde{U}_1) \right| \leq \widetilde{C}_k(M)\|U_1\|_{L^\infty},
$$
$$
\left| A_k(\widetilde{\partial_t U}_1) \right| \leq \widetilde{C}_k(M)\|\partial_t U_1\|_{L^\infty}.
$$

If we couple these facts derived from the beginning of the argument with (7.146), (7.164), $\|U_1\|_{L^\infty} \le C\epsilon^{3/4}$, and

$$\frac{\widetilde{C}_k(M)}{D_k(M)} \sim C_k(M) \le C\epsilon^{-1/4},$$

then it follows that

$$D_{k,1}(M) - \frac{1}{k}A_k(\widetilde{U}_1) \sim D_k(M), \tag{7.192}$$

$$\left|A_k(\widetilde{\partial_t U_1})\right| \le C_k^2(M)D_k(M)\frac{\mathbb{D}_\uparrow^{1/2} + \epsilon^{3/4}b}{\tau}, \tag{7.193}$$

$$\left|D_{k,2}(M) + \left(2 + \frac{1}{k}\right)A_k(\widetilde{U}_1) + B_k(\widetilde{U}_1)\right| \tag{7.194}$$
$$\le CM^{-2k+2} + C_k(M)D_k(M)\|U_1\|_{L^\infty}.$$

All we are left to do is estimating $Z^{\mathrm{bdry}}(\phi_\uparrow)$ and $Z^{\mathrm{bulk}}(\phi_\uparrow)$. For the former, directly from its formula (i.e., (7.107)), we deduce

$$\left|Z^{\mathrm{bdry}}(\phi_\uparrow)\right| \le \|\phi_\uparrow\|_{L^\infty(\frac{r}{\tau}\le 2M)}\left\|y\partial_y\chi_M\frac{\sin(Q_\tau)}{r}\right\|_{L^1(rdr)}.$$

In the case when $k = 1$, an application of (7.162) and (7.164) yields

$$\|\phi_\uparrow\|_{L^\infty(\frac{r}{\tau}\le 2M)} \le C(\log(M))^{1/2}\frac{\mathbb{D}_\uparrow^{1/2}}{\tau},$$

whereas if $k > 1$, then, using (7.130), (7.133), and (7.164), one derives the stronger bound

$$\|\phi_\uparrow\|_{L^\infty(\frac{r}{\tau}\le 2M)} \le C\frac{\mathbb{D}_\uparrow^{1/2}}{\tau}.$$

On the other hand, the previous facts on $y\partial_y\chi_M$ imply

$$\left\|y\partial_y\chi_M\frac{\sin(Q_\tau)}{r}\right\|_{L^1(rdr)} \le C\tau\|\sin(Q)\|_{L^1(dy;M\le y\le 2M)} \le CM^{-k+1}\tau.$$

Putting together the last four estimates and (7.191), we obtain

$$\left|Z^{\mathrm{bdry}}(\phi_\uparrow)\right| \le Z_k(M)\mathbb{D}_\uparrow^{1/2}. \tag{7.195}$$

Now, we address the $Z^{\mathrm{bulk}}(\phi_\uparrow)$ term, for which its formula (7.106), (7.84), and (7.91) lead to

$$\left|Z^{\mathrm{bulk}}(\phi_\uparrow)\right|$$

$$\le C\left(\left\|\chi_M\sin(Q_\tau)U_1\,\phi_\downarrow(Q_\tau)\phi_\uparrow\right\|_{L^1(rdr)} + \left\|\chi_M\sin(Q_\tau)U_1\frac{U_1\phi_\uparrow}{r}\right\|_{L^1(rdr)}\right)$$

$$\le C\|\chi_M\sin(Q_\tau)\|_{L^2(rdr)}\|U_1\|_{L^\infty}\left(\|\phi_\downarrow(Q_\tau)\phi_\uparrow\|_{L^2(rdr)} + \left\|\frac{U_1\phi_\uparrow}{r}\right\|_{L^2(rdr)}\right).$$

Based on (7.98), we notice immediately that

$$\|\chi_M \sin(Q_\tau)\|_{L^2(rdr)} \sim D_k^{1/2}(M)\tau.$$

Using (7.133), (7.142), and (7.164), we infer

$$\|\phi_\downarrow(Q_\tau)\phi_\uparrow\|_{L^2(rdr)} \leq C\frac{D_\uparrow^{1/2}}{\tau}.$$

If $k = 1$, then (7.135), (7.142), and (7.164) imply

$$\left\|\frac{U_1\phi_\uparrow}{r}\right\|_{L^2(rdr)} \leq C\frac{D_\uparrow^{1/2}}{\tau}.$$

When $k > 1$, applying (7.131), (7.133), and (7.164), we deduce the stronger estimate

$$\left\|\frac{U_1\phi_\uparrow}{r}\right\|_{L^2(rdr)} \leq C\|U_1\|_{L^\infty}\frac{D_\uparrow^{1/2}}{\tau}.$$

Collecting all these facts connected to $Z^{\mathrm{bulk}}(\phi_\uparrow)$, we derive

$$|Z^{\mathrm{bulk}}(\phi_\uparrow)| \leq D_k^{1/2}(M)\|U_1\|_{L^\infty}D_\uparrow^{1/2}. \tag{7.196}$$

Finally, the desired estimate (7.190) appears as the joint consequence of the last bound together with (7.192), (7.193), (7.194), and (7.195). □

## 7.8 The last step in the argument for collapse

We finally arrive at the last step of our argument proving the collapse for wave maps, in which the main ingredients are the differential inequalities derived in the previous section, involving the second-order energy, the speed, and the acceleration. In fact, we will work with simplified version of these inequalities, that are obtained from (7.169), (7.170), and (7.190) by taking advantage of $C_k(M) \leq C\epsilon^{-1/4}$, $\|U_1\|_{L^\infty} \leq C\epsilon^{3/4}$, and $|b| \leq C\epsilon$ (see (7.122)):

$$|I| \leq \frac{C\left(b^3 D_\uparrow^{1/2} + \epsilon^{1/2}b D_\uparrow(t) + \epsilon^{-1/4}D_\uparrow^{3/2}\right) + C^3\epsilon^{-3/4}D_\uparrow^2}{\tau^3}, \tag{7.197}$$

$$|N| \leq C\left(b^2 D_\uparrow^{1/2} + \epsilon D_\uparrow\right) + C^2\epsilon^{-1/2}D_\uparrow^{3/2}, \tag{7.198}$$

$$|b'| \leq \delta_k b^2 + \epsilon_k D_\uparrow^{1/2}. \tag{7.199}$$

In the above, $C$, $\delta_k$, and $\epsilon_k$ are absolute constants, with the first one being independent of any of the variables in the analysis, whereas $\delta_k$ and $\epsilon_k$ can be made arbitrarily small by choosing $\epsilon$ sufficiently small and $M$ sufficiently large[9], related, however, by $C_k(M) \leq C\epsilon^{-1/4}$.

It will also be important to upgrade the preliminary assumptions on the initial data made in (7.92) and (7.121), i.e.,

$$\mathbb{E} = \mathbb{E}_\uparrow + \mathbb{E}_\downarrow = 4k + \epsilon^2, \qquad \mathbb{E}_\uparrow < \epsilon^2, \qquad \|\widehat{\mathcal{L}}_k U_1(0)\|_{L^2(rdr)} \leq C\epsilon, \quad (7.200)$$

by additionally asking for

$$b(0) > 0, \qquad \mathbb{D}_\uparrow(0) \leq b^4(0), \quad (7.201)$$

to be true too.

These are all the necessary facts we need in order to demonstrate a trapping regime, which leads to collapse in finite time.

**Theorem 7.5.** *Under the assumptions (7.200) and (7.201), if $\epsilon$ is sufficiently small, then there exist constants $K \gg 1$ and $\delta \ll 1$ such that the estimates*

$$\mathbb{D}_\uparrow \leq Kb^4 \qquad and \qquad |b'| \leq \delta b^2 \quad (7.202)$$

*hold in any time interval of existence for a classical solution[10] $R = R(t,r)$ to (7.3), of the type (7.72) (or, equivalently, (7.73)). Moreover, there exists $0 < T < \infty$ such that*

$$\lim_{t \to T} \tau(t) = 0,$$

*thus implying collapse of $R$ at time $t = T$.*

**Proof.** We start the argument by choosing $K \gg 1$ to satisfy

$$\delta \stackrel{\text{def}}{=} \delta_k + \epsilon_k K^{1/2} \ll 1, \quad (7.203)$$

$$\beta \stackrel{\text{def}}{=} \max\left\{ C(K^{-1/2} + \epsilon) + C^4 \epsilon^{3/2} K^{1/2}, \atop C(K^{-1/2} + \epsilon^{1/2}) + C^2 \epsilon^{3/4} K^{1/2} + C^6 \epsilon^{9/4} K \right\} \ll 1, \quad (7.204)$$

and proving, by contradiction, that the first bound in (7.202) is true.

Therefore, we consider $[0, T)$ to be the time interval of existence and assume that (7.202) does not hold for all $t \in [0, T)$. As $\mathbb{D}_\uparrow(0) \leq b^4(0)$, by

---

[9]The ratios $M^{-k+1}/D_k(M)$ and $Z_k(M)/D_k(M)$ in (7.190) can be made arbitrarily small by picking $M \gg 1$.

[10]This is guaranteed by invoking any local existence result of classical solutions for the semilinear equation (7.3).

the continuity of the classical solution, it follows that there exists a first time $t_* \in (0, T)$ such that

$$\mathbb{D}_\uparrow(t) < Kb^4(t), \quad \forall t \in [0, t_*), \qquad \mathbb{D}_\uparrow(t_*) = Kb^4(t_*). \qquad (7.205)$$

We will achieve a contradiction by showing, in fact, that $\mathbb{D}_\uparrow(t_*) < Kb^4(t_*)$.

In the above scenario, using (7.199) and (7.203), we infer that

$$|b'| \le (\delta_k + \epsilon_k K^{1/2}) b^2 = \delta b^2 \qquad (7.206)$$

is true on $[0, t_*]$. This implies that, on the same interval, we have

$$\frac{d}{dt} \left\{ \frac{b}{\tau^{1/2}} \right\} = \frac{b' + \frac{1}{2} b^2}{\tau^{3/2}} \ge \left( \frac{1}{2} - \delta \right) \frac{b^2}{\tau^{3/2}} \ge 0.$$

Hence, due to (7.201), we obtain

$$\frac{b(t)}{\tau^{1/2}(t)} \ge \frac{b(0)}{\tau^{1/2}(0)} > 0, \qquad \forall t \in [0, t_*]. \qquad (7.207)$$

In particular, $b$ is positive throughout $[0, t_*]$. Next, we use (7.204), (7.205), and $b = |b| \le C\epsilon$, in the context of (7.197) and (7.198), to deduce that on $[0, t_*]$ one has

$$
\begin{aligned}
|I| &\le \frac{C \left( K^{1/2} b^5 + \epsilon^{1/2} K b^5 + \epsilon^{-1/4} K^{3/2} b^6 \right) + C^3 \epsilon^{-3/4} K^2 b^8}{\tau^3} \\
&\le \frac{\left[ C(K^{-1/2} + \epsilon^{1/2}) + C^2 \epsilon^{3/4} K^{1/2} + C^6 \epsilon^{9/4} \right] K b^5}{\tau^3} \qquad (7.208) \\
&\le \beta K \frac{b^5}{\tau^3}
\end{aligned}
$$

and

$$
\begin{aligned}
|N| &\le C \left( K^{1/2} b^4 + \epsilon K b^4 \right) + C^2 \epsilon^{-1/2} K^{3/2} b^6 \\
&\le \left[ C \left( K^{1/2} + \epsilon K \right) + C^4 \epsilon^{3/2} K^{1/2} \right] b^4 \qquad (7.209) \\
&\le \beta K b^4.
\end{aligned}
$$

Following this, we integrate the differential identity (7.167) over the interval $[0, t_*]$ and, based on (7.168), we derive

$$\frac{\mathbb{D}_\uparrow(t_*) + N(t_*)}{\tau^2(t_*)} + \int_0^{t_*} \left\{ \frac{3b}{4\tau} \| \phi_\downarrow(Q_\tau) \phi_\uparrow \|_{L^2(rdr)}^2 \right\} dt$$

$$= \frac{\mathbb{D}_\uparrow(0) + N(0)}{\tau^2(0)} + \int_0^{t_*} \{ I(t) \} dt,$$

which implies

$$\mathbb{D}_\uparrow(t_*) \le -N(t_*) + \frac{\tau^2(t_*)}{\tau^2(0)} \left( \mathbb{D}_\uparrow(0) + N(0) \right) + \tau^2(t_*) \int_0^{t_*} \{ I(t) \} dt.$$

If we couple this estimate with $\mathbb{D}_\uparrow(0) \le b^4(0)$, (7.207), (7.208), and (7.209), then we obtain

$$\mathbb{D}_\uparrow(t_*)$$

$$\le \beta K b^4(t_*) + (1 + \beta K)\frac{b^4(0)}{\tau^2(0)}\tau^2(t_*) + \beta K \tau^2(t_*)\int_0^{t_*}\left\{\frac{b^5}{\tau^3}\right\}dt \quad (7.210)$$

$$\le (1 + 2\beta K)b^4(t_*) + \beta K \tau^2(t_*)\int_0^{t_*}\left\{\frac{b^5}{\tau^3}\right\}dt.$$

Thus, the key to this argument is to deduce a favorable bound for the integral in the above estimate. Relying on the identities $b(t) = -\dot\tau(t)$ and $b' = \dot b(t)\tau(t)$, we can perform the following integration by parts:

$$\int_0^{t_*}\left\{\frac{b^5}{\tau^3}\right\}dt = \frac{1}{2}\int_0^{t_*}\left\{\frac{d}{dt}\left(\frac{1}{\tau^2}\right)b^4\right\}dt$$

$$= \frac{1}{2}\left(\frac{b^4(t_*)}{\tau^2(t_*)} - \frac{b^4(0)}{\tau^2(0)}\right) - 2\int_0^{t_*}\left\{\frac{b^3 b'}{\tau^3}\right\}dt.$$

For the last integral, we use (7.206) to infer

$$\left|\int_0^{t_*}\left\{\frac{b^3 b'}{\tau^3}\right\}dt\right| \le \delta\int_0^{t_*}\left\{\frac{b^5}{\tau^3}\right\}dt,$$

which allows us to absorb it on the left-hand side of the previous relation. As a consequence, we derive

$$\int_0^{t_*}\left\{\frac{b^5}{\tau^3}\right\}dt \le \frac{1}{2(1 - 2\delta)}\frac{b^4(t_*)}{\tau^2(t_*)}.$$

If we apply this bound in connection to (7.210) and take into account that $K \gg 1$, while both $\delta \ll 1$ and $\beta \ll 1$, then we conclude that

$$\mathbb{D}_\uparrow(t_*) \le \left[1 + \left(2 + \frac{1}{2(1 - 2\delta)}\right)\beta K\right]b^4(t_*) < K b^4(t_*),$$

which achieves the desired contradiction. Hence, we proved that the inequality $\mathbb{D}_\uparrow \le K b^4$ holds in the entire interval of existence $[0, T)$ and, as a byproduct of the previous argument, the same can be said about (7.206) and, subsequently, (7.207). This finishes the proof of (7.202).

In proving the collapse, we proceed by integrating (7.207), written in the form

$$\frac{-\dot\tau(t)}{\tau^{1/2}(t)} = \frac{b(t)}{\tau^{1/2}(t)} \ge \frac{b(0)}{\tau^{1/2}(0)} \stackrel{\text{def}}{=} \delta_* > 0, \qquad \forall t \in [0, T),$$

on the whole interval $[0, T)$. It follows that

$$\tau^{1/2}(t) + \frac{\delta_*}{2}t \le \tau^{1/2}(0), \qquad \forall t \in [0, T),$$

which yields an upper bound for $T$ equal to

$$\frac{2}{\delta_*}\tau^{1/2}(0),$$

and, hence, demonstrates the existence of such a finite time at which $\lim_{t \to T} \tau(t) = 0$. $\qquad\square$

This theorem achieves our modest goal of proving finite time collapse for solutions of the equivariant equation (7.3) and the next natural question is: how can one improve this method to obtain a sharp description of the blow-up profile, like in Theorem 7.1?

The first observation is that, instead of working with estimates for the second-order energy defined as in (7.164), we can use local versions of these bounds, valid only in certain regions of the time slice $\{(t, r) | 0 \le r < \infty\}$, which provide more qualitative information. This is because we describe time through the dynamic variable $\tau = \tau(t)$ and, due to (7.202), we know that the surface

$$S \overset{\text{def}}{=} \left\{ (t, r) \mid r - (1 + 2\delta)\frac{\tau}{b} = 0 \right\}$$

is spacelike. We can tie this to the energy being better behaved inside a ball of the type $\{r \le \overline{C}\tau/b\}$ (for some arbitrary constant $\overline{C}$) as, at least when $k > 1$, (7.202) implies

$$\int\limits_{r \le \overline{C}\tau/b} \{\phi_\uparrow^2\} \, r dr \le \frac{\overline{C}^2 \tau^2}{2b^2} \|\phi_\uparrow\|_{L^\infty}^2 \le \frac{C^2}{b^2} \mathbb{D}_\uparrow \le C^2 K b^2.$$

Another improvement can be made by working, from the very beginning, with $M_1$ (and, consequently, $M$) replaced by suitable expressions depending on $b$. However, this creates problems, especially for the cases when $k = 1$ or 2, because $M$ plays a critical role in most of the bounds.

Finally, one can try and derive a sharper version for the differential equation (7.103) satisfied by $b$, which could lead to better asymptotics. The best tool for achieving this goal is by integrating the conformal identity (7.62) over balls like the ones discussed above. For this idea and many others, we invite the reader to consult the paper of Raphaël-Rodnianski [136].

# Appendix A

# Tools from geometry: invariance, symmetry, and curvature

The purpose of the present appendix is to introduce, for the convenience of the reader, the concepts from differential geometry that are used in this book. The first part consists of background in differential geometry with emphasis on tools that are relevant for geometric partial differential equations. The second part contains a discussion on the concepts of invariance and symmetry; in particular, we outline the derivation of the energy-momentum tensor, which is one of our main objects of interest. This development follows unpublished notes by Demetrios Christodoulou. The last part presents the curvature tensor and focuses on its relevance to general relativity, our interest being the connection to the wave map problem. We do not strive for completeness, although the appendix is essentially self-contained. Instead, we simply concentrate on those geometrical aspects that are needed in the present work. There are many books that the reader can consult on differential geometry, the standard textbook by do Carmo [38] being one example. On general relativity, the excellent book by Wald [183] is a good reference. The reader should be aware that there is a wealth of books discussing various aspects of general relativity.

Computations in differential geometry are presented in various forms. The abstract formulation avoids indices and thus places emphasis on the coordinate independence of the quantities involved. It is the most elegant approach, but one cannot perform any computations beyond the ones that are well-understood. The index notation expresses all quantities with respect to some coordinate system and, as a result, is more convenient for calculations once the barrier of handling indices is overcome. The method of Élie Cartan [22], which uses differential forms and orthogonal frames, is particularly elegant and suited for a certain type of computations. In general, the various methods complement each other. Here, we adopt the use of indices as the most appropriate method for the calculations that we have in mind.

## A.1 Background in differential geometry

Throughout this appendix, we mainly work with an $n + 1$-dimensional Lorentzian manifold $(M^{n+1}, g_{\mu\nu})$ (with the metric being symmetric, i.e., $g_{\mu\nu} = g_{\nu\mu}$), which means that, for every fixed point $p \in M$, there exists an invertible linear transformation that maps the metric $g_{\mu\nu}(p)$ into

$$m_{\mu\nu} \overset{\text{def}}{=} \mathrm{diag}(-1, 1, \ldots, 1),$$

also known as the Minkowski metric. Thus, the metric can be seen as an invertible, symmetric matrix, having exactly one negative eigenvalue and $n$ positive ones, unlike the case of a Riemannian metric which is strictly positive definite. Furthermore, from here on, we adopt the Einstein summation convention over repeated indices.

On a general manifold, we consider objects generalizing vectors and matrices, called tensors of valence $(a, b)$ (where $a$ and $b$ are nonnegative integers), which are identified in some local coordinate system by their components,

$$T^{\nu_1 \ldots \nu_a}_{\mu_1 \ldots \mu_b}.$$

In particular, the metric $g_{\mu\nu}$ is a tensor of valence $(0, 2)$. The vectors, which are tensors of valence $(1, 0)$ and whose components are denoted by $X^\mu$, where $0 \le \mu \le n$, "live" on the tangent bundle that is formed by attaching to each point on the manifold its tangent space, i.e.,

$$TM = \bigcup_{p \in M} T_p M.$$

The metric $g_{\mu\nu}$ gives rise to the notion of duality and this idea is expressed via the operation of lowering/raising indices of tensors. Thus, a vector $X^\mu \in T_p M$ has a dual covector $X_\mu := g_{\mu\alpha} X^\alpha$ living on the dual space $(T_p M)^*$. In a similar spirit, covectors are tensors of valence $(0, 1)$ living on the cotangent bundle, which is formed by coupling every point on the manifold to the dual to its tangent space. A covector is described by $Y_\mu$, with $0 \le \mu \le n$ (notice the lower index convention), and its dual is given by $Y^\mu := g^{\mu\alpha} Y_\alpha$, where we use $g^{\mu\alpha}$ to denote the inverse[1] of $g_{\mu\alpha}$.

---

[1] One has the fundamental relation $g^{\mu\alpha} g_{\alpha\nu} = \delta^\mu{}_\nu$.

If one transitions from a set of coordinates $(x^\mu)$ to a new one $(y^{\mu'})$ via a local invertible transformation $x^\mu = x^\mu(y^{\mu'})$ whose derivative is denoted by

$$l_{\mu'}^{\ \mu} \overset{\text{def}}{=} \frac{\partial x^\mu}{\partial y^{\mu'}},$$

then the vectors and covectors must transform according to the rules

$$X^{\mu'} = l_\mu^{\ \mu'} X^\mu \qquad \text{and} \qquad Y_{\mu'} = l_{\mu'}^{\ \mu} Y_\mu. \tag{A.1}$$

In the above,

$$l_\mu^{\ \mu'} \overset{\text{def}}{=} \frac{\partial y^{\mu'}}{\partial x^\mu}$$

and one notices that $l_\mu^{\ \mu'}$ and $l_{\mu'}^{\ \mu}$ are inverses of each other. We say that covectors are covariant, while vectors are contravariant. This discussion also gives us an idea about how general tensors transform, namely

$$T_{\mu_1' \ldots \mu_b'}^{\nu_1' \ldots \nu_a'} = l_{\mu_1'}^{\ \mu_1} \ldots l_{\mu_b'}^{\ \mu_b} l_{\nu_1}^{\ \nu_1'} \ldots l_{\nu_a}^{\ \nu_a'} T_{\mu_1 \ldots \mu_b}^{\nu_1 \ldots \nu_a}. \tag{A.2}$$

Specializing this formula to the case of the metric $g_{\mu\nu}$, we deduce

$$g_{\mu'\nu'} = l_{\mu'}^{\ \mu} l_{\nu'}^{\ \nu} g_{\mu\nu}. \tag{A.3}$$

Alternatively, this could have been seen arising by observing that the quadratic expression

$$g_{\mu\nu} dx^\mu dx^\nu$$

is invariant under a change of coordinates, meaning

$$g_{\mu\nu} dx^\mu dx^\nu = g_{\mu'\nu'} dx^{\mu'} dx^{\nu'}.$$

If we denote

$$g \overset{\text{def}}{=} \sqrt{|\det(g_{\mu\nu})|} \tag{A.4}$$

and consider the natural measure of integration on $M$ to be

$$d\mu_g \overset{\text{def}}{=} g\, dx^0 \ldots dx^n, \tag{A.5}$$

then this measure turns out to be independent of the choice of coordinates due to

$$\det(l_{\mu'}^{\ \mu}) dy^0 \ldots dy^n = dx^0 \ldots dx^n.$$

Another benefit of working with the metric is that we can use it to construct the inner product of two vectors according to

$$\langle X, Y \rangle \overset{\text{def}}{=} g_{\mu\nu} X^\mu Y^\nu = X^\mu (g_{\mu\nu} Y^\nu) = X^\mu Y_\mu, \tag{A.6}$$

which is another coordinate independent quantity. This allows for the following classification of vectors: $X^\mu$ is timelike if $\langle X, X \rangle < 0$, null if $g\langle X, X \rangle = 0$, and spacelike if $\langle X, X \rangle > 0$. We can further split timelike vectors into past and future oriented ones. As the metric also introduces the concept of infinitesimal distance through

$$\pm ds^2 = g_{\mu\nu}(x(s)) \frac{dx^\mu}{ds} \frac{dx^\nu}{ds},$$

where $x^\mu = x^\mu(s)$ describes a curve in local coordinates, then we say that the curve is timelike if

$$g_{\mu\nu}(x(s))\dot{x}^\mu(s)\dot{x}^\nu(s) < 0, \qquad \forall\, s.$$

Accordingly, a timelike geodesic minimizes the distance between two points that can be joined with a timelike curve. Moreover, for each point $p$ of an orientable manifold $M$, we can distinguish three sets. The future consists of all points $q \in M$ that can be connected to $p$ with a future oriented timelike geodesic. The past is made up of points $q \in M$ that can be joined with $p$ by a past oriented timelike geodesic, whereas the rest consists of points that are inaccessible from $p$ through timelike geodesics.

## A.2 Calculus on manifolds

The next topic we are interested in is how to differentiate tensors on a manifold. For a scalar function $\phi : M \mapsto \mathbb{R}$, the ordinary partial derivative $\partial_\mu \phi$ is a covector, as it can be easily checked. Similarly, $\partial^\mu \phi = g^{\mu\alpha} \partial_\alpha \phi$ defines a vector. A problem arises when one considers derivatives of vectors since, for a vector field $X^\nu$, the ordinary derivative $\partial_\mu X^\nu$ is not a tensor (i.e., it does not transform according to (A.2)). To fix this issue, the idea is to add an appropriate correction to the ordinary derivative and, for this purpose, one introduces the Christoffel symbols, which are determined by three indices and are denoted as $\Gamma^\alpha_{\beta\gamma}$. With their help, we define the following covariant derivatives,

$$\nabla_\mu X_\nu \stackrel{\text{def}}{=} \partial_\mu X_\nu - \Gamma^\alpha_{\mu\nu} X_\alpha, \tag{A.7}$$

$$\nabla_\mu X^\nu \stackrel{\text{def}}{=} \partial_\mu X^\nu + \Gamma^\nu_{\mu\alpha} X^\alpha, \tag{A.8}$$

$$\nabla_\lambda T_{\mu\nu} \stackrel{\text{def}}{=} \partial_\lambda T_{\mu\nu} - \Gamma^\alpha_{\mu\lambda} T_{\alpha\nu} - \Gamma^\alpha_{\lambda\nu} T_{\mu\alpha}, \tag{A.9}$$

and, in this way, we say that the Christoffel symbols induce a connection on the manifold. If $\Gamma^\alpha_{\beta\gamma} = \Gamma^\alpha_{\gamma\beta}$, then the connection is said to be symmetric.

There is an obvious benefit if one requires that $\nabla_\lambda g_{\mu\nu} = 0$, which can be interpreted as the metric being constant with respect to covariant differentiation. The fundamental theorem of Riemannian geometry is the observation that this condition determines the Christoffel symbols in a unique manner.

**Theorem A.1.** *Suppose that, for a smooth manifold $(M, g_{\mu\nu})$, the equation (A.9) represents the rule of covariant differentiation for tensors of valence $(0, 2)$. If the associated connection is symmetric and $\nabla_\lambda g_{\mu\nu} = 0$, then the Christoffel symbols $\Gamma^\lambda_{\mu\nu}$ are uniquely determined by the formulas*

$$\Gamma_{\mu\nu;\lambda} = \frac{1}{2}\left\{\partial_\mu g_{\nu\lambda} + \partial_\nu g_{\mu\lambda} - \partial_\lambda g_{\mu\nu}\right\}, \tag{A.10}$$

$$\Gamma^\lambda_{\mu\nu} = g^{\lambda\alpha}\Gamma_{\mu\nu;\alpha}. \tag{A.11}$$

**Proof.** Using (A.9) and the fact that $\nabla_\lambda g_{\mu\nu} = 0$, we deduce

$$\nabla_\mu g_{\nu\lambda} = \partial_\mu g_{\nu\lambda} - \Gamma^\alpha_{\mu\nu}g_{\alpha\lambda} - \Gamma^\alpha_{\mu\lambda}g_{\nu\alpha} = 0,$$
$$\nabla_\nu g_{\lambda\mu} = \partial_\nu g_{\lambda\mu} - \Gamma^\alpha_{\nu\lambda}g_{\alpha\mu} - \Gamma^\alpha_{\nu\mu}g_{\lambda\alpha} = 0,$$
$$\nabla_\lambda g_{\mu\nu} = \partial_\lambda g_{\mu\nu} - \Gamma^\alpha_{\lambda\mu}g_{\alpha\nu} - \Gamma^\alpha_{\lambda\nu}g_{\mu\alpha} = 0.$$

By adding the first two equations and then subtracting the last one, we derive

$$\partial_\mu g_{\nu\lambda} + \partial_\nu g_{\nu\lambda} - \partial_\lambda g_{\mu\nu} = 2\Gamma^\alpha_{\mu\nu}g_{\alpha\lambda}.$$

If we define $\Gamma_{\mu\nu;\lambda} := g_{\alpha\lambda}\Gamma^\alpha_{\mu\nu}$, then (A.10) and (A.11) follow immediately. $\square$

A number of relevant remarks are now in order.

**Remark A.1.** Based on (A.7) and (A.8), we infer the Leibniz's rule

$$\partial_\mu\langle X, Y\rangle = \nabla_\mu\langle X, Y\rangle = \langle\nabla_\mu X, Y\rangle + \langle X, \nabla_\mu Y\rangle.$$

In fact, $\nabla g = 0$ implies that the same rule also applies to covariant derivatives of arbitrary products of tensors (e.g., partial contraction of products like $T_\mu{}^\alpha T_\alpha{}^\nu$), with the general scheme being

$$\nabla\{T^{(1)}T^{(2)}\} = \{\nabla T^{(1)}\}T^{(2)} + T^{(1)}\{\nabla T^{(2)}\}.$$

**Remark A.2.** The Christoffel symbols are not tensors. If we make the change of coordinates $x^\rho = x^\rho(y^\lambda)$, then, after a long computation, we discover that

$$\tilde{\Gamma}^\lambda_{\mu\nu} = \frac{\partial y^\lambda}{\partial x^\rho}\frac{\partial x^\sigma}{\partial y^\mu}\frac{\partial x^\tau}{\partial y^\nu}\Gamma^\rho_{\sigma\tau} + \frac{\partial y^\lambda}{\partial x^\rho}\frac{\partial^2 x^\rho}{\partial y^\mu \partial y^\nu},$$

which clearly does not obey rule (A.2). However, as we will see later, certain combinations of Christoffel symbols and their derivatives yield the curvature tensor.

**Remark A.3.** The following is an observation that is useful when one considers perturbations of a metric, which is a subject of interest in general relativity. Assume that for a parameter $s$ we have a family of metrics $g_{\mu\nu}(s)$, with corresponding Christoffel symbols $\Gamma^\lambda_{\mu\nu}(s)$. For each $s$, there is a derivation rule $\nabla^{(s)}$ which can be written abbreviately as $\nabla^{(s)} = \partial \pm \Gamma(s)$. The goal is to determine the "perturbation" $\Gamma^\lambda_{\mu\nu}(s) - \Gamma^\lambda_{\mu\nu}(0)$ and a calculation similar to the one in the previous theorem produces

$$\Gamma^\lambda_{\mu\nu}(s) - \Gamma^\lambda_{\mu\nu}(0)$$
$$= \frac{1}{2}g^{\lambda\rho}(s)\left\{\nabla^{(0)}_\mu g_{\nu\rho}(s) + \nabla^{(0)}_\nu g_{\nu\rho}(s) - \nabla^{(0)}_\rho g_{\mu\nu}(s)\right\}. \qquad (A.12)$$

**Remark A.4.** This remark is important when we consider integrals. If we consider the partial contraction $\Gamma^\beta_{\alpha\beta}$ in (A.11), then, due to (A.4), we arrive at

$$\Gamma^\beta_{\alpha\beta} = \frac{1}{2}g^{\beta\gamma}\partial_\alpha g_{\beta\gamma} = \frac{1}{2g^2}A^{\beta\gamma}\partial_\alpha g_{\beta\gamma} = \frac{\partial_\alpha g}{g}, \qquad (A.13)$$

where $A^{\beta\gamma}$ is the cofactor corresponding to the entry $g_{\beta\gamma}$. As a consequence, for the divergence of a vector field, one has

$$g\nabla_\mu X^\mu = g\partial_\mu X^\mu + g\Gamma^\nu_{\mu\nu}X^\mu = g\partial_\mu X^\mu + \partial_\mu g X^\mu = \partial_\mu\left\{gX^\mu\right\}.$$

Together with (A.5), this equation can be applied to integrate the divergence of a vector field over a region $\Omega \subset M$, thus leading to Gauss's formula

$$\int_\Omega \{\nabla_\mu X^\mu\}d\mu_g = \int_{\partial\Omega} \{X^\mu n_\mu\}dS. \qquad (A.14)$$

In the above, $n_\mu$ is a covector normal to the boundary of $\Omega$ (if there is any). This observation has many implications, in particular to the derivation of conservation laws.

## A.3   The wave operator and the Lagrangian of a scalar field

The wave equation arises as the Euler-Lagrange equation of an appropriate Lagrangian. Let us consider a scalar field $\phi : M \mapsto \mathbb{R}$ and the following type of Lagrangian,

$$\mathbb{L}[\phi] \stackrel{\text{def}}{=} \frac{1}{2} \int_M \{g^{\mu\nu}\nabla_\mu\phi\nabla_\nu\phi\}\, d\mu_g = \frac{1}{2} \int_M \{\nabla_\mu\phi\nabla^\mu\phi\}\, d\mu_g. \tag{A.15}$$

The functional derivative of the above expression is

$$\frac{d}{ds}\Big|_{s=0} \mathbb{L}[\phi + s\psi] = \int_M \{g^{\mu\nu}\nabla_\mu\phi\nabla_\nu\psi\}\, d\mu_g = \int_M \{(-\nabla_\mu\nabla^\mu\phi)\psi\}\, d\mu_g,$$

where $\psi$ is an arbitrary function with compact support. Hence, a formal critical point for $\mathbb{L}$ must satisfy the wave equation

$$\Box\phi \stackrel{\text{def}}{=} -\nabla_\mu\nabla^\mu\phi = 0. \tag{A.16}$$

One notices the importance of the condition $\nabla g = 0$, since it allows us to integrate by parts with covariant derivatives, thus bypassing the metric tensor. In the above, we introduced the wave operator or the d'Alembertian, which, in the case of the Minkowski spacetime $(\mathbb{R}^{n+1}, \text{diag}(-1, 1, \ldots, 1))$, takes the familiar form

$$\Box = \partial_t^2 - \Delta,$$

with $x^0 := t$ denoting the time coordinate.

## A.4   The concept of Lie derivative

We would like to investigate now the symmetries of the Lagrangian $\mathbb{L}[\phi]$. For this reason, we introduce the concept of Lie derivative associated to a vector field. We start by considering a vector field denoted by $\xi^\mu = \xi^\mu(x)$ and define its flow on the manifold via the solution of the following system of ordinary differential equations:

$$\frac{dx^\mu}{ds} = \xi^\mu(x(s)), \qquad x^\mu(0) = x_0^\mu. \tag{A.17}$$

This can be understood as a continuous change of coordinates by writing the solution of the system like $x^\mu = x^\mu(s, x_0^\alpha)$, where $x^\mu(0, x_0^\alpha) = x_0^\mu$. Differentiation of tensors along this flow gives rise to the notion of Lie derivative, which is given by

$$\mathcal{L}_\xi X^\mu = \xi^\alpha \nabla_\alpha X^\mu - X^\alpha \nabla_\alpha \xi^\mu, \tag{A.18}$$

$$\mathcal{L}_\xi X_\mu = \xi^\alpha \nabla_\alpha X_\mu + X_\alpha \nabla_\mu \xi^\alpha. \tag{A.19}$$

A simple verification shows that these expressions are, in fact, independent of the connection induced on the manifold. This is to be expected since the differentiation of a vector field along the previous flow should be connection-free.

In order to motivate (A.18) and (A.19), let us consider the change of coordinates $x_0^{\mu'} \mapsto x^\mu(s, x_0^{\mu'})$ and set

$$l_{\mu'}^{\mu} \overset{\text{def}}{=} \frac{\partial x^\mu}{\partial x_0^{\mu'}}.$$

One notices that the inverse of the matrix $l_{\mu'}^{\mu}$, denoted as $l_{\mu}^{\mu'}$, has the formula

$$l_{\mu}^{\mu'} = \frac{\partial x_0^{\mu'}}{\partial x^\mu}.$$

If we differentiate (A.17) with respect to the initial data, then we derive that the evolutions of the matrices $l_{\mu'}^{\mu}(s)$ and $l_{\mu}^{\mu'}(s)$ are described by

$$\frac{dl_{\mu'}^{\mu}}{ds} = (\partial_\alpha \xi^\mu) l_{\mu'}^{\alpha}, \qquad \frac{dl_{\mu}^{\mu'}}{ds} = -l_\alpha^{\mu'}(\partial_\mu \xi^\alpha). \tag{A.20}$$

The associated initial conditions are $l_{\mu'}^{\mu}(0) = \delta_{\mu'}^{\mu}$ and $l_{\mu}^{\mu'}(0) = \delta_{\mu}^{\mu'}$, respectively. The Lie derivative of a vector field $X^\mu$ is then the derivative of $X^{\mu'} := X^\mu l_\mu^{\mu'}$ with respect to the parameter $s$, evaluated at $s = 0$, and a direct computation yields

$$\frac{dX^{\mu'}}{ds}\Big|_{s=0} = \frac{dX^\mu}{ds}\Big|_{s=0} l_\mu^{\mu'} + X^\mu \frac{dl_\mu^{\mu'}}{ds}\Big|_{s=0} = \xi^\alpha \partial_\alpha X^\mu - X^\alpha \partial_\alpha \xi^\mu. \tag{A.21}$$

One can easily check the identity

$$\xi^\alpha \partial_\alpha X^\mu - X^\alpha \partial_\alpha \xi^\mu = \xi^\alpha \nabla_\alpha X^\mu - X^\alpha \nabla_\alpha \xi^\mu,$$

which certifies (A.18) and the previous heuristics stating that the covariant derivatives do not play any role in this context. In a similar manner, one motivates (A.19).

Next, we can extend (A.18) and (A.19) to tensors of valence $(0, 2)$ and $(2, 0)$ in the following way:

$$\mathcal{L}_\xi T_{\mu\nu} = \xi^\alpha \nabla_\alpha T_{\mu\nu} + T_{\alpha\nu} \nabla_\mu \xi^\alpha + T_{\mu\alpha} \nabla_\nu \xi^\alpha, \tag{A.22}$$

$$\mathcal{L}_\xi T^{\mu\nu} = \xi^\alpha \nabla_\alpha T^{\mu\nu} - T^{\alpha\nu} \nabla_\alpha \xi^\mu - T^{\mu\alpha} \nabla_\alpha \xi^\nu. \tag{A.23}$$

In particular, for the metric tensor, we obtain

$$\mathcal{L}_\xi g_{\mu\nu} = \nabla_\mu \xi_\nu + \nabla_\nu \xi_\mu, \qquad \mathcal{L}_\xi g^{\mu\nu} = -\nabla^\mu \xi^\nu - \nabla^\nu \xi^\mu, \tag{A.24}$$

with the first formula measuring the deformation of the metric along the flow of $\xi^\mu$. If $\mathcal{L}_\xi g_{\mu\nu} = 0$, then the metric is invariant under the flow and we say that $\xi^\mu$ is a Killing vector field. Such vector fields are important in deriving conservation laws, as well as in symmetry reductions.

Following this, we would like to analyze how the measure of integration $d\mu_g$ defined by (A.5) behaves under the flow generated by $\xi^\mu$. To simplify matters, we focus on the case of a $3 + 1$-dimensional manifold and write $d\mu_g$ in the form

$$d\mu_g = \frac{1}{4!}\epsilon_{\alpha\beta\gamma\delta}dx^\alpha dx^\beta dx^\gamma dx^\delta, \tag{A.25}$$

which is more amenable to our computations. In the above, $\epsilon_{\alpha\beta\gamma\delta}$ denotes the totally antisymmetric or Levi-Civita tensor on the manifold $M$, which is given by

$$\epsilon_{\alpha\beta\gamma\delta} \overset{\text{def}}{=} \sqrt{g}\,\widehat{\epsilon}_{\alpha\beta\gamma\delta},$$

with

$$\widehat{\epsilon}_{\alpha\beta\gamma\delta} \overset{\text{def}}{=} \text{sign}\left\{0 \mapsto \alpha,\ 1 \mapsto \beta,\ 2 \mapsto \gamma,\ 3 \mapsto \delta\right\}.$$

Thus, $\widehat{\epsilon}_{\alpha\beta\gamma\delta}$ is equal to either zero if there is a repeated index, or the sign of $(\alpha, \beta, \gamma, \delta)$, seen as a permutation of $(1, 2, 3, 4)$.

In order to compute the Lie derivative of $d\mu_g$, we start by arguing that

$$\begin{aligned}
\mathcal{L}_\xi\left(\epsilon_{\alpha\beta\gamma\delta}\right) =&\ \xi^\lambda\nabla_\lambda\epsilon_{\alpha\beta\gamma\delta} + \epsilon_{\lambda\beta\gamma\delta}\nabla_\alpha\xi^\lambda + \epsilon_{\alpha\lambda\gamma\delta}\nabla_\beta\xi^\lambda \\
&+ \epsilon_{\alpha\beta\lambda\delta}\nabla_\gamma\xi^\lambda + \epsilon_{\alpha\beta\gamma\lambda}\nabla_\delta\xi^\lambda.
\end{aligned} \tag{A.26}$$

For the covariant derivative of the Levi-Civita tensor we have

$$\begin{aligned}
\nabla_\lambda\epsilon_{\alpha\beta\gamma\delta} =&\ \left(\partial_\lambda g\right)\widehat{\epsilon}_{\alpha\beta\gamma\delta} - \epsilon_{\rho\beta\gamma\delta}\Gamma^\rho_{\lambda\alpha} - \epsilon_{\alpha\rho\gamma\delta}\Gamma^\rho_{\lambda\beta} \\
&- \epsilon_{\alpha\beta\rho\delta}\Gamma^\rho_{\lambda\gamma} - \epsilon_{\alpha\beta\gamma\rho}\Gamma^\rho_{\lambda\delta}.
\end{aligned}$$

In particular, using the antisymmetric properties of this tensor, we deduce

$$\begin{aligned}
\nabla_\lambda\epsilon_{0123} =&\ \left(\partial_\lambda g\right)\widehat{\epsilon}_{0123} - \epsilon_{\rho 123}\Gamma^\rho_{\lambda 0} - \epsilon_{0\rho 23}\Gamma^\rho_{\lambda 1} \\
&- \epsilon_{01\rho 3}\Gamma^\rho_{\lambda 2} - \epsilon_{012\rho}\Gamma^\rho_{\lambda 3} \\
=&\ \left[\frac{\partial_\lambda g}{g} - \Gamma^\rho_{\lambda\rho}\right]\epsilon_{0123} = 0.
\end{aligned}$$

Similarly, one derives that $\nabla_\lambda\epsilon_{\alpha\beta\gamma\delta} = 0$, which means that the Levi-Civita tensor is constant with respect to covariant differentiation. If we apply this fact to (A.26), then we infer that

$$\begin{aligned}
\mathcal{L}_\xi\epsilon_{0123} =&\ \epsilon_{\lambda 123}\nabla_0\xi^\lambda + \epsilon_{0\lambda 23}\nabla_1\xi^\lambda \\
&+ \epsilon_{01\lambda 3}\nabla_2\xi^\lambda + \epsilon_{012\lambda}\nabla_3\xi^\lambda \\
=&\ \epsilon_{0123}\left(\nabla_\lambda\xi^\lambda\right)
\end{aligned}$$

and, subsequently,

$$\mathcal{L}_\xi(\epsilon_{\alpha\beta\gamma\delta}) = \epsilon_{\alpha\beta\gamma\delta}(\nabla_\lambda \xi^\lambda).$$

This implies

$$\mathcal{L}_\xi(d\mu_g) = (\nabla_\mu \xi^\mu) d\mu_g \qquad (A.27)$$

and, for the case when $\nabla_\mu \xi^\mu = 0$, the integration measure is preserved under the flow of $\xi^\mu$. On a related note, the reader may recall that a divergence-free vector field defines a volume-preserving transformation.

## A.5    The energy-momentum tensor

In this section, we follow up on the computations involving Lie derivatives and look to derive the energy-momentum tensor associated with solutions to (A.16). First, we define

$$g_{\mu\nu}(s) \overset{\text{def}}{=} g_{\mu\nu}(x^\alpha(s, x_0^{\alpha'})), \qquad \phi(s) \overset{\text{def}}{=} \phi(x^\alpha(s, x_0^{\alpha'})),$$

to be the family of metrics and scalar maps, respectively, generated by the flow (A.17). We observe immediately that the Lagrangian $\mathbb{L}[\phi(s); g_{\mu\nu}(s)]$, obtained by replacing in (A.15) $\phi$ and $g_{\mu\nu}$ with $\phi(s)$ and $g_{\mu\nu}(s)$, is independent of $s$ and, as a result, we deduce

$$\frac{d}{ds}\bigg|_{s=0} \mathbb{L}[\phi(s); g_{\mu\nu}(s)] = 0.$$

If we differentiate inside the integral representing the Lagrangian and rely on (A.27), we infer that

$$\int_M \bigg\{ g^{\mu\nu}\nabla_\mu\phi(\mathcal{L}_\xi\nabla_\nu\phi)$$

$$+ \frac{1}{2}\left[(\mathcal{L}_\xi g^{\mu\nu})\nabla_\mu\phi\nabla_\nu\phi + (\nabla_\lambda\xi^\lambda)g^{\mu\nu}\nabla_\mu\phi\nabla_\nu\phi\right]\bigg\}d\mu_g = 0. \qquad (A.28)$$

For the first integral term, we notice that the Lie derivative and the covariant derivative commute, i.e.,

$$\mathcal{L}_\xi(\nabla_\nu\phi) = \xi^\alpha\nabla_\alpha\nabla_\nu\phi + \nabla_\alpha\phi\nabla_\nu\xi^\alpha = \nabla_\nu(\xi^\alpha\nabla_\alpha\phi) = \nabla_\nu(\mathcal{L}_\xi\phi),$$

and, subsequently, we have

$$\int_M \left\{ g^{\mu\nu}\nabla_\mu\phi(\nabla_\nu\mathcal{L}_\xi\phi) \right\} d\mu_g = \int_M \left\{ (-\nabla_\nu\nabla^\nu\phi)\mathcal{L}_\xi\phi \right\} d\mu_g = 0,$$

with the last equality being justified by (A.16). Therefore, if we couple this fact with (A.24) and (A.28), then we obtain the identity

$$\int_M \left\{ -\left(\nabla^\mu \xi^\nu + \nabla^\nu \xi^\mu\right)\left(\nabla_\mu \phi \nabla_\nu \phi\right) + \left(\nabla_\lambda \xi^\lambda\right)\left(g^{\mu\nu}\nabla_\mu \phi \nabla_\nu \phi\right)\right\} d\mu_g = 0.$$

Next, assuming that the vector field $\xi^\mu$ has compact support and integrating by parts, we derive

$$\int_M d\mu_g \left\{ \xi^\nu \nabla_\mu \left(\nabla^\mu \phi \nabla_\nu \phi - \frac{1}{2}\delta^\mu_{\ \nu}\left(\nabla_\alpha \phi \nabla^\alpha \phi\right)\right)\right\} = 0.$$

At this point, we introduce the energy-momentum tensor to be given by

$$T_{\mu\nu} \overset{\text{def}}{=} \nabla_\mu \phi \nabla_\nu \phi - \frac{1}{2}g_{\mu\nu}\left(\nabla_\alpha \phi \nabla^\alpha \phi\right) \tag{A.29}$$

and argue that the last integral identity implies the following set of structure equations:

$$\nabla_\mu T^\mu_{\ \nu} = 0. \tag{A.30}$$

Thus, we have proved the ensuing result which can be seen as a particular case of the celebrated Noether's theorem.

**Theorem A.2.** *If $\phi$ is a formal critical point for the Lagrangian (A.15), meaning that it satisfies (A.16), then the energy-momentum tensor defined by (A.29) satisfies the identities (A.30).*

The structure equations express the invariance of the Lagrangian under a continuous change of coordinates defined via the flow of a compactly supported vector field. These are not conservation laws because the contraction $\nabla_\mu T^\mu_{\ \nu}$ does not represent the divergence of a certain vector field. However, by contracting (A.30) with an arbitrary vector field $X^\nu$, we deduce

$$\nabla_\mu \left\{T^\mu_{\ \nu} X^\nu\right\} = T^{\mu\nu}\left(\nabla_\mu X_\nu\right) = \frac{1}{2}T^{\mu\nu}\left(\mathcal{L}_X g_{\mu\nu}\right),$$

which leads to the conservation law

$$\nabla_\mu \left\{T^\mu_{\ \nu} X^\nu\right\} = 0,$$

if $X^\nu$ is Killing. In the case when $X^\nu$ is conformal Killing, i.e.,

$$\mathcal{L}_X g_{\mu\nu} = c g_{\mu\nu}$$

where $c$ is the conformal factor, and $M^{n+1}$ is a Lorentzian manifold, we obtain in a similar manner that

$$\nabla_\mu \left\{T^\mu_{\ \nu} X^\nu\right\} = \frac{1}{2}c T^{\mu\nu} g_{\mu\nu} = c\frac{1-n}{4}\langle \nabla\phi, \nabla\phi\rangle,$$

with the notation convention $\langle \nabla \phi, \nabla \phi \rangle = \nabla_\mu \phi \nabla^\mu \phi$. Therefore, we produce a conservation law only if $n = 1$, which is to say that the $1 + 1$-dimensional wave equation is conformally invariant.

To conclude this section, Killing vector fields generate conserved quantities which are important in physics. For us, the energy-momentum tensor is a source of a priori estimates because of the following observation. Let us consider a timelike vector field $X^\mu$ and a spacelike hypersurface $\Sigma$, meaning that its normal vector $n^\mu$ is timelike. If we additionally assume that $\Sigma$ is orientable, such that we can choose $n^\mu$ to be consistently future-oriented, then the quantity $T_{\mu\nu} X^\mu n^\nu$ has a definite sign. This is essential for obtaining useful a priori bounds, as is explained with some detail in Chapter 1.

### A.6   Maxwell's equations

Another set of equations, which can be derived from a Lagrangian and are relevant for our discussion in this appendix, are Maxwell's equations. In order to present them, we start by considering a gauge field $A_\mu$, with $0 \le \mu \le 3$, and construct the antisymmetric tensor

$$F_{\mu\nu}(A) \stackrel{\text{def}}{=} \nabla_\mu A_\nu - \nabla_\nu A_\mu, \tag{A.31}$$

where we write $F(A)$ to denote its dependence on $A$. Due to its antisymmetry, the tensor $F_{\mu\nu}$ has only six germane components, which form the electric and magnetic fields. The former is given by

$$(F_{01}, F_{02}, F_{03}),$$

whereas the latter is represented by

$$\left( \epsilon_1{}^{kl} F_{kl}, \epsilon_2{}^{kl} F_{kl}, \epsilon_3{}^{kl} F_{kl} \right),$$

with $\epsilon_{jkl}$ being the Levi-Civita tensor in $\mathbb{R}^3$. It is a simple observation that this construction does not depend on the (symmetric) connection, since a direct calculation based on $\Gamma^\gamma_{\alpha\beta} = \Gamma^\gamma_{\beta\alpha}$ yields $F_{\mu\nu} = \partial_\mu A_\nu - \partial_\nu A_\mu$.

Next, we can raise the indices of $F_{\mu\nu}$ to form another antisymmetric tensor,

$$F^{\mu\nu} \stackrel{\text{def}}{=} g^{\mu\alpha} g^{\nu\beta} F_{\alpha\beta},$$

and, with its help, we build the scalar quantity

$$\langle F, F \rangle \stackrel{\text{def}}{=} F_{\mu\nu} F^{\mu\nu}.$$

This allows us to define the Lagrangian of the electromagnetic field by

$$\mathbb{L}[A] \stackrel{\text{def}}{=} \frac{1}{4} \int_M \{F^{\mu\nu} F_{\mu\nu}\} \, d\mu_g = \frac{1}{4} \int_M \{\langle F, F \rangle\} \, d\mu_g, \qquad \text{(A.32)}$$

for which a variational argument leads to

$$\frac{d}{ds}\bigg|_{s=0} \mathbb{L}(A + sB) = \frac{1}{2} \int_M \{\langle F(A), F(B) \rangle\} \, d\mu_g = 0.$$

An integration by parts shows that

$$\frac{1}{2} \int_M \{\langle F(A), F(B) \rangle\} \, d\mu_g = - \int_M \{(\nabla_\mu F^{\mu\nu}) B_\nu\} \, d\mu_g$$

and, consequently, we obtain

$$\nabla_\mu F^{\mu\nu} = 0, \qquad \text{(A.33)}$$

which are Maxwell's equations.

An important aspect of Maxwell's equations is duality. For the particular case when $(M, g_{\mu\nu})$ is the $3 + 1$-dimensional Minkowski spacetime, one has the obvious identity

$$\nabla_\alpha F_{\beta\gamma} + \nabla_\beta F_{\gamma\alpha} + \nabla_\gamma F_{\alpha\beta} = 0, \qquad \text{(A.34)}$$

as covariant derivatives coincide with ordinary derivatives in this setting. If we define the dual tensor $F^*_{\mu\nu}$ by

$$F^*_{\mu\nu} \stackrel{\text{def}}{=} \epsilon_{\mu\nu}{}^{\alpha\beta} F_{\alpha\beta}, \qquad \text{(A.35)}$$

then, due to (A.34), we infer

$$\nabla_\mu F^{*\mu}{}_\nu = -\epsilon_\nu{}^{\mu\alpha\beta} \nabla_\mu F_{\alpha\beta} = 0. \qquad \text{(A.36)}$$

The same property also holds in the general case of a Lorentzian manifold. However, to see how this comes about, we have to introduce the curvature tensor on the manifold and study its symmetries. The reason for this is that $A_\mu$ is a covector and iterated second-order derivatives of a vector or covector do not commute.

## A.7    The Riemann curvature tensor

As motivated above, we formalize here the curvature tensor and discuss some of its essential properties. Moreover, we explore its relation to general relativity. The following result serves as the definition for the curvature tensor.

**Theorem A.3.** *Let $(M, g_{\mu\nu})$ be a smooth manifold endowed with a symmetric connection induced by (A.7)-(A.8). There exists a unique tensor of valence $(1,3)$, denoted by $R_{\alpha\beta\gamma}{}^{\delta}$, such that the formulas[2]*

$$(\nabla_\alpha \nabla_\beta - \nabla_\beta \nabla_\alpha) X_\gamma \overset{\text{def}}{=} - R_{\alpha\beta\gamma}{}^{\delta} X_\delta, \qquad (A.37)$$

$$(\nabla_\alpha \nabla_\beta - \nabla_\beta \nabla_\alpha) X^\gamma \overset{\text{def}}{=} R_{\alpha\beta\delta}{}^{\gamma} X^\delta, \qquad (A.38)$$

*hold for any covector $X_\gamma$ and vector $X^\gamma$, respectively. Furthermore, one can entirely express this tensor in terms of Christoffel symbols via the equation*

$$R_{\alpha\beta\gamma}{}^{\delta} = \partial_\alpha \Gamma^\delta_{\beta\gamma} - \partial_\beta \Gamma^\delta_{\alpha\gamma} - \left( \Gamma^\rho_{\alpha\gamma} \Gamma^\delta_{\beta\rho} - \Gamma^\rho_{\beta\gamma} \Gamma^\delta_{\alpha\rho} \right). \qquad (A.39)$$

**Proof.** We first notice that (A.37) and (A.38) are not independent of each other. Indeed, for a symmetric connection and any scalar field $\phi$, we have $(\nabla_\alpha \nabla_\beta - \nabla_\beta \nabla_\alpha)\phi = 0$, which implies

$$(\nabla_\alpha \nabla_\beta - \nabla_\beta \nabla_\alpha)(X^\delta Y_\delta) = 0.$$

This obviously shows that (A.38) follows as a consequence of (A.37) and, thus, we focus on proving (A.37).

Using (A.7), (A.9), and the symmetry of the connection, we deduce

$$\nabla_\alpha \nabla_\beta X_\gamma = \partial_\alpha (\nabla_\beta X_\gamma) - \Gamma^\rho_{\alpha\beta} \nabla_\rho X_\gamma - \Gamma^\rho_{\alpha\gamma} \nabla_\beta X_\rho$$
$$= \partial_\alpha \left( \partial_\beta X_\gamma - \Gamma^\delta_{\beta\gamma} X_\delta \right)$$
$$- \Gamma^\rho_{\alpha\beta} \left( \partial_\rho X_\gamma - \Gamma^\delta_{\rho\gamma} X_\delta \right) - \Gamma^\rho_{\alpha\gamma} \left( \partial_\beta X_\rho - \Gamma^\delta_{\beta\rho} X_\delta \right)$$

and

$$\nabla_\beta \nabla_\alpha X_\gamma = \partial_\beta (\nabla_\alpha X_\gamma) - \Gamma^\rho_{\beta\alpha} \nabla_\rho X_\gamma - \Gamma^\rho_{\beta\gamma} \nabla_\alpha X_\rho$$
$$= \partial_\beta \left( \partial_\alpha X_\gamma - \Gamma^\delta_{\alpha\gamma} X_\delta \right)$$
$$- \Gamma^\rho_{\beta\alpha} \left( \partial_\rho X_\gamma - \Gamma^\delta_{\rho\gamma} X_\delta \right) - \Gamma^\rho_{\beta\gamma} \left( \partial_\alpha X_\rho - \Gamma^\delta_{\alpha\rho} X_\delta \right).$$

If we subtract term by term the two equations, then we derive

$$(\nabla_\alpha \nabla_\beta - \nabla_\beta \nabla_\alpha) X_\gamma$$
$$= \left\{ \partial_\alpha \Gamma^\delta_{\beta\gamma} - \partial_\beta \Gamma^\delta_{\alpha\gamma} - \left( \Gamma^\rho_{\alpha\gamma} \Gamma^\delta_{\beta\rho} - \Gamma^\rho_{\beta\gamma} \Gamma^\delta_{\alpha\rho} \right) \right\} X_\delta,$$

which justifies both (A.37) and (A.39).    $\square$

---

[2]The sign convention is the opposite of the one in Wald [183].

**Remark A.5.** By adopting the antisymmetrization conventions

$$S_{[\alpha\beta]} \stackrel{\text{def}}{=} \frac{1}{2}\left(S_{\alpha\beta} - S_{\beta\alpha}\right),$$

$$S_{[\alpha|\gamma|\beta]} \stackrel{\text{def}}{=} \frac{1}{2}\left(S_{\alpha\gamma\beta} - S_{\beta\gamma\alpha}\right),$$

we can write the curvature tensor in a slightly more compact formula:

$$\frac{1}{2}R_{\alpha\beta\gamma}{}^{\delta} = \partial_{[\alpha}\Gamma^{\delta}_{\beta]\gamma} - \Gamma^{\rho}_{[\alpha|\gamma|}\Gamma^{\delta}_{\beta]\rho}.$$

**Remark A.6.** If we recall that the Christoffel symbols have the schematic form

$$\Gamma^{\delta}_{\beta\gamma} = \Gamma^{\delta}_{\beta\gamma}(g, \partial g),$$

then we can argue that

$$R_{\alpha\beta\gamma}{}^{\delta} = R_{\alpha\beta\gamma}{}^{\delta}(g, \partial g, \partial^2 g).$$

The curvature tensor enjoys certain symmetries which are essential in understanding the nature of this important geometric concept. It is more convenient to formulate these properties in terms of the purely covariant version of the curvature tensor, which is defined by

$$R_{\alpha\beta\gamma\delta} \stackrel{\text{def}}{=} R_{\alpha\beta\gamma}{}^{\rho}g_{\delta\rho}. \tag{A.40}$$

**Lemma A.1.** *The tensor $R_{\alpha\beta\gamma\delta}$ given by (A.40) satisfies the following identities*[3]:

$$R_{\alpha\beta\gamma\delta} = -R_{\beta\alpha\gamma\delta}, \tag{A.41}$$

$$R_{\alpha\beta\gamma\delta} = -R_{\alpha\beta\delta\gamma}, \tag{A.42}$$

$$R_{\alpha\beta\gamma\delta} = R_{\gamma\delta\alpha\beta}, \tag{A.43}$$

$$R_{\alpha\beta\gamma\delta} + R_{\beta\gamma\alpha\delta} + R_{\gamma\alpha\beta\delta} = 0, \tag{A.44}$$

$$\nabla_{\alpha}R_{\beta\gamma\mu\nu} + \nabla_{\beta}R_{\gamma\alpha\mu\nu} + \nabla_{\gamma}R_{\alpha\beta\mu\nu} = 0, \tag{A.45}$$

*where the last two are known in literature as the Ricci and Bianchi identities, respectively.*

---

[3] Collectively, (A.41)-(A.44) are referred to as Ricci's identities, with (A.44) being the most prominent one.

**Proof.** The first property, (A.41), is the immediate result of either (A.37) or (A.38), coupled with (A.40). To see why (A.42) is true, we observe that

$$0 = 2\nabla_{[\alpha}\nabla_{\beta]}g_{\gamma\delta} = -R_{\alpha\beta\gamma}{}^{\rho}g_{\rho\delta} - R_{\alpha\beta\delta}{}^{\rho}g_{\gamma\rho} = -\left(R_{\alpha\beta\gamma\delta} + R_{\alpha\beta\delta\gamma}\right).$$

For proving the Ricci identity, (A.44), we start by deriving from (A.37) that, for a scalar field $\phi$, we have

$$\left(\nabla_{\alpha}\nabla_{\beta} - \nabla_{\beta}\nabla_{\alpha}\right)\nabla_{\gamma}\phi = -R_{\alpha\beta\gamma}{}^{\delta}\nabla_{\delta}\phi.$$

If we adopt the antisymmetrization convention

$$S_{[\alpha\beta\gamma]} \stackrel{\text{def}}{=} \frac{1}{3!}\left(S_{\alpha\beta\gamma} + S_{\beta\gamma\alpha} + S_{\gamma\alpha\beta} - S_{\beta\alpha\gamma} - S_{\gamma\beta\alpha} - S_{\alpha\gamma\beta}\right),$$

then the previous equation implies

$$R_{[\alpha\beta\gamma]}{}^{\delta}\nabla_{\delta}\phi = \frac{1}{3}\left(\nabla_{\alpha}\nabla_{[\beta}\nabla_{\gamma]}\phi + \nabla_{\beta}\nabla_{[\gamma}\nabla_{\alpha]}\phi + \nabla_{\gamma}\nabla_{[\alpha}\nabla_{\beta]}\phi\right).$$

However, $\nabla_{[\alpha}\nabla_{\beta]}\phi = 0$ and, due to (A.41), we infer

$$R_{[\alpha\beta\gamma]\delta} = \frac{1}{3}\left(R_{\alpha\beta\gamma\delta} + R_{\beta\gamma\alpha\delta} + R_{\gamma\alpha\beta\delta}\right).$$

Hence, (A.44) is the consequence of these last three facts.

The third identity, (A.43), follows by applying together (A.41), (A.42), and (A.44). First, using (A.44) and (A.41), we deduce

$$R_{\alpha\beta\gamma\delta} = -R_{\beta\gamma\alpha\delta} - R_{\gamma\alpha\beta\delta} = R_{\gamma\beta\alpha\delta} - R_{\gamma\alpha\beta\delta},$$
$$-R_{\alpha\beta\delta\gamma} = R_{\beta\delta\alpha\gamma} + R_{\delta\alpha\beta\gamma} = R_{\delta\alpha\beta\gamma} - R_{\delta\beta\alpha\gamma}.$$

If we add these equations and we also factor in (A.42), then we derive

$$R_{\alpha\beta\gamma\delta} = -2R_{[\gamma|\alpha\beta|\delta]} + 2R_{[\gamma|\beta\alpha|\delta]} = -4R_{[\gamma|[\alpha\beta]|\delta]}.$$

Similarly, we obtain

$$R_{\alpha\beta\gamma\delta} = -4R_{[\alpha|[\gamma\delta]|\beta]}.$$

Thus, the previous two equations imply

$$R_{\gamma\delta\alpha\beta} = -4R_{[\gamma|[\alpha\beta]|\delta]} = R_{\alpha\beta\gamma\delta},$$

which is the desired identity.

Finally, in order to prove the Bianchi identity (A.45), we argue that for an arbitrary vector field $X^{\delta}$ we have

$$\left(\nabla_{\alpha}\nabla_{\beta} - \nabla_{\beta}\nabla_{\alpha}\right)\nabla_{\gamma}X^{\delta} = -R_{\alpha\beta\gamma}{}^{\rho}\nabla_{\rho}X^{\delta} + R_{\alpha\beta\rho}{}^{\delta}\nabla_{\gamma}X^{\rho}.$$

By antisymmetrization, this leads to

$$2\nabla_{[\alpha}\nabla_{\beta}\nabla_{\gamma]}X^{\delta} = R_{[\alpha\beta|\rho|}{}^{\delta}\nabla_{\gamma]}X^{\rho}. \tag{A.46}$$

In a similar spirit, we discover that

$$\nabla_\alpha \left( \nabla_\beta \nabla_\gamma - \nabla_\gamma \nabla_\beta \right) X^\delta = \nabla_\alpha R_{\beta\gamma\rho}{}^\delta X^\rho + R_{\beta\gamma\rho}{}^\delta \nabla_\alpha X^\rho$$

and, subsequently,

$$2\nabla_{[\alpha} \nabla_\beta \nabla_{\gamma]} X^\delta = \nabla_{[\alpha} R_{\beta\gamma]\rho}{}^\delta X^\rho + R_{[\beta\gamma|\rho|}{}^\delta \nabla_{\alpha]} X^\rho. \tag{A.47}$$

By comparing the right-hand sides of (A.46) with (A.47), we obtain

$$\nabla_{[\alpha} R_{\beta\gamma]\rho}{}^\delta = 0,$$

which implies, based on (A.41), that

$$\nabla_\alpha R_{\beta\gamma\rho}{}^\delta + \nabla_\beta R_{\gamma\alpha\rho}{}^\delta + \nabla_\gamma R_{\gamma\alpha\rho}{}^\delta = 0.$$

Thus, the argument for (A.45) is complete.      □

## A.8    The energy-momentum tensor for Maxwell's equations

With this new knowledge on the curvature tensor, we are going back to Maxwell's equations with two goals in mind. The first one is to provide an argument for the dual tensor defined by (A.35) to satisfy (A.36) in the general case of a Lorentzian manifold. The second goal is to derive the energy-momentum tensor associated to Maxwell's equations (A.33) and explore its properties relative to Killing and conformal Killing vector fields.

For the former, if we rely on (A.35), (A.37), and (A.44), then we deduce

$$\nabla_\mu F^{*,\mu\nu} = 2\epsilon^{\mu\nu\alpha\beta} \nabla_\mu \nabla_\alpha A_\beta = \epsilon^{\mu\nu\alpha\beta} \left( \nabla_\mu \nabla_\alpha - \nabla_\alpha \nabla_\mu \right) A_\beta$$
$$= -\epsilon^{\mu\nu\alpha\beta} R_{\mu\alpha\beta}{}^\gamma A_\gamma = -\epsilon^{\mu\alpha\beta\nu} R_{\mu\alpha\beta}{}^\gamma A_\gamma = 0,$$

which proves (A.36). Thus, we have an extra set of equations that complements the original Maxwell's equations (A.33).

In what concerns the energy-momentum tensor in this setting, we adopt similar to the one that led us to (A.29). Specifically, we work with the flow (A.17) generated by a vector field $\xi^\mu$, which induces the family of gauge fields $A(s)$ and metrics $g_{\mu\nu}(s)$. Next, we modify the Lagrangian (A.32) by replacing $A$ and $g_{\mu\nu}$ with $A(s)$ and $g_{\mu\nu}(s)$, respectively, and argue that

$$\frac{d}{ds}\bigg|_{s=0} \mathbb{L}[A(s); g_{\mu\nu}(s)] = 0,$$

on the basis of the new Lagrangian being independent of $s$. Hence, using Lie derivatives, we derive the identity

$$\int_M \left\{ \langle F, \mathcal{L}_\xi F \rangle - \left[ (\nabla^\mu \xi^\alpha + \nabla^\alpha \xi^\mu) g^{\nu\beta} F_{\mu\nu} F_{\alpha\beta} \right. \right. \tag{A.48}$$
$$\left. \left. + \frac{1}{2} \nabla_\mu \xi^\mu \langle F, F \rangle \right] \right\} d\mu_g = 0.$$

We claim that the first integral term is irrelevant moving forward, i.e., its integral is zero. For this purpose, we start by applying (A.19) and (A.22) to infer

$$\mathcal{L}_\xi A_\beta = \xi^\gamma \nabla_\gamma A_\beta + A_\gamma \nabla_\beta \xi^\gamma,$$

$$\mathcal{L}_\xi (\nabla_\alpha A_\beta) = \xi^\gamma \nabla_\gamma \nabla_\alpha A_\beta + \nabla_\gamma A_\beta \nabla_\alpha \xi^\gamma + \nabla_\alpha A_\gamma \nabla_\beta \xi^\gamma.$$

By differentiating the first equation, we obtain

$$\nabla_\alpha (\mathcal{L}_\xi A_\beta) = \nabla_\alpha \xi^\gamma \nabla_\gamma A_\beta + \xi^\gamma \nabla_\alpha \nabla_\gamma A_\beta + \nabla_\alpha A_\gamma \nabla_\beta \xi^\gamma + A_\gamma \nabla_\alpha \nabla_\beta \xi^\gamma$$
$$= \xi^\gamma \nabla_\gamma \nabla_\alpha A_\beta + \nabla_\alpha A_\gamma \nabla_\beta \xi^\gamma + \nabla_\alpha \xi^\gamma \nabla_\gamma A_\beta$$
$$+ \xi^\gamma (\nabla_\alpha \nabla_\gamma - \nabla_\gamma \nabla_\alpha) A_\beta + A_\gamma \nabla_\alpha \nabla_\beta \xi^\gamma.$$

If we compare this formula to the one for $\mathcal{L}_\xi (\nabla_\alpha A_\beta)$, then it follows that

$$F(\mathcal{L}_\xi A) = \mathcal{L}_\xi F(A) - R_{\alpha\gamma\beta\delta} A^\delta \xi^\gamma + R_{\beta\gamma\alpha\delta} A^\delta \xi^\gamma + R_{\alpha\beta\gamma\delta} A^\delta \xi^\gamma = \mathcal{L}_\xi F(A),$$

with the last equality being motivated by (A.41) and (A.44). Therefore, integrating by parts and using (A.33), we deduce

$$\int_M \{\langle F(A), \mathcal{L}_\xi F(A)\rangle\} \, d\mu_g = \int_M \{\langle F(A), F(\mathcal{L}_\xi A)\rangle\} \, d\mu_g$$
$$= -2 \int_M \{(\nabla_\mu F^{\mu\nu}) \mathcal{L}_\xi A_\nu\} \, d\mu_g = 0,$$

which yields the claim and implies

$$\int_M \left\{ (\nabla^\mu \xi^\alpha + \nabla^\alpha \xi^\mu) g^{\nu\beta} F_{\mu\nu} F_{\alpha\beta} + \frac{1}{2} \nabla_\mu \xi^\mu \langle F, F\rangle \right\} d\mu_g = 0.$$

Following this, we introduce the energy-momentum tensor to be given by

$$T_{\mu\nu} \overset{\text{def}}{=} D_{\mu\nu} - \frac{1}{4} g_{\mu\nu} \langle F, F\rangle, \tag{A.49}$$

where

$$D_{\mu\nu} \overset{\text{def}}{=} g^{\alpha\beta} F_{\mu\alpha} F_{\nu\beta}.$$

If we perform another set of integrations by parts in the last integral identity, then we derive

$$\int_M \{(\nabla_\mu T^\mu_{\ \nu}) \xi^\nu\} \, d\mu_g = 0,$$

and, subsequently,

$$\nabla_\mu T^\mu_{\ \nu} = 0. \tag{A.50}$$

We can summarize these findings in the next theorem.

**Theorem A.4.** *Let $(M, g_{\mu\nu})$ be a $3 + 1$-dimensional smooth manifold and assume $A_\mu$ to be a gauge field. If the antisymmetric tensor $F_{\mu\nu}$ defined by (A.31) satisfies Maxwell's equations (A.33), then the energy-momentum tensor described by (A.49) obeys the structure equations (A.50).*

Like in Section A.5, we obtain

$$\nabla_\mu \left\{ T^\mu_{\ \nu} X^\nu \right\} = 0,$$

for an arbitrary Killing vector field $X^\nu$. In the case of a conformal Killing vector field, we have

$$\nabla_\mu \left\{ T^\mu_{\ \nu} X^\nu \right\} = \frac{1}{2} c\, g_{\mu\nu} T^{\mu\nu} = \frac{1}{2} c \left( D^\mu_{\ \mu} - \frac{1}{4} \delta^\mu_{\ \mu} \langle F, F \rangle \right) = 0.$$

Thus, both Killing and conformal Killing vector fields generate conserved quantities in this framework.

## A.9 The Ricci tensor and general relativity

In this final section, we follow up on the description of the Riemann curvature tensor and, with its help, introduce the Ricci tensor and the Ricci scalar curvature. These are then used in formulating the celebrated Einstein equations of general relativity.

We start by contracting the curvature tensor in the $(2, 4)$ positions and thus define the Ricci tensor by

$$R_{\alpha\beta} \stackrel{\text{def}}{=} R_{\alpha\delta\beta}^{\ \ \ \delta} = g^{\gamma\delta} R_{\alpha\gamma\beta\delta}. \tag{A.51}$$

Subsequently, the Ricci scalar curvature is given by

$$R \stackrel{\text{def}}{=} R_\alpha^{\ \alpha} = g^{\alpha\beta} R_{\alpha\beta} = g^{\alpha\beta} g^{\gamma\delta} R_{\alpha\gamma\beta\delta}. \tag{A.52}$$

We first notice that the Ricci tensor is symmetric and this is immediate by contracting (A.43) with $g^{\beta\delta}$:

$$R_{\alpha\gamma} = g^{\beta\delta} R_{\alpha\beta\gamma\delta} = g^{\delta\beta} R_{\gamma\delta\alpha\beta} = R_{\gamma\alpha}.$$

Next, we derive an identity which is very important in general relativity. We start by invoking (A.45), (A.41), and (A.42) to claim that

$$\nabla_\alpha R_{\beta\gamma\delta\rho} + \nabla_\gamma R_{\beta\alpha\rho\delta} - \nabla_\beta R_{\alpha\gamma\delta\rho} = 0.$$

If we contract this equation with $g^{\alpha\delta}g^{\gamma\rho}/2$ and rely on $\nabla g = 0$, then we obtain

$$\nabla_\alpha \left\{ R^\alpha_{\ \beta} - \frac{1}{2}\delta^\alpha_{\ \beta}R \right\} = \nabla_\alpha R^\alpha_{\ \beta} - \frac{1}{2}\nabla_\beta R$$

$$= \frac{1}{2}\nabla_\alpha \left\{ g^{\alpha\delta}g^{\gamma\rho}R_{\beta\gamma\delta\rho} \right\} + \frac{1}{2}\nabla_\gamma \left\{ g^{\gamma\rho}g^{\alpha\delta}R_{\beta\alpha\rho\delta} \right\} \qquad (A.53)$$

$$- \frac{1}{2}\nabla_\beta \left\{ g^{\alpha\delta}g^{\gamma\rho}R_{\alpha\gamma\delta\rho} \right\} = 0.$$

Following this, for the remainder of this section, we focus on Einstein equations of general relativity, whose physical motivation is well-explained in various textbooks (e.g., Wald [183]). This is why we bypass this issue here and, instead, begin by defining the Einstein tensor according to

$$G_{\mu\nu} \overset{\text{def}}{=} R_{\mu\nu} - \frac{1}{2}g_{\mu\nu}R. \qquad (A.54)$$

In the presence of an external field with energy-momentum tensor labeled by $T_{\mu\nu}$, the Einstein equations are given by

$$G_{\mu\nu} = cT_{\mu\nu}, \qquad (A.55)$$

where $c$ is an appropriate constant. One can see that if the energy-momentum tensor originates from a coordinate-independent Lagrangian (as is always the case), then $\nabla_\mu T^\mu_{\ \nu} = 0$. This is compatible with $\nabla_\mu G^\mu_{\ \nu} = 0$, which holds due to (A.53) and

$$G^\alpha_{\ \beta} = R^\alpha_{\ \beta} - \frac{1}{2}\delta^\alpha_{\ \beta}R.$$

In the absence of the field, we have $T_{\mu\nu} = 0$ and, thus, (A.55) become $G_{\mu\nu} = 0$. From (A.54), it follows that $\text{tr}(G_{\mu\nu}) = -R$ and, consequently, $R = 0$. Hence, in this case, (A.55) further reduce to

$$R_{\mu\nu} = 0, \qquad (A.56)$$

which are called the Einstein vacuum equations. These have been studied extensively and are known to have physically-relevant solutions like the flat Minkowski spacetime, the Schwarzschild solution, and the Kerr solution. The last two represent stationary spherically symmetric and axisymmetric solutions, respectively, being prototypes for gravitational collapse (i.e., solutions with a certain type of singularities, also known as black holes). The following is an important observation due to Hilbert, which connects the Einstein vacuum equations to a Lagrangian.

**Theorem A.5.** *The Einstein vacuum equations* (A.56) *are the Euler-Lagrange equations of the Hilbert functional*

$$\mathbb{L}[g] \overset{\text{def}}{=} \int_M \{R\}\, d\mu_g. \qquad (A.57)$$

**Proof.** The challenge in proving this result consists in the fact that the metric is an unknown and we are supposed to compute the variation of the above functional with respect to this metric, which is actually part of the manifold on which we integrate.

We start by considering, as in Remark A.3, a family of metrics $g_{\mu\nu}(s)$ parametrized by $s$, with the variation $\dot{g}_{\mu\nu}(0) := \gamma_{\mu\nu}$. For example, we can work with $g_{\mu\nu}(s) = g_{\mu\nu} + s\gamma_{\mu\nu}$, $g_{\mu\nu}$ being the background metric. We recall the formula (A.25) for the measure of integration on $M$, which can be written as

$$d\mu_g = g\,dx,$$

where

$$dx := \frac{1}{4!}\widehat{\epsilon}_{\alpha\beta\gamma\delta}dx^\alpha dx^\beta dx^\gamma dx^\delta.$$

Hence, the Hilbert functional (A.57) takes the form

$$\mathbb{L}[g] = \int_\Omega \left\{ g g^{\mu\nu} R_{\mu\nu} \right\} dx,$$

for some $\Omega \subset M$. Using the formal notational convention for the variation of the metric (i.e., $\delta g_{\mu\nu} = \dot{g}_{\mu\nu} = \gamma_{\mu\nu}$), we can infer

$$\frac{\delta\mathbb{L}[g]}{\delta g_{\mu\nu}} = \int_\Omega \left\{ \dot{g}g^{\mu\nu}R_{\mu\nu} + g\dot{g}^{\mu\nu}R_{\mu\nu} + gg^{\mu\nu}\dot{R}_{\mu\nu} \right\} dx. \tag{A.58}$$

For the first integral term, if $A^{\mu\nu}$ denotes the cofactor corresponding to the matrix entry $g_{\mu\nu}$, then, due to $A^{\mu\nu} = -g^2 g^{\mu\nu}$, we deduce

$$\dot{g} = -\frac{1}{2}\frac{\dot{g}_{\mu\nu}A^{\mu\nu}}{g} = \frac{1}{2}g\gamma_{\mu\nu}g^{\mu\nu}.$$

Concerning the second integral term, we take advantage of the identity

$$g^{\mu\alpha}(s)\,g_{\alpha\nu}(s) = \delta^\mu_\nu,$$

which, after differentiation with respect to $s$, yields

$$\dot{g}^{\mu\nu} = -g^{\mu\alpha}\dot{g}_{\alpha\beta}g^{\beta\nu} = -g^{\mu\alpha}\gamma_{\alpha\beta}g^{\beta\nu}.$$

Thus, inserting into (A.58) these two equations for $\dot{g}$ and $\dot{g}^{\mu\nu}$, respectively, we find that

$$\begin{aligned}
\frac{\delta\mathbb{L}[g]}{\delta g_{\mu\nu}} &= \int_\Omega \left\{ g\left[ \gamma_{\mu\nu}\left(\frac{1}{2}g^{\mu\nu}R - R^{\mu\nu}\right) + g^{\mu\nu}\dot{R}_{\mu\nu} \right] \right\} dx \\
&= \int_\Omega \left\{ g\left( \gamma_{\mu\nu}G^{\mu\nu} + g^{\mu\nu}\dot{R}_{\mu\nu} \right) \right\} dx.
\end{aligned} \tag{A.59}$$

Next, we argue that $g^{\mu\nu}\dot{R}_{\mu\nu}$ is the divergence of some vector field $P^\alpha$, i.e.,

$$g^{\mu\nu}\dot{R}_{\mu\nu} = \nabla_\alpha P^\alpha,$$

and, hence, this integral term in the above variational derivative contributes only a boundary term after integration. For this purpose, using (A.39), we obtain

$$\dot{R}_{\mu\alpha\nu}{}^\beta = \partial_\mu C_{\alpha\nu}^\beta - \partial_\alpha C_{\mu\nu}^\beta - \left(\Gamma_{\mu\nu}^\rho C_{\alpha\rho}^\beta + C_{\mu\nu}^\rho \Gamma_{\alpha\rho}^\beta - \Gamma_{\alpha\nu}^\rho C_{\mu\rho}^\beta - C_{\alpha\nu}^\rho \Gamma_{\mu\rho}^\beta\right),$$

where

$$C_{\mu\nu}^\lambda \overset{\text{def}}{=} \dot{\Gamma}_{\mu\nu}^\lambda(0).$$

If we recall the formula for the covariant derivative

$$\nabla_\mu C_{\alpha\nu}^\beta = \partial_\mu C_{\alpha\nu}^\beta - \Gamma_{\mu\alpha}^\rho C_{\rho\nu}^\beta - \Gamma_{\mu\nu}^\rho C_{\alpha\rho}^\beta + \Gamma_{\mu\rho}^\beta C_{\alpha\nu}^\rho,$$

then we arrive at the expression

$$\dot{R}_{\mu\alpha\nu}{}^\beta = \nabla_\mu C_{\alpha\nu}^\beta - \nabla_\alpha C_{\mu\nu}^\beta$$

and, subsequently,

$$\dot{R}_{\mu\nu} = \nabla_\mu C_{\alpha\nu}^\alpha - \nabla_\alpha C_{\mu\nu}^\alpha.$$

According to (A.12), we can write

$$C_{\mu\nu}^\alpha = \frac{1}{2}\left(\nabla_\mu \gamma_\nu{}^\alpha + \nabla_\nu \gamma_\mu{}^\alpha - \nabla^\alpha \gamma_{\mu\nu}\right),$$

with $\nabla$ denoting now the derivation with respect to the background metric $g_{\mu\nu}(0) := g_{\mu\nu}$ and $\gamma_\mu{}^\alpha := g^{\alpha\rho}\gamma_{\mu\rho}$. Applying this formula to the previous equation for $\dot{R}_{\mu\nu}$, we discover that

$$\dot{R}_{\mu\nu} = \frac{1}{2}\left\{\nabla_\mu \nabla_\nu \gamma + (\nabla_\alpha \nabla^\alpha)\gamma_{\mu\nu} - \left(\nabla_\alpha \nabla_\mu \gamma_\nu{}^\alpha + \nabla_\alpha \nabla_\nu \gamma_\mu{}^\alpha\right)\right\}$$

where we also introduced $\gamma := g^{\alpha\rho}\gamma_{\alpha\rho}$. It follows that

$$g^{\mu\nu}\dot{R}_{\mu\nu} = -\Box\gamma - \nabla_\alpha \nabla_\beta \gamma^{\beta\alpha} = \nabla_\alpha\left\{\nabla^\alpha \gamma - \nabla_\beta \gamma^{\beta\alpha}\right\} = \nabla_\alpha P^\alpha,$$

with $P^\alpha := \nabla^\alpha \gamma - \nabla_\beta \gamma^{\beta\alpha}$.

Therefore, if the variation has compact support, then we can upgrade (A.59) to read

$$\frac{\delta\mathbb{L}[g]}{\delta g_{\mu\nu}} = \int_\Omega \left\{g\gamma_{\mu\nu} G^{\mu\nu}\right\} dx,$$

which, set equal to zero, leads to the vanishing of the Einstein tensor and, consequently, (A.56).                                                                    $\Box$

**Remark A.7.** With this result in mind, if we revisit the case when an external field $\phi : M \mapsto \mathbb{R}$ is present, we recognize that the original Einstein equations (A.55) are also described by a Lagrangian,

$$\mathbb{L}[g_{\mu\nu}, \phi] \stackrel{\text{def}}{=} \int\limits_{M} \{R - cg_{\mu\nu}\nabla^{\mu}\phi\nabla^{\nu}\phi\}\, d\mu_g,$$

which couples the metric and the field. Indeed, the variational derivative of this Lagrangian with respect to $g_{\mu\nu}$ gives (A.55), whereas the variation of the field produces the wave equation

$$\Box_g \phi = 0,$$

where the d'Alembertian is with respect to the metric (see (A.16)). Together, these form a nonlinear coupled system since the metric has to be constructed from the Einstein equations, which depend on the field. The radially-symmetric model was studied by Christodoulou in a series of seminal papers.

In the book, we do not consider problems that arise in general relativity, except to notice the connection of a symmetry reduction for the Einstein vacuum equations to a system, part of which is the wave map problem with the hyperbolic upper-half plane as its target manifold. This serves as motivation for the study of wave maps, as well as a source of open problems, and the reader can consult Chapter 1 for a discussion of these issues.

# Bibliography

[1] Adkins, G. and Nappi, C. (1984). Stabilization of chiral solitons via vector mesons, *Phys. Lett. B* **137**, 3-4, pp. 251–256.

[2] Adkins, G., Nappi, C. and Witten, E. (1983). Static properties of nucleons in the Skyrme model, *Nucl. Phys. B* **228**, 3, pp. 552–566.

[3] Anderson, L., Gudapati, N. and Szeftel, J. (2015). Global regularity for the 2 + 1 dimensional equivariant Einstein-wave map system, Preprint, arXiv:1501.00616.

[4] Aratyn, H., Ferreira, L. A. and Zimerman, A. H. (1999). Exact static soliton solutions of (3+1)-dimensional integrable theory with nonzero Hopf numbers, *Phys. Rev. Lett.* **83**, 9, pp. 1723–1726.

[5] Bach, R. and Weyl, H. (1922). Neue lösungen der einsteinschen gravitationsgleichungen, *Mathematische Zeitschrift* **13**, 1, pp. 134–145.

[6] Bahouri, H., Chemin, J.-Y. and Danchin, R. (2011). *Fourier analysis and nonlinear partial differential equations*, Grundlehren der Mathematischen Wissenschaften [Fundamental Principles of Mathematical Sciences], Vol. 343 (Springer, Heidelberg).

[7] Bahouri, H. and Gérard, P. (1999). High frequency approximation of solutions to critical nonlinear wave equations, *Amer. J. Math.* **121**, 1, pp. 131–175.

[8] Balachandran, A. P., Nair, V. P., Rajeev, S. G. and Stern, A. B. (1982). Exotic levels from topology in the quantum-chromodynamic effective lagrangian, *Phys. Rev. Lett.* **49**, 16, pp. 1124–1127.

[9] Balachandran, A. P., Nair, V. P., Rajeev, S. G. and Stern, A. B. (1983). Soliton states in the quantum-chromodynamic effective lagrangian, *Phys. Rev. D* **27**, 5, pp. 1153–1164.

[10] Battye, R. A. and Sutcliffe, P. M. (1997). Symmetric skyrmions, *Phys. Rev. Lett.* **79**, 3, pp. 363–366.

[11] Battye, R. A. and Sutcliffe, P. M. (1998). Knots as stable soliton solutions in a three-dimensional classical field theory, *Phys. Rev. Lett.* **81**, 22, pp. 4798–4801.

[12] Battye, R. A. and Sutcliffe, P. M. (1999). Solitons, links and knots, *R. Soc. Lond. Proc. Ser. A Math. Phys. Eng. Sci.* **455**, 1992, pp. 4305–4331.

[13] Beals, M. (1983). Self-spreading and strength of singularities for solutions to semilinear wave equations, *Ann. of Math. (2)* **118**, 1, pp. 187–214.

[14] Belavin, A. A. and Polyakov, A. M. (1975). Metastable states of two-dimensional isotropic ferromagnets, *JETP Lett.* **22**, 10, pp. 245–247.

[15] Benson, K. and Bucher, M. (1993). Skyrmions and semilocal strings in cosmology, *Nucl. Phys. B* **406**, 1-2, pp. 355–376.

[16] Bergh, J. and Löfström, J. (1976). *Interpolation spaces. An introduction* (Springer-Verlag, Berlin), Grundlehren der Mathematischen Wissenschaften, No. 223.

[17] Bizoń, P., Chmaj, T. and Rostworowski, A. (2007). Asymptotic stability of the skyrmion, *Phys. Rev. D* **75**, 12, pp. 121702–121706.

[18] Bizoń, P., Chmaj, T. and Tabor, Z. (2001). Formation of singularities for equivariant $(2 + 1)$-dimensional wave maps into the 2-sphere, *Nonlinearity* **14**, 5, pp. 1041–1053.

[19] Born, M. and Infeld, L. (1934). Foundations of the new field theory, *Proc. Roy. Soc. London Ser. A* **144**, 852, pp. 425–451.

[20] Bourgain, J. (1993a). Fourier transform restriction phenomena for certain lattice subsets and applications to nonlinear evolution equations. I. Schrödinger equations, *Geom. Funct. Anal.* **3**, 2, pp. 107–156.

[21] Bourgain, J. (1993b). Fourier transform restriction phenomena for certain lattice subsets and applications to nonlinear evolution equations. II. The KdV-equation, *Geom. Funct. Anal.* **3**, 3, pp. 209–262.

[22] Cartan, E. (2013). *Exterior differential systems and its applications*, translated from French by Mehdi Nadjafikhah.

[23] Carter, B. (1973). Black hole equilibrium states, in *Black holes/Les astres occlus (École d'Été Phys. Théor., Les Houches, 1972)* (Gordon and Breach, New York), pp. 57–214.

[24] Cazenave, T. (2003). *Semilinear Schrödinger equations, Courant Lecture Notes in Mathematics*, Vol. 10 (New York University, Courant Institute of Mathematical Sciences, New York; American Mathematical Society, Providence, RI).

[25] Cazenave, T., Shatah, J. and Tahvildar-Zadeh, A. (1998). Harmonic maps of the hyperbolic space and development of singularities in wave maps and Yang-Mills fields, *Ann. Inst. H. Poincaré Phys. Théor.* **68**, 3, pp. 315–349.

[26] Cho, Y. and Ozawa, T. (2009). Sobolev inequalities with symmetry, *Commun. Contemp. Math.* **11**, 3, pp. 355–365.

[27] Christ, M. and Kiselev, A. (2001). Maximal functions associated to filtrations, *J. Funct. Anal.* **179**, 2, pp. 409–425.

[28] Christodoulou, D. (2000). *The action principle and partial differential equations, Annals of Mathematics Studies*, Vol. 146 (Princeton University Press, Princeton, NJ).

[29] Christodoulou, D. and Tahvildar-Zadeh, A. (1993). On the regularity of spherically symmetric wave maps, *Comm. Pure Appl. Math.* **46**, 7, pp. 1041–1091.

[30] Côte, R. (2005). Instability of nonconstant harmonic maps for the $(1 + 2)$-dimensional equivariant wave map system, *Int. Math. Res. Not.*, 57, pp. 3525–3549.

[31] Creek, M. (2013). Large-data global well-posedness for the $(1 + 2)$-dimensional equivariant Faddeev model, Preprint, arXiv:1310.4708.

[32] Creek, M., Donninger, R., Schlag, W. and Snelson, S. (2016). Linear stability of the skyrmion, Preprint, arXiv:1603.03662.

[33] Crutchfield, W. Y. and Bell, J. B. (1994). Instabilities of the Skyrme model, *J. Comput. Phys.* **110**, 2, pp. 234–241.

[34] Dafermos, C. M. (2005). *Hyperbolic conservation laws in continuum physics*, Grundlehren der Mathematischen Wissenschaften [Fundamental Principles of Mathematical Sciences], Vol. 325, 2nd edn. (Springer-Verlag, Berlin).

[35] D'Ancona, P. and Georgiev, V. (2004). On the continuity of the solution operator to the wave map system, *Comm. Pure Appl. Math.* **57**, 3, pp. 357–383.

[36] D'Ancona, P. and Georgiev, V. (2005). Wave maps and ill-posedness of their Cauchy problem, in *New trends in the theory of hyperbolic equations*, *Oper. Theory Adv. Appl.*, Vol. 159 (Birkhäuser, Basel), pp. 1–111.

[37] Derrick, G. H. (1964). Comments on nonlinear wave equations as models for elementary particles, *J. Mathematical Phys.* **5**, pp. 1252–1254.

[38] do Carmo, M. P. (1992). *Riemannian geometry*, Mathematics: Theory & Applications (Birkhäuser Boston, Inc., Boston, MA), translated from the second Portuguese edition by Francis Flaherty.

[39] Dolbeault, J. (1990). Existence de solutions symétriques pour un modèle de champs de mésons: le modèle d'Adkins et Nappi, *Comm. Partial Differential Equations* **15**, 12, pp. 1743–1786.

[40] Eells, J. and Lemaire, L. (1978). A report on harmonic maps, *Bull. London Math. Soc.* **10**, 1, pp. 1–68.

[41] Ernst, F. J. (1968). New formulation of the axially symmetric gravitational field problem, *Phys. Rev.* **167**, 5, p. 1175.

[42] Esteban, M. J. (1986). A direct variational approach to Skyrme's model for meson fields, *Comm. Math. Phys.* **105**, 4, pp. 571–591.

[43] Faddeev, L. D. (1975). Quantization of solitons, Preprint IAS print-75-QS70 (Inst. Advanced Study, Princeton, NJ), 32 pp.

[44] Faddeev, L. D. (1976). Some comments on the many-dimensional solitons, *Lett. Math. Phys.* **1**, 4, pp. 289–293.

[45] Faddeev, L. D. and Niemi, A. J. (1997). Stable knot-like structures in classical field theory, *Nature* **387**, pp. 58–61.

[46] Fang, D. and Wang, C. (2006). Some remarks on Strichartz estimates for homogeneous wave equation, *Nonlinear Anal.* **65**, 3, pp. 697–706.

[47] Foster, D. J. and Sutcliffe, P. M. (2009). Baby skyrmions stabilized by vector mesons, *Phys. Rev. D* **79**, 12, pp. 125026, 5 pp.

[48] Friedrichs, K. O. (1954). Symmetric hyperbolic linear differential equations, *Comm. Pure Appl. Math.* **7**, pp. 345–392.

[49] Friedrichs, K. O. (1958). Symmetric positive linear differential equations, *Comm. Pure Appl. Math.* **11**, pp. 333–418.

[50]  Geba, D.-A. and da Silva, D. (2013). On the regularity of the 2 + 1 dimensional equivariant Skyrme model, *Proc. Amer. Math. Soc.* **141**, 6, pp. 2105–2115.

[51]  Geba, D.-A., Nakanishi, K. and Rajeev, S. G. (2012). Global well-posedness and scattering for Skyrme wave maps, *Commun. Pure Appl. Anal.* **11**, 5, pp. 1923–1933.

[52]  Geba, D.-A., Nakanishi, K. and Zhang, X. (2015). Sharp global regularity for the 2 + 1-dimensional equivariant Faddeev model, *Int. Math. Res. Not.*, 22, pp. 11549–11565.

[53]  Geba, D.-A. and Rajeev, S. G. (2010a). A continuity argument for a semilinear Skyrme model, *Electron. J. Differential Equations*, 86, pp. 1–9.

[54]  Geba, D.-A. and Rajeev, S. G. (2010b). Nonconcentration of energy for a semilinear Skyrme model, *Ann. Physics* **325**, 12, pp. 2697–2706.

[55]  Gell-Mann, M. and Lévy, M. M. (1960). The axial vector current in beta decay, *Nuovo Cimento (10)* **16**, pp. 705–726.

[56]  Gibbons, G. W. (2003). Causality and the Skyrme model, *Phys. Lett. B* **566**, 1-2, pp. 171–174.

[57]  Ginibre, J. and Velo, G. (1995). Generalized Strichartz inequalities for the wave equation, *J. Funct. Anal.* **133**, 1, pp. 50–68.

[58]  Glassey, R. T. (1977). On the blowing up of solutions to the Cauchy problem for nonlinear Schrödinger equations, *J. Math. Phys.* **18**, 9, pp. 1794–1797.

[59]  Grillakis, M. G. (1990). Regularity and asymptotic behaviour of the wave equation with a critical nonlinearity, *Ann. of Math. (2)* **132**, 3, pp. 485–509.

[60]  Grillakis, M. G. (1991). Classical solutions for the equivariant wave map in 1 + 2 dimensions, Preprint.

[61]  Grillakis, M. G. (1992). Regularity for the wave equation with a critical nonlinearity, *Comm. Pure Appl. Math.* **45**, 6, pp. 749–774.

[62]  Grillakis, M. G. (1998). Energy estimates and the wave map problem, *Comm. Partial Differential Equations* **23**, 5-6, pp. 887–911.

[63]  Grillakis, M. G. (2000). On the wave map problem, in *Nonlinear wave equations (Providence, RI, 1998), Contemp. Math.*, Vol. 263 (Amer. Math. Soc., Providence, RI), pp. 71–84.

[64]  Gürsey, F. (1960). On the symmetries of strong and weak interactions, *Nuovo Cimento (10)* **16**, pp. 230–240.

[65]  Gürsey, F. (1961). On the structure and parity of weak interaction currents, *Ann. Physics* **12**, 1, pp. 91–117.

[66]  Hadamard, J. (1902). Sur les problèmes aux dérivées partielles et leur signification physique, *Princeton University Bulletin* **13**, pp. 49–52.

[67]  Hawking, S. W. and Ellis, G. F. R. (1973). *The large scale structure of space-time* (Cambridge University Press, London-New York), Cambridge Monographs on Mathematical Physics, No. 1.

[68]  Hélein, F. (2002). *Harmonic maps, conservation laws and moving frames, Cambridge Tracts in Mathematics*, Vol. 150, 2nd edn. (Cambridge University Press, Cambridge), translated from the 1996 French original, with a foreword by James Eells.

[69] Hopf, H. (1926). Zum Clifford-Kleinschen Raumproblem, *Math. Ann.* **95**, 1, pp. 313–339.

[70] Hörmander, L. (1960). Estimates for translation invariant operators in $L^p$ spaces, *Acta Math.* **104**, pp. 93–140.

[71] Hörmander, L. (1976). *Linear partial differential operators* (Springer Verlag, Berlin-New York).

[72] Hörmander, L. (1988). $L^1$, $L^\infty$ estimates for the wave operator, in *Analyse mathématique et applications* (Gauthier-Villars, Montrouge), pp. 211–234.

[73] Hörmander, L. (1997). *Lectures on nonlinear hyperbolic differential equations, Mathématiques & Applications (Berlin) [Mathematics & Applications]*, Vol. 26 (Springer-Verlag, Berlin).

[74] Hughes, T. J. R., Kato, T. and Marsden, J. E. (1976). Well-posed quasi-linear second-order hyperbolic systems with applications to nonlinear elastodynamics and general relativity, *Arch. Rational Mech. Anal.* **63**, 3, pp. 273–294.

[75] Hunt, R. A. (1965). *Operators acting on Lorentz spaces* (ProQuest LLC, Ann Arbor, MI), thesis (Ph.D.)–Washington University in St. Louis.

[76] Jost, J. (2005). *Riemannian geometry and geometric analysis*, Fourth edn., Universitext (Springer-Verlag, Berlin).

[77] Kapitanskiĭ, L. V. and Ladyzhenskaya, O. A. (1983). The Coleman principle for finding stationary points of invariant functionals, *Zap. Nauchn. Sem. Leningrad. Otdel. Mat. Inst. Steklov. (LOMI)* **127**, pp. 84–102.

[78] Keel, M. and Tao, T. (1998a). Endpoint Strichartz estimates, *Amer. J. Math.* **120**, 5, pp. 955–980.

[79] Keel, M. and Tao, T. (1998b). Local and global well-posedness of wave maps on $\mathbb{R}^{1+1}$ for rough data, *Int. Math. Res. Not.*, 21, pp. 1117–1156.

[80] Kenig, C. E. and Merle, F. (2006). Global well-posedness, scattering and blow-up for the energy-critical, focusing, non-linear Schrödinger equation in the radial case, *Invent. Math.* **166**, 3, pp. 645–675.

[81] Kenig, C. E. and Merle, F. (2008). Global well-posedness, scattering and blow-up for the energy-critical focusing non-linear wave equation, *Acta Math.* **201**, 2, pp. 147–212.

[82] Kerr, R. P. (1963). Gravitational field of a spinning mass as an example of algebraically special metrics, *Phys. Rev. Lett.* **11**, pp. 237–238.

[83] Killing, W. (1891). Ueber die Clifford-Klein'schen Raumformen, *Math. Ann.* **39**, 2, pp. 257–278.

[84] Klainerman, S. (1980). Global existence for nonlinear wave equations, *Comm. Pure Appl. Math.* **33**, 1, pp. 43–101.

[85] Klainerman, S. (2000). PDE as a unified subject, *Geom. Funct. Anal.*, Special Volume, Part I, pp. 279–315, GAFA 2000 (Tel Aviv, 1999).

[86] Klainerman, S. and Machedon, M. (1994). Smoothing estimates for null forms and applications, *Int. Math. Res. Not.*, 9, pp. 383–389.

[87] Klainerman, S. and Machedon, M. (1995). Smoothing estimates for null forms and applications, *Duke Math. J.* **81**, 1, pp. 99–133, a celebration of John F. Nash, Jr.

[88] Klainerman, S. and Rodnianski, I. (2001). On the global regularity of wave maps in the critical Sobolev norm, *Int. Math. Res. Not.*, 13, pp. 655–677.

[89] Klainerman, S. and Selberg, S. (1997). Remark on the optimal regularity for equations of wave maps type, *Comm. Partial Differential Equations* **22**, 5-6, pp. 901–918.

[90] Klainerman, S. and Selberg, S. (2002). Bilinear estimates and applications to nonlinear wave equations, *Commun. Contemp. Math.* **4**, 2, pp. 223–295.

[91] Koch, H., Tataru, D. and Visan, M. (2014). *Dispersive equations and nonlinear waves*, Oberwolfach Seminars, Vol. 45 (Birkhäuser).

[92] Krieger, J. (2003). Global regularity of wave maps from $\mathbb{R}^{3+1}$ to surfaces, *Comm. Math. Phys.* **238**, 1-2, pp. 333–366.

[93] Krieger, J. (2004). Global regularity of wave maps from $\mathbb{R}^{2+1}$ to $H^2$. Small energy, *Comm. Math. Phys.* **250**, 3, pp. 507–580.

[94] Krieger, J. (2008). Global regularity and singularity development for wave maps, in *Surveys in differential geometry. Vol. XII. Geometric flows, Surv. Differ. Geom.*, Vol. 12 (Int. Press, Somerville, MA), pp. 167–201.

[95] Krieger, J. and Schlag, W. (2009). Concentration compactness for critical wave maps, Preprint, arXiv:0908.2474.

[96] Krieger, J. and Schlag, W. (2012). *Concentration compactness for critical wave maps*, EMS Monographs in Mathematics (European Mathematical Society (EMS), Zürich).

[97] Krieger, J., Schlag, W. and Tataru, D. (2008). Renormalization and blow up for charge one equivariant critical wave maps, *Invent. Math.* **171**, 3, pp. 543–615.

[98] Landman, M. J., LeMesurier, B. J., Papanicolaou, G. C., Sulem, C. and Sulem, P.-L. (1989). Singular solutions of the cubic Schrödinger equation, in *Integrable systems and applications (Île d'Oléron, 1988), Lecture Notes in Phys.*, Vol. 342 (Springer, Berlin), pp. 207–217.

[99] Lawrie, A. (2015). Conditional global existence and scattering for a semilinear Skyrme equation with large data, *Comm. Math. Phys.* **334**, 2, pp. 1025–1081.

[100] Lee, B. W. (1972). Chiral dynamics, in *Cargèse lectures in physics, Vol. 5* (Gordon and Breach, New York), pp. 1–117.

[101] Lee, D.-H. and Kane, C. L. (1990). Boson-vortex-skyrmion duality, spin-singlet fractional quantum Hall effect, and spin-$\frac{1}{2}$ anyon superconductivity, *Phys. Rev. Lett.* **64**, 12, pp. 1313–1317.

[102] Leese, R. A., Peyrard, M. and Zakrzewski, W. J. (1990). Soliton scatterings in some relativistic models in $(2 + 1)$ dimensions, *Nonlinearity* **3**, 3, pp. 773–807.

[103] Lei, Z., Lin, F. and Zhou, Y. (2011). Global solutions of the evolutionary Faddeev model with small initial data, *Acta Math. Sin. (Engl. Ser.)* **27**, 2, pp. 309–328.

[104] LeMesurier, B., Papanicolaou, G., Sulem, C. and Sulem, P.-L. (1987). The focusing singularity of the nonlinear Schrödinger equation, in *Directions in partial differential equations (Madison, WI, 1985), Publ. Math. Res. Center Univ. Wisconsin*, Vol. 54 (Academic Press, Boston, MA), pp. 159–201.

[105] LeMesurier, B. J., Papanicolaou, G., Sulem, C. and Sulem, P.-L. (1988a). Focusing and multi-focusing solutions of the nonlinear Schrödinger equation, *Phys. D* **31**, 1, pp. 78–102.

[106] LeMesurier, B. J., Papanicolaou, G. C., Sulem, C. and Sulem, P.-L. (1988b). Local structure of the self-focusing singularity of the nonlinear Schrödinger equation, *Phys. D* **32**, 2, pp. 210–226.

[107] Leray, J. (1953). *Hyperbolic differential equations* (The Institute for Advanced Study, Princeton, N. J.).

[108] Li, D. (2012). Global wellposedness of hedgehog solutions for the $(3 + 1)$ Skyrme model, Preprint, arXiv:1208.4977.

[109] Lin, F. and Yang, Y. (2004a). Existence of energy minimizers as stable knotted solitons in the Faddeev model, *Comm. Math. Phys.* **249**, 2, pp. 273–303.

[110] Lin, F. and Yang, Y. (2004b). The Faddeev knots as stable solitons: existence theorems, *Sci. China Ser. A* **47**, 2, pp. 187–197.

[111] Machihara, S., Nakanishi, K. and Tsugawa, K. (2010). Well-posedness for nonlinear Dirac equations in one dimension, *Kyoto J. Math.* **50**, 2, pp. 403–451.

[112] Manton, N. S. (1987). Geometry of skyrmions, *Comm. Math. Phys.* **111**, 3, pp. 469–478.

[113] Manton, N. S. and Sutcliffe, P. M. (2004). *Topological solitons*, Cambridge Monographs on Mathematical Physics (Cambridge University Press, Cambridge).

[114] McLeod, J. B. and Troy, W. C. (1991). The Skyrme model for nucleons under spherical symmetry, *Proc. Roy. Soc. Edinburgh Sect. A* **118**, 3-4, pp. 271–288.

[115] Merle, F. and Raphaël, P. (2003). Sharp upper bound on the blow-up rate for the critical nonlinear Schrödinger equation, *Geom. Funct. Anal.* **13**, 3, pp. 591–642.

[116] Merle, F. and Raphaël, P. (2004). On universality of blow-up profile for $L^2$ critical nonlinear Schrödinger equation, *Invent. Math.* **156**, 3, pp. 565–672.

[117] Merle, F. and Raphaël, P. (2005a). The blow-up dynamic and upper bound on the blow-up rate for critical nonlinear Schrödinger equation, *Ann. of Math. (2)* **161**, 1, pp. 157–222.

[118] Merle, F. and Raphaël, P. (2005b). Profiles and quantization of the blow up mass for critical nonlinear Schrödinger equation, *Comm. Math. Phys.* **253**, 3, pp. 675–704.

[119] Merle, F. and Raphaël, P. (2006). On a sharp lower bound on the blow-up rate for the $L^2$ critical nonlinear Schrödinger equation, *J. Amer. Math. Soc.* **19**, 1, pp. 37–90 (electronic).

[120] Morawetz, C. S. (1961). The decay of solutions of the exterior initial-boundary value problem for the wave equation, *Comm. Pure Appl. Math.* **14**, pp. 561–568.

[121] Morawetz, C. S. (1968). Time decay for the nonlinear Klein-Gordon equations, *Proc. Roy. Soc. Ser. A* **306**, pp. 291–296.

[122] Morawetz, C. S. and Strauss, W. A. (1972). Decay and scattering of solutions of a nonlinear relativistic wave equation, *Comm. Pure Appl. Math.* **25**, pp. 1–31.

[123] Nahmod, A., Stefanov, A. and Uhlenbeck, K. (2003). On the well-posedness of the wave map problem in high dimensions, *Comm. Anal. Geom.* **11**, 1, pp. 49–83.

[124] Nash, J. (1956). The imbedding problem for Riemannian manifolds, *Ann. of Math. (2)* **63**, pp. 20–63.

[125] O'Neil, R. (1963). Convolution operators and $L(p, q)$ spaces, *Duke Math. J.* **30**, pp. 129–142.

[126] O'Neill, B. (1983). *Semi-Riemannian geometry, Pure and Applied Mathematics*, Vol. 103 (Academic Press, Inc. [Harcourt Brace Jovanovich, Publishers], New York), with applications to relativity.

[127] Opic, B. and Kufner, A. (1990). *Hardy-type inequalities, Pitman Research Notes in Mathematics Series*, Vol. 219 (Longman Scientific & Technical, Harlow).

[128] Ovchinnikov, Y. N. and Sigal, I. M. (2011). On collapse of wave maps, *Phys. D* **240**, 17, pp. 1311–1324.

[129] Perelman, G. (2001). On the formation of singularities in solutions of the critical nonlinear Schrödinger equation, *Ann. Henri Poincaré* **2**, 4, pp. 605–673.

[130] Peyrard, M., Piette, B. and Zakrzewski, W. J. (1992a). Soliton scattering in the Skyrme model in $(2+1)$ dimensions. I. Soliton-soliton case, *Nonlinearity* **5**, 2, pp. 563–583.

[131] Peyrard, M., Piette, B. and Zakrzewski, W. J. (1992b). Soliton scattering in the Skyrme model in $(2 + 1)$ dimensions. II. More general systems, *Nonlinearity* **5**, 2, pp. 585–600.

[132] Piette, B., Schroers, B. J. and Zakrzewski, W. J. (1995). Multisolitons in a two-dimensional Skyrme model, *Z. Phys.* **C65**, 1, pp. 165–174.

[133] Pohlmeyer, K. (1976). Integrable Hamiltonian systems and interactions through quadratic constraints, *Comm. Math. Phys.* **46**, 3, pp. 207–221.

[134] Ponce, G. and Sideris, T. C. (1993). Local regularity of nonlinear wave equations in three space dimensions, *Comm. Partial Differential Equations* **18**, 1-2, pp. 169–177.

[135] Qing, J. (1993). Boundary regularity of weakly harmonic maps from surfaces, *J. Funct. Anal.* **114**, 2, pp. 458–466.

[136] Raphaël, P. and Rodnianski, I. (2012). Stable blow up dynamics for the critical co-rotational wave maps and equivariant Yang-Mills problems, *Publ. Math. Inst. Hautes Études Sci.*, pp. 1–122.

[137] Rauch, J. and Reed, M. (1982). Nonlinear microlocal analysis of semilinear hyperbolic systems in one space dimension, *Duke Math. J.* **49**, 2, pp. 397–475.

[138] Rodnianski, I. and Sterbenz, J. (2010). On the formation of singularities in the critical O(3) $\sigma$-model, *Ann. of Math. (2)* **172**, 1, pp. 187–242.

[139] Schoen, R. and Yau, S. T. (1976). Harmonic maps and the topology of stable hypersurfaces and manifolds with non-negative Ricci curvature, *Com-*

*ment. Math. Helv.* **51**, 3, pp. 333–341.

[140] Schwarzschild, K. (1916). Über das gravitationsfeld eines massenpunktes nach der einsteinschen theorie, *Sitzungsberichte der Königlich Preußischen Akademie der Wissenschaften (Berlin), 1916, Seite 189-196* **1**, pp. 189–196.

[141] Selberg, S. (1999). Multilinear spacetime estimates and applications to local existence theory for nonlinear wave equations, Ph.D. Thesis, Princeton University.

[142] Selberg, S. (2001). Graduate course on nonlinear wave equations, Lecture notes, Johns Hopkins University.

[143] Shatah, J. (1988). Weak solutions and development of singularities of the SU(2) σ-model, *Comm. Pure Appl. Math.* **41**, 4, pp. 459–469.

[144] Shatah, J. and Strauss, W. (1996). Breathers as homoclinic geometric wave maps, *Phys. D* **99**, 2-3, pp. 113–133.

[145] Shatah, J. and Struwe, M. (1998). *Geometric wave equations, Courant Lecture Notes in Mathematics*, Vol. 2 (New York University Courant Institute of Mathematical Sciences, New York).

[146] Shatah, J. and Struwe, M. (2002). The Cauchy problem for wave maps, *Int. Math. Res. Not.*, 11, pp. 555–571.

[147] Shatah, J. and Tahvildar-Zadeh, A. (1992). Regularity of harmonic maps from the Minkowski space into rotationally symmetric manifolds, *Comm. Pure Appl. Math.* **45**, 8, pp. 947–971.

[148] Shatah, J. and Tahvildar-Zadeh, A. (1994). On the Cauchy problem for equivariant wave maps, *Comm. Pure Appl. Math.* **47**, 5, pp. 719–754.

[149] Sickel, W. and Skrzypczak, L. (2000). Radial subspaces of Besov and Lizorkin-Triebel classes: extended Strauss lemma and compactness of embeddings, *J. Fourier Anal. Appl.* **6**, 6, pp. 639–662.

[150] Sickel, W. and Skrzypczak, L. (2012). On the interplay of regularity and decay in case of radial functions II. Homogeneous spaces, *J. Fourier Anal. Appl.* **18**, 3, pp. 548–582.

[151] Sickel, W., Skrzypczak, L. and Vybiral, J. (2012). On the interplay of regularity and decay in case of radial functions I. Inhomogeneous spaces, *Commun. Contemp. Math.* **14**, 1, pp. 1250005, 60.

[152] Skyrme, T. H. R. (1961a). A non-linear field theory, *Proc. Roy. Soc. London Ser. A* **260**, pp. 127–138.

[153] Skyrme, T. H. R. (1961b). Particle states of a quantized meson field, *Proc. Roy. Soc. Ser. A* **262**, pp. 237–245.

[154] Skyrme, T. H. R. (1962). A unified field theory of mesons and baryons, *Nucl. Phys.* **31**, pp. 556–569.

[155] Smith, H. F. and Sogge, C. D. (2000). Global Strichartz estimates for nontrapping perturbations of the Laplacian, *Comm. Partial Differential Equations* **25**, 11-12, pp. 2171–2183.

[156] Sogge, C. D. (1995). *Lectures on nonlinear wave equations*, Monographs in Analysis, II (International Press, Boston, MA).

[157] Speck, J. (2012). The nonlinear stability of the trivial solution to the Maxwell-Born-Infeld system, *J. Math. Phys.* **53**, 8, pp. 083703, 83.

[158] Stein, E. M. (1993). *Harmonic analysis: real-variable methods, orthogonality, and oscillatory integrals*, Princeton Mathematical Series, Vol. 43 (Princeton University Press, Princeton, NJ), with the assistance of Timothy S. Murphy, Monographs in Harmonic Analysis, III.

[159] Sterbenz, J. (2005). Angular regularity and Strichartz estimates for the wave equation, *Int. Math. Res. Not.*, 4, pp. 187–231, with an appendix by Igor Rodnianski.

[160] Sterbenz, J. and Tataru, D. (2010a). Energy dispersed large data wave maps in 2 + 1 dimensions, *Comm. Math. Phys.* **298**, 1, pp. 139–230.

[161] Sterbenz, J. and Tataru, D. (2010b). Regularity of wave-maps in dimension 2 + 1, *Comm. Math. Phys.* **298**, 1, pp. 231–264.

[162] Strauss, W. A. (1977). Existence of solitary waves in higher dimensions, *Comm. Math. Phys.* **55**, 2, pp. 149–162.

[163] Strichartz, R. S. (1977). Restrictions of Fourier transforms to quadratic surfaces and decay of solutions of wave equations, *Duke Math. J.* **44**, 3, pp. 705–714.

[164] Struwe, M. (1985). On the evolution of harmonic mappings of Riemannian surfaces, *Comment. Math. Helv.* **60**, 4, pp. 558–581.

[165] Struwe, M. (2003). Equivariant wave maps in two space dimensions, *Comm. Pure Appl. Math.* **56**, 7, pp. 815–823, dedicated to the memory of Jürgen K. Moser.

[166] Talanov, V. I. and Vlasov, S. N. (1989). Distributed wave collapse in the nonlinear Schrödinger equation, in *Nonlinear waves, 1*, Res. Rep. Phys. (Springer, Berlin), pp. 224–234.

[167] Tao, T. (1998). Counterexamples to the $n = 3$ endpoint Strichartz estimate for the wave equation, Preprint.

[168] Tao, T. (2000). Spherically averaged endpoint Strichartz estimates for the two-dimensional Schrödinger equation, *Comm. Partial Differential Equations* **25**, 7-8, pp. 1471–1485.

[169] Tao, T. (2001a). Global regularity of wave maps. I. Small critical Sobolev norm in high dimension, *Int. Math. Res. Not.*, 6, pp. 299–328.

[170] Tao, T. (2001b). Global regularity of wave maps. II. Small energy in two dimensions, *Comm. Math. Phys.* **224**, 2, pp. 443–544.

[171] Tao, T. (2006). *Nonlinear dispersive equations*, CBMS Regional Conference Series in Mathematics, Vol. 106 (Published for the Conference Board of the Mathematical Sciences, Washington, DC; by the American Mathematical Society, Providence, RI), Local and global analysis.

[172] Tao, T. (2008a). Global regularity of wave maps III. Large energy from $\mathbb{R}^{1+2}$ to hyperbolic spaces, Preprint, arXiv:0805.4666.

[173] Tao, T. (2008b). Global regularity of wave maps IV. Absence of stationary or self-similar solutions in the energy class, Preprint, arXiv:0806.3592.

[174] Tao, T. (2008c). Global regularity of wave maps V. Large data local well-posedness and perturbation theory in the energy class, Preprint, arXiv:0808.0368.

[175] Tao, T. (2009a). Global regularity of wave maps VI. Abstract theory of minimal-energy blowup solutions, Preprint, arXiv:0906.2833.

[176] Tao, T. (2009b). Global regularity of wave maps VII. Control of delocalised or dispersed solutions, Preprint, arXiv:0908.0776.

[177] Tataru, D. (1998). Local and global results for wave maps. I, *Comm. Partial Differential Equations* **23**, 9-10, pp. 1781–1793.

[178] Tataru, D. (2001). On global existence and scattering for the wave maps equation, *Amer. J. Math.* **123**, 1, pp. 37–77.

[179] Tataru, D. (2004). The wave maps equation, *Bull. Amer. Math. Soc. (N.S.)* **41**, 2, pp. 185–204.

[180] Tataru, D. (2005). Rough solutions for the wave maps equation, *Amer. J. Math.* **127**, 2, pp. 293–377.

[181] Turok, N. and Spergel, D. (1990). Global texture and the microwave background, *Phys. Rev. Lett.* **64**, 23, pp. 2736–2739.

[182] Vakulenko, A. F. and Kapitanskiĭ, L. V. (1979). Stability of solitons in $S^2$ of a nonlinear $\sigma$-model, *Dokl. Akad. Nauk SSSR* **246**, 4, pp. 840–842.

[183] Wald, R. M. (1984). *General relativity* (University of Chicago Press, Chicago, IL).

[184] Weidig, T. (1999). The baby Skyrme models and their multi-skyrmions, *Nonlinearity* **12**, 6, pp. 1489–1503.

[185] Weinstein, G. (1990). On rotating black holes in equilibrium in general relativity, *Comm. Pure Appl. Math.* **43**, 7, pp. 903–948.

[186] Weinstein, G. (1992). The stationary axisymmetric two-body problem in general relativity, *Comm. Pure Appl. Math.* **45**, 9, pp. 1183–1203.

[187] Weinstein, M. I. (1989). The nonlinear Schrödinger equation—singularity formation, stability and dispersion, in *The connection between infinite-dimensional and finite-dimensional dynamical systems (Boulder, CO, 1987)*, Contemp. Math., Vol. 99 (Amer. Math. Soc., Providence, RI), pp. 213–232.

[188] Wilkins, D. (2005). A course in Riemannian geometry, Lecture notes, Trinity College.

[189] Witten, E. (1983a). Current algebra, baryons, and quark confinement, *Nucl. Phys. B* **223**, 2, pp. 433–444.

[190] Witten, E. (1983b). Global aspects of current algebra, *Nucl. Phys. B* **223**, 2, pp. 422–432.

[191] Wong, W. W.-Y. (2011). Regular hyperbolicity, dominant energy condition and causality for Lagrangian theories of maps, *Classical Quantum Gravity* **28**, 21, pp. 215008, 23.

# Index

471

Printed in the United States
By Bookmasters